ANALYSIS AND MODELLING OF NON-STEADY FLOW IN PIPE AND CHANNEL NETWORKS

ANALYSIS AND MODELLING OF NON-STEADY FLOW IN PIPE AND CHANNEL NETWORKS

Vinko Jović

University of Split, Croatia

A John Wiley & Sons, Ltd., Publication

This edition first published 2013
© 2013 John Wiley & Sons, Ltd

Registered office
John Wiley & Sons Ltd, The Atrium, Southern Gate, Chichester, West Sussex, PO19 8SQ, United Kingdom

For details of our global editorial offices, for customer services and for information about how to apply for permission to reuse the copyright material in this book please see our website at www.wiley.com.

Library of Congress Cataloging-in-Publication Data

Jovic, Vinko.
  Analysis and modelling of non-steady flow in pipe and channel networks / Vinko Jovic.
    pages cm
  Includes bibliographical references and index.
  ISBN 978-1-118-53214-0 (hardback : alk. paper) – ISBN 978-1-118-53686-5 (mobi) – ISBN 978-1-118-53687-2 (ebook/epdf) – ISBN 978-1-118-53688-9 (epub) – ISBN 978-1-118-53689-6 (wiley online library)
  1. Pipe–Hydrodynamics. 2. Hydrodynamics. I. Title.
  TC174.J69 2013
  621.8′672–dc23

                                                                                2012039412

A catalogue record for this book is available from the British Library.

ISBN: 978-1-118-53214-0

Typeset in 9/11pt Times by Aptara Inc., New Delhi, India
Printed and bound in Malaysia by Vivar Printing Sdn Bhd

Contents

Preface

This book deals with flows in pipes and channel networks from both the standpoint of hydraulics and of modelling techniques and methods. These classical engineering problems occur in the course of the design and construction of hydroenergy plants, water-supplies, and other systems. The author presents his experience in solving these problems from the early 1970s to the present. During this period new methods of solving hydraulic problems have evolved, primarily due to the development of electronic (analog and digital) computers, that is the development of numerical methods. The publication of this book is closely connected to the history and impact of the author's software package for solving non-steady pipe flow using the finite element method, called ***Simpip*** which is an abbreviation of ***simulation of pipe flow***. Initially, the program was intended for solving flows in pipe networks; however, it was soon expanded to flows in channels (see paper[1]), though the name was retained. This program package can be found at www.wiley.com/go/jovic and has been used and is currently used for the solution of a great number of engineering problems in Croatia. It also has international references (it has been in the international market since 1992, but was withdrawn from the market for the author's private reasons).

Chapter 1 – Hydraulic Networks. Many numerical methods result from the property of the scalar product of functions in Hilbert space, that is from the fundamental lemma of the variation calculus. This class of numerical methods includes the finite element method or – more precisely – the finite element technique. This means that the introduction of localized coordinate functions, both for the base of an approximate solution and for the test space, leads to the finite elements and topological properties which are connected into a network.

By assembling finite elements into a union which forms an entire domain, it is possible to assemble a global system of equations, which determines the modeled problem by using the same topological properties from the elemental equations. The finite element technique does not depend upon the mathematical method used for deriving the element equations.

It can be noted that hydraulic networks possess the same topological properties of the finite elements mesh and they are predetermined for problem solving using the finite element technique. These are unified networks. Unified networks can include various types of hydraulic branches such as pipes, valves, pumps, and other elements from the pressure systems, natural and artificial channels/canals, rivers, underground natural channels, and all other elements of channel systems. Unified hydraulic networks enable modelling of superficially quite different flows, such as modelling the water hammer with simultaneous modelling of the wave phenomena in the channel. *The basis for solving unified networks is the numerical interpretation of the basic physical laws of mass and energy conservation.* The first chapter presents a universal procedure for developing a matrix and vectors of the finite element from the elemental equations which are assembled into a global matrix and a vector of the equations system.

---

[1] Jović V. (1997) Non-steady Flow in Pipes and Channels by Finite Element Method. Proceedings of XVII Congress of the IAHR, **2**, pp. 197–204, Baden-Baden, 1977.

This is a *fundamental system of equations* which cannot be solved since the global matrix is singular so that the system *has to be completed by natural and essential boundary conditions*.

Chapter 2 – Modelling of Incompressible Fluid Flow. This chapter presents the derivation of a matrix and vector of a pipe finite element and a typical example for modelling the steady flow using the finite elements technique. The solution is iterative using the Newton–Raphson method in the assembled *banded matrix* of the system. It also states the drawbacks of the chosen method for assembling and solving the system of equations and it introduces a *frontal technique*[2] which eliminates the unknowns already in the assembling phase. Since the frontal technique is "natural" for several reasons, and since it has been adopted as the basis for the program solution `SimpipCore` (which can be found at www.wiley.com/go/jovic), all the program phases of modelling the steady flow of incompressible fluid have been explained.

```
use GlobalVars
if(OpenSimpipInputOutputFiles()) then
if Input() then
  if BuildMesh() then ; finite element mesh
    if Steady(t0) then ; solve steady solution, t=t0
      call Output
    endif
  endif
endif
endif
```

Modelling non-steady incompressible fluid flow is a logical continuation of modelling the steady flow by expanding it with a time loop, within which a respective matrix and vector of the non-steady flow of the pipe finite element are recalled. The initial conditions of the non-steady flow are the previously computed steady flows. Non-steady flow of the incompressible fluid can be divided into a quasi non-steady (temporally gradually varying) and non-steady (rigid) flow.

Matrices and vectors of the finite element of the quasi non-steady and rigid flows can be easily obtained from the basic laws of mass and energy conservation.

Chapter 3 – Natural Boundary Conditions Objects. In the fundamental system, the external nodal discharge is a natural condition which completes the nodal equation. It is a natural communication of the hydraulic network with the other systems, which is realized by using objects such as various *valves, water tanks, vessels, surge tanks, and other objects*. Generally, the external discharge depends upon the nodal piezometric height so that both a vector and matrix of the fundamental system are updated with a respective derivation. Special attention is paid to modelling the surge tanks as complex structures in a hydroenergetic system.

Chapter 4 – Water Hammer – Classic Theory. Modelling of non-steady phenomena cannot be imagined without a respective physical interpretation of the phenomenon, in this case the classic theory of the water hammer. Special attention has been paid to the relative motion of the water hammer and its phases as well as to the sudden acceleration and column separation of the water body. This chapter presents some classical computation methods and the principle of protection from the water hammer. By using the kinematic characteristics of wave front and linear relations of the water hammer (superposition principle) it is possible to obtain the wave functions and a general solution of the water hammer determined by a classic theory.

Chapter 5 – Equations of Non-steady Flow in Pipes. The beginning of this chapter presents the equation of the state of matter in the form of a p-V-T surface and in the form of phase projections, followed by equations of the water state under various flow conditions. Subsequently, the differential equations of

---

[2]Irons, B.M. (1970) A frontal solution program, Int. J. Num. Meth. **2**: 5–32.

a one-dimensional non-steady pipe flow are derived in a less typical way beginning with the principle of the mechanics of a material point. These are the continuity equation and the dynamic equation which follow from the law of mass conservation and the mechanical energy of the fluid particle. It has been shown that a precise analysis of a one-dimensional flow is not possible without simplification of the members resulting from the kinetic energy flow; this should be remembered when explaining the results of numerical modelling. Furthermore, various models of steady and non-steady flows of compressible and incompressible fluids in elastic and rigid pipelines are considered. Thus, it was possible to obtain equations of characteristics for the flow of elastic liquid by the simple transformation of the continuity equation and a dynamic equation; however, a more general, *R. Courant and K.O. Friedrichs,* procedure was employed for the compressible fluid. Finally, some analytical solutions of linearized equations for the non-steady water flow are presented.

Chapter 6 – Modelling of Non-steady Flow of Compressible Liquid. This chapter presents the numerical solution of the pipe flow with and without friction by employing a method of characteristics using discrete coordinates of the spatial and temporal variable. The solution can also be expressed as discrete values of primitive variables p, v or wave functions Γ^+, Γ^-. The computation uses recursion. It is interesting that for the pipe flow without friction a simple recursive program can be made without a mesh of characteristics. For modelling the non-steady flow of the elastic fluid in hydraulic networks, matrices and vectors of the pipe finite element have been developed as follows: by the direct numerical interpretation of the continuity equation and dynamic equation and by applying the method of characteristics. The program solution SimpipCore (available at www.wiley.com/go/jovic) enables an *optional* choice of one or two procedures for integrating a matrix and vector of the pipe finite element.

Chapter 7 – Valves and Joints. Valves and various transient objects are relatively short branches which are used as joint elements connecting other branches in a hydraulic network. These are also finite elements used in modelling the hydraulic network, and therefore respective matrices and vectors have been determined for each valve type or connecting object. A non-return valve, either open or closed, should be treated as a special case wherein the *valve status* for the steady flow is given in advance, whereas for the non-steady flow it is computed from the hydraulic state of the system.

Chapter 8 – Pumping Units. Pumping units are a branch of the hydraulic network which consist of a pump and an asynchronous electromotor. This chapter presents the elements of turbo machines. Successful modelling of a non-steady flow with the functioning of pumping units is possible if we know the detailed characteristics of the pumps, i.e. the four-quadrant characteristics. The producers of pumps deliver pumps of serial production after performing standard tests (part of the first quadrant) while complete characteristics (four quadrants) are made only for special orders. In order to model abnormal operating conditions of pumping units it is necessary to reconstruct the detailed characteristic of the pump from normal characteristics. Furthermore, the program SimpipCore (available at www.wiley.com/go/jovic) reconstructs an approximate momentum characteristic of the electromotor according to the type of the declared pump working point and the number of rotations (for an electromotor 50 Hz).

The optional parameter SpeedTransients, with a false default, controls the *nominal rotation velocity of the pumping units*. The finite element matrix and vector are derived from three equations: the continuity equation, dynamic equation, and dynamic equation of the machine rotation, that is the computations of the discharge, manometric height, and the angular velocity of the pumping units, depending upon the SpeedTransients *status and the operating variables of the voltage and frequency*.

Chapter 9 – Open Channel Flow. Modelling of the non-steady channel flow is exceptionally complex since the flow can be subcritical or supercritical, that is during the modelling phases it can change from a subcritical to a supercritical state, or vice versa. Consequently, the program solution SimpipCore is restricted to modelling phases with an advance-determined flow state. The flow and the channel type are declared in the *input* phase, that is in the *initialization* of the channel section. The channel stretch is a branch of the hydraulic network which consists of a series of *channel finite elements*. The spatial position is defined by the coordinates of the points of the flow axes in which the *cross-sectional profiles* of the riverbed have been assembled. The continuity equation and the dynamic equation of the channel flow are formally equal to equations obtained for the pipe flow since they result from the same laws of mass and energy conservation.

The matrix and the vector of the channel finite element, obtained by the integration of the continuity equation and the dynamic (energy) equation, retain all the properties of the mass and energy conservation law both for the channel stretch and the entire hydraulic network, which is a necessary condition for acceptable engineering modelling. Generally, it is possible to form the finite element matrix and vector using the interpolation of the boundary characteristics in a similar way as in pipe finite elements. However, experience in modelling has shown that this seriously threatens the energy and mass conservation law for the channel stretch, which can be explained by the errors caused by necessary interpolations.

Chapter 10 – Numerical Modelling in Karst. Approximately 50% of the soil in the Republic of Croatia is covered by Dinaric karst and significant karst terrains are densely populated, especially the coastal areas. The circulation of the groundwater in the Dinaric karst takes place within a well-developed channel system. Precipitation rapidly sinks underground through a system of fractures, flows through underground channels, and is drained in karst springs and submerged springs. This chapter is written according to the PhD. thesis of Davor Bojanić: Hydrodynamic modelling of karst aquifers (Faculty of Civil Engineering-Architecture and Surveying, University of Split), in 2011, which developed a matrix and vector of a karst channel finite element, a karst channel stretch and an elementary catchment – surface "Karst" for collecting effective rainfall and for defining spatial porosity. Thus, the karst channel stretch is one of the branches of a hydraulic network in the SimpipCore program solution (available at www.wiley.com/go/jovic).

Chapter 11 – Convective-Dispersive Flows. In unified hydraulic networks, apart from primary flows, other secondary processes can be coupled or not coupled to the basic flow. Secondary processes behave according to the law of extensive field conservation, for example the transfer of heat and substances. This chapter presents a solution for a convective-dispersive heat transfer in the hydraulic network; however, the derivation of the finite element matrix and vector in the first chapter is universal and is also valid for modelling other secondary processes.

Chapter 12 – Hydraulic Vibrations in Networks. Forced vibrations are a well developed harmonic state in the hydraulic network resulting from harmonic excitation. The vibration excitation can be any source in the hydraulic network which periodically changes the pressure or the discharge during its normal functioning. Vibration modelling is solved in the frequency domain, that is in the complex area of numbers.

Appendix A – Program solutions. The Appendix can be found at www.wiley.com/go/jovic and presents the program solution SimpleSteady – a typical educational program for modelling steady flow which uses a banded matrix for solving the system of equations. Furthermore, it includes the sources of the Fortran modulus ODE for solving ordinary differential equations with initial conditions and the program tests for the surge tank and a vessel.

Appendix B – SimpipCore. This appendix can also be found at www.wiley.com/go/jovic and presents the SimpipCore project which was developed by an integrated developmental environment *Microsoft Developer Studio* and *Compaq Visual Fortran Versison 6.6*.

The accompanying website www.wiley.com/go/jovic contains, apart from the SimpipCore project (the Fortran sources and Project Workspace, that is the makefile), an independent Windows installation, a user manual, and examples with a series of Fortran sources and tests.

1

Hydraulic Networks

1.1 Finite element technique

1.1.1 Functional approximations

Let us observe a class of methods that can be generated from the property of the scalar product of functions in a *Hilbert*[1] space

$$(\varepsilon, w) = \int_{\Omega} \varepsilon(x)w(x)d\Omega. \tag{1.1}$$

The following *lemma* is a direct consequence of the scalar product (1.1) property: if, for a continuous function $\varepsilon\colon \Omega \to \mathbb{R}$ and for each continuous function $w\colon \Omega \to \mathbb{R}, w \neq 0; \Omega \subset \mathbb{R}^n$

$$\int_{\Omega} \varepsilon(x)w(x)d\Omega = 0; \quad x \in \Omega, \tag{1.2}$$

then $\varepsilon(x) \equiv 0$ for each $x \in \Omega$. The lemma (1.2) is often called the fundamental lemma of the variational calculus. The fundamental lemma will not be proved here since its validity is intuitive. The following can be considered; since $\varepsilon(x)$ and $w(x)$ are the vectors while Eq. (1.2) is a scalar product of vectors, a scalar product of any vector $w(x)$ and vector $\varepsilon(x)$ will always be equal to zero only if vector $\varepsilon(x)$ is the null vector.

The fundamental lemma is widely applied in numerical analysis. Procedures derived from the fundamental lemma can be either approximate[2] or exact.[3] Approximate procedures arise from the meaning of the functional approximation regardless of whether the function was set directly or as a solution of differential equations.

An approximation of a function is sought in the form of the n-dimensional vector. Let the

$$f(x)\colon \Omega \to \mathbb{R} \tag{1.3}$$

---

[1] David Hilbert, German mathematician (1862–1943).
[2] Approximate in the analytical sense.
[3] Exact in the analytical sense.

Analysis and Modelling of Non-Steady Flow in Pipe and Channel Networks, First Edition. Vinko Jović.
© 2013 John Wiley & Sons, Ltd. Published 2013 by John Wiley & Sons, Ltd.

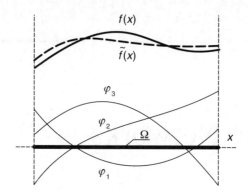

Figure 1.1   Approximation of a function.

be a function for which approximation is sought in the form of a vector $\tilde{f}(x)$. If an n-dimensional vector space is selected, then $\tilde{f}(x)$ is sought in the form of a linear combination of basis vectors

$$\tilde{f}(x): \sum_{i=1}^{n} \alpha_i \varphi_i(x); \quad \varphi_i(x): \Omega \to \mathbb{R}, \tag{1.4}$$

where φ_i are the basis or coordinate vectors, that is linearly independent functions, while α_i are the unknown parameters of a linear combination.

One meaning of an approximation is illustrated in Figure 1.1. Furthermore, the sum sign will be omitted because the Einstein[4] summation convention and other rules of indices will be applied. The difference between the function and its approximation is called a residual

$$\varepsilon(x) = f(x) - \tilde{f}(x). \tag{1.5}$$

There is a question of the criteria for calculation of the unknown parameters of a linear combination in terms of the error minimization $\varepsilon(x)$. If fundamental lemma is applied, then

$$\int_{\Omega} (f - \alpha_i \varphi_i) w_j d\Omega = 0$$

$$i, j = 1, 2, 3, \ldots n \tag{1.6}$$

Should Eq. (1.6) be valid for each continuous function, then the function $f(x)$ will be developed in a convergent series (1.4) for base φ_i. However, in order to calculate n unknown parameters of a linear combination, it will be enough to set n independent conditions, which is achieved by selection of n linearly independent w functions. If n is a finite number, then the fundamental lemma will be satisfied in an approximate sense while functions will be developed with n members from the convergent series.

Since, according to Eq. (1.5), vector $\varepsilon(x)$ is expressed by n coordinate vectors, that is an approximation base $\varphi_i(x)$, it is also obvious that n linearly independent functions $w_j(x)$ will form a base of a certain space, which is called the test space, while vector $w_j(x)$ is a coordinate vector of the test space.

The unknown approximation parameters α_i are obtained by developing Eq. (1.6):

$$\alpha_i \int_{\Omega} \varphi_i w_j d\Omega = \int_{\Omega} f w_j d\Omega$$

$$i, j = 1, 2, 3, \ldots n \tag{1.7}$$

---

[4] Albert Einstein, world famous physicist (1879–1955).

that is, following the calculation of integrals, from the equation system

$$a_{ij}\alpha_i = b_j$$
$$i, j = 1, 2, 3, \ldots n$$

(1.8)

where

$$a_{ij} = \int_{\Omega} \varphi_i w_j d\Omega; \quad b_j = \int_{\Omega} f w_j d\Omega.$$

(1.9)

Since the approximation base and the test base can be selected from the wide range of functions, in general there are certain dilemmas regarding that selection. Details are given in Jovic (1993)0, while the most important approximation methods will be listed hereinafter:

- least squares integral method $w_j(x) = \varphi_j(x)$,
- approximation with orthogonal basis (Legendre[5] and Chebyshev[6] polynomials, harmonic functions),
- the collocation method, algebraic, in particular Lagrange[7] polynomials,
- the finite element method and spline approximations,
- a transfinite mapping method.

1.1.2 Discretization, finite element mesh

Discretization. A problem of function approximation, with the area Ω discretized into sufficiently small finite elements e according to the one-dimensional concept, will be analyzed, see Figure 1.2. A finite element is selected so the function can be approximated by simple functions such as polynomials. A finite element mesh forms a compatible configuration, refer to Figure 1.2b, which provides a union without overlapping

$$\Omega = \bigcup_{i=1}^{m} e_i.$$

(1.10)

Table of element connections. A connection between finite elements to form a compatible configuration is written in the table of element connections such as the following:

	Global nodes	
Element e	1 local	2 local
1	1	2
2	2	3
3	3	4
4	4	5
5	5	6
6	. . .	. . .

where information is written for each finite element e regarding the correspondence between the local nodes and the global ones. A table of element connections is the basic topologic feature of the finite element mesh.

---

[5] Adrien-Marie Legendre, French mathematician (1752–1833).
[6] Pafnuty Lvovich Chebyshev, Russian mathematician (1821–1894).
[7] Joseph-Louis Lagrange, mathematician and astronomer (1736–1813).

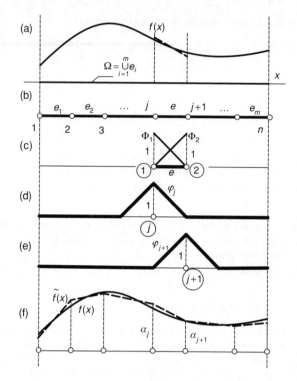

Figure 1.2   Localized basis functions.

Shape or interpolation functions. Over each finite element $e = 1, 2, 3, \ldots m$ a function is approximated by *interpolation* functions, which form a base of solutions over a finite element. These functions shape the solution over an element and are, therefore, called the *shape* or interpolation functions. They are either normalized polynomials of the Lagrange class, or Hermite[8] polynomials attached to the nodes or some other interpolation polynomials. Figure 1.2c shows a two-nodal finite element with the shape functions Φ_1 and Φ_2 attached to local nodes 1 and 2. The shape functions are the basis vectors of the finite element. These basis vectors are generating a linear form of the function over a finite element, while approximation of a function over an area Ω will be a polygonal one. Higher order approximation is achieved by elements with more nodes.

Localized global functions. A localized base, as shown in Figures 1.2d and 1.12e for the nodes j and $j + 1$, is built from the shape functions of the adjacent elements. Hence, one localized base φ_i, with the definition area Ω, is appointed to each global node i. A localized base is intensive within a limited area, that is elements that contain the respective node, while outside it is equal to zero.

A sought approximation of the function is a linear combination of localized coordinate functions $\tilde{f}(x) = \alpha_i \varphi_i(x)$. The value of the localized base equals 1 in the nodes that it is appointed to. Thus, the unknown parameters become equal to the nodal values of an approximate solution $\alpha_i = \tilde{f}_i$.

Continuity of an approximation. Linear shape functions secure continuity of a function, though not its derivations. Thus, it is said that an approximation belongs to the class $\mathbb{C}^0$.

---

[8]Charles Hermite, French mathematician (1822–1901).

```
n,m ; no nodes and elements
connect(m,2) ; element connectivities
Ag(n,n),Bg(n) ; global matrix and vector
Ae(2,2),Be(2) ; elemental matrix and vector
; loop over finite elements:
for e = 1 to m do
  call matrix(Ae,Be) ; compute matrix and vector
  for k = 1 to 2
    r = connect(e,1) ; first global node
    Bg(r)=Bg(r)+Be(k)
    For l = 1 to 2
      s = connect(e,2) ; second global node
      Ag(r,s)=Ag(r,s)+Ae(k,l)
    End loop l
  end loop k
end loop e
```

Figure 1.3 Assembling algorithm.

Finite element matrix and vector. If the least squares integral approximation procedure is applied, with the test base equal to the approximations base $w_j(x) = \varphi_j(x)$, the matrix and vector members are obtained

$$
a_{ij} = \int_{\Omega} \varphi_i \varphi_j dx = \sum_{e=1}^{m} \int_e \varphi_i \varphi_j dx \quad \text{and} \quad b_j = \int_{\Omega} f \varphi_j dx = \sum_{e=1}^{m} \int_e f \varphi_j dx. \tag{1.11}
$$

These are the global matrix and vector, which are integrated from the contribution from individual finite elements. If in integrals (1.11) global functions φ are replaced by local ones Φ, then the finite element matrix and vector are obtained[9]

$$
a_{kl}^e = \int_e \Phi_k \Phi_l de \quad \text{i} \quad b_l^e = \int_e f \Phi_l de. \tag{1.12}
$$

System assembling. A procedure of global equation system generation will be presented using an algorithm written in the pseudo-language, as shown in Figure 1.3. The assembling procedure starts with an empty global matrix and an empty global vector.

For each element e the finite element matrix $a_{k,l}^e$, $k, l = 1, 2$ and vector b_k^e, $l = k, 2$ are calculated and superimposed into the global matrix and vector using the connectivity table, see Figure 1.4. Note that calculation over finite elements is independent and can be processed in parallel.[10] Also note that the element assembling schedule is irrelevant.

Figure 1.5 shows the global matrix and vector separately for each element and their final form.

---

[9]It is often referred to as the stiffness matrix and the load vector with respect to physical features occurring in the solving of the problem of elastic body equilibrium.

[10]This property is suitable for computers with parallel processors.

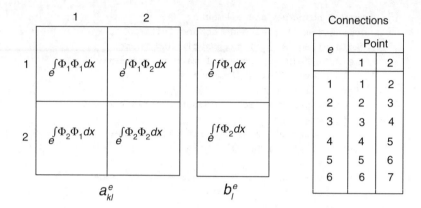

Figure 1.4   Matrix, vector, and table of connectivity of the finite element.

1.1.3   Approximate solution of differential equations

Strong formulation

Let us focus on differential equations obtained by description of natural phenomena and their approximate solutions. Generally, a mathematical model of a natural phenomenon is formally written in the following form

$$F(X, U, D^k U) = 0, \qquad (1.13)$$

where $X = (x_1, x_2, x_3, \ldots x_p)$ are the coordinates of the space where the phenomenon takes place, p is the space dimension, $U = (u_1, u_2, u_3, \ldots u_s)$ is the intensive field describing the phenomenon – namely

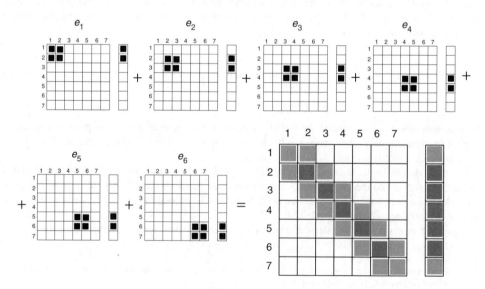

Figure 1.5   Assembling.

solution of the equation, s is the degree of freedom of a system, for a vector function it can be 1, 2, or 3 depending on the spatial dimension, and $D^k U$ is the generalized partial derivation of the k-th order.

A solution of Eq. (1.13) is sought for the initial and boundary conditions. It is understood that the solution exists; it is unique and can be expressed as a vector

$$\tilde{U} = \alpha_i \varphi_i(X); \quad i = 1, 2, 3, \ldots n \tag{1.14}$$

with the unknown α_i parameters and φ_i basis vectors, functions selected from a class that are derivable enough.

By introducing Eq. (1.14) into Eq. (1.13) and applying the fundamental lemma, n independent equations are obtained for defining the parameters[11]

$$\int_\Omega F(X, \alpha_i \varphi_i, D^k \alpha_i \varphi_i) W_j d\Omega = 0 \tag{1.15}$$

$$i, j = 1, 2, 3, \ldots n$$

The aforementioned equation system does not have to be regular because a solution without the initial and boundary conditions is not unique. Initial conditions are the known values of the function at a certain time. Thus, satisfying the initial conditions refers to the problem of approximation of the set function. Let the boundary conditions be set forth by the expression

$$G(X, U, D^k U) = 0; \quad X \in \Gamma, \tag{1.16}$$

where Γ is the boundary of the domain Ω. The order of partial derivations in Eq. (1.16) is, in general, lower than the partial derivations order in Eq. (1.15). Since an approximate solution shall also satisfy the boundary conditions, the fundamental lemma will be applied again to boundary conditions. Thus

$$\int_\Omega F(X, \alpha_i \varphi_i, D^k \alpha_i \varphi_i) W_j d\Omega = \int_\Gamma G(X, \alpha_i \varphi_i, D^k \alpha_i \varphi_i) W_j d\Gamma \tag{1.17}$$

$$i, j = 1, 2, 3, \ldots n$$

A solution is sought as an approximation by linear combination of the basis functions, which are derivable enough and satisfy the boundary conditions. These are the strong conditions. Strong formulation procedures play an important role in engineering that shall also not be negligible in the future. An approximate solution is sought in a linear combination of the global basis functions that are derivable enough

$$\tilde{U} = \alpha_k \varphi_k. \tag{1.18}$$

If an approximate solution shall satisfy the boundary conditions accurately, a procedure of boundary condition homogenization is applied by transformation

$$U = \Psi + V \tag{1.19}$$

---

[11]Which are the weight factors; this is often called the weighted residuals method, in particular in terms of differential equations' approximate solutions.

so that, after introducing Eq. (1.19) into Eq. (1.15), the strong formulation becomes

$$\int_{\Omega} F^{(1)}(X, \alpha_i \varphi_i, D^k \alpha_i \varphi_i) W_j d\Omega = \int_{\Omega} F^{(2)}(X, \Psi, D^k \Psi) W_j d\Gamma$$

$$i, j = 1, 2, 3, \ldots n$$

(1.20)

where the right side in Eq. (1.17) is eliminated due to homogeneous boundary conditions. If the equation set by a linear operator is observed

$$L(u) = 0 \qquad (1.21)$$

with the solution sought based on the mixed boundary conditions: natural $q = q(x); x \in \Gamma_1$ and essential $u = g(x), x \in \Gamma_2$, by application of the fundamental lemma, the strong formulation can be written in the following form

$$\int_{\Omega} L(\tilde{u}) w_j d\Omega = \int_{\Gamma_1} (\tilde{q} - q) \, w_j d\Gamma + \int_{\Gamma_2} (\tilde{u} - g) \, w_j d\Gamma, \qquad (1.22)$$

where the boundary integrals are divided into two parts. Methods for approximate solving of differential equations can be classified according to procedure:

- according to the selection of the test space base: the moment method, the point collocation method, the least squares method, the least squares collocation method, the subdomain method or the subdomain collocation method (Biezeno and Koch[12]), the Galerkin[13]–Bubnov[14] method, and other methods;
- according to the selection of an approximate solution base (basis separation, basis localization, minimization of the solution variation leads to the Galerkin procedure);
- operator methods for discrete and continuous parameters.

Weak formulation

Unlike the strong formulation, the problem is solved by integral transformations to decrease the derivation order. An integral formulation is solved instead of a differential equation, and weaker conditions are set for an approximate solution. The finite element technique and the Galerkin method (variational *procedures*) are the most commonly used for calculation of matrices and vectors. The numerical form of conservation law is one of the very important weak formulations. It is also referred to as the finite volume method or method of subdomain.

For easier understanding of the weak formulation procedures, a typical solution of the Boussinesq[15] equation will be presented. The Boussinesq equation is a parabolic equation used for the description of the heat conduction problem in physics, seepage problem in hydraulics, and other problems in electrical

---

[12]C. B. Biezeno, J. J. Koch, Dutch engineers.

[13]Boris Grigoryevich Galerkin, Russian/Soviet an engineer and mathematician (1871–1945).

[14]Ivan Grigoryevich Bubnov, Russian marine engineer (1872–1919).

[15]J. V. Boussinesq, French physicist and mathematician (1842–1929).

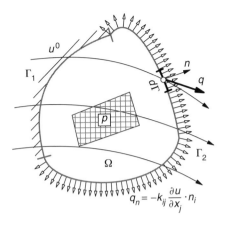

Figure 1.6 Domain or control volume.

engineering and so on. The steady form of the equation has an elliptical form. Starting from the heat conservation law over a control volume Ω, the following is obtained

$$\frac{\partial}{\partial t}\int_\Omega cu d\Omega + \int_\Gamma q_i n_i d\Gamma = \int_\Omega p d\Omega, \tag{1.23}$$

where the first integral is the rate of change of heat inside the control volume, the second integral is the change occurring due to heat flux through the surface Γ enclosing the control volume, while the third integral is the heat production within the control volume, see Figure 1.6.

Application of the GGO Theorem[16] on the surface interval is written as

$$\frac{\partial}{\partial t}\int_\Omega cu d\Omega + \int_\Omega \frac{\partial q_i}{\partial x_i} d\Omega = \int_\Omega p d\Omega. \tag{1.24}$$

After grouping under one integral, the following is obtained

$$\int_\Omega \left(c\frac{\partial u}{\partial t} + \frac{\partial q_i}{\partial x_i} - p \right) d\Omega = 0. \tag{1.25}$$

This particular integral vanishes for each area, which means that the sub-integral function must be equal to zero. Then, the heat continuity equation is obtained in the form

$$c\frac{\partial u}{\partial t} + \frac{\partial q_i}{\partial x_i} - p = 0. \tag{1.26}$$

---

[16]General integral transformation theorem using the projection for $\mathbb{R}^n$:

$$\int_\Gamma f n_i d\Gamma = \int_\Omega \frac{\partial f}{\partial x_i} d\Omega$$

discovered independently by Gauss, German mathematician and astronomer (1777–1855), Green, English mathematician and physicist (1793–1844), and Ostrogradski, Russian mathematician and physicist (1801–1862) and thus named after them.

A simple dynamic equation – generalized as Fourier's[17] law can be applied to the thermal conduction processes

$$q_i = -k_{ij}\frac{\partial u}{x_j}, \tag{1.27}$$

where q_i is the thermal flux and k_{ij} is the thermal conduction tensor. Introducing Fourier's law into the conservation law

$$c\frac{\partial u}{\partial t} = \frac{\partial}{\partial x_i}k_{ij}\frac{\partial u}{x_j} + p \tag{1.28}$$

a Boussinesq equation is obtained. When the fundamental lemma is applied to the Boussinesq equation, an extended form is obtained

$$\int_\Omega c\frac{\partial u}{\partial t}wd\Omega - \int_\Omega w\frac{\partial}{\partial x_i}k_{ij}\frac{\partial u}{x_j}d\Omega - \int_\Omega pwd\Omega = 0. \tag{1.29}$$

Partial integration[18] will be applied to the second integral; thus, expression (1.29) will become

$$\int_\Omega c\frac{\partial u}{\partial t}wd\Omega + \int_\Omega k_{ij}\frac{\partial c}{x_j}\frac{\partial w}{\partial x_i}d\Omega = \int_\Gamma wk_{ij}\frac{\partial u}{\partial x_i}n_i d\Gamma + \int_\Omega pwd\Omega. \tag{1.30}$$

Integral equation (1.30) is a weak formulation of the Boussinesq equation. An approximate solution will be sought in the form of a linear combination of basis functions

$$u = u_r(t)\varphi_r(x_i), \tag{1.31}$$

where the values of linear combination of time-dependent function are the unknowns (nodal temperatures). They are determined from the equation system, with test functions $w = \varphi_s(x_i)$. If the finite element technique is applied (localized approximation base and test space) the following is obtained

$$\frac{du_r}{dt}\int_\Omega c\varphi_r\varphi_s d\Omega + u_r\int_\Omega k_{ij}\frac{d\varphi_r}{dx_j}\frac{d\varphi_s}{dx_i}d\Omega = \int_{\Gamma_2}\varphi_s q_n d\Gamma + \int_\Omega \varphi_s p d\Omega, \tag{1.32}$$

where integrals are matrices and vectors. A boundary $\Gamma = \Gamma_1 \cup \Gamma_2$ consists of two parts. The first is an integral over the boundary Γ_1, with the known value of solution u^0, which does not have to be calculated; and the integral over the boundary Γ_2 with the known prescribed discharge q_n in the direction of the normal. If the following marked

capacitive matrix:
$$C_{rs} = \int_\Omega c\varphi_r\varphi_s d\Omega, \tag{1.33}$$

divergence matrix:
$$D_{rs} = \int_\Omega k_{ij}\frac{d\varphi_r}{dx_j}\frac{d\varphi_s}{dx_i}d\Omega, \tag{1.34}$$

---

[17]Fourier, French mathematician and physicist (1768–1830).
[18]Partial integration $\int_\Omega u\frac{\partial v}{\partial x_i}d\Omega = \int_\Gamma uvn_i d\Gamma - \int_\Omega \frac{\partial u}{\partial x_i}\frac{\partial v}{\partial x_i}d\Omega$ is obtained by the GGO transformation theorem.

vector of boundary thermal fluxes:
$$Q_s = \int_{d\Gamma} \varphi_s q_n d\Gamma, \tag{1.35}$$

heat production vector:
$$P_s = \int_{\Omega} \varphi_s \, p \, d\Omega, \tag{1.36}$$

where the expressions have a physical meaning, a *discrete global system* is obtained in the form

$$C_{rs} \frac{du_r}{dt} + D_{rs} u_r = Q_s + P_s. \tag{1.37}$$

Ordinary differential equations are obtained with nodal functions to be solved.

Figure 1.7a shows a discrete system. As can be observed, the boundary nodal discharge Q_s consists of the concentrated contributions of the adjacent elements. A production vector P_s is interpreted similarly, as a contribution from the adjacent elements with the common node.

Each finite element can be observed separately as an isolated discrete system, see Figure 1.7b. Elemental discrete equations can also be applied to

$$C^e_{rs} \frac{du_r}{dt} + D^e_{rs} u_r = Q^e_s + P^e_s. \tag{1.38}$$

For steady flow, nodal discharges will be

$$Q^e_s = D^e_{rs} u_r. \tag{1.39}$$

Note that, besides the table of finite element connections, there are other topologic properties of the finite element method.

Figure 1.8 shows the generation of finite element configuration around the node s.

The same assembling procedure can be applied to nodal continuity equations, because the thermal flux conservation law is valid for node s

$$\sum_{e=1}^{p} Q^e_s = 0 \tag{1.40}$$

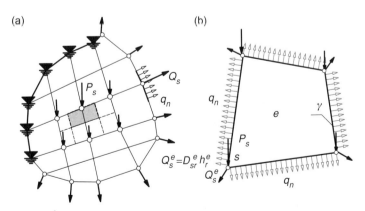

Figure 1.7 Discrete system (a) global system and (b) finite element.

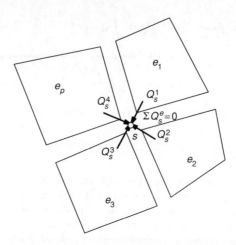

Figure 1.8   Generation of the finite element configuration and nodal equation.

that consists of nodal discharge Q_s^e contributions from the adjacent elements. If the nodal discharge vector over an element can be expressed via a nodal temperature vector, the global nodal equation can be written in the following form

$$\sum_{e=1}^{P} Q_s^e = \sum_{e=1}^{P} D_{rs}^e u_r = 0, \tag{1.41}$$

where the matrix D_{rs}^e is equal to the finite element matrix. That matrix can be generated using the thermal flux over an element that is, in the presented example, numerically completely equal to the finite element matrix obtained from the weak formulation. Numerical equivalence is completely understandable because a linear differential equation was analyzed; also, a weak formulation gives the thermal flux conservation law, as can be observed if test function values in Eq. (1.30) are adopted as constant.

   First example. A steady thermal flux will be observed on a bar of the length L, see Figure 1.9a, which consists of three segments with different thermal conduction coefficients k_e and the same length ΔL.

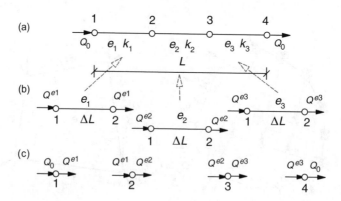

Figure 1.9   Steady thermal flux along the bar. (a) bar, (b) decomposition of bar on segments (finite elements), (c) nodal flux continuity.

A steady thermal flux Q_0 and temperature distribution u_r shall be determined in characteristic points $r = 1,2,3,4$ along the bar. Heat conduction is described by the following thermal flux equations

continuity equation
$$\frac{dQ}{dl} = 0, \tag{1.42}$$

dynamic equation
$$k\frac{du}{dl} + Q = 0, \tag{1.43}$$

where Q is the thermal flux, u is temperature, while k is the thermal conduction coefficient, constant for a respective segment. A solution can be sought for the following boundary conditions:

(a) Known thermal flux Q_0 on the one end $(l = L)$ and the known temperature on the same or the opposite end of a bar $(l = 0)$; or

(b) Known boundary temperatures $u(0)$ and $u(L)$; thus the unknown thermal flux and the remaining temperature distribution are to be defined.

ad (a)
The task is trivial, because integration of the thermal flux continuity equation (1.42) from the known boundary flux Q_0, for example starting from the left edge, to any distance, gives:

$$\int_0^l \frac{dQ}{dl} dl = Q(l) - Q_0 = 0, \tag{1.44}$$

that is the thermal flux along the bar is constant and equal to Q_0.

Temperature distribution in nodes is determined by the dynamic equation's (1.43) integration over a segment e with the constant k_e, starting from the edge with the unknown temperature:

$$\int_{l_1}^{l_2} \left(k_e \frac{du}{dl} + Q^e \right) dl = 0. \tag{1.45}$$

The following is obtained

$$u_2 = u_1 + \frac{Q_0}{k_e} \int_1^2 dl = u_1 + \frac{Q_0}{k_1} \frac{\Delta L}{2}$$

$$u_3 = u_2 + \frac{Q_0}{k_2} \frac{\Delta L}{2} \tag{1.46}$$

$$u_4 = u_3 + \frac{Q_0}{k_3} \frac{\Delta L}{2}$$

ad (b)
The task is implicit. In general, four unknown nodal temperatures u_r, $r = 1,2,3,4$ and three thermal fluxes Q^e, $e = 1,2,3$, are sought on a bar; namely, seven unknowns. Thus, there will be seven independent equations in the calculations. If four thermal flux nodal equations and three dynamic equations

are written for three segments, the system will be formally closed. Integration of the heat continuity equation gives

$$\int \frac{dQ}{dl} dl = Q = const \qquad (1.47)$$

thus the thermal flux will be constant along the bar and each of its sections e: $Q^e = Q$. Nodal flux continuity equations are written for each point $r = 1, 2, 3, 4$, thus connecting the thermal fluxes from the adjacent sections, as shown in Figure 1.9c

$$
\begin{array}{ll}
1: & Q_0 - Q^{e_1} = 0 \\
2: & Q^{e_1} - Q^{e_2} = 0 \\
3: & Q^{e_2} - Q^{e_3} = 0 \\
4: & Q^{e_3} - Q_0 = 0
\end{array}
\qquad (1.48)
$$

or written as a sum of nodal fluxes $\sum_{p=1}^{2} {}^{r}Q^{e_p} = 0$. The expression is obviously read as a sum of boundary fluxes for separate segments e_p around the node r, where p is the number of adjacent sections. Then, the dynamic equation is integrated, and the following is valid for each segment e, see Figure 1.9b,

$$\int_e \left(k_e \frac{du}{dl} + Q^e \right) dl = 0 \qquad (1.49)$$

from which

$$k_e (u_s - u_r) + Q^e \Delta L = 0 \qquad (1.50)$$

thus, for each segment $e = 1, 2, 3$ the same number of equations is obtained

$$F^e(u_r, u_s, Q^e) = 0 \qquad (1.51)$$

with the thermal flux equal to

$$Q^e = -k^e \frac{u_2 - u_1}{\Delta L}. \qquad (1.52)$$

When Eq. (1.52) flux is introduced into nodal equation (1.48), nodal fluxes are eliminated, thus, nodal equations are expressed only by nodal temperatures

$$Q_0 + k^{e_1} \frac{u_2 - u_1}{\Delta l} = 0,$$

$$-k^{e_1} \frac{u_2 - u_1}{\Delta l} + k^{e_2} \frac{u_3 - u_2}{\Delta l} = 0,$$

$$-k^{e_2} \frac{u_3 - u_2}{\Delta l} + k^{e_3} \frac{u_4 - u_3}{\Delta l} = 0,$$

$$-k^{e_3} \frac{u_4 - u_3}{\Delta l} - Q_0 = 0.$$

As written in matrix form

$$
\begin{bmatrix} Q_0 - Q^{e1} \\ Q^{e1} - Q^{e2} \\ Q^{e2} - Q^{e3} \\ Q^{e3} - Q_0 \end{bmatrix} =
\begin{bmatrix}
-\dfrac{k_1}{\Delta L} & \dfrac{k_1}{\Delta L} & & \\
\dfrac{k_1}{\Delta L} & -\dfrac{k_1}{\Delta L} - \dfrac{k_2}{\Delta L} & \dfrac{k_2}{\Delta L} & \\
& \dfrac{k_2}{\Delta L} & -\dfrac{k_2}{\Delta L} - \dfrac{k_3}{\Delta L} & \dfrac{k_3}{\Delta L} \\
& & \dfrac{k_3}{\Delta L} & -\dfrac{k_3}{\Delta L}
\end{bmatrix} \cdot
\begin{bmatrix} u_1 \\ u_2 \\ u_3 \\ u_4 \end{bmatrix} +
\begin{bmatrix} +Q_0 \\ 0 \\ 0 \\ -Q_0 \end{bmatrix} =
\begin{bmatrix} 0 \\ 0 \\ 0 \\ 0 \end{bmatrix}.
$$

$$(1.53)$$

When boundary temperatures $u_1 = u(0)$ and $u_4 = u(L)$ are introduced into previous equations, and the system is solved by the unknowns; the unknown temperatures u_2 and u_3 are obtained. Then, the thermal fluxes will be calculated Q^{e1}, Q^{e2}, $Q^{e3} = Q_0$ according to Eq. (1.52).

Finite element matrix and vector from the conservation law (first example). If bar discretization into finite elements is visualized in a manner such that each finite element corresponds to a respective bar segment, then all properties of the finite element technique can be used in problem solving; such as the following connectivity table:

	Global nodes	
Element e	1 local	2 local
1	1	2
2	2	3
3	3	4

and finite element matrix to form the global equation system. The finite element matrix is generated from the elemental equation (1.50) when elemental flux is calculated:

as the first local node
$$
-Q^e = \frac{k^e}{\Delta l} \begin{bmatrix} -1 & +1 \end{bmatrix} \begin{bmatrix} u_1 \\ u^2 \end{bmatrix},
\tag{1.54}
$$

and

as the second local node
$$
Q^e = \frac{k^e}{\Delta l} \begin{bmatrix} +1 & -1 \end{bmatrix} \begin{bmatrix} u_1 \\ u^2 \end{bmatrix}.
\tag{1.55}
$$

Finally, the finite element matrix comes from equations

$$
\begin{bmatrix} -Q^e \\ +Q^e \end{bmatrix} = \underbrace{\frac{k_e}{\Delta L} \begin{bmatrix} -1 & +1 \\ +1 & -1 \end{bmatrix}}_{\text{finte elemet matrix A}} \cdot \begin{bmatrix} u_1 \\ u_2 \end{bmatrix}.
\tag{1.56}
$$

The global system of equations has the meaning of the nodal equation of continuity (1.48). It can be assembled from contributions from individual finite elements.

$$\sum_{p} {}^{r}Q^e = D_{rs}u_s = 0.$$

(1.57)

The equation system (1.57) shall be extended with the natural boundary conditions. Natural boundary conditions are the boundary thermal flux Q_0, see Figure 1.9a, which is added to the nodal equations vector

$$\sum_{p} {}^{r}Q^e + Q_0^r = D_{rs}u_s + Q_0^r = 0,$$

(1.58)

thus

$$
\begin{bmatrix}
-\dfrac{k_1}{\Delta L} & \dfrac{k_1}{\Delta L} & & \\[2ex]
\dfrac{k_1}{\Delta L} & -\dfrac{k_1}{\Delta L} - \dfrac{k_2}{\Delta L} & \dfrac{k_2}{\Delta L} & \\[2ex]
& \dfrac{k_2}{\Delta L} & -\dfrac{k_2}{\Delta L} - \dfrac{k_3}{\Delta L} & \dfrac{k_3}{\Delta L} \\[2ex]
& & \dfrac{k_3}{\Delta L} & -\dfrac{k_3}{\Delta L}
\end{bmatrix}
\cdot
\begin{bmatrix} u_1 \\ u_2 \\ u_3 \\ u_4 \end{bmatrix}
+
\begin{bmatrix} +Q_0 \\ 0 \\ 0 \\ -Q_0 \end{bmatrix}
=
\begin{bmatrix} 0 \\ 0 \\ 0 \\ 0 \end{bmatrix}.
$$

(1.59)

External thermal flux is positive if it increases the system heat, otherwise it is negative. A natural boundary condition can be generalized as a concentric external thermal load in each node; and the vector Q_0^r is named the external load vector.

The system (1.58) is still unsolvable since the matrix is singular. A temperature shall be known in at least one node, which is achieved by the introduction of the known prescribed value $u_r = U_0$. Thus, the matrix becomes singular and the system solvable.

Finite element matrix and vector from weak formulation (first example). It will be shown that the finite element matrix (1.56) obtained from the nodal continuity equations is formally equal to the standard derivation of matrix. After introducing the dynamic equation (1.43) into the continuity equation (1.42), a thermal conduction differential equation is obtained in the following form

$$\frac{d}{dl}\left(k\frac{du}{dl}\right) = 0.$$

(1.60)

A fundamental lemma will be applied to the obtained equation

$$\int_L \frac{d}{dl}\left(k\frac{du}{dl}\right) w\, dl = 0$$

(1.61)

after which it is partially integrated according to the partial integration rules[19]

$$\int_L \frac{d}{dl}\left(k\frac{du}{dl}\right) w\, dl = \left(wk\frac{du}{dl}\right)\Big|_0^L - \int_L k\frac{dw}{dl}\frac{du}{dl}\, dl = 0,$$

(1.62)

---

[19]Partial integration: $\int_L u\,dv = (uv)\big|_0^L - \int_L v\,du.$

from which

$$\int_L k \frac{du}{dl} \frac{dw}{dl} dl = \left(wk \frac{du}{dl} \right) \Big|_0^L. \tag{1.63}$$

A weak formulation of differential equation (1.60) is obtained. Natural boundary conditions, namely boundary thermal fluxes $Q_0 = -kdu/dl$, are on the right side.

The bar will be divided into finite elements that correspond to bar segments with the constant thermal conduction coefficients k_e. A localized base and Galerkin's selection of the test functions will be used

$$u = u_r \varphi_r(l), \quad w = \varphi_s(l) \tag{1.64}$$

when introduced into Eq. (1.63), a discrete system is obtained

$$u_r \int_L k \frac{d\varphi_r}{dl} \frac{d\varphi_s}{dl} dl = (-\varphi_s Q_0)|_0^L \tag{1.65}$$

and is written in the following form

$$D_{rs} u_s + Q_s = 0, \tag{1.66}$$

where D_{rs} is the global matrix and Q_s is the global vector. Vector Q_s contains all natural boundary conditions. A global matrix will be generated from the matrices of individual finite elements obtained by integration of the finite element shape functions

$$D_{rs}^e = \int_{\Delta L} k_e \frac{d\Phi_r}{dl} \frac{d\Phi_s}{dl} dl = \frac{k_e}{\Delta L} \begin{bmatrix} -1 & +1 \\ +1 & -1 \end{bmatrix}. \tag{1.67}$$

Second example. The previously analyzed example shows the finite element matrix and vector generation when the problem is described with one elemental discharge, since only the dynamic equation was used for problem solving over a finite element. A similar problem will be analyzed where both elemental equations will be used for problem solution on a finite element

thermal flux continuity equation: $\quad\quad\quad\quad \dfrac{dQ}{dl} = p, \tag{1.68}$

dynamic equation: $\quad\quad\quad\quad k \dfrac{du}{dl} + Q = 0. \tag{1.69}$

Apart from the constant thermal load p along the bar in the continuity equation, all other parameters are completely equal to the parameters from the previous example.

Finite element matrix and vector from the conservation law (second example). The integration of the continuity equation and dynamic equations on a finite element

$$\int_{\Delta L} \frac{dQ}{dl} dl = \int_{\Delta L} p\, dl, \quad \int_{\Delta L} k_e \frac{du}{dl} dl + \int_{\Delta L} Q\, dl = 0 \tag{1.70}$$

leads to two algebraic equations

$$Q_2 - Q_1 = p\Delta L, \tag{1.71}$$

$$k_e(u_2 - u_1) + (Q_1 + Q_2)\frac{\Delta L}{2} = 0, \tag{1.72}$$

where the last integral will be calculated by application of the mean value integral theorem.[20]
Elemental equations will be written in the matrix form

$$\frac{2k_e}{\Delta L}\begin{bmatrix} 0 & 0 \\ -1 & +1 \end{bmatrix}\begin{bmatrix} u_1 \\ u_2 \end{bmatrix} + \underbrace{\begin{bmatrix} -1 & +1 \\ +1 & +1 \end{bmatrix}}_{[\underline{Q}]}\begin{bmatrix} Q_1 \\ Q_2 \end{bmatrix} = p\Delta L\begin{bmatrix} 1 \\ 0 \end{bmatrix}. \tag{1.73}$$

The matrix equation will be multiplied by the inverse matrix[21]

$$[\underline{Q}]^{-1} = \frac{1}{2}\begin{bmatrix} -1 & +1 \\ +1 & +1 \end{bmatrix}:$$

$$\frac{2k_e}{\Delta L}\frac{1}{2}\begin{bmatrix} -1 & +1 \\ +1 & +1 \end{bmatrix}\begin{bmatrix} 0 & 0 \\ -1 & +1 \end{bmatrix}\begin{bmatrix} u_1 \\ u_2 \end{bmatrix} + \begin{bmatrix} Q_1 \\ Q_2 \end{bmatrix} = \frac{1}{2}\begin{bmatrix} -1 & +1 \\ +1 & +1 \end{bmatrix}\begin{bmatrix} p\Delta L \\ 0 \end{bmatrix}. \tag{1.74}$$

After rearranging, two elemental discharges in the matrix form will be obtained

$$\begin{bmatrix} Q_1 \\ Q_2 \end{bmatrix} = \frac{p\Delta L}{2}\begin{bmatrix} -1 \\ +1 \end{bmatrix} + \frac{k_e}{\Delta L}\begin{bmatrix} +1 & -1 \\ +1 & -1 \end{bmatrix}\begin{bmatrix} u_1 \\ u_2 \end{bmatrix}. \tag{1.75}$$

The finite element matrix and vector will be obtained by alteration of the algebraic sign in the first row

$$\begin{bmatrix} -Q_1 \\ Q_2 \end{bmatrix} = \underbrace{\frac{k_e}{\Delta L}\begin{bmatrix} -1 & +1 \\ +1 & -1 \end{bmatrix}}_{A^e}\begin{bmatrix} u_1 \\ u_2 \end{bmatrix} + \underbrace{\frac{p\Delta L}{2}\begin{bmatrix} +1 \\ +1 \end{bmatrix}}_{B^e}, \tag{1.76}$$

that is,

$$[\underline{A}] = \frac{k_e}{\Delta L}\begin{bmatrix} -1 & +1 \\ +1 & -1 \end{bmatrix}, \quad [\underline{B}] = \frac{p\Delta L}{2}\begin{bmatrix} +1 \\ +1 \end{bmatrix}. \tag{1.77}$$

Finite element matrix and vector from the weak formulation (second example). The continuity equation and the dynamic equation can be written together in the single thermal conduction equation

$$\frac{d}{dl}\left(-k\frac{du}{dl}\right) = p. \tag{1.78}$$

---

[20]Let f(x) be continuous on [a, b]. Set $F(x) = \int_a^x f(t)dt$, the first mean value theorem for integrals implies $\int_a^x f(t)dt = f(c)(b-a)$, where the point $f(c)$ is called the average value of $f(x)$ on [a, b].

[21]The formula for 2×2 matrix inversion: $A = \begin{bmatrix} a & b \\ c & d \end{bmatrix}$, $A^{-1} = \frac{1}{ad-bc}\begin{bmatrix} d & -b \\ -c & a \end{bmatrix}$.

A fundamental lemma will be applied to the obtained equation

$$\int_L \frac{d}{dl}\left(k\frac{du}{dl} + p\right)wdl = 0 \tag{1.79}$$

then it will be partially integrated according to the partial integration rules

$$\int_L \frac{d}{dl}\left(k\frac{du}{dl}\right)wdl + \int_L wpdl = \left(wk\frac{du}{dl}\right)\Big|_0^L - \int_L k\frac{dw}{dl}\frac{du}{dl}dl + \int_L wpdl = 0, \tag{1.80}$$

from which a weak formulation will be obtained

$$\int_L k\frac{du}{dl}\frac{dw}{dl}dl = \left(wk\frac{du}{dl}\right)\Big|_0^L + p\int_L wpdl. \tag{1.81}$$

When localized basis functions $u = u_r\varphi_r(l)$ and Galerkin's test base $w = \varphi_s(l)$ are applied, a global equation system is obtained that will be generated from the contribution of separate finite elements. The finite element matrix

$$D_{rs}^e = \int_{\Delta L} k_e\frac{d\Phi_r}{dl}\frac{d\Phi_s}{dl}dl = \frac{k_e}{\Delta L}\begin{bmatrix} -1 & +1 \\ +1 & -1 \end{bmatrix} \tag{1.82}$$

and vector

$$F_s = p\int_{\Delta l} \Phi_s dl = p\frac{\Delta L}{2}\begin{bmatrix} +1 \\ +1 \end{bmatrix} \tag{1.83}$$

are integrated in a manner such that the global basis functions are replaced with the shape functions.

Numerical form of the conservation law

Let us observe the wave equation that occurs in different fields of physics, such as acoustics, solid mechanics, fluid mechanics, electricity, and other fields. One kind of wave equation is the linear form of non-steady flow in pipes and channels. The equation can be generated from two equations

continuity equation: $\qquad\qquad \dfrac{gA}{c^2}\dfrac{\partial h}{\partial t} + \dfrac{\partial Q}{\partial x} = 0,$ $\qquad\qquad$ (1.84)

dynamic equation: $\qquad\qquad \dfrac{1}{gA}\dfrac{\partial Q}{\partial t} + \dfrac{\partial h}{\partial x} = 0.$ $\qquad\qquad$ (1.85)

If the continuity equation is partially derived in time, the dynamic equation is partially derived by the x variable, and when added together a linear wave equation is obtained

$$\frac{\partial^2 h}{\partial t^2} = c^2\frac{\partial^2 h}{\partial x^2}. \tag{1.86}$$

A solution describes two waves: a pressure wave expressed by the piezometric head and the velocity wave described by discharge. An approximate solution will be obtained by application of the fundamental lemma; thus, the continuity equation will be written as

$$\int_L \left(\frac{gA}{c^2} \frac{\partial h}{\partial t} + \frac{\partial Q}{\partial x} \right) \delta h d\Omega = 0, \tag{1.87}$$

where the test function is equal to the piezometric head wave variation δh. Similarly, a fundamental lemma is written for the dynamic equation where the test function is equal to the discharge wave variation δQ

$$\int_L \left(\frac{1}{gA} \frac{\partial Q}{\partial t} + \frac{\partial h}{\partial x} \right) \delta Q d\Omega = 0. \tag{1.88}$$

The piezometric head $h(x,t)$ and discharge $Q(x,t)$ will be expressed by a linear combination of the basis functions

$$h = h_r(t), \varphi_r(x), \quad Q = Q_r(t), \varphi_r(x), \tag{1.89}$$

where linear combination parameters, that is nodal values, are time-dependent functions. According to Galerkin's procedure, variations are equal to the variations of basis vectors $\delta h = \delta Q = \varphi_s(x)$. Thus, after introducing them into Eqs (1.87) and (1.88)

$$\frac{gA}{c^2} \frac{dh_r}{dt} \int_L \varphi_s \varphi_r dx + Q_p \int_L \varphi_s \frac{d\varphi_p}{dx} dx = 0, \tag{1.90}$$

$$\frac{1}{gA} \frac{dQ_p}{dt} \int_L \varphi_s \varphi_p dx + h_r \int_L \varphi_s \frac{d\varphi_r}{dx} dx = 0, \tag{1.91}$$

written in the form of a discrete system of ordinary differential equations

for the continuity equation: $\dfrac{gA}{c^2} \underline{H}_{sr} \dfrac{dh_r}{dt} + \underline{Q}_{sp} Q_p = 0,$ \hfill (1.92)

for the dynamic equation: $\dfrac{1}{gA} \underline{H}_{sp} \dfrac{dQ_p}{dt} + \underline{Q}_{sr} h_r = 0.$ \hfill (1.93)

Indicated global system integrals are marked as the matrices $\underline{H}$ and $\underline{Q}$. The global system matrices are assembled using the finite element matrices. If a global base φ is replaced with the base Φ in Eqs (1.90) and (1.91) then, the elemental matrices for linear two-noded elements are obtained

$$\underline{H}^e_{sr} = \int_{\Delta L} \Phi_s \Phi_r dx = \Delta L \begin{bmatrix} \dfrac{1}{3} & \dfrac{1}{6} \\ \dfrac{1}{6} & \dfrac{1}{3} \end{bmatrix} \quad \text{and} \quad \underline{Q}^e_{sr} = \int_{\Delta L} \Phi_s \frac{d\Phi_r}{dx} dx = \begin{bmatrix} -\dfrac{1}{2} & +\dfrac{1}{2} \\ -\dfrac{1}{2} & +\dfrac{1}{2} \end{bmatrix}, \tag{1.94}$$

where ΔL is the finite element length and we obtain

$$\frac{gA\Delta L}{c^2}\begin{bmatrix} \dfrac{1}{3} & \dfrac{1}{6} \\[2mm] \dfrac{1}{6} & \dfrac{1}{3} \end{bmatrix} \cdot \begin{bmatrix} \dfrac{dh_1}{dt} \\[2mm] \dfrac{dh_2}{dt} \end{bmatrix} + \begin{bmatrix} -\dfrac{1}{2} & +\dfrac{1}{2} \\[2mm] -\dfrac{1}{2} & +\dfrac{1}{2} \end{bmatrix} \cdot \begin{bmatrix} Q_1 \\[2mm] Q_2 \end{bmatrix} = 0, \tag{1.95}$$

$$\frac{\Delta L}{gA}\begin{bmatrix} \dfrac{1}{3} & \dfrac{1}{6} \\[2mm] \dfrac{1}{6} & \dfrac{1}{3} \end{bmatrix} \cdot \begin{bmatrix} \dfrac{dQ_1}{dt} \\[2mm] \dfrac{dQ_2}{dt} \end{bmatrix} + \begin{bmatrix} -\dfrac{1}{2} & +\dfrac{1}{2} \\[2mm] -\dfrac{1}{2} & +\dfrac{1}{2} \end{bmatrix} \cdot \begin{bmatrix} h_1 \\[2mm] h_2 \end{bmatrix} = 0. \tag{1.96}$$

If both equations in Eq. (1.95) are added together, a numerical form of the continuity equation is obtained

$$\frac{gA\Delta L}{c^2}\begin{bmatrix} \dfrac{1}{2} & \dfrac{1}{2} \end{bmatrix} \cdot \begin{bmatrix} \dfrac{dh_1}{dt} \\[2mm] \dfrac{dh_2}{dt} \end{bmatrix} + \begin{bmatrix} -1 & +1 \end{bmatrix} \cdot \begin{bmatrix} Q_1 \\[2mm] Q_2 \end{bmatrix} = 0. \tag{1.97}$$

Similarly, if both equations in Eq. (1.96) are added, a numerical form of the dynamic equation is obtained

$$\frac{\Delta L}{gA}\begin{bmatrix} \dfrac{1}{2} & \dfrac{1}{2} \end{bmatrix} \cdot \begin{bmatrix} \dfrac{dQ_1}{dt} \\[2mm] \dfrac{dQ_2}{dt} \end{bmatrix} + \begin{bmatrix} -1 & +1 \end{bmatrix} \cdot \begin{bmatrix} h_1 \\[2mm] h_2 \end{bmatrix} = 0 \tag{1.98}$$

over a finite element.

1.2   Unified hydraulic networks

A hydraulic network is a system of linear hydraulic branches connected in nodes, such as the water supply network shown in Figure 1.10a. Hydraulic network braches can be pipelines, channels, pumps/turbines, different valves, and similar structures.

The pipe finite element mesh shown in Figure 1.10b and the channel finite element mesh shown in Figure 1.10c are also hydraulic networks. Although, in general, hydraulic networks can be made of multi-dimensional branches – finite elements – only linear branches will be analyzed here.

Hence, each branch is a finite element with one upstream and one downstream boundary discharge, see Figure 1.11. Each local node is associated with one global node. A positive discharge is defined in the direction from the upstream (local index 1) to the downstream (local index 2) node.

A finite element is an isolated part of an area over which a problem solution is known as either *accurate* or *approximate*. A solution is expressed in the parametric form as a function of boundary conditions. Provided that compatibility conditions are respected, a global system of equations can be obtained by assembling a system of elemental equations using the superposition principles. That procedure is termed

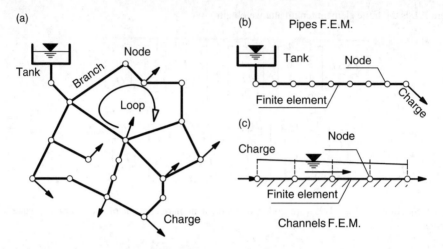

Figure 1.10   Hydraulic networks, an example.

a finite element technique. It has the property of universality, because it is applicable to different problems and, therefore, to flow modelling problems in hydraulic networks.

A hydraulic network configuration is defined by topological data, which requires marking of all network nodes and branches by a unique numerical mark or label, as shown in the example of Figure 1.12.

If these data are organized in the elemental connections table, which contains an index or label of the upstream and downstream nodes for each element, the system configuration can be assembled by the global system assembly algorithm using its constituents – finite elements. Thus, for example, Figure 1.13 shows an algorithm for a system configuration plot, written in pseudo programming language, where a plot can be drawn by superposition of several finite element plots like Figure 1.12.

The same superposition principle can be applied to the assembling of the equation system that defines hydraulic states – a global system is also assembled by superposition of respective equations for each finite element.

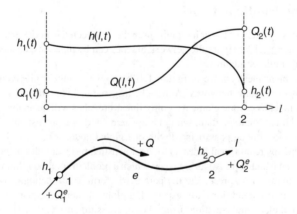

Figure 1.11   Branch – a finite element.

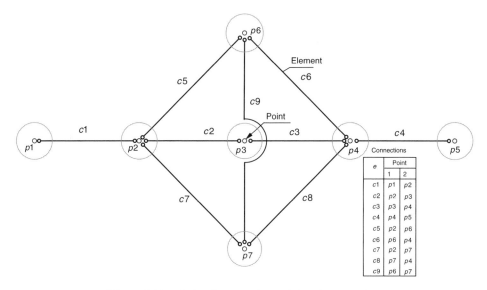

Figure 1.12 Hydraulic network as a finite element union.

1.3   Equation system

1.3.1   Elemental equations

Each branch of the hydraulic network is a *finite element* on which hydraulic states are defined by a solution of flow described by the continuity equation and the dynamic equation. A solution is sought for the known initial and boundary conditions and can be either accurate or approximate (analytical or numerical).

In general, two algebraic equations can be written from the known solution over an element, with the *boundary conditions as parameters*; namely, the upstream and downstream piezometric head h_r, h_s and the upstream and downstream boundary discharges Q_1^e, Q_2^e

$$F_1^e\left[h_r(t), h_s(t), Q_1^e(t), Q_2^e(t)\right] = 0$$
$$F_2^e\left[h_r(t), h_s(t), Q_1^e(t), Q_2^e(t)\right] = 0 \qquad (1.99)$$

```
n,m ; number of nodes and elements
connections(m,2) ; elemental connections
xy(n)    ; nodal xy coordinates

;loop over all elements:
for e = 1 to m do
        r = connection (e,1) ; first global node
        s = connection (e,2) ; second global node
        call MoveTo(xy(r)) ; move to point
        call LineTo(xy(s)) ; draw line to point
end loop e
```

Figure 1.13   Configuration plotting algorithm.

1.3.2   Nodal equations

Hydraulic states in a hydraulic network are defined by the piezometric head in N nodes $h_r(t)$, $r = 1, 2, 3, \ldots, N$, that can be used for the calculation of piezometric heads or discharges in every point of the network. In order to calculate the N unknown nodal piezometric heads, N independent equations shall be introduced; namely, the continuity equations in nodes

$$F_r = \sum_p {}^r Q^e_{k(p)} = 0, \tag{1.100}$$

where $Q^e_{k(p)}$ is the elemental boundary discharge and p is the number of finite elements – branches in a node.

Figure 1.14 shows an example of a node with three branches ($p = 3$), with the respective nodal equation

$$F_r = +Q^{e_1}_2 - Q^{e_2}_1 + Q^{e_3}_2 = 0. \tag{1.101}$$

In a hydraulic system there are $2M$ unknown elemental boundary discharges

$$Q^e_k = \left[Q^e_1, Q^e_2 \right]; \quad e = 1, 2, 3, \ldots, M \tag{1.102}$$

and N unknown nodal piezometric heads

$$h_r; \quad r = 1, 2, 3, \ldots, N. \tag{1.103}$$

In order to determine $N + 2M$ unknowns, $2M$ elemental equations of the kind as Eq. (1.99) and N nodal continuity equations shall be formed as Eq. (1.100) so that the system is formally closed.

Similar to the assembling of a hydraulic network configuration plot from separate finite element plots using the table of elemental connections, a nodal equation vector can be generated.

An algorithm for assembling the nodal equations vector, written in pseudo programming language, is shown in Figure 1.15.

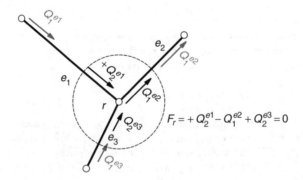

Figure 1.14   Nodal continuity equation.

```
n,m ; number of nodes and elements
connections (m,2) ; elemental connections
F(n)=0 ;empty nodal equations
; loop over all elements:
for e = 1 to m do
    ;loop over all elemental nodes:
    for k = 1 to 2 do
        r = connections(e,k) ; k-th global node
        ; complement r-th nodal equation:
        F(r) = F(r)+(-1)^kQ(k) ; k-th elemental equation
    end loop k
end loop e
```

Figure 1.15   The basis system nodal equations assembly algorithm.

Note that, in the nodal sum, the upstream (first) elemental discharge Q_1 refers to outflow from a node while the downstream (second) elemental discharge Q_2 is added to the node, see Figure 1.16. The algebraic sign of the elemental discharge in a nodal sum is defined by $(-1)^k$.

1.3.3   Fundamental system

Elemental (1.99) and nodal (1.100) equations form the global fundamental system of equations, which can be written in the following form

$$\Phi_i(U_j) = 0$$
$$i, j = 1, 2, 3, \ldots (N + 2M)$$,

(1.104)

where U_j is the vector of the unknowns $[h, Q^e]$, where the first N members are the unknown piezometric heads, while the remaining $2M$ are the unknown boundary elemental discharges. The *Newton–Raphson*

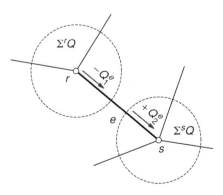

Figure 1.16   Elemental discharge contribution in nodal equation.

iterative procedure for the non-linear system of algebraic equations (1.104) is used in the following form

$$U_j^{(k+1)} = U_j^{(k)} + \Delta U_j, \tag{1.105}$$

where k represents the iterative step. Increments of the unknown ΔU_j are calculated by successive solving of the linear system of equations

$$\frac{\partial \Phi_i^{(k)}}{\partial U_j} \Delta U_j = -\Phi_i^{(k)}, \quad i, j = 1, 2, 3, \ldots (N + 2M), \tag{1.106}$$

that is as a solution of the matrix equation

$$J \Delta U = -\Phi \Rightarrow \Delta U = -J^{-1}\Phi. \tag{1.107}$$

Figure 1.17 presents the global equation system, where the Jacobian matrix **J** of the fundamental system consists of the following block matrices

$$\mathbf{J} = \begin{bmatrix} \mathbf{G} & \mathbf{Q} \\ \mathbf{HH} & \mathbf{QQ} \end{bmatrix}. \tag{1.108}$$

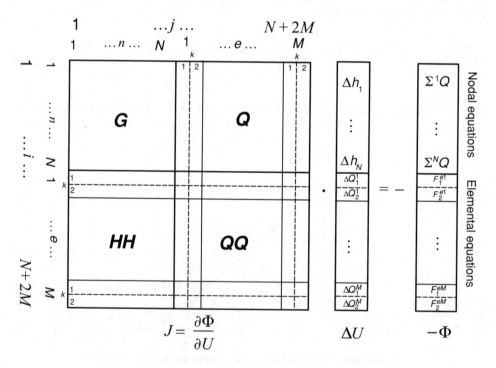

Figure 1.17　Graphic presentation of the global equation system.

Block matrices **G**, **Q** are the matrices of nodal equations, while **HH** and **QQ** are the matrices of elemental equations. Block **G** is written in the following form

$$\mathbf{G}_{rs} = \frac{\partial}{\partial h_s} \sum_p {}^r Q_k^e = 0$$

(1.109)

$$r, s = 1, 2, 3, \ldots, N$$

and contains a derivation of the nodal sum of the discharges $\sum_p {}^r Q_k^e$ by a variable h and is equal to the zero matrix. Block **Q** from the nodal equations contains derivations of the nodal sum of the discharges $\sum_p {}^r Q_k^e$ by elemental discharges. It is equal to

$$\mathbf{Q}_{rek} = \frac{\partial}{\partial Q_e} \sum_p {}^r Q_k^e = \begin{cases} -1 & k = 1 \\ +1 & k = 2 \end{cases}$$

$$r = 1, 2, 3, \ldots, N$$
$$e = 1, 2, 3, \ldots, M$$
$$k = 1, 2$$

(1.110)

and contains either -1 or $+1$ depending on the discharge algebraic sign. If the Newton–Raphson procedure is applied to elemental equations (1.99), then we obtain:

$$\begin{bmatrix} \dfrac{\partial F_1^e}{\partial h_r} & \dfrac{\partial F_1^e}{\partial h_s} \\ \dfrac{\partial F_2^e}{\partial h_r} & \dfrac{\partial F_2^e}{\partial h_s} \end{bmatrix} \cdot \begin{bmatrix} \Delta h_r \\ \Delta h_s \end{bmatrix} + \begin{bmatrix} \dfrac{\partial F_1^e}{\partial Q_1} & \dfrac{\partial F_1^e}{\partial Q_2} \\ \dfrac{\partial F_2^e}{\partial Q_1} & \dfrac{\partial F_2^e}{\partial Q_2} \end{bmatrix} \cdot \begin{bmatrix} \Delta Q_1 \\ \Delta Q_2 \end{bmatrix} = - \begin{bmatrix} F_1^e \\ F_2^e \end{bmatrix}.$$

(1.111)

Block **HH** from the elemental equations contains derivations of elemental equations by nodal piezometric heads

$$\mathbf{HH}_{k,s}^e = \frac{\partial F_k^e}{\partial h_s}$$

$$k = 1, 2$$
$$e = 1, 2, 3, \ldots, M$$
$$s = 1, 2, 3, \ldots, N$$

(1.112)

Block **QQ** from the elemental equations contains derivations of elemental equations by elemental discharges

$$\mathbf{QQ}_{k,s}^e = \frac{\partial F_k^e}{\partial Q_l}$$

$$k, l = 1, 2$$

$$e = 1, 2, 3, \ldots, M$$

(1.113)

Note that there are several possible modifications of the Newton–Raphson procedure, because the accurate partial gradient calculations are not necessary for convergence. It can easily be proved with an example of the calculation of a zero point of a non-linear equation with one unknown, see Figure 1.18.

In the system of non-linear equations that is common in hydraulic network solving, convergence is usually quadratic type convergence. However, in general, a convergence of an iterative procedure cannot

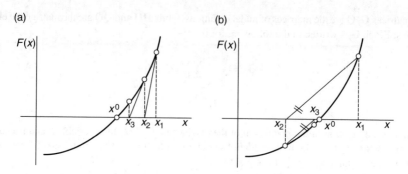

Figure 1.18   An illustration of the Newton–Raphson procedure: (a) tangent method (b) fixed secant method.

be guaranteed without detailed analysis. Divergence usually appears in the weak formulation of the Jacobian matrix. In general, according to the *Banach*[22] *fixed point theorem*, an iterative procedure will converge if the mapping is a contraction.

1.4   Boundary conditions

1.4.1   Natural boundary conditions

A fundamental system of equations is unsolvable because it is irregular without particular conditions, namely, boundary conditions.

Nodes could have different functions; for example a hydraulic network contains other hydraulic objects or structures such as water tanks, surge tanks, air tanks, relief valves, and similar.

These are the places where the system communicates with the external space; namely, nodal functions are the *natural boundary conditions* of a hydraulic system. Boundary conditions extend the fundamental system by an additional discharge member

$$Q_r^0 = Q_r^0 \left(t, \frac{dh_r}{dt}, h_r \right),$$ (1.114)

See Figure 1.19, which is positive for external inflow. Thus, the extended nodal continuity function has the following form

$$\overbrace{\sum_p {}^r Q^e}^{\text{fundamental part}} + \overbrace{Q_r^0}^{\text{boundary conditions}} = 0.$$ (1.115)

The global equation system obtains a form as shown in Figure 1.20:

$$\frac{\partial}{\partial U_j} \left(\Phi_i^{(k)} + Q_i^0 \right) \Delta U_j = - \left(\Phi_i^{(k)} + Q_i^0 \right)$$ (1.116)

---

[22]Stefan Banach, a Polish mathematician (1892–1945).

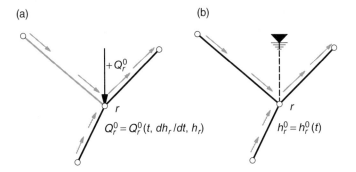

Figure 1.19 Boundary conditions, extension or modification of the nodal equation.

it is modified only in part nodal equations and as an addition to the right hand side vector

$$F_r^{new} = F_r^{old} - Q_r^0 \tag{1.117}$$

and an addition to the matrix

$$G_{r,r}^{new} = G_{r,r}^{old} + \frac{\partial Q_r^0}{\partial h_r}. \tag{1.118}$$

Steady state

Updating of the global vector (the right hand side in the system of equations) in the steady state is simple, due to the additional property of discharge, and has the form of Eq. (1.117).

If the discharge Q_r^0 depends on the nodal piezometric head h_r, the global system matrix shall also be updated by a respective partial derivative (1.118).

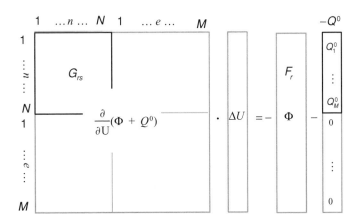

Figure 1.20 Fundamental system extension by natural boundary condition.

In a steady flow for $t = t_0$, the fundamental system is updated by natural boundary conditions in the form of an additional discharge term $Q_r^0 (t_0, h_r)$, which is positive in case of an external inflow into node r. Two types of external inflow will be considered:

(a) in the *explicit* form $Q_r^0(t_0) = Q_0 + f G(t_0)$, where Q_0 is the constant, $G(t)$ is the graph object (see Chapter 2, Section 2.2 Gradually varied flow in time) , f is the factor that the graph value for $t = t_0$ is multiplied by,
(b) in the *implicit* form $Q_r^0(t_0) = A + B h_r$, where A and B are the constants. This boundary condition depends on the piezometric head in node r.

Non-steady state

In non-steady flow, the initial fundamental system refers to the volume balance in a given time step; thus the system is updated by the respective nodal volume

$$\Delta V_r^0 = \int_{\Delta t} Q_r^0 dt = (1 - \vartheta) \Delta t\, Q_r^0 + \vartheta\, \Delta t\, Q_r^{0+} \tag{1.119}$$

which is added to the global system vector

$$F_r^{new} = F_r^{old} - \Delta V_r^0.$$

Respective updating of the global matrix has the following form

$$G_{r,r}^{new} = G_{r,r}^{old} + \frac{\partial \Delta V_r^0}{\partial h_r^+}.$$

1.4.2   Essential boundary conditions

Even when natural boundary conditions are added, they still do not ensure regularity of the equation system; for example if $Q_r^0 = const$ or $Q_r^0 = Q_r^0(t)$, then all the solutions that differ in a constant will satisfy the elemental equations. Thus, an *essential* boundary condition shall be added, such as the piezometric head h_r^0 prescribed in at least one node; see Figure 1.19b.

If Q_r^0 is a function of the nodal piezometric head $Q_r^0(h_r)$, then respective derivatives shall be added to the Jacobian matrix.

An essential boundary condition, such as the prescribed piezometric head $h_r = h_r^0$, is introduced by a modification of the r-th row of the global system of nodal equations. Since the solution for asymmetric systems is by simple replacement of the new equation, the existing r-th row is erased, the main diagonal is set to 1, the r-th vector member is set to 0, and the solution (Δh_r increment) will be 0; thus, the prescribed value remains unchanged.

1.5   Finite element matrix and vector

The solution of the global equation system (a fundamental system extended with boundary conditions) in the full matrix form $(N + 2M) \times (N + 2M)$ is neither appropriate nor efficient due to unfavorable filling in of the matrix. However, note that the discharge increment ΔQ^e can be eliminated from the nodal equations prior to the filling of the fundamental system matrix. Then, the nodal equations will contain only the unknown increments of nodal piezometric heads; namely, the equation system will be reduced to the filled **G** matrix solving.

Thus, hydraulic network problem solving can be reduced to the standard procedure that is efficiently applied in the finite element technique:

- calculation of the finite element matrix and vector,
- filling in of the system global matrix,
- system extension with boundary conditions,
- equation system solving.

Not only that, it is shown that the same finite element matrix and vector generation procedure can be applied to different types of hydraulic branches, both for steady and non-steady states.

Generalized elemental equations will be written again in the following form

$$
\begin{aligned}
F_1^e\left[h_r(t), h_s(t), Q_1^e(t), Q_2^e(t)\right] = 0 \\
F_2^e\left[h_r(t), h_s(t), Q_1^e(t), Q_2^e(t)\right] = 0
\end{aligned}
\tag{1.120}
$$

where the first local node of an element corresponds to the r-th global node while the second local node corresponds to the s-th global node.

If the Newton–Raphson method is applied to the elemental equations, then

$$
\begin{bmatrix} \dfrac{\partial F_1^e}{\partial h_r} & \dfrac{\partial F_1^e}{\partial h_s} \\[2ex] \dfrac{\partial F_2^e}{\partial h_r} & \dfrac{\partial F_2^e}{\partial h_s} \end{bmatrix} \cdot \begin{bmatrix} \Delta h_r \\[1ex] \Delta h_s \end{bmatrix} + \begin{bmatrix} \dfrac{\partial F_1^e}{\partial Q_1} & \dfrac{\partial F_1^e}{\partial Q_2} \\[2ex] \dfrac{\partial F_2^e}{\partial Q_1} & \dfrac{\partial F_2^e}{\partial Q_2} \end{bmatrix} \cdot \begin{bmatrix} \Delta Q_1 \\[1ex] \Delta Q_2 \end{bmatrix} = - \begin{bmatrix} F_1^e \\[1ex] F_2^e \end{bmatrix}
\tag{1.121}
$$

and formally written using matrix-vector operations

$$
[H] \cdot [\Delta h] + [Q] \cdot [\Delta Q] = [F].
\tag{1.122}
$$

Figure 1.21 shows the global system filling following the addition of the first element e. Note that the nodal discharge sum vector is filled by the pseudo code algorithm shown in Figure 1.15, while in the Jacobian matrix (block **Q**), which refers to the derivation of nodal equations by discharge, there are values $+1$ or -1, depending on whether the discharge is of an inflow or outflow. Part of the Jacobian matrix (block **G**), which refers to the derivation of nodal equations by piezometric heads, remains empty.

When the elemental equations (1.122) are multiplied by the inverse matrix $\left[Q\right]^{-1}$

$$
\left[Q\right]^{-1}[H] \cdot [\Delta h] + \left[Q\right]^{-1}[Q] \cdot [\Delta Q] = \left[Q\right]^{-1}[F],
\tag{1.123}
$$

that is, after arranging, the following is obtained

$$
[A] \cdot [\Delta h] + [\Delta Q] = [B],
\tag{1.124}
$$

where

$$
[A] = \left[Q\right]^{-1}[H],
\tag{1.125}
$$

$$
[B] = \left[Q\right]^{-1}[F].
\tag{1.126}
$$

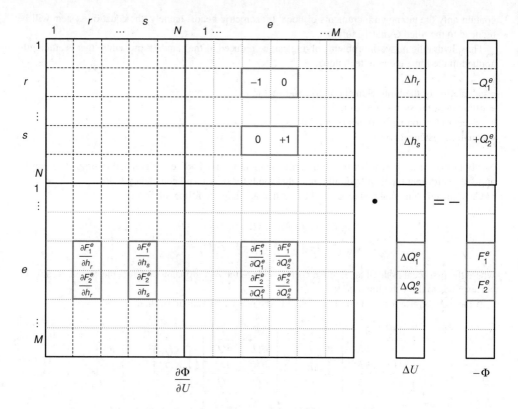

Figure 1.21   Global system filling width one element.

Figure 1.22 shows the global system following the operations carried out on elemental equations of the added single element e.

If the algebraic sign is altered in the second row of the elemental equation, then elemental discharge increments in nodal equations can be eliminated by adding the first row of the elemental equation with the r-th nodal equation and the second row of the elemental equation with the s-th nodal equation.

Accordingly, the finite element matrix A^e and vector B^e can be generated in the following form:

$$A^e = \begin{bmatrix} +\underline{A}_{11} & +\underline{A}_{12} \\ -\underline{A}_{21} & -\underline{A}_{22} \end{bmatrix}^e, \tag{1.127}$$

$$B^e = \underbrace{\begin{bmatrix} +Q_1^e \\ -Q_2^e \end{bmatrix}}_{(1)} + \underbrace{\begin{bmatrix} +\underline{B}_1 \\ -\underline{B}_2 \end{bmatrix}^e}_{(2)}. \tag{1.128}$$

This will fill the block matrix **G** by the assembling procedure, while the block matrix **Q** will become an empty matrix, as shown for one element in Figure 1.23. Member (1) is the vector part before, while member (2) is the vector part after, elimination of elemental equations from the nodal sums.

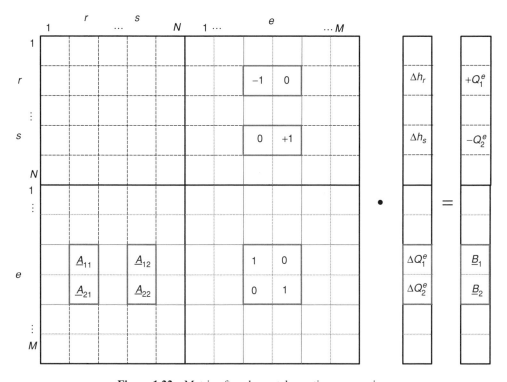

Figure 1.22   Matrix after elemental equation rearranging.

When the described procedure is applied to all finite elements, a fundamental system is modified into nodal equations without the elemental discharge increments in the Newton–Raphson form

$$\frac{\partial F_r^{(k)}}{\partial h_s}\Delta h_s = -\Gamma_r^{(k)}$$

$$r, s = 1, 2, 3, \ldots N$$

(1.129)

and the elemental equation (1.124)

$$\left[\underline{A}^e\right]\cdot\left[\Delta h^e\right]+\left[\Delta Q^e\right]=\left[\underline{B}^e\right].$$

(1.130)

After boundary conditions are introduced, nodal equations become solvable by the unknown increments of piezometric heads since they do not depend on elemental equations.

Unknown increments of elemental discharges can be calculated from the modified elemental equations (1.130) after the unknown nodal increments of piezometric heads are calculated

$$\left[\Delta Q^e\right]=\left[\underline{B}^e\right]-\left[\underline{A}^e\right]\cdot\left[\Delta h^e\right].$$

(1.131)

A procedure of finite element matrix and vector generation when two elemental equations are used was presented in the previous text. When an incompressible liquid $\rho = cons$ is being modeled, the upstream

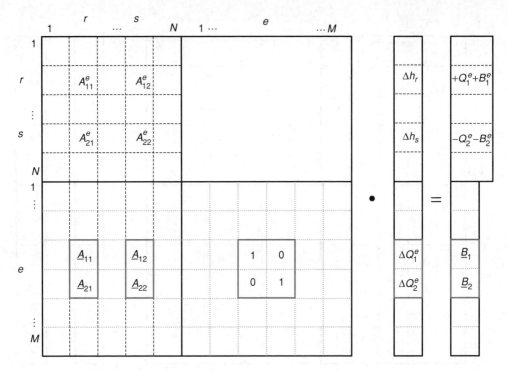

Figure 1.23   Global system after elimination of the first element elemental discharges.

and downstream boundary discharges are equal $Q_1^e = Q_2^e = Q^e$; thus, only a dynamic equation is used on a finite element that leads to the following form

$$F^e \left[h_r, h_s, Q^e \right] = 0. \tag{1.132}$$

Figure 1.24 shows the extended global Jacobian matrix at the moment of the first element assembling. Then, the Newton–Raphson iterative form of the element e will be

$$\left[\frac{\partial F^e}{\partial h_r} \quad \frac{\partial F^e}{\partial h_s} \right] \cdot \left[\begin{matrix} \Delta h_r \\ \Delta h_s \end{matrix} \right] + \frac{\partial F^e}{\partial Q^e} \Delta Q^e = -F^e. \tag{1.133}$$

This is formally written using the matrix-vector operations

$$\left[\underline{H} \right] \cdot [\Delta h] + \left[\underline{Q} \right] \cdot [\Delta Q] = \left[\underline{F} \right]. \tag{1.134}$$

When the previous expression is multiplied by the inverse member $\left[\underline{Q} \right]^{-1}$ then

$$\left[\underline{A} \right] \cdot [\Delta h] + [\Delta Q] = \left[\underline{B} \right], \tag{1.135}$$

from which an increase in the elemental discharge can be calculated as

$$[\Delta Q] = \left[\underline{B} \right] - \left[\underline{A} \right] [\Delta h], \tag{1.136}$$

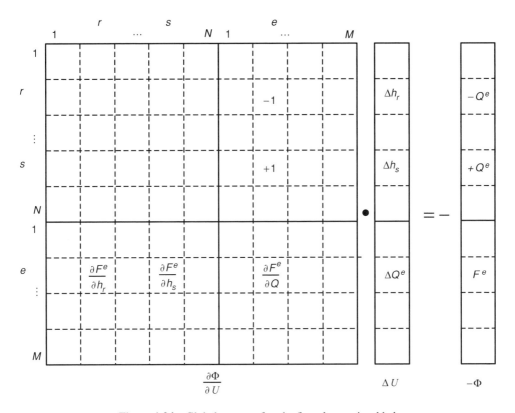

Figure 1.24 Global system after the first element is added.

where

$$[\underline{A}] = [\underline{Q}]^{-1}[\underline{H}],$$ (1.137)

$$[\underline{B}] = [\underline{Q}]^{-1}[\underline{F}].$$ (1.138)

In previous expressions, $[\underline{A}]$ is a two-member vector while $[\underline{B}]$ is a scalar. A process of elimination of elemental discharge increments from the nodal continuity equations defines the structure of the finite element matrix

$$A^e = \begin{bmatrix} +\underline{A} \\ -\underline{A} \end{bmatrix}$$ (1.139)

and vector

$$B^e = \underbrace{\begin{bmatrix} +Q^e \\ -Q^e \end{bmatrix}}_{(1)} + \underbrace{\begin{bmatrix} +\underline{B} \\ -\underline{B} \end{bmatrix}}_{(2)}.$$ (1.140)

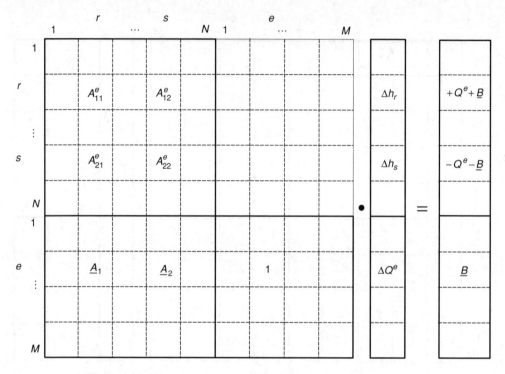

Figure 1.25   Situation after elemental discharge increases elimination.

Member (1) is an existing member on the right hand side before elimination, while (2) is the residual after elemental discharge elimination from the nodal equation.

Figure 1.25 shows the global system after elimination of the elemental discharge increments from nodal equations; namely, after the first element matrix and vector assembling.

Reference

Jović, V. (1993) *Introduction to Numerical Engineering Modelling* (in Croatian). Aquarius Engineering, Split.

Further reading

Connor, J.C. and Brebbia, C.A. (1976) *Finite Element Techniques for Fluid Flow*. Butterworths, London.

Hinton, E., Owen, D.R.J. (1977) *Finite Element Programming*. Academic Press, London.

Hinton, E., Owen, D.R.J. (1979) *An Introduction to Finite Element Computation*. Pineridge Press Ltd, Swansea.

Irons, B.M. (1970) A frontal solution program. *Int. J. Num. Meth.* 2, 5–32.

2

Modelling of Incompressible Fluid Flow

2.1 Steady flow of an incompressible fluid

2.1.1 Equation of steady flow in pipes

Due to small velocities and relatively long pipeline length, it is assumed that the velocity head and all local losses are negligible when compared to linear friction resistance. Accordingly, the continuity and dynamic equations will be

$$Q = Av = const, \tag{2.1}$$

$$\frac{dh}{dl} + \frac{\lambda}{D}\frac{v^2}{2g} = 0. \tag{2.2}$$

The dynamic equation integrated over a finite element of the length L has the following form

$$h_2 - h_1 + \lambda\frac{L}{D}\frac{v^2}{2g} = 0, \tag{2.3}$$

where h_1, h_2 are the piezometric heads at the upstream and downstream ends of the pipe, $\lambda(R_e, \varepsilon/D)$ is the Darcy[1]–Weisbach[2] friction factor, L is the pipe length, D is the pipe diameter, v is the mean velocity, and g is the gravity acceleration. Figure 2.1 shows hydraulic heads and losses on the pipe element.

Coefficient λ depends on the Reynolds[3] number R_e and relative roughness ε/D where ε is the absolute hydraulic roughness.

---

[1] Henry Darcy, French engineer (1803–1853).
[2] Julius Weisbach, German mathematician and engineer (1806–1871).
[3] Osborne, Reynolds, British professor of engineering (1848–1912).

Analysis and Modelling of Non-Steady Flow in Pipe and Channel Networks, First Edition. Vinko Jović.
© 2013 John Wiley & Sons, Ltd. Published 2013 by John Wiley & Sons, Ltd.

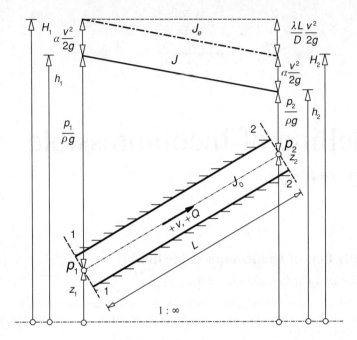

Figure 2.1 Pipe finite element.

This is determined from the Moody[4] chart, shown in Figure 2.2, which represents the synthesis of the tests carried out by Nikuradze[5] and the Colebrook–White[6] analyses of measurements of the resistance to flow in technical pipes.

If $R_e < 2320$, the flow is laminar and the Hagen[7]–Poisseuille[8] law can be applied

$$\lambda = \frac{64}{R_e},$$
(2.4)

while the Colebrook–White equation is used for larger Reynolds numbers

$$\frac{1}{\sqrt{\lambda}} = 1,14 - 2\log\left(\frac{k}{D} + \frac{9,35}{R_e\sqrt{\lambda}}\right),$$
(2.5)

where k is the absolute hydraulic roughness. The Colebrook–White equation asymptotically comprises both the fully turbulent rough and smooth flow in conduits. For calculation of the Darcy–Weisbach resistance coefficient λ, a functional procedure from the module Hydraulics.f90 will be used

```
real*8 function aMoody(Re,rr),
```

---

[4]Lewis Ferry Moody (1875–1953).
[5]Johan Nikuradze, Georgian hydraulic engineer, worked in Goettingen, Germany.
[6]C. M. White and C. F. Colebrook: Fluid friction in roughened pipes, *Proceeding of the Royal Society, London*, 1937.
[7]Gotthilf Heinrich Ludwig Hagen (1799–1884).
[8]Jean Lous Poisseuille (1799–1869).

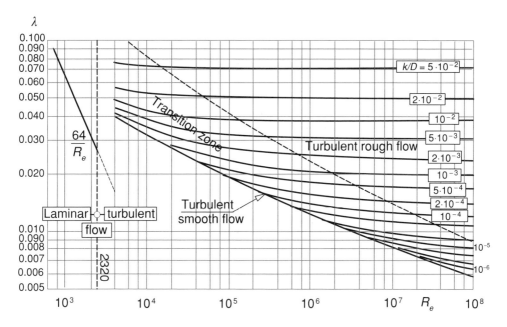

Figure 2.2   Moody chart.

which calculates the coefficient λ for the prescribed Reynolds number and relative roughness `rr`. If in expression (2.3) mean velocity is expressed using the discharge $v = Q/A$, where A is the area of the pipe cross-section, the following expression is obtained

$$h_2 - h_1 + \lambda \frac{L}{2gDA^2} Q^2 = 0 \tag{2.6}$$

that can be written in the form

$$F^e: \quad h_2 - h_1 + \beta \, |Q| \, Q = 0 \tag{2.7}$$

with the resistance term

$$\beta(Q) = \frac{L}{2gDA^2} \lambda(Q). \tag{2.8}$$

Expression (2.7) is *the dynamic equation of the steady flow in pipes*, which determines the correct algebraic sign of the piezometric head difference between the "upstream" and "downstream" pipe end. Namely, a positive discharge Q is defined as the flow in the direction from the first towards the second end of a pipe, and resistances always resist the flow. Thus, in the resistance term an absolute value of discharge Q is used, so the resistance term β is always positive.

2.1.2   Subroutine `SteadyPipeMtx`

Computation procedure

The pipe finite element matrix and vector for the steady flow are calculated in a subroutine

<code>subroutine SteadyPipeMtx(ielem).</code>

Although in the implementation of the subroutine `SteadyPipeMtx` the pipe finite element matrix and vector for the steady flow are calculated by a numerical procedure, their extended form will also be written here. Thus, for the elemental equation:

$$F^e: \quad h_2 - h_1 + \beta \, |Q| \, Q = 0 \tag{2.9}$$

the Newton–Raphson iterative form is written

$$\begin{bmatrix} \dfrac{\partial F^e}{\partial h_1} & \dfrac{\partial F^e}{\partial h_2} \end{bmatrix} \cdot \begin{bmatrix} \Delta h_1 \\ \Delta h_2 \end{bmatrix} + \dfrac{\partial F^e}{\partial Q} \Delta Q^e = -F^e, \tag{2.10}$$

which is formally written using the matrix-vector operations

$$[\underline{H}] \cdot [\Delta h] + [\underline{Q}] \cdot [\Delta Q] = [\underline{F}]. \tag{2.11}$$

When the previous equation is multiplied by the inverse term $[\underline{Q}]^{-1}$, then

$$[\underline{A}] \cdot [\Delta h] + [\Delta Q] = [\underline{B}] \tag{2.12}$$

from which the value of the elemental discharge increment can be calculated

$$[\Delta Q] = [\underline{B}] - [\underline{A}] [\Delta h]. \tag{2.13}$$

Scalar value $[\underline{F}]$ is equal to

$$[\underline{F}] = -(h_2 - h_1 + \beta \, |Q| \, Q), \tag{2.14}$$

while vector $[\underline{H}]$ has the form

$$[\underline{H}] = \begin{bmatrix} \dfrac{\partial F^e}{\partial h_1} & \dfrac{\partial F^e}{\partial h_2} \end{bmatrix} = \begin{bmatrix} -1 & 1 \end{bmatrix}. \tag{2.15}$$

Scalar value $[\underline{Q}]$ is equal to

$$[\underline{Q}] = \dfrac{\partial F^e}{\partial Q} = 2\beta \, |Q| + |Q| \, Q \dfrac{d\beta}{dQ}, \tag{2.16}$$

where the second term in the equation appears due to the dependence of the resistance term on the Reynolds number. The iterative process will still converge even when that term is omitted. The inverse value $[\underline{Q}]^{-1}$ equals to

$$[\underline{Q}]^{-1} = \dfrac{1}{2\beta \, |Q|}. \tag{2.17}$$

Vector value $[\underline{A}]$ is equal to

$$[\underline{A}] = [\underline{Q}]^{-1} [\underline{H}] = \left[-\frac{1}{2\beta\,|Q|} \quad \frac{1}{2\beta\,|Q|} \right] \tag{2.18}$$

and the scalar value $[\underline{B}]$ is

$$[\underline{B}] = [\underline{Q}]^{-1} [\underline{F}] = -\frac{h_2 - h_1}{2\beta\,|Q|} - \frac{Q}{2}. \tag{2.19}$$

The pipe finite element matrix has the following form

$$A^e = \begin{bmatrix} +\underline{A} \\ -\underline{A} \end{bmatrix} = \begin{bmatrix} -\dfrac{1}{2\beta\,|Q|} & \dfrac{1}{2\beta\,|Q|} \\[2ex] \dfrac{1}{2\beta\,|Q|} & -\dfrac{1}{2\beta\,|Q|} \end{bmatrix}. \tag{2.20}$$

The pipe finite element vector has the following form

$$B^e = \underbrace{\begin{bmatrix} +Q \\ -Q \end{bmatrix}}_{(1)} + \underbrace{\begin{bmatrix} -\dfrac{h_2 - h_1}{2\beta\,|Q|} - \dfrac{Q}{2} \\[2ex] \dfrac{h_2 - h_1}{2\beta\,|Q|} + \dfrac{Q}{2} \end{bmatrix}}_{(2)}. \tag{2.21}$$

Term (1) is the existing term on the right hand side before elimination, while term (2) is the contribution following the elimination of the elemental discharge from the nodal equation.

In the matrix and vector expressions, there is a division with the absolute value of discharge; thus, a division with zero shall also be considered. In that case, one shall take into account that for small flow rates the flow is laminar, the resistances are proportional to the discharge, and the elemental equation has the linear form

$$F^e: \quad h_2 - h_1 + \beta^* Q = 0, \tag{2.22}$$

where

$$\beta^* = \frac{64\nu}{2g\,D^2 A}L. \tag{2.23}$$

According to this, a derivation shall be applied for $Q = 0$

$$[\underline{Q}] = \frac{\partial F^e}{\partial Q} = \beta^*, \tag{2.24}$$

that is in expressions (2.20) and (2.21) the term $2\beta\,|Q|$ shall be replaced by β^*. Actually, the linear hydraulic resistance law can be applied when the Reynolds number ≤ 2320. The implemented function aMoody will provide accurate values for each $|Q| \neq 0$.

```
n,m          ; number of nodes and elements
connect(m,2) ; element connectivities
F(n)=0       ; fundamental  vector
G(n,n)=0     ; fundamental  matrix
Ae(2,2)      ; elemental matrix
Be(2)        ; elemental vector
; loop over all elements:
for e = 1 to m do
     call SteadyPipeMtx(Ae,Be) ; compute Ae, Be
     ; loop over all elemental nodes:
     for k = 1 to 2 do
       r = connect(e,k)        ; k-th global node
       F(r) = F(r)+Be(k)       ; fill in the vector
       For l = 1 to 2 do
         s = connect(e,l)     ; l-th global node
         G(r,s) = G(r,s)+Ae(k,l)  ; fill in the matrix
       end loop l
     end loop k
end loop e
```

Figure 2.3 Assembly: Algorithm of the fundamental system assembling.

2.1.3   Algorithms and procedures

Fundamental system assembling

Figure 2.3 shows the algorithm for the fundamental system assembling, written in the pseudo program language. The matrix G and vector F of the fundamental system are filled in with the contributions of respective finite element matrices and vectors; namely, a connectivity table and the principle of superposing are used. At www.wiley.com/go/jovic, in Appendix A Program solutions, there is a Fortran implementation of the procedure subroutine Assembly.

Banded system of equations

A block matrix G of the fundamental system has a banded form, see Figure 2.4a. Matrices of the equation systems that originate from the finite element technique (this also applies to the finite difference technique) are symmetrical and banded matrices; namely, they are filled in the narrow band between the two end diagonals.

The number of diagonals p beside the main diagonal is called the band width.[9] It is determined as the largest of all differences of nodal indexes over all elements and is easily calculated from the connectivity table connect(e,c).

As can be seen, values of the global matrix outside of the end diagonals are equal to zero and do not participate in the elimination algorithm. Thus, it is recommended to use the economical memory organization of the matrix and the elimination algorithm adapted to the economical form, see Figure 2.4b. In that case, matrix filling shall be modified to suit the economical form. The only difference from the

---

[9]it is actually a half band width.

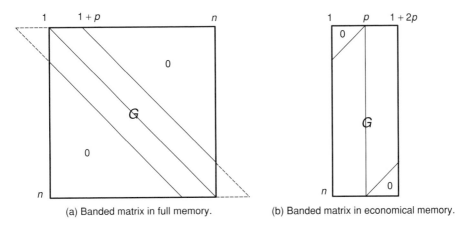

(a) Banded matrix in full memory.   (b) Banded matrix in economical memory.

Figure 2.4 Full and economical memory organization of the system matrix.

filling of the full matrix is that the target columns in the economical form are calculated from the formula:

$$s' = s - r + p + 1. \tag{2.25}$$

Note that the position of the main diagonal in the economical form is equal to $1 + p$. For the purposes of band width calculation, the respective program solution shall be written such as the function procedure:

```
integer function lbandw(),
```

see www.wiley.com/go/jovic – Appendix A Program solutions. The aforementioned appendix also contains the subroutine:

```
subroutine bandsol(a,b,n,m),
```

which solves the system of equations written in the economical form of the memory organization.

Algorithm of the solution

Figure 2.5 shows an overview of the algorithm for solving the problem of a hydraulic network by the finite element technique, written in a pseudo program language.

The solution is sought by an iterative procedure. Namely, from the system of nodal equations, increments of nodal piezometric heads are calculated first:

$$h_r^{(k+1)} = h_r^{(k)} + \Delta h_r \tag{2.26}$$

and then the elemental discharge increments from the expression (2.13):

$$\left[\Delta Q^e \right] = \left[\underline{B}^e \right] - \left[\underline{A}^e \right] \cdot \left[\Delta h^e \right]. \tag{2.27}$$

Accuracy shall be tested in each iterative step. If the accuracy is satisfactory, iteration stops. An educational software for modelling steady flow in pipe hydraulic networks, named SimpleSteady

```
n,m          ; number of nodes and elements
connect(m,2) ; element connectivities
F(n),U(n)=0  ; F fundamental  vector, in the return procedure equal to the solution U
G(n,n)=0     ; fundamental  matrix
H(n)         ; nodal piezometric heads
Q(m)         ; elemental discharges
dQ           ; elemental discharge increment
; iterative loop
for iter = 1 to maxiter do
     call assembly(G,F) ; assembly fundamental  matrix G and vector F
     call applyBDC(G,F) ; apply boundary conditions
     call solver(G,F)    ; F is a solution of nodal increments U
     for r=1 to n
        H(r) = H(r) + U(r)   ; add piezometric head increment
     end loop r
     dh_max = max_U      ; max H increment standard
     ; loop over all elements:
     for e = 1 to m do
        r = connect(e,1)     ; first global node
        s = connect(e,2)     ; second global node
        dh = [U(r),U(s)]     ; vector of elemental nodal increments
        dQ = B - A*dh        ; elemental discharge increment
        Q(e) = Q(e)+dQ       ; add elemental discharge increment
        dQ_max = max_dQ      ; max Q increment standard
     end loop e
     if dh_max <= epsH and dQ_max <= epsQ then exit iter
end loop iter
call output  ; print output data
```

Figure 2.5 SimpleSteady – algorithm of the solution obtained by the finite element technique.

(www.wiley.com/go/jovic) is based on the presented algorithms and program pseudo code of the procedures:

```
program SimpleSteady
use SSglobalVars  ! memory module
  call input              ! read data file
  call SolveSteady        ! steady solution
  call output             ! print results
end program SimpleSteady
```

It is written in Fortran, the implementation of which can be found in Appendix A – Program solution, and the accompanying website (www.wiley.com/go/jovic). Apart from the Fortran sources, a small test example can also be found in Appendix A.

Drawbacks of the solution with the banded matrix

Modelling a hydraulic network requires respective numeration, i.e. node numeration in a continuous numerical sequence. The numeration order shall be in the shortest possible direction so as to obtain the least possible band width of the global matrix. All finite elements shall also be numerated. The order of pipe element numeration is irrelevant; however, it still has to be a continuous numerical sequence.

Input data are organized in the input file, which contains the data on the number of nodes and elements and information on branches, that is elements from a previously numerated network (pipe indexes and connections of pipes with nodes, pipe diameter, absolute pipe roughness, and pipe length). Then there are boundary conditions: node number, nodal type, and boundary condition value.

A numerical model of a hydraulic network can be used for different purposes; thus, depending on the particulars, some branches shall be added and others annulled, which results in either an increase or decrease in the number of elements and nodes. In that case a new topological scheme of the network, new numbering, and a new input file shall be elaborated.

An educational example of the numerical model is obviously not flexible enough for large hydraulic networks. Therefore, it should be substantially modified in terms of data inputs; for example it should incorporate automatic mesh re-numeration with the minimization of the matrix band width and similar.

Instead of an educational example being updated and improved to make it a professional one, based on the solution of nodal equations using the banded matrix, this book will present the program solution `SimpipCore` originating from the frontal technique for assembling and solving the system of equations.

2.1.4   Frontal procedure

Assembly, boundary conditions, and elimination

The frontal procedure is a unique method for assembling and eliminating nodal equations in the finite element technique. It was developed by B. M. Irons (1970), and is well documented in Hinton and Owens (1977).

The basic idea arises from the fact that each node in a mesh can be related to one row and one column in the global system of equations, which are assembled by superposition of the contributions of elemental matrices and vectors of finite elements. Superposition of the row and column of a node is completed when contribution of the last element connected with that node is added. At that moment, a nodal equation can be eliminated.

Let us generate a finite element mesh of a hydraulic system (*one-dimensional two-nodal elements*), as in the example shown in Figure 2.6, consisting of six finite elements.

Let each element have a matrix and vector of 2 x 2 in size, which are, according to the connectivity table (see Figure 2.6) transferred into the global matrix of 5 x 5 in size. Thus, one matrix row corresponds to one node.

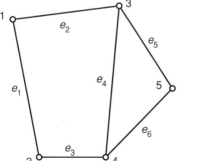

e	p_1	p_2
1	1	2
2	1	3
3	2	4
4	3	4
5	3	5
6	4	5

Figure 2.6   An example of the finite element mesh and connectivities.

Table 2.1   Global matrix and vector of the right hand side after adding the first
three elements. Following the elimination of the 1st and 2nd node, only the shaded
part of the global matrix remains active.

Node	1	2	3	4	5	
1	(1)+(2)	(1)	(2)	0		(1) + (2)
2	(1)	(1) + (3)	0	(3)		(1)
3	(2)	0	(2)	0		(2)
4	0	(3)	0	(3)		
5						

Table 2.1 shows filling of the global matrix after processing the first three elements, where the number in parenthesis denotes the element that generates the contribution.

If further filling of the matrix is imagined, note that the row 1 and column 1, as well as the respective right hand side, remain unchanged. The reason is that none of the remaining elements contain node 1, and their matrices do not contribute to the row and column of the node 1. Thus, the first row and the first column, as well as the first term of the right hand side vector already have the final form, and there is no reason not to immediately eliminate the nodal unknown by elimination procedure from all equations, that is it can be eliminated from the first column. Following the elimination of the node 1, memory space occupied by that row and column becomes free.

It is understood that, before reusing that space, the eliminated equation was recorded somewhere, that is saved for the subsequent substitution procedure for the computation of unknowns.

This example shows that everything that applies to the first node can also be applied to the second, since the second row and the second column already have a final form and can be eliminated too. Since the elemental matrices and vectors are being added according to the sequence order from the connectivity table, note that elimination of the nodal equation becomes possible at the moment when the global index of the node appears for the last time in the connectivity table. Thus, the last appearance of the node in the connectivity table is marked with, for example, the negative index of a node, such as is shown in the connectivity table in Figure 2.7.

A basic concept is to put the equation for the first subsequent node in the place of the eliminated node, unless it is already in the list of active nodes. Thus, by adding the matrix of the fourth element, none of the new nodes are introduced and the matrix is assembled on the positions of nodes 3 and 4. By adding the matrix of this element, none of the equations has its final form and the assembly process continues, i.e. the matrix of the element 5 is added. Element 5 introduces the new node 5, and the contribution of this node will be redirected to the first free space, that is the first row and the first column. Thus, after adding the matrix of the fifth element, the frontal processor array will look as shown in Figure 2.7.

Following adding of the fifth element, it can be observed that the equation for node 3 already has its final form and can be eliminated; with only nodes 4 and 5 still being active, that is their equations are not yet complete. Only after adding the sixth element will these two remaining equations be complete and they can then be eliminated by the same procedure.

Note that the algorithm of simultaneous assembling and elimination of nodal unknowns is also possible if book keeping of active and eliminated nodes is introduced. Likewise, book keeping shall be applied to free positions where newly-introduced nodes can be placed.

The elimination of nodal equations is obviously possible at the moment when the node appears in the list of elemental nodes for the last time, which shall be somehow marked in the list. It is done in a manner such that the appearance of the list of elemental nodes is marked by the negative index of the node. Since indexes of all nodes are positive integers, the node index algebraic sign is altered.

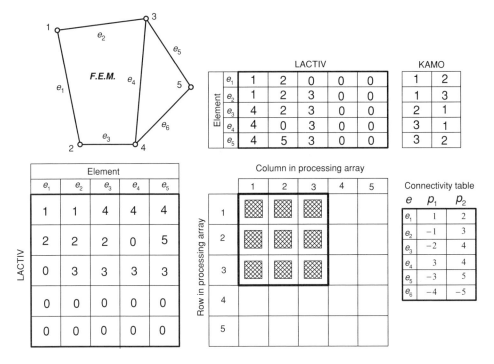

Figure 2.7 The front after the fifth element (*NACTIV*=3).

Furthermore, a list of active nodes **LACTIV** is introduced, which shows whether the matrix row or column belongs to an active node or is free.

In the very beginning, the list of active nodes is empty, and the list length is equal to zero:

LACTIV: 0,0,0,...; list length: **NACTIV** = 0.

Filling of the matrix with nodal contributions is organized according to free spaces in the list of active nodes; thus, a destination row (and column) in the matrix is defined for each node from the elemental list, for example using the destination list **KAMO**. If there are no free spaces in the list of active nodes, the list is extended.

Thus, the destination list before adding the matrix of the first element will be:

KAMO: 1,2

because those positions are free.

After the matrix of the first element is added, the list of active nodes has the form:

LACTIV: 1,2,0,0,0,...; i.e. the list length is **NACTIV**=2.

For the second element, nodes 1 and 3, the destination list will be:

KAMO: 1,3

because node 1 is active and there is no free space for node 3. Thus, the list of active nodes shall be extended in the following form:

LACTIV: 1,2,3,0,0,...; **NACTIV**=3.

Because node 1 can now be eliminated, the list of active nodes is altered and has the following form:

LACTIV: 0,2,3,0,0,...; **NACTIV**=3

that is the list is shorter, although it has free space in the first position.

In order to add the third element, nodes 2 and 4, the destination list will be:

KAMO: 2,1

that is the first free space in the list of active nodes will be used for node 4, and the active space 2 for node 2. The list of active nodes has not been extended and has the form:

LACTIV: 4,2,3,0,0,0,...; **NACTIV**=3.

Now, node 2 can be eliminated and the remaining active nodes in the list are:

LACTIV: 4,0,3,0,0,0,...; **NACTIV**=3

thus freeing the second space in the list etc.

Assembling the global system of equations and the elimination of the last nodal equation ends with the last element in the process. Nodal values of the solution are calculated by the reverse substitution procedure.

Clearly, before elimination of any node, it should be checked whether the node was under the restriction of essential boundary conditions, or whether natural boundary conditions should be added to the element; and act accordingly.

Variable **NACTIV,** that is the length of the active nodes list which shows the portion of the working (processing) matrix taken by active equations, is called the *front length*. It should as small as possible in order to minimize dimensions of the frontal working matrix.

Unlike the band width of the banded matrix, which depends on the sequence of node numbering, the *length of the front depends on the sequence of element numbering*. For the frontal procedure, elements should be numbered in the shortest direction possible.[10]

An algorithm for the frontal procedure of assembling and elimination, written by a pseudo program language, is shown in Figure 2.8.

Note that the moment of assembling the last finite element also ends the elimination process and starts the reverse phase or computation of the unknowns.

Back substitution

The reverse procedure of computation of the unknowns begins with downloading the eliminated equation from the disc,[11] starting from the last equation and moving towards the first one; which is a standard back substitution. It uses the same memory space as the frontal procedure matrix for computation of the unknowns.

Program module **Front.f90**

Nowadays, there are numerous program solutions for the frontal procedure, starting from the procedure for symmetrical matrices, a procedure with the complete record of eliminated equations in RAM,

---

[10]It is better to talk about the order of introducing elements into the frontal procedure than element indexes. Elements can be marked with names (labels) instead of integers, which allows greater flexibility in practical modelling, as will be shown later in developing the programming solution SimpipCore.

[11]Or from the other mass memory.

- mark the last appearance of each node in the list of elemental nodes
  (connectivity table) CONECT using the negative index of the node
- initiate lists ACTIVE, KAMO to zero

FOR each element e DO

- compute finite element matrix and vector
- prepare the list of elemental nodes LELNOD
- prepare the list of active nodes LACTIV

  FOR each node r from the elemental nodes list DO

    IF node is active in the list THEN
- it is destination KAMO

    ELSE IF there is a free space, it is
- destination KAMO

    ELSE extend the list of active nodes by one,
- and it is destination KAMO

    END IF
- assemble the elemental matrix and vector elements into frontal
matrix according to destinations from the list KAMO

    FOR each active node from the list of active nodes DO

     IF node is the last one negative) THEN

      IF there are boundary conditions, apply;
- carry out elimination
- write the equation and other information on disc
- deactivate the node by deleting the row and column in the
  frontal matrix and list of active nodes,
- reduce the length of the list of active nodes, if possible.

      END IF

     ENDIF

    END loop s

  END loop r

END loop e

Figure 2.8   Frontal algorithm of nodal equation assembling and elimination.

combined records, etc. These are, more or less, variations of the program solution printed in 1970; namely, modification of Irons' original solution.

For the needs of the program SimpipCore, subroutines and functional procedures of the frontal solution of hydraulic networks have been written, see (Jovic, 1993), attached in the file FRONT.F90.

Memory space and parameters

Memory space and the parameters of the frontal solver are shown in the following lines and explained using Fortran comments:

```
!--------------------------Frontal solver------------------------
!
!    mxactv - maximum length of the front (maximum number of active nodes)
!    mxfron - dimensions of the frontal matrix for processing
!    fprmem - frontal matrix for processing
!    fprvec - frontal vector of the right hand side
!             in the return phase of the front it serves as the vector
!             of active unknowns
```

```
!      tmpvec - auxiliary vector, in the elimination phase it serves for
!               copying of the elimination factor. It is written for purposes
!               of reconstruction of the right hand side. In the phase of
!               computation of the unknowns, it serves for
!               downloading of an active row of the matrix
!      lactiv - list of active nodes
!      kamo   - destination of nodes in frontal matrix
!      nactiv - number of active nodes in the process (length of the front)
!      numequ - record counter, output data for the front
!
!        front_len - front length
!        LunFrn - logical unit for recording equations after elimination
!        Front_Order - array of elements appearance order in the front
!        EquaNormMx  - maximal norm of Frontal Equations, abs(fprvec)
!
integer,parameter:: mxactv=250
integer,parameter:: mxfron=mxactv
real*8 fprmem(mxfron,mxfron),fprvec(mxfron),tmpvec(mxfron)
integer lactiv(mxactv),kamo(mxactv),nactiv,numequ
!
integer front_len
integer,parameter:: LunFrn=77
integer,allocatable:: Front_Order(:)
real*8 EquaNormMx/0/
!-------------------------End Frontal solver----------------------------
```

This is located in the global Fortran module `module GlobalVars`, and available when the instruction `use GlobalVars` is used.

Prefrontal procedure and initialization

In the phase before initialization of the front, the first thing to do is to record the last appearance of the node in the list of elemental nodes. This phase is called the prefrontal phase and it is implemented as the function procedure:

```
logical function PreFrontal.
```

A possible algorithm for recording the last appearance of a node in the list of elemental nodes could be a direct one, for example take the node, search the list, and test if the node is the last one, if yes then mark it with negative number. This algorithm has a major drawback; namely, the entire list must be searched in order to determine the last appearance in the list. It is a large number of searches and, therefore, very expensive for a large number of elements and nodes. Thus, a far better solution is applied here.

    The last appearance of the node will be determined more easily if the list is searched in reverse order, that is from the back. Then, the first appearance of the node will be what it searched for. This procedure is implemented in the subroutine:

```
subroutine MarkLastNode(istat).
```

This algorithm is fast; however, it requires one auxiliary array for registry of recorded nodes. In the subroutine `MarkLastNode` it is the array `NODMAR`. This array, which has a size equal to the total number of nodes, is only a temporary one, that is it is dynamically allocated and de-allocated within the subroutine.

Together with the aforementioned subroutines, the same file contains the functional procedure:

```
integer function FrontLength()
```

which is used for calculation of the maximum front length for the prescribed list of elemental nodes. It is used to determine the length of a record in a direct file during elimination of nodal equations.

Before starting the loop over elements, memory space shall be initialized by the subroutine:

```
subroutine IniFrnt(),
```

that is all variables of the frontal procedure memory are set to zero.

Lists of active nodes and destinations

The algorithm for determining the destination list **KAMO**, the active nodes list **LACTIV,** and the length of the active nodes list **NACTIV** is separated as a subroutine:

```
subroutine flact(noelnd,lelnod,nactiv,mxactv,kamo,lactiv).
```

Similarly, deactivation of the nodes in the lists of active nodes and possible reduction of the list length was also made an individual subroutine:

```
subroutine fldea(nactiv,lactiv).
```

Subroutines **FLACT** and **FLDEA** are used in subroutines **FrontLength** and **FRONTU.**

Assembling, elimination, and substitution

Part of the frontal procedure algorithm for assembling and elimination has been made an individual subroutine:

```
subroutine FrontU(ielem,lun)
```

to be called in a loop over all elements. The reverse phase of computation of the unknowns has been written as an individual subroutine:

```
subroutine FrontB(lun)
```

to be called after elimination is completed. Input, output, and auxiliary parameters are described in the each subroutine.

2.1.5   Frontal solution of steady problem

`SimpipCore` – the main procedure

Frontal solution of the steady problem consists of organizing the memory and several procedures. Modern Fortran compilers enable user-defined types of data and dynamic allocation of memory, for example node data are defined as the data type in the following form:

```
type Point_t
  integer kind       ; node type
  real*8 x,y,z        ; coordinates
  real*8 Hp,Hk        ; piez. head
  ...
endtype
```

```
Use GlobalVars
if(OpenSimpipInputOutputFiles()) then
 if Input() then
  if BuildMesh() then ; finite element mesh
   if Steady(t0) then ; solve steady solution, t=t0
     call Output
   endif
  endif
 endif
Endif
```

Figure 2.9   `SimpipCore` – frontal solution.

where all relevant nodal data, for example coordinates, nodal piezometric heads etc. are listed in the structure. A finite element is defined individually in the following form:

```
type Element_t
 integer kind      ; element type
 integer n1,n2     ; connectivities
 real*8 Q          ; discharge
 real*8 Qp(2),Qk(2)
 ...
endtype
```

with the structure containing connectivities, elemental discharges etc.

In the program solution `SimpipCore`, memory requirements are organized in the module `GlobalVars`, which is made available within any procedure by:

```
use GlobalVars
```

It contains all types of data for different objects as well as different parameters and variables. All types refer to both steady and unsteady flow.

Figure 2.9 shows the main procedure `SimpipCore` of the frontal solution of the steady problem, from which all procedures required for the sizing of the configuration and solution are called.

Procedure Input

Procedure `Input` downloads the system data that are written in the textual (ASCII) file, e.g. global indexes of nodes, their coordinates, data on hydraulic branches such as pipe finite elements, etc. Implementation of the data input procedure can be either trivial or very sophisticated such as the one that, for each type of object, dynamically allocates the memory space, carries out control of data accuracy, and calls respective procedures. In an implemented version of the procedure `Input` of the program solution `SimpipCore`, *parameter definition*, can be used, for example:

```
Parameters
 D = 1.5      ; diameter of tunnel (this is comment)
 A = D^2*Pi/4 ; cross-sectional area, Pi is implemented in the program
 L = 5003
```

that is *algebraic expressions* as input data, such as:

```
Points       ; x y z
 0.1*L 45.02 292.04
 0.2*L 32.87 282.00
```

```
Use GlobalVars
if AllocateMesh() then ; allocate the finite element mesh
 if FeMesh() then        ; assemble the finite element mesh
  if SortElements() then ; sort elements for frontal procedure
   if PreFrontal()then ; call prefrontal procedure
    return true          ; BuildMesh returns status of mesh
   endif
  endif
 endif
endif
```

Figure 2.10   Functional procedure `BuildMesh`.

that enables very high flexibility of the program solution, see www.wiley.com/go/jovic Appendix B, section Input/Output syntax.

Procedure BuildMesh

Successfully completed data input allows calling of the procedure `BuildMesh` for assembling the finite elements configuration, see Figure 2.10. This procedure first allocates the dynamic memory for finite elements, and then allocates the other data to the elements depending on the type of hydraulic branches in use.[12] When the procedure `FeMesh` is successfully completed, the procedure `SortElements` for finite elements sorting in a contiguous connected front is called. Actually, it does not have an optimum but a reasonable length. The frontal procedure enables very high modelling flexibility since node and finite element numbering can be in the form of *labels – alphanumerical marks* (names).

The numbering order is not relevant. Apart from sorting elements for the frontal procedure (global array `FrontOrder`), procedure `SortElements` can also detect the existence of mesh discontinuity in the downloaded configuration.

Following the successfully completed procedure `SortElements`, the `PreFrontal` procedure starts, which records the last appearance of nodes.

When all parts are completed, procedure `BuildMesh` returns the logical value *true*, otherwise it returns *false*.

Subroutine Steady

Following the successfully completed procedure `BuildMesh`, see Figure 2.9, a procedure `Steady`, for the iterative solution of a steady problem, is called.

The pseudo code of the procedure is shown in Figure 2.11. At the beginning of the procedure `Steady`, *boundary conditions* are initialized by calling the procedure `GetBdc`. At this moment, only natural and essential boundary conditions, such as the prescribed nodal discharge Q_r^0 and the prescribed piezometric head h_r^0, will be observed.

Within the iterative loop, a frontal solver is initialized first by calling `IniFrnt` then the array `dU` that contains nodal piezometric head increments is emptied. It is followed by a loop over all elements within which, for the current element from the `FrontOrder`, based on the finite element type, a respective subroutine for computation of the finite element matrix and vector is called.

If the finite element type is a pipe, it will be called the subroutine `SteadyPipeMtx`. After that, a frontal procedure `FrontU` for assembly and elimination is called. Let us repeat once again that

---

[12]For now, hydraulic branches are the pipe finite elements.

```
Use GlobalVars      ; program parameters and memory space
CurrentTime = T0    ; time
call GetBdc         ; initialize boundary conditions

Do while(iter < niter)
  call IniFrnt()           ; initiate front
  dU = 0         ; empty solution array
  For k = 1 to NumElements Do   ; loop over all elements
    ielem = FrontOrder(k)         ; index of sorted element
    select ElementType(ielem)    ; branching based on the element type
      case PipeElement
            call SteadyPipeMtx(ielem) ; compute Aᵉ,Bᵉ
      case ...
          ...
    end select
    call FrontU(ielem,lun) ; assembly, boundary conditions and elimination
  end loop k
  call FrontB(lun) ; back substitution phase - calculate the increment dU
  call IncVar       ; increment variables
  if maxNorm <= epsilon then exit loop
end do
```

Figure 2.11   Procedure Steady.

previously prepared boundary conditions are introduced in the frontal procedure FrontU immediately before elimination of the nodal equation. With the end of the loop, the last unknown is also eliminated and the reverse back phase FrontB of the computation of the unknown increments dU can start.

Following the computation of the nodal value increments, respective piezometric head increments are added to the nodal piezometric heads in a subroutine IncVar

$$h_r^{(k+1)} = h_r^{(k)} + \Delta h_r \tag{2.28}$$

while the elemental discharge increments are obtained from the expression

$$[\Delta Q] = [\underline{B}] - [\underline{A}][\Delta h]. \tag{2.29}$$

Respective elemental discharge increments are

$$Q^{(k+1)} = Q^{(k)} + \Delta Q, \tag{2.30}$$

where iteration steps are marked in parentheses.

Convergence and accuracy testing

Accuracy shall be tested in each iterative step. Norms for the maximum increment are calculated within the subroutine IncVar. If the accuracy is satisfied, iteration stops. The following norms can be used for accuracy testing:

(a) The maximum piezometric head increment norm
$$\max_{r=1,2,3,...N} |\Delta h_r| \le \varepsilon_h, \text{ where } \varepsilon_h \text{ is the selected criterion for piezometric head accuracy.}$$

(b) The maximum elemental discharge increment norm

$$\max_{e=1,2,3,...M} |\Delta Q_e| \leq \varepsilon_Q,$$ where ε_Q is the selected criterion for elemental discharge accuracy.

(c) The norm for nodal discharge equilibrium

$$\max_{r=1,2,3,...N} \left| \sum_p Q_e + Q_r^0 \right| \leq \varepsilon_{\sum Q},$$ where $\varepsilon_{\sum Q}$ is the selected criterion for the accuracy of the

discharge sum in a node.

(d) The accuracy norm for equation solution.

In the Newton–Raphson system of equations, the right hand side vector approaches the zero vector when the system converges towards the solution; namely, each vector component approaches zero, which can be used for the testing of solution accuracy:

$$\max_{r=1,2,3,...N} |\Delta F_r| \leq \varepsilon_e,$$ where ε_e is the selected criterion for accuracy of nodal equation solution.

An iterative process can converge either monotonously or alternating.

Computer accuracy

Integers can be written using one or several bytes (one byte is a memory location that consists of eight bits). The most important bit is used for the algebraic sign, while others represent an integer exactly, within the range shown below:

integer*1	$-2^7=-128$	$2^7-1=127$
integer*2	$-2^{15}=-32\,768$	$2^{15}-1=+32\,767$
integer*4	$-2^{32}=-2\,147\,483\,648$	$2^{31}-1=+2\,147\,483\,647$

Real numbers are written in the floating point format; for example for a real number of the real*4 type (or single precision) four bytes are used, in a manner that bits from 0 to 22 are used for the mantissa, bits from 23 to 30 for the exponent, and the 31st bit for the algebraic sign. There are different floating point format standards.

The smallest and the largest real number for real*4 are within the range from 10^{-38} to 10^{+38}. A precision order is 2^{23}, that is 7 decimal digits. If a real number cannot be accurately written in the floating point format, then it will be written using the closest one, which introduces the *rounding error*. Thus, computer presentation of real numbers is inaccurate, i.e. accurate with the error of number rounding. It can be assumed that a real number which takes four bytes is accurate to 6 decimal digits. A floating number format is generally used in computer science to write real numbers regardless of the program language, that is not only in Fortran.

Modern Fortran compilers contain different *intrinsic* functions for determining precision of representation of real and other types of numbers, the minimum and maximum values, and similar (it is necessary to browse the manuals or help system). Integer*4 and real*4 are the standards of modern Fortran, that is if Integer a or real b are written, they refer to integer a and real number b of four bytes length.[13]

Computer operations with real numbers always introduce rounding errors that are not taken into account in standard calculations. However, it shall be kept in mind if, for example, the equality of two real numbers is tested; or if in an iterative calculation iteration should be stopped at the moment when the required accuracy is achieved.

---

[13]The Fortran compiler can optionally change this default version to eight bytes.

For example, if the equality of two real numbers a and b is to be tested by software, application of the logic expression

$$if \quad a - b = 0 \quad then \quad a = b$$

can give the correct result by accident. Otherwise it will not, since in the floating point format, the difference $a - b \neq 0$ will not be equal to the representation of the number "0." Note that "*zero*" depends on the number size, that is a number of significant digits of a real number. A source of a small Fortran test program is shown in the following text. Numbers $a = 150 \cdot 10^6$ and $b = 100$ have been prescribed. Let $c = a + b$ and $d = c - a$, which should be equal to number b.

```
Program Test Accuracy
implicit none
real a/150.0e6/,b/100.0/
real c,d,Toleranca
    write(*,*) 'Problem of difference of the two close real numbers:'
    write(*,*)
    write(*,*) 'a,b',a,b
    c=a+b
    write(*,*) 'c=a+b',c
    d=c-a
    write(*,*)
    write(*,*) 'd=c-a',d
    write(*,*)
    write(*,*) 'Should be b-(c-a)=0, but it is not, it is:',d-b
    write(*,*)
    write(*,*) 'Tolerance of the number is "a" on 6 digits is :',
    Toleranca(a,6)
    write(*,*)
contains
    real function Toleranca(x,nDigits)
    implicit none
    real x
    integer nDigits
        if (x.eq.0) then
                Toleranca=5*10**(-nDigits-1)
        elseif (abs(x).ge.1) then
                Toleranca=5*10**(aint(Log10(Abs(x)))-nDigits)
        elseif (abs(x).lt.1) then
                Toleranca=5*10**(aint(Log10(Abs(x)))-nDigits-1)
        endif
    end function Toleranca
End Program
```

The difference d-b should be zero; however, the screen print:

```
Problem of difference of the two close real numbers:
 a,b  1.5000000E+08    100.0000
 c=a+b  1.5000010E+08
 d=c-a   96.00000
 Should be b-(c-a)=0, but it is not, it is:  -4.000000
 Toleranca of the number is "a" on 6 digits is :   500.0000
```

shows that it is not; instead, it is equal to -4! Note that number c was calculated accurately in addition while subtraction caused an error because number c was calculated out of the precision of 6 significant digits.

Program testing of the equality of the two real numbers shall be

$$if \quad |a - b| \leq \varepsilon \quad then \quad a = b,$$

where ε is a small real number called a real number *tolerance* and depends on the size of the number a or b. The tolerance ε shall be determined based on the required accuracy, that is the required number of digits, based on the applied precision of real numbers. Calculation of a tolerance for a real number `real*4` $a = 150 \cdot 10^6$ on 6 significant digits from the example gives a value of 500. Thus, every number smaller than 500 compared to number a will be equal to zero.

For the purposes of calculation of the real number tolerance of the length `real*8`, a functional procedure `Toleranca(x,nZnamenaka)` was written in the program solution *SimpipCore*, and can be found in the Fortran module `Toleranca.f90`.

Fortran sources of the aforementioned and other ancillary procedures, subroutines and functional procedures are given at www.wiley.com/go/jovic, in the file *SimpipCore program*. Input data syntax can also be found there in the file *SimpipCore User Manual*.

2.1.6   Steady test example

Figure 2.12 shows a small hydraulic network that consists of nine pipe branches and seven nodes (points). A piezometric head of $h = 100$ m was prescribed to the one end of the network while a consumption of 250 l/s was prescribed to the other end.

Other data are given in the input file, see Figure 2.13. The same figure also shows the output data. In this test example, accuracy of the iteration procedure of 15 significant digits is required (although satisfactory results can be obtained with the accuracy of 4 significant digits).

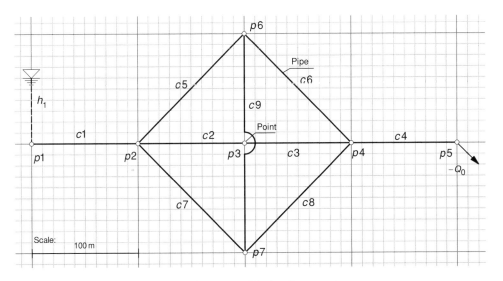

Figure 2.12   Small hydraulic network.

Input file:	Output file:
Small network.simpip	Small network.prt

```
;                              Front length:    4
; small network test          Finite element mesh O.K.
;                              Steady initial conditions
Options                        St 0 iter 11 Hn 0.617E-14 Qn 0.388E-16 Eq 0.261E-14
    input short                Steady O.K
    digits 15                  Stage: 0        Time:   0.00000
Parameters                         Point   Piez.head      SumQ
    Qo = 250/1000                  p1      100.000 -0.250000
    D = 0.300                      p2      95.9271 0.00000
    eps = 0.25/1000                p3      95.3405 0.00000
Points                             p4      94.7539 0.00000
    p1 0 0 0                       p5      90.6810 0.250000
    p2 100 0 0                     p6      95.3405 0.00000
    p3 200 0 0                     p7      95.3405 0.00000
    p4 300 0 0                     El.     Name    Q
    p5 400 0 0                     1       c1      0.250000
    p6 200 100 0                   2       c2      -0.934703E-01
    p7 200 -100 0                  3       c3      0.934703E-01
Pipes                              4       c4      0.250000
    c1 p1 p2 D eps                 5       c5      0.782649E-01
    c2 p3 p2 D eps                 6       c6      0.782649E-01
    c3 p3 p4 D eps                 7       c7      0.782649E-01
    c4 p4 p5 D eps                 8       c8      0.782649E-01
    c5 p2 p6 D eps                 9       c9      -0.284237E-17
    c6 p6 p4 D eps
    c7 p2 p7 D eps
    c8 p7 p4 D eps
    c9 p6 p7 D eps
Charge p5 -Qo
Piezo  p1 100
Steady 0
Print
    solStage 0
```

Figure 2.13   Small hydraulic network – input and output data.

Basic variables are given in parametric form and are used in further definition of the network. Pipes and points are alphanumeric labels – names. The system recharge *Charge* is positive by default. This consumption has a negative algebraic sign. To briefly mention the convention default discharge algebraic sign; a discharge is positive for the flow in the direction from the first to the second point of an element. The first node is the first one appointed to the input list; thus, for example for pipe *c2* the first node will be *p3* and the second will be *p2*. The results show that discharge in the pipe is negative because the flow is not directed from point *p3* to *p2* but otherwise. The significance of this definition of the algebraic sign will be shown later in the analysis of unsteady flow. Namely, in unsteady flow, the algebraic sign of

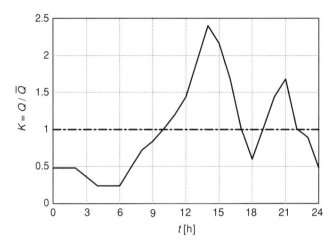

Figure 2.14   Variability of consumption.

discharge depends on acceleration and cannot be related to the algebraic sign of the difference in nodal piezometric heads.

The essential boundary condition is prescribed as *Piezometric* data. Input data syntax is intuitive and the details are given at www.wiley.com/go/jovic Appendix B, see Input/Output syntax.

2.2   Gradually varied flow in time

2.2.1   Time-dependent variability

Consumption in a water supply network is variable during the day and can be expressed as

$$Q_r^0(t) = \bar{Q}_r^0 K(t), \tag{2.31}$$

where $Q_r^0(t)$ is the time-dependent consumption, $\bar{Q}_r^0$ is the mean daily consumption in node r, and $K(t)$ is the graph of water consumption variability. A typical graph of consumption variability is shown in Figure 2.14 and can be drawn at the level of nodal consumption or the system consumption.

This is a polygonal graph for which the program solution SimpipCore defines an special object Graph of general purpose; namely, as the table function $G(x)$ of data pairs (x,G). For example:

```
Graph   G(x)    ;(G(x)=name of graph)
  -4      9.8    ;  x₁,G₁
  0.2     5.4    ;  x₂,G₂
  6.42   -24     ;  x₃,G₃
  ...     ...           ...
```

Similarly, other boundary conditions, for example prescribed time-dependent nodal piezometric heads, can be expressed in the form of a polygonal graph.

The object `Graph` allows operations such as the computation of the tabular function values (polygonal graph) for the prescribed value of an argument

$$y = G(x) \tag{2.32}$$

using the function procedure

```
real*8 Function PgVal(Ndata, x, y, xval).
```

Different functions can be defined on a polygonal graph; namely, functional procedures for the calculation of the surface area, wetted parameter, and the static moment of a channel area, which are used in channel flow modelling:

```
real*8 Function PgArea(Ndata, x, y, xval),
real*8 Function PgPerim(Ndata, x, y, xval),
real*8 Function PgStatic(Ndata, x, y, xval).
```

These functional procedures are also implemented in the FFortran module `Graphs.f90`.
    A polygonal graph allows operations in the linear form:

$$U(t) = U_c + f G(t), \tag{2.33}$$

where $U(t)$ is the value of a time-dependent datum, U_c is its constant part, and f is the factor that the graph value is multiplied by. In the example of nodal time-dependent consumption, expression (2.31), a constant part is equal to zero, the factor is equal to the mean daily consumption $\bar{Q}_r^0(t)$, while the graph is equal to the variation of consumption $K(t)$.

2.2.2   Quasi non-steady model

Variation of daily consumption is balanced by one or several tanks in a hydraulic system. A small hydraulic network with a tank connected in node $p1$ and time-dependent consumption in node $p5$, as shown in Figure 2.15, will be analyzed.

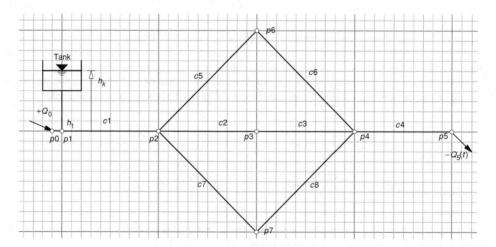

Figure 2.15   Tank in water supply network.

Time variations are slow; thus, it can be considered that flow in the network is steady in every moment. A hydraulic network state, described by basic variables (h, Q) in a selected time $t_0 = t$ can be determined by modelling a series of steady states for boundary conditions that correspond to that time. These are the prescribed piezometric head in node $p1$ that corresponds to the water level in a tank $h_1 = h_k(t)$ and the nodal consumption $Q_r^0(t) = \bar{Q}_r^0 K(t)$.

Instead of several individual calculations of a series of steady states (using the steady model), a model adequate for gradually variable boundary conditions can be developed. This model is called a *quasi non-steady* model – non-steady because it deals with time-dependent flow and quasi because the flow is not unsteady but steady.

In time-dependent flow, nodal load $Q_r^0(t)$ is a time-dependent function described by the time-dependent polygonal graph, according to expression (2.31).

2.2.3   Subroutine `QuasiUnsteadyPipeMtx`

The matrix and vector of the pipe finite element for the quasi-unsteady flow are calculated by the subroutine

```
subroutine QuasiUnsteadyPipeMtx(ielem).
```

Modelling of time-dependent flow requires integration of all equations in time. Integration of nodal equations is carried out in a time step Δt and the *obtained expressions have the meaning of the volume balance in time interval*. Numerical integration of ordinary differential equations and integrals is given at www.wiley.com/go/jovic Appendix A.

Integration of the initial, basic term of the nodal continuity equation is written as

$$\Delta V_r = \int_{\Delta t} \sum_p {}^r Q^e(t) dt = (1 - \vartheta) \Delta t \sum_p {}^r Q^e + \vartheta \Delta t \sum_p {}^r Q^{e+} = 0, \qquad (2.34)$$

where the "+" sign marks the unknown state at the end of the time interval, and the absence of the sign the known state at the beginning of the interval. For purposes of modelling of time-dependent states, a value of the integration parameter $\vartheta = 0.5$ will be adopted, based on the experience.

Natural boundary conditions are also integrated in time. An initial fundamental system is updated with nodal loads in such a manner that respective volume is added to the system

$$\Delta V_r^0 = \int_{\Delta t} Q_r^0 dt = (1 - \vartheta) \Delta t Q_r^0 + \vartheta \Delta t Q_r^{0+}. \qquad (2.35)$$

Preparation of the boundary conditions perform routine `GetBDC`. Nodal equations are updated in the procedure `FrontU` immediately before elimination of the nodal unknown.

The respective matrix and vector of the quasi-unsteady flow are obtained from the elemental equation. The equation of the quasi-unsteady flow in a pipe is equal to the steady equation

$$h_s - h_r + \beta |Q| Q = 0, \qquad (2.36)$$

where r is the upstream (first) and s is the downstream (second) global index of the nodes on an element, while variables are time-dependent. By integration of the elemental equation in time interval Δt the following is obtained

$$F^e = \int_{\Delta t} [h_s - h_r + \beta |Q| Q] dt = \qquad (2.37)$$

$$(1 - \vartheta) \Delta t (h_r - h_s + \beta |Q| Q) + \vartheta \Delta t (h_r^+ - h_s^+ + \beta^+ |Q^+| Q^+) = 0.$$

The Newton–Raphson iterative form for the element e is

$$\left[\frac{\partial F^e}{\partial h_r^+} \quad \frac{\partial F^e}{\partial h_s^+}\right] \cdot \left[\begin{array}{c}\Delta h_r \\ \Delta h_s\end{array}\right] + \frac{\partial F^e}{\partial Q^+}\Delta Q = -F^e, \tag{2.38}$$

which is formally written using the matrix-vector operations

$$[\underline{H}] \cdot [\Delta h] + [\underline{Q}] \cdot [\Delta Q] = [\underline{F}] \tag{2.39}$$

with the vector term equal to

$$[\underline{H}] = \left[-\vartheta\,\Delta t \quad +\vartheta\,\Delta t\right] = \vartheta\,\Delta t \left[-1 \quad +1\right] \tag{2.40}$$

and the scalar term

$$[\underline{Q}] = 2\vartheta\,\Delta t\beta^+\,|Q^+|, \tag{2.41}$$

that is inverse value equal to

$$[\underline{Q}]^{-1} = \frac{1}{2\vartheta\,\Delta t\beta^+\,|Q^+|}. \tag{2.42}$$

When expression (2.39) is multiplied by the inverse term $\left[\underline{Q}\right]^{-1}$

$$[\underline{A}] \cdot [\Delta h] + [\Delta Q] = [\underline{B}], \tag{2.43}$$

from which value of the elemental discharge increment can be calculated

$$[\Delta Q] = [\underline{B}] - [\underline{A}][\Delta h], \tag{2.44}$$

where

$$[\underline{A}] = [\underline{Q}]^{-1}[\underline{H}], \tag{2.45}$$

$$[\underline{B}] = [\underline{Q}]^{-1}[\underline{F}]. \tag{2.46}$$

In the aforementioned expressions $[\underline{A}]$ is a two-term vector while $[\underline{B}]$ is a scalar. Integrated nodal equation (2.34) shows that elemental discharges are multiplied by the factor $\vartheta\,\Delta t$; thus the sought structure of the finite element matrix for the quasi-unsteady flow will be

$$A^e = \vartheta\,\Delta t \left[\begin{array}{c}+\underline{A} \\ -\underline{A}\end{array}\right]. \tag{2.47}$$

The finite element vector

$$B^e = (1 - \vartheta\,\Delta t)\underbrace{\left[\begin{array}{c}+Q \\ -Q\end{array}\right]}_{(1)} + \vartheta\,\Delta t\left[\begin{array}{c}+Q^+ \\ -Q^+\end{array}\right] + \vartheta\,\Delta t\underbrace{\left[\begin{array}{c}+\underline{B} \\ -\underline{B}\end{array}\right]}_{(2)} \tag{2.48}$$

contains term (1) which is the contribution of the element to the nodal equation and term (2) which is the contribution of the elimination of the nodal discharge increment from nodal equations.

If $Q = 0$, similar to the steady flow, then

$$F^e = \int_{\Delta t} \left[h_r - h_s + \beta^* Q \right] dt =$$

$$(1 - \vartheta)\Delta t \left(h_r - h_s + \beta^* Q \right) + \vartheta \, \Delta t \left(h_r^+ - h_s^+ + \beta^* Q^+ \right) = 0,$$

(2.49)

where

$$\beta^* = \frac{64\nu}{2g \, D^2 A} L.$$

(2.50)

According to this, for $Q = 0$ a value of the derivation shall be applied

$$[\underline{Q}] = \frac{\partial F^e}{\partial Q^+} = \vartheta \, \Delta t \beta^*.$$

(2.51)

The rest of the procedure remains the same.

2.2.4 Frontal solution of unsteady problem

The solution of time-dependent changes in a hydraulic network can be monitored through several hydraulic states, starting from the prescribed initial state, that is initial conditions.

Initial state is usually the steady state at the time $t = t_0$, which is obtained by calling the procedure Steady(t0). The time stage is a set of all time-dependent variables. The time stage is prescribed at the beginning of the time interval Δt and the unknown state at its end. The unknown state is defined by the time integration of the elemental and nodal equations for the prescribed boundary conditions; namely, by determining the increments of variables Δh and ΔQ, which is accomplished within the time loop.

A frontal solution of the quasi-unsteady flow, that is a problem gradually varied in time, is a logical extension of the frontal solution of the steady problem. An iterative loop is repeated within the time loop, see pseudo program solution Unsteady shown in Figure 2.16. Variable kstage is a counter of time stages. At the beginning of the loop, a subroutine RstVar is called, which sets the calculated state from the previous interval as the initial state. Then, explicitly set boundary conditions are prepared by calling the subroutine GetBDC and a new value for the time variable is set.

An iterative loop for the solution of the state at the end of the time interval is modified in comparison with the steady problem only in the part when the respective finite element matrix and vector are called. Instead of calling the subroutine SteadyPipeMtx, the QuasiUnsteadyPipeMtx subroutine shall be called.

Following the end of the iterative loop, and before transferring to the next time state, the calculated state is stored by calling the subroutine SaveSolutionStage.

Figure 2.17 shows the abridged form of the main procedure *SimpipCore*, extended to solving time-dependent problems; namely, the unsteady problem modelling. The full Fortran source can be found at www.wiley.com/go/jovic.

```
call Steady(t0)

for kstage = 1 to maxstage do
      Equalize the state at the end and the beginning of the time step:
      call RstVar
      Prepare boundary conditions:
      call GetBDC
      Set new value of the time variable:
      t = t + Dt
      Do while(iter < niter)
         call IniFrnt()     ; initialize front
         dU = 0             ; empty solution array
         For k = 1,NumElements Do        ; loop over all elements
           ielem = FrontOrder(k); index of sorted element
           select ElementType(ielem) ; branching based on the element type
              case PipeElement
                 if runmode = QuasiUnsteady then
                    call QuasiUnsteadyPipeMtx(ielem) ; compute Aᵉ,Bᵉ
                 else
                    ...
              case ...
                 ...
           end select
         call FrontU(ielem,lun) ; assembling and elimination
         end loop k
         call FrontB(lun) ; back substitution phase - increments calc. dU
         call IncVar()
      end Do                ; end of iterative loop
      call SaveSolutionStage ; store results
end loop kstage            ; end of time loop
```

Figure 2.16   Procedure Unsteady in frontal solution.

```
   Use GlobalVars
       if OpenSimpipInputOutputFiles() then
           if Input() then
             if BuildMesh() then    ; finite element mesh
              if Steady(t0) then   ; steady solution for time t=t0
                 if Unsteady then ; unsteady solution
                 endif
                 call Output       ; print results
               endif
             endif
           endif
       endif
```

Figure 2.17   The main procedure SimpipCore, an extension of the steady to unsteady model.

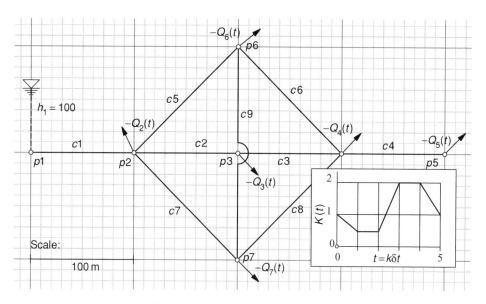

Figure 2.18   Small hydraulic network.

2.2.5   Quasi-unsteady test example

Figure 2.18 shows a small hydraulic network with a nodal consumption gradually varied in time. The figure shows a graph of consumption $K(t)$ variation in time, with the time prescribed as parametric time unit $\delta t = 3600$ s.

Consumption in nodes has the following form:

$$Q_r(t) = -Q^0 K(t), \tag{2.52}$$

where the value Q^0 is equal to 50 l/s in every node. Variation of consumption in the input file has been prescribed as the Graph object named K(t). The time data are prescribed as algebraic expressions which are transferred into numerical data by the *SimpipCore Parser (compiler of input data)*.

Expression (2.33) is used for assigning nodal consumption, see www.wiley.com/go/jovic Appendix B, Input/Output syntax, where the constant part of the consumption equals zero and the factor that the graph is multiplied by equals $-Q^0$ (the minus sign refers to consumption).

Figure 2.19 shows the input file on the left with all the input data and the request for computation of the steady state at time $t = 0$ and five unsteady states at $\delta t = 3600$ s time intervals. Calculated piezometric heads in all network nodes are shown in the graph on the right.

2.3   Unsteady flow of an incompressible fluid

2.3.1   Dynamic equation

If the flow is gradually varied in time, the fluid inertia can be neglected and the flow will be steady at every moment. Otherwise, the flow will be unsteady. Let us observe the unsteady flow of an incompressible fluid in a circular pipe of constant cross-section area. Because of incompressibility, that is constant

```
;small Quasi non-steady
     network test
Options
   run quasy
   input short
   digits 15
Parameters
   Qo = 50/1000
   dt = 3600
   D = 0.300
   eps = 0.25/1000
Points
   p1 0 0 0
   p2 100 0 0
   p3 200 0 0
   p4 300 0 0
   p5 400 0 0
   p6 200 100 0
   p7 200 -100 0
Pipes
   c1    p1 p2  D  eps
   c2    p2 p3  D  eps
   c3    p3 p4  D  eps
   c4    p4 p5  D  eps
   c5    p2 p6  D  eps
   c6    p6 p4  D  eps
   c7    p2 p7  D  eps
   c8    p7 p4  D  eps
   c9    p6 p7  D  eps
Graph K(t)
   0 1
   1*dt 0.5
   2*dt 0.5
   3*dt 2
   4*dt 2
   5*dt 1.0
Charge
   p2 0 -Qo K(t)
   p3 0 -Qo K(t)
   p4 0 -Qo K(t)
   p5 0 -Qo K(t)
   p6 0 -Qo K(t)
   p7 0 -Qo K(t)
Piezo  p1 100
Steady 0
Unsteady 5 dt
Print
   solStage 0 3 5
```

Figure 2.19 Small hydraulic network – input data and output graph.

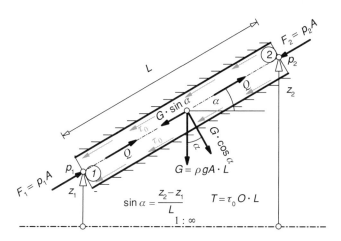

Figure 2.20   Pipe flow.

density, the flow rate does not change along the pipe and the fluid moves as if rigid. If mass movement in the entire pipe is observed, a dynamic equation can be written for that type of movement, as shown in Figure 2.20.

Starting from the second Newton law of motion, equilibrium between the inertia and the acting force, a balance of forces in the flow direction can be written as

$$\rho A L \frac{dv}{dt} = F_1 - F_2 - G \sin \alpha - T. \tag{2.53}$$

where forces F_1, F_2 are the pressure forces, G is weight, and T is the force due to friction on the pipe wetted perimeter O, as shown in Figure 2.20:

$$\rho A L \frac{dv}{dt} = (p_1 - p_2) A - \rho g (z_2 - z_1) A - \tau_0 O L. \tag{2.54}$$

After division by $\rho g A$

$$\frac{L}{g} \frac{dv}{dt} = \underbrace{z_1 + \frac{p_1}{\rho g}}_{h_1} - \underbrace{z_2 - \frac{p_2}{\rho g}}_{h_2} - \underbrace{\frac{\tau_0}{\rho g R} L}_{\beta |Q| Q} \tag{2.55}$$

a dynamic equation of incompressible (rigid) fluid flow in a pipe of constant cross-section area is obtained in the form of Bernoulli's equation in head form

$$h_2 - h_1 + \beta |Q| Q + \frac{L}{g A} \frac{dQ}{dt} = 0, \tag{2.56}$$

where h is the piezometric head and the flow resistances $\Delta h = \beta |Q| Q$ are equal to the flow resistances for the same steady flow discharge.

Modelling of unsteady flow of an incompressible fluid, that is a constant density fluid, is also called a *rigid fluid flow modelling*. In the case of an incompressible (rigid) fluid and a rigid pipeline, discharge is constant along the pipe at every moment

$$Q(l, t) = Q(t). \tag{2.57}$$

Note that the gradient of energy and the piezometric line are equal due to the existence of a developed boundary layer and constant cross-section area.

2.3.2 Subroutine RgdUnsteadyPipeMtx

The finite element matrix and vector for unsteady flow of an incompressible fluid are obtained by integration of the elemental equation

$$
\begin{aligned}
F^e = \int_{\Delta t} &\left[h_s - h_r + \beta \, |Q| \, Q + \frac{L}{gA} \frac{dQ}{dt} \right] dt = \\
&(1 - \vartheta)\Delta t \, (h_s - h_r + \beta \, |Q| \, Q) + \\
&+ \vartheta \, \Delta t \left(h_s^+ - h_r^+ + \beta^+ \left| Q^+ \right| Q^+ \right) + \\
&+ \frac{L}{gA} (Q^+ - Q) = 0.
\end{aligned}
\tag{2.58}
$$

The pipe finite element matrix and vector can be obtained directly from an integrated elemental equation (2.58), using a universal procedure. The procedure starts from the Newton–Raphson iterative form

$$
\left[\frac{\partial F^e}{\partial h_r^+} \quad \frac{\partial F^e}{\partial h_s^+} \right] \cdot \left[\begin{array}{c} \Delta h_r \\ \Delta h_s \end{array} \right] + \frac{\partial F^e}{\partial Q^+} \Delta Q = -F^e
\tag{2.59}
$$

that is formally written using the matrix-vector operations

$$
[\underline{H}] \cdot [\Delta h] + [\underline{Q}] \cdot [\Delta Q] = [\underline{F}]
\tag{2.60}
$$

with the vector term

$$
[\underline{H}] = [-\vartheta \, \Delta t \quad +\vartheta \, \Delta t] = \vartheta \, \Delta t \left[-1 \quad +1 \right]
\tag{2.61}
$$

and the scalar term

$$
[\underline{Q}] = \frac{L}{gA} + 2\vartheta \, \Delta t \beta^+ \left| Q^+ \right|
\tag{2.62}
$$

and the inverse value equal to

$$
[\underline{Q}]^{-1} = \frac{1}{\dfrac{L}{gA} + 2\vartheta \, \Delta t \beta^+ \left| Q^+ \right|}.
\tag{2.63}
$$

Other steps of the procedure are carried out numerically in the subroutine

```
subroutine RgdUnsteadyPipeMtx,
```

according to the formulas valid both for the steady and time-dependent flow. Equation (2.63) shows that the inverse term value can also be calculated for $Q = 0$. However, it is recommended to apply the previously described procedure for that value too. Namely, the initial iterative values of all elemental discharges are equal to zero, which generates a linear system of nodal equations, the solution of which provides a logical discharge increment already in the first iteration.

For unsteady flow modelling, the elemental loop in subroutine **Unsteady** shall be updated by calling the subroutine **UnsteadyPipeMtx** for the pipe finite element matrix and vector computation:

```
...
select ElementType(ielem) ; branching based on to the pipe element type
   case PipeElement
      ; compute Aᵉ,Bᵉ
      call UnsteadyPipeMtx(ielem)
   case ...
      ...
end select
...
```

while the subroutine **UnsteadyPipeMtx** calls subroutines **QuasiUnsteadyPipeMtx** or **RgdUnsteadyPipeMtx** for the finite element matrix and vector computation:

```
...
; compute Aᵉ,Bᵉ
if runmode = QuasiUnsteady then
   call QuasiUnsteadyPipeMtx(ielem)
else if runmode = RigidUnsteady then
   call RgdUnsteadyPipeMtx(ielem)
endif
...
```

depending on the unsteady flow computation type, the `runmode` parameter.

2.3.3   Incompressible fluid acceleration

Two classical examples of incompressible fluid acceleration from the state at rest to the steady flow will be observed:

(a) instantaneous opening of the gate at the pipe end as shown in Figure 2.21a, and
(b) instantaneous rising of the hydraulic grade line at the left pipe end, as shown in Figure 2.21b.

In case (a) gate opening is immediately followed by flow acceleration. Available flow energy is defined by the pressure head difference, h_0, of which one part pertains to transformation into kinetic energy, the other to friction resistance, and the remaining part to acceleration. Something similar happens in case (b), where the forced inflow head h_0 transforms into kinetic energy, the other to friction resistance and acceleration.

Thus, the Bernoulli equation can be applied

$$h_0 = \frac{v^2}{2g} + \zeta \frac{v^2}{2g} + \frac{L}{g} \frac{dv}{dt}, \qquad (2.64)$$

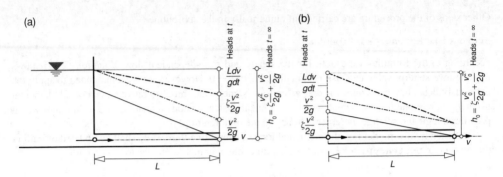

Figure 2.21   Incompressible fluid acceleration.

where the coefficient of resistance due to friction ζ equals

$$\zeta = \lambda \frac{L}{D}$$

and is constant during acceleration (turbulent flow in rough conduits). In a steady flow there is no acceleration, thus

$$h_0 = (1 + \zeta)\frac{v_0^2}{2g}, \tag{2.65}$$

so the dynamic equation will be

$$L\frac{dv}{dt} = (1 + \zeta)\frac{v_0^2 - v^2}{2} \tag{2.66}$$

and will be solved by separation of variables

$$dt = \frac{2L}{1 + \zeta}\frac{dv}{v_0^2 - v^2}. \tag{2.67}$$

After integration

$$\int_0^t dt = \frac{2L}{1 + \zeta} \int_0^v \frac{dv}{v_0^2 - v^2} \tag{2.68}$$

the following is obtained

$$t = \frac{2L}{(1 + \zeta)v_0} \operatorname{artgh} \frac{v}{v_0} \tag{2.69}$$

or expressed in logarithmic form

$$t = \frac{L}{(1 + \zeta)v_0} \ln \frac{v_0 + v}{v_0 - v}. \tag{2.70}$$

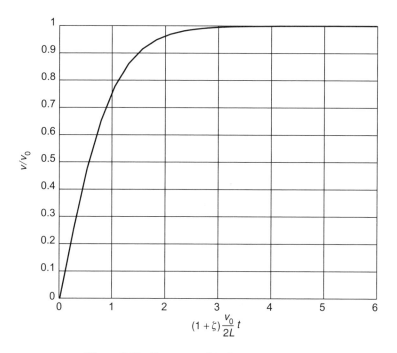

Figure 2.22   Incompressible fluid acceleration.

The obtained result can be written in the following form

$$\frac{v}{v_0} = \tanh \frac{v_0}{2L}(1 + \zeta)t \tag{2.71}$$

as shown graphically in Figure 2.22 with the abscissa values

$$x = \frac{v_0}{2L}(1 + \zeta)t. \tag{2.72}$$

Note that after $x = 3$, the flow can be considered steady.

Thus, for example the time required for water acceleration from the state at rest to the steady discharge of $Q = 92$ m^3/s (corresponding to the head $h_0 \approx 30.55$ m) in the pipeline of 5 m diameter, 2000 m length, and 1 mm roughness ($\lambda = 0.0138$) is

$$t = \frac{6L}{\left(1 + \lambda \dfrac{L}{D}\right) v_0} \approx 394 \text{ s.}$$

If a velocity head is negligible, which is a regular assumption in hydraulic network models, then the acceleration equation will be

$$\frac{v}{v_0} = \tanh \frac{v_0}{2L}\zeta t. \tag{2.73}$$

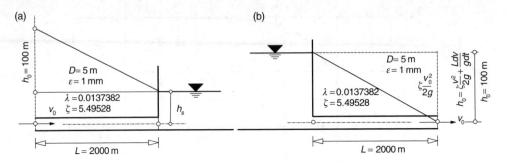

Figure 2.23   Acceleration at (a) forced inflow and (b) gate opening, test example.

2.3.4   Acceleration test

Software *SimpipCore*, option run rigid, shall be used to solve acceleration of incompressible water in pipes for:

(a) forced inflow at instantaneous piezometric head elevation, and
(b) sudden instantaneous gate opening

as shown in Figure 2.23. Initial conditions are hydrostatic at time $t = 0$. Sudden opening and forced inflow are given by polygonal Graph h(t).

The numerical solution shall be compared with the analytical according to expression (2.73). On the left side of Figure 2.24, there are input data that define the hydraulic system. A solution is sought in the time of 250 seconds with the time step of one second. Note that both acceleration cases are the same, only the kind and points of boundary conditions are altered.

Figure 2.24 shows the time graph of velocity development in the pipe according to the numerical and analytical solution on the right hand side. There is correspondence of numerical and analytical solution graphs and the difference cannot be seen! Note that the steady state is achieved in about 125 seconds.

2.3.5   Rigid test example

Figure 2.25 shows a small hydraulic network with nodal consumption variable in time, which has already been analyzed as quasi-unsteady flow in Section 2.2.5 with the time increment $\delta t = 3600$s from the graph of variation in consumption $K(t)$.

The same problem will be solved as unsteady flow of incompressible (rigid) fluid, first with the time increment (a) $\delta t = 3600$s then (b) $\delta t = 36$ s.

Figure 2.26 shows the input file on the left hand side with all input data and the request for computation of the steady state at the time $t = 0$ and 500 unsteady states at time increments $0.01 \cdot \delta t$. Attention should be paid to the option run rigid, required for unsteady computation of incompressible fluid flow.

According to alternative (a) $\delta t = 3600$s calculated piezometric heads in all mesh nodes are shown in the graph in the top right of the Figure 2.26. If the results are compared with the quasi-unsteady model, notable differences can be observed.

According to alternative (b) $\delta t = 36$s calculated piezometric heads in all mesh nodes are shown in the graph in the bottom right of the Figure 2.26. If the results are compared with the quasi-unsteady model, differences can be observed caused by water inertia, that is there are piezometric head oscillations at changes in discharge (break points of graph $K(t)$).

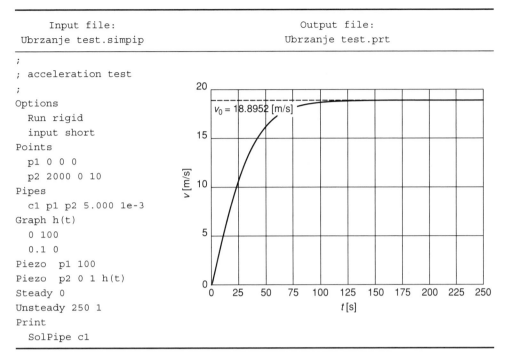

```
Input file:                          Output file:
Ubrzanje test.simpip                 Ubrzanje test.prt
```

```
;
; acceleration test
;
Options
  Run rigid
  input short
Points
  p1 0 0 0
  p2 2000 0 10
Pipes
  c1 p1 p2 5.000 1e-3
Graph h(t)
  0 100
  0.1 0
Piezo  p1 100
Piezo  p2 0 1 h(t)
Steady 0
Unsteady 250 1
Print
  SolPipe c1
```

Figure 2.24   Acceleration test.

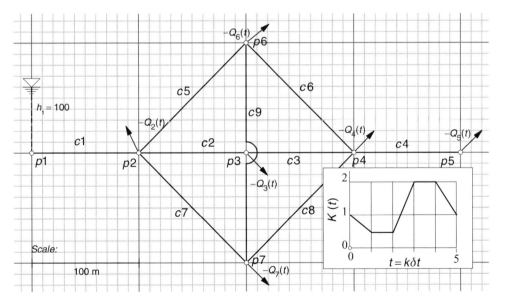

Figure 2.25   Small hydraulic network.

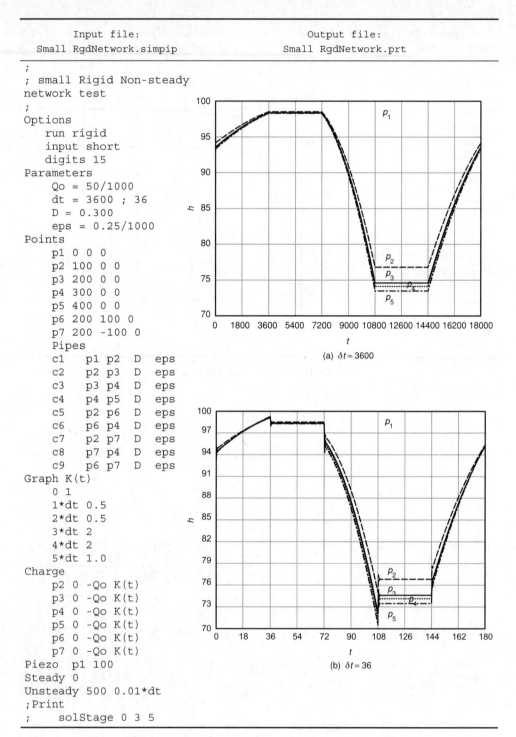

Input file:
Small RgdNetwork.simpip

Output file:
Small RgdNetwork.prt

```
;
; small Rigid Non-steady
network test
;
Options
    run rigid
    input short
    digits 15
Parameters
    Qo = 50/1000
    dt = 3600 ; 36
    D = 0.300
    eps = 0.25/1000
Points
    p1 0 0 0
    p2 100 0 0
    p3 200 0 0
    p4 300 0 0
    p5 400 0 0
    p6 200 100 0
    p7 200 -100 0
    Pipes
    c1    p1 p2   D   eps
    c2    p2 p3   D   eps
    c3    p3 p4   D   eps
    c4    p4 p5   D   eps
    c5    p2 p6   D   eps
    c6    p6 p4   D   eps
    c7    p2 p7   D   eps
    c8    p7 p4   D   eps
    c9    p6 p7   D   eps
Graph K(t)
    0 1
    1*dt  0.5
    2*dt  0.5
    3*dt  2
    4*dt  2
    5*dt  1.0
Charge
    p2 0 -Qo K(t)
    p3 0 -Qo K(t)
    p4 0 -Qo K(t)
    p5 0 -Qo K(t)
    p6 0 -Qo K(t)
    p7 0 -Qo K(t)
Piezo  p1 100
Steady 0
Unsteady 500 0.01*dt
;Print
;    solStage 0 3 5
```

Figure 2.26   Small hydraulic network – input data and the output graph.

References

Hinton, E. and Owen, D.R.J. (1977) *Finite Element Program*, Academic Press, London.

Irons, B.M. (1970) A frontal solution program. *Int. J. Num. Meth.* 2, 5–32.

Jović, V. (1993) *Introduction to Numerical Engineering Modelling* (in Croatian), Aquarius Engineering, Split.

Further Reading

Allen, T. Jr. and Ditsworth, R.L. (1972) *Fluid Mechanics, International Student Edition*. McGraw-Hill Kogakusha Ltd. Tokio.

Douglas, J.F., Gasiorek, J.M., and Swaffield, J.A. (1979) *Fluid Mechanics*. Pitman, Bath.

Fox, R.W. and McDonald, A.T. (1973) *Introduction to Fluid Mechanics*. John Wiley & Sons, New York.

Hinton, E. and Owen, D.R.J. (1979) *An Introduction to Finite Element Computations*. Pineridge Press Ltd, Swansea.

Jović, V. (2006) *Fundamentals of Hydromechanics* (in Croatian: Osnove hidromehanike). Element, Zagreb.

Rouse, H. (1962) *Fluid Mechanics for Hydraulic Engineers*. Dover Pub. Inc., New York.

Rouse, H. (1969) *Hydraulics* (prijevod na srpski Tehnička hidraulika). Građevinska knjiga, Beograd.

Streeter, V.L. (1958) *Fluid Mechanics*. McGraw-Hill Book Co., New York.

Vallentine, H.R. (1969) *Applied Hydrodynamics*. Butterworths, London.

Vennard, J.K. (1961) *Elementary Fluid Mechanics*, 4th edn. John Wiley & Sons, New York.

Walshaw, A.C. and Jobson, D.A. (1962) *Mechanics of Fluids*. Longmans, London.

Wilson, D.H. (1959) *Hydrodynamics*. E. Arnold Pub. Ltd, London.

3

Natural Boundary Condition Objects

3.1 Tank object

3.1.1 Tank dimensioning

Figure 3.1a shows a scheme of a classical *simple* tank. Its work volume is defined by the bottom elevation Z_b, top elevation Z_t, and the constant plan area A_k, which is filled from above from the external system.

In general, the tank may be equipped with the *safety weir* at an elevation Z_p, which overflows the discharge Q_p when water level h_k in the tank exceeds the weir elevation.

Figure 3.1b shows a scheme of a *compensation* tank in node *r*. Simple tank functioning can be presented as case (b) in the way that the branch filled by discharge Q_{in}, marked by a dashed line in figure, is added in node *r*. In both cases, the tank is filled by discharge

$$Q_k = Q_{in} - Q_{out}. \tag{3.1}$$

In water supply systems, daily consumed water quantities are equal to daily supplied water quantities, and this is called the *daily volume balance* V_0. In cycle $T = 24$ h of the daily flow balance, the mean value of the inflow and outflow discharge is equal to the mean daily balance discharge $\bar{Q} = V_0/T$, where V_0 is the daily balance volume.

Consumption discharge $Q_{out}(t)$ varies around the daily mean value and can be expressed as

$$Q_{out}(t) = \bar{Q} K_{out}(t), \tag{3.2}$$

where $\bar{Q}$ is the mean daily discharge of the system and $K_{out}(t)$ is the graph of consumption variation. Similarly, the daily supply discharge $Q_{in}(t)$ can be expressed as

$$Q_{in}(t) = \bar{Q} K_{in}(t), \tag{3.3}$$

where $K_{in}(t)$ is the graph of supply variability. The equation of continuity for the tank with a weir will be

$$\frac{dV}{dt} + Q_p = Q_k, \tag{3.4}$$

Analysis and Modelling of Non-Steady Flow in Pipe and Channel Networks, First Edition. Vinko Jović.
© 2013 John Wiley & Sons, Ltd. Published 2013 by John Wiley & Sons, Ltd.

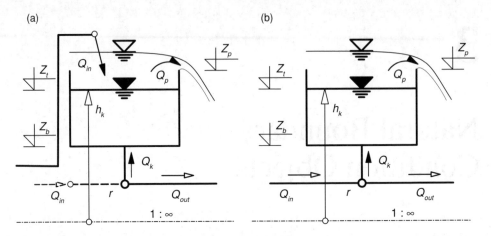

Figure 3.1   Tank.

where Q_k is the tank inflow, positive for tank filling and negative for tank emptying, Q_p is the overflow discharge, and V is the volume of liquid in the tank. A tank is usually sized so that there is no overflow during the balance cycle.

The necessary tank volume for input and output volume balance is defined by integration of the continuity equation

$$\frac{dV}{dt} = Q_{in} - Q_{out}. \tag{3.5}$$

Let the initial volume be equal to $V(t_0)$ at time $t = t_0$; thus, integration of Eq. (3.5) gives the accumulated volume at any time

$$V(t) - V(t_0) = \bar{Q} \int_{t_0}^{t} (K_{in} - K_{out})dt = \frac{V_0}{T} \int_{t_0}^{t} (K_{in} - K_{out})dt, \tag{3.6}$$

that is the accumulated volume can be expressed as a portion of the daily water consumption

$$\frac{V(t) - V(t_0)}{V_0} = \frac{1}{T} \int_{t_0}^{t} (K_{in} - K_{out})dt. \tag{3.7}$$

The required tank volume for daily balance of consumption is equal to the difference between the maximum maximorum and the minimum minimorum of the integral value on the right hand side of expression (3.6) $V_{var} = V_{max}^{max} - V_{min}^{min}$.

Expressed as portion of daily water consumption it will be

$$\frac{V_{var}}{V_0} = \frac{V(t)_{max}^{max} - V(t_0)}{V_0} - \frac{V(t)_{min}^{min} - V(t_0)}{V_0} = \frac{V_{max}^{max} - V_{min}^{min}}{V_0}. \tag{3.8}$$

Note that the initial volume $V(t_0)$ can be arbitrary. Integral (3.7) can be calculated by some of the numerical methods.

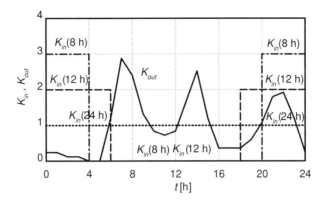

Figure 3.2 Tank filling during 24, 12, and 8 hours.

Figure 3.2 shows the graph of tank emptying variability $K_{out}(t)$ and three different graphs of filling variability $K_{in}(t)$; namely, tank filling during 24, 12, and 8 hours. For each of the tank filling graphs, integral (3.7) and their extremes were calculated. The results are shown in Table 3.1.

Table 3.1

	4 24 h	5 12 h	6 8 h
V_{var}/V_0	22.5%	66.25%	74.25%

Note the great impact of tank filling type on the volume required for daily balance of consumption; it is something that the design engineer should take into account during the water supply system design process. Overall construction costs also depend on several other factors. Thus, an adequate cost–benefit analysis should be carried out.

3.1.2 Tank model

Let the tank be connected to the hydraulic network in node r. Discharge from the node to the tank is equal to the unbalanced sum of discharges of all connected elements

$$Q_k = \sum_p{}^r Q^e. \tag{3.9}$$

Namely, it is equal to the accumulation discharge. The nodal equation of the node that included the tank has the following form

$$\sum_p{}^r Q^e = \frac{dV}{dt} + Q_p. \tag{3.10}$$

The overflow discharge is equal to

$$\begin{aligned} h_k > Z_p : \quad & Q_p = a(h_k - Z_p)^b \\ h_k \leq Z_p : \quad & Q_p = 0 \end{aligned}, \tag{3.11}$$

where a and b are the weir constants. In general, the water level in the tank differs from the piezometric head at the connecting node. Let there be asymmetric losses at the tank entrance $\Delta h = \beta^{\pm} Q_k$, where sign "+" denotes the loss coefficient for positive, and sign "−" for negative, Q_k. Due to entrance losses there is a difference between the piezometric head at the inlet node and the level in the tank

$$h_k = h - \beta^{\pm} Q_k^2. \tag{3.12}$$

For objects such as the tank with the weir and the throttle additional *secondary variables* are introduced such as the water level in the tank h_k, the tank discharge Q_k, and the overflow discharge Q_p. Secondary variables are entirely defined by the state of the fundamental primary variables. The iterative condition of secondary variables is calculated in the phase of calculation of the primary variable increments, see subroutine `IncVar`. Calculation of tank secondary variables is implemented in a subroutine

subroutine CalcTankVars(inode),

while updating the tank initial fundamental system as a separate boundary condition is given in a subroutine

subroutine TankNode(inode,iactiv,nactiv),

which is called in the frontal procedure `FrontU` before elimination of the nodal equation, see program module `Front.f90`. These subroutines are found in the program module `Tanks.f90`.

The hydraulic state of the tank is described by an ordinary differential equation that can be solved if the initial condition is prescribed. The initial condition is either prescribed volume or fluid level in the tank. Since the steady state is the most commonly adopted initial condition in the hydraulic network, the tank behavior in the steady state shall be analyzed.

The steady state of the tank is defined by the fact that all time-dependent changes are equal to zero; thus there is equality of the storage discharge Q_k and the overflow discharge Q_p

$$\underbrace{\frac{dV}{dt}}_{=0} + Q_p = Q_k. \tag{3.13}$$

Hence, when the tank is not fitted with the safety weir, the storage discharge is equal to zero and, in the steady state, the node at which the tank is connected acts as a simple connection. The water level in tank h_k, according to Eq. (3.12), will be equal to the piezometric head in the node of the tank connection.

Figure 3.3 shows possible steady states of the tank.

Figure 3.3a shows the tank with the safety weir in node r with the piezometric head $h_r > Z_p$. For piezometric heads $Z_b < h_r \le Z_p$ tank elevation makes sense since it will be filled with water. For piezometric heads $h_r < Z_b$ the tank will be empty.

Figure 3.3b shows possible solutions for the tank without a safety weir; namely, when $Q_k = 0$. When the solution is within the range $Z_b < h_r \le Z_t$, the computed tank elevation is adequate and the solution makes sense; otherwise it does not.

Fundamental system, vector, and matrix updating

The nodal equation of the node with the tank is

$$\sum_p {}^r Q^e = Q_p, \tag{3.14}$$

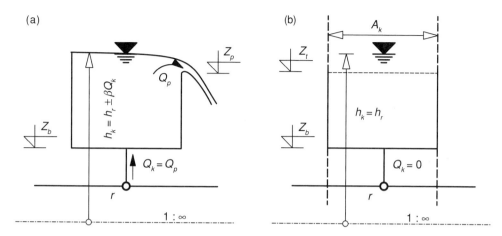

Figure 3.3 Tank elevation.

where the overflowing discharge is equal to

$$
\begin{aligned}
h_k > Z_p : \quad & Q_p = a(h_k - Z_p)^b \\
h_k \le Z_p : \quad & Q_p = 0
\end{aligned}
\tag{3.15}
$$

Vector updating in the steady state is

$$
F_r^{new} = F_r^{old} + Q_p
\tag{3.16}
$$

which is added to the global system vector (positive discharge $Q_k = Q_p$ refers to water withdrawal from the hydraulic network – the fundamental system).

Terms that update the fundamental system vector are functions of fundamental variables. Thus, the fundamental system matrix will also be updated by the respective derivatives

$$
G_{r,r}^{new} = G_{r,r}^{old} - \frac{\partial Q_p}{\partial h_r}
\tag{3.17}
$$

that are calculated as

$$
\frac{\partial Q_p}{\partial h_r} = \frac{\partial Q_p}{\partial h_k} \frac{dh_k}{dh_r},
\tag{3.18}
$$

where

$$
\frac{dh_k}{dh_r} \cong 1
\tag{3.19}
$$

which does not impact significantly on the stability of iteration. Depending on the water level in the tank, that is the weir operation

$$
\begin{aligned}
h_k > Z_p : & \quad \frac{\partial Q_p}{\partial h_k} = Q_p \frac{b}{h_k - Z_p} \\
h_k \leq Z_p : & \quad \frac{\partial Q_p}{\partial h_k} = 0
\end{aligned}
$$
(3.20)

In the steady state, an explicitly prescribed initial condition in the tank could be required. The water level in the tank indirectly defines the piezometric state in the node, see expression (3.12); thus, it is the natural boundary condition for the steady state.

In the non-steady state the increment of volume is equal to

$$
\Delta V_k = \int_{\Delta t} Q_k dt = \int_{\Delta t} \left(\frac{dV}{dt} + Q_p \right) dt.
$$
(3.21)

Following integration in the time step, the following is obtained

$$
\Delta V_k = (1 - \vartheta)\Delta t\, Q_p + \vartheta \Delta t\, Q_p^+ + (V^+ - V)
$$
(3.22)

and added to the global system vector (positive volume increment ΔV_k refers to water withdrawal from the system)

$$
F_r^{new} = F_r^{old} + \Delta V_k.
$$
(3.23)

Values for the overflowing discharge and the tank volume at the beginning and the end of the time step are calculated using the respective water levels in the tank. The tank weir is active for water levels above the weir elevation and is calculated according to the expressions

$$
\begin{aligned}
h_k > Z_p : & \quad Q_p = a(h_k - Z_p)^b \\
h_k \leq Z_p : & \quad Q_p = 0
\end{aligned}
$$
(3.24)

for the overflowing discharge at the beginning and the overflowing discharge at the end of the time interval

$$
\begin{aligned}
h_k > Z_p : & \quad Q_p^+ = a(h_k^+ - Z_p)^b \\
h_k \leq Z_p : & \quad Q_p^+ = 0
\end{aligned}
$$
(3.25)

Terms that update the fundamental system vector are functions of the fundamental variables; thus, the fundamental system matrix will also be updated by respective derivatives

$$
G_{r,r}^{new} = G_{r,r}^{old} - \frac{\partial \Delta V_k}{\partial h_r^+}
$$
(3.26)

that are calculated as follows

$$
\frac{\partial \Delta V_k}{\partial h_r^+} = \frac{\partial \Delta V_k}{\partial h_k^+} \frac{dh_k^+}{dh_r^+} = \left(A_k + \vartheta \Delta t \frac{\partial Q_p^+}{\partial h_k^+} \right) \frac{dh_k^+}{dh_r^+}.
$$
(3.27)

where the second member in the parentheses, depending on the water level in the tank, that is the weir operation, equals

$$h_k^+ > Z_p : \quad \frac{\partial Q_p^+}{\partial h_k^+} = Q_p^+ \frac{b}{h_k^+ - Z_p} ,$$

$$h_k^+ \leq Z_p : \quad \frac{\partial Q_p^+}{\partial h_k^+} = 0 \tag{3.28}$$

Derivation of the water level in the tank by piezometric level at node r

$$\frac{dh_k^+}{dh_r^+} \cong 1 \tag{3.29}$$

does not impact significantly on the stability of iteration.

3.1.3 Tank test examples

Steady state

Figure 3.4 shows the tests that the tank model should satisfy in steady modelling.

(a) Test

In the example, the boundary piezometric conditions are such as to expect a full tank. The tank is fitted with a safety weir, see input and output data shown in Figure 3.5.

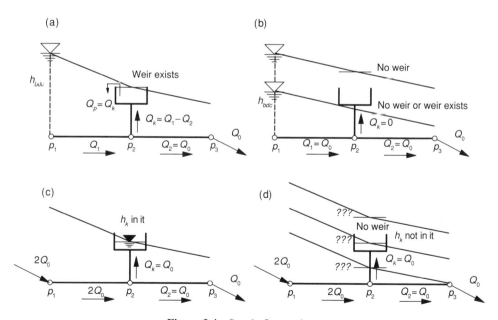

Figure 3.4 Steady flow tank tests.

Input file:	Output file:
Tank Steady a) test.simpip	Tank Steady a) test.prt

```
;                              Front length:    2
; Tank Steady a)test           Finite element mesh O.K.
;                              Steady initial conditions
Options                        St    0 iter   12   Hn 0.595E-14    Qn
   input short                 0.297E-16   Eq 0.577E-14
   run quasi                   Steady O.K.
   digits 15                   Stage:    0        Time:        0.00000
Parameters                            Point    Piez.head             SumQ
   Qo = 50/1000                       p1       110.000 -0.269654
   D = 0.300                          p2       105.26  0.219654
   eps = 0.25/1000                    p3       105.094 0.500000E-01
   Vtank = 600                        El.      Name     Q
Points                                1        c1       0.269654
   p1 0 0 0                           2        c2       0.500000E-01
   p2 100 0 0
   p3 200 0 0
Pipes
   c1    p1 p2   D   eps
   c2    p2 p3   D   eps
Tank tnk p2 Vtank/5 95 105 Qo/(0.1)^1.5 1.5
Piezom p1 110
Charge
   p3 -Qo
Steady 0
Print
   SolStage 0
```

Figure 3.5   Tank Steady (a) test.

A solution is one in which the tank overflow discharge is equal to the difference between the pipes *c1* and *c2*, which is 0.269654 – 0.05 = 0.219664 m^3/s and is equal to the negative sum of discharges in node *p2*. Since the weir parameters, according to the input data, are $a = 1.58113883$ and $b = 1.5$ the overflow height is 0.268 m, which gives the overflow level of 105 + 0.268 = 105.268 m.

(b)   Test
The tank has no weir. Since the input data do not give an explicit initial condition for the tank (discharge to the tank is zero), the steady state model will give the level in the tank node as there is no tank at all.

(c)   Test
This refers to an explicitly prescribed tank initial condition. At the initial time, tank volume, that is the water level in the tank, is prescribed. Note that, apart from the explicitly prescribed tank level, there is no other explicitly prescribed natural boundary condition, that is prescribed piezometric head, in the modeled system.

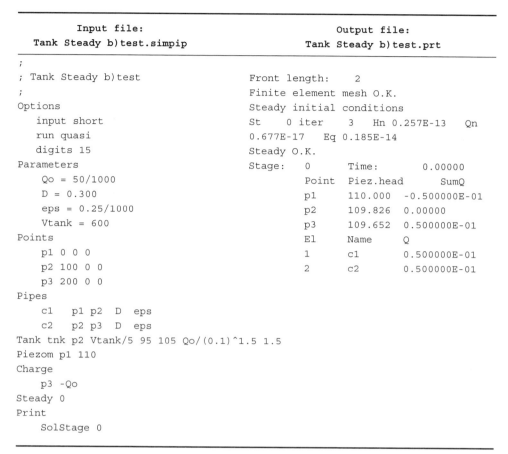

Input file: Tank Steady b)test.simpip	Output file: Tank Steady b)test.prt

```
;
; Tank Steady b)test
;
Options
    input short
    run quasi
    digits 15
Parameters
    Qo = 50/1000
    D = 0.300
    eps = 0.25/1000
    Vtank = 600
Points
    p1 0 0 0
    p2 100 0 0
    p3 200 0 0
Pipes
    c1    p1 p2   D   eps
    c2    p2 p3   D   eps
Tank tnk p2 Vtank/5 95 105 Qo/(0.1)^1.5 1.5
Piezom p1 110
Charge
    p3 -Qo
Steady 0
Print
    SolStage 0
```

```
Front length:    2
Finite element mesh O.K.
Steady initial conditions
St    0 iter    3   Hn 0.257E-13    Qn
0.677E-17   Eq 0.185E-14
Steady O.K.
Stage:   0      Time:       0.00000
         Point  Piez.head      SumQ
         p1     110.000   -0.500000E-01
         p2     109.826   0.00000
         p3     109.652   0.500000E-01
         El     Name       Q
         1      c1         0.500000E-01
         2      c2         0.500000E-01
```

Figure 3.6 Tank Steady (b) test.

The initial water level in the tank is given by the `SimpipCore` instruction:

$$\text{Initialize Tank name hValue,}$$

see the input file on the left hand side of Figure 3.7

(d)   Test
In this example, the unknowns are the piezometric boundary conditions. An explicit initial condition for the tank has also not been prescribed. As expected, the system is unsolvable since it is not regular.

Quasi unsteady condition

Figure 3.9 shows a tank in a small water supply network with consumption varying according to the diagram $K_{out}(t)$, shown in Figure 3.2.

The daily water requirement is $V_0 = 25920$ m^3 or, expressed as uniform discharge, 300 l/s that is added to the system in front of the tank at point $p0$. The tank size was defined based on the criterion of uniform all-day filling, see Table 3.1; namely, as 22.5% of the total required water volume, which is $V_{var} = 5832$ m^3.

Input file: Tank Steady c)test.simpip	Output file: Tank Steady c)test.prt

```
;                                    Front length:    2
; Tank Steady c)test                 Finite element mesh O.K.
;                                    Steady initial conditions
Options                              St    0 iter    3   Hn 0.312E-13   Qn
    input short                      0.678E-17    Eq 0.233E-14
    run quasi                        Steady O.K.
    digits 15                        Stage:    0       Time:      0.00000
Parameters                                  Point   Piez.head   SumQ
    Qo = 50/1000                            p1      100.669 -0.100000
    D = 0.300                               p2      100.000 0.500000E-01
    eps = 0.25/1000                         p3      99.8260 0.500000E-01
    Vtank = 600                             El.     Name    Q
Points                                      1       c1      0.100000
    p1 0 0 0                                2       c2      0.500000E-01
    p2 100 0 0
    p3 200 0 0
Pipes
    c1    p1 p2   D   eps
    c2    p2 p3   D   eps
Tank tnk p2 Vtank/5 95 105 Qo/(0.1)^1.5 1.5
Init TANK tnk 100
Charge
    p1 2*Qo
    p3 -Qo
Steady 0
Print
    SolStage 0
```

Figure 3.7　Tank Steady (c) test.

At time $t = 0$, the initial condition in the tank is prescribed.

Figure 3.10 shows the input data for the water supply network and the results of operation modelling throughout 24 hours. Variation of daily consumption is prescribed by the graph $K_{out}(t)$. Graph data included in the file "K(t).inc" is inserted into input data by "@includefilename", as the graph object.

The right hand side of the figure shows some of the results of the water supply system simulation in one day; namely, the piezometric heads in some system nodes and changes in the water level in the tank.

Note that all water levels at the end of 24 h are returned to the initial value, as is expected due to the character of the variation graph, which is balanced for cyclic repetition on a 24 hour interval.

The final tank condition is equal to the initial condition, that is the initialized value 100 m. The maximum level is 104.22 and the minimum 99.20 m. The tank level has a 5 m variation. The tank cross-section area is $A_k = 1166.4 \text{ m}^2$. The volume required for daily balance of consumption is

$$V_{var} = 5A_k = 5 \cdot 1166.4 = 5832 \text{ m}^3$$

that is, equal to the one obtained through application of Table 3.1.

Input file: Tank Steady d)test.simpip	Output file: Tank Steady d)test.prt
Options	Front length: 2
input short	Finite element mesh O.K.
run quasi	Steady initial conditions
digits 15	*** Irregular system found. Impossible
Parameters	to solve! ***
Qo = 50/1000	Point: p2 Tank: tnk
D = 0.300	*** Frontal procedure fatal error! ***
eps = 0.25/1000	
Vtank = 600	
Points	
p1 0 0 0	
p2 100 0 0	
p3 200 0 0	
Pipes	
c1 p1 p2 D eps	
c2 p2 p3 D eps	
Tank tnk p2 Vtank/5 95 105	
Charge	
p1 2*Qo	
p3 -Qo	
Steady 0	
Print	
SolStage 0	

Figure 3.8   Tank Steady (d) test.

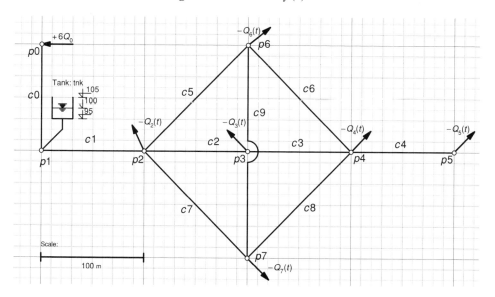

Figure 3.9   Tank in a small water supply network.

<table>
<tr><td>

Input file:
Tank QuNetwork.simpip

</td><td>

Output file:
Tank QuNetwork.prt

</td></tr>
</table>

```
;
; Tank Quasi Non-steady
  network test
;
Options
    input short
    run quasi
    digits 15
Parameters
    Qo = 50/1000
    D = 0.300
    eps = 0.25/1000
    Vtank = 0.225*6*Qo*24*3600
Points
    p0 0 100 0
    p1 0 0 0
    p2 100 0 0
    p3 200 0 0
    p4 300 0 0
    p5 400 0 0
    p6 200 100 0
    p7 200 -100 0
Pipes
    c0   p0 p1   D   eps
    c1   p1 p2   D   eps
    c2   p2 p3   D   eps
    c3   p3 p4   D   eps
    c4   p4 p5   D   eps
    c5   p2 p6   D   eps
    c6   p6 p4   D   eps
    c7   p2 p7   D   eps
    c8   p7 p4   D   eps
    c9   p6 p7   D   eps
Tank tnk p1 Vtank/5 95 105
Init TANK tnk 100
@K(t).inc
Charge
    p0 6*Qo
    p2 0 -Qo K(t)
    p3 0 -Qo K(t)
    p4 0 -Qo K(t)
    p5 0 -Qo K(t)
    p6 0 -Qo K(t)
    p7 0 -Qo K(t)
Steady 0
Unsteady 24 3600
Print
  SolTank tnk
  SolPoint p0 p1 p2 p3 p4
  p5 p6 p7
```

(a) Piezometric heads.

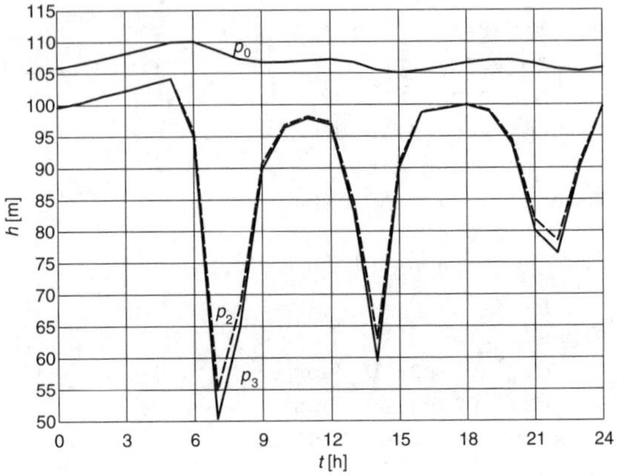

(b) Lavel in tank.

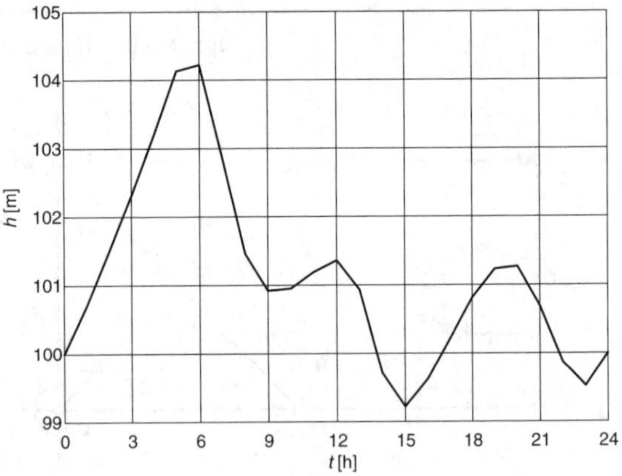

Figure 3.10   Tank QuNetwork test.

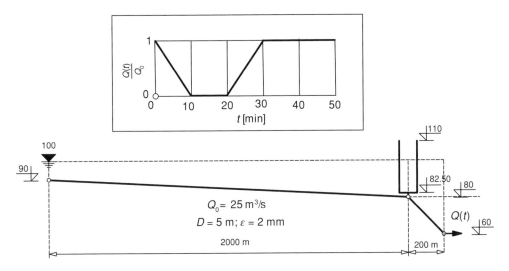

Figure 3.11   Tank quasi unsteady – rigid flow test.

Rigid fluid

Application of the quasi unsteady model out of the usual modelling of the daily operation of a water supply system can be very questionable because the influence of inertia is hard to assess.

This can be shown by an example.

Figure 3.11 shows a system consisting of the supply pipeline, the tank, and the penstock. Prescribed values are the piezometric head at the beginning of the supply pipeline and variable consumption at the end of the outlet pipeline. Variable consumption is given in the form of the consumption diagram (graph), also shown in the figure.

The tank cross-section area is $A_k = 100$ m^2. Other data are given in the figure and input file.

Consumption changes are relatively slow; for example closing and discharge rising takes 10 minutes, which seems slow enough.

Initial conditions are prescribed as a steady solution, that is a steady state for the consumption $Q_0 = 25$ m^3/s.

The problem thus described will be solved first as quasi unsteady flow and then as non-steady flow of an incompressible fluid (rigid model) by application of the option run.

The obtained results are shown on the right hand side of Figure 3.12, for the tank level. The result is more than surprising for someone who is unfamiliar with the problem because it seems that the results are incorrect or refer to two different systems.

For someone who is well acquainted with the pipe flow and modelling problems, it is clear that the cardinal differences in tank behavior are the result of the inertia factors in the supply pipeline, which are obviously not negligible.

The influence of inertia is manifested by the piezometric head and discharge oscillations in the system supply pipeline–tank. Oscillations occur in both tank filling and tank emptying phases since the liquid in the supply pipeline cannot suddenly slow down or speed up.

Thus, in hydraulic networks with inertia phenomena that are not negligible, special tanks – surge tanks – are introduced with the particular role to receive or discharge water at relatively fast changes in system consumption. Surge tanks will be discussed in Section 3.3 Surge tank.

Input file:	Output file:
Tank Qu-Rgd test.simpip	Tank Qu-Rgd test.prt

```
; Quasi/Rigid unsteady models
;
; tests of inertia influence on tank solution
;

Options
        input short
        Run rigid ; quasi
        digits 8
Parameters
        Qo = 25
        D = 5
        eps = 2.e-3
        Ak = 100
        Ha = 100
Points
        p1 0 0 90
        p2 2000 0 80
        p3 2200 0 60
Pipes
   c1    p1 p2    D   eps
   c2    p2 p3    D   eps
Graph Q(t)
        0 1
        600 0
        1200 0
        1800 1
Charge p3 0 -Qo Q(t)
Piezo p1 Ha
Tank Komora p2 Ak 82.5 110
Steady 0
Unsteady 600 10
Print
        SolTank Komora
```

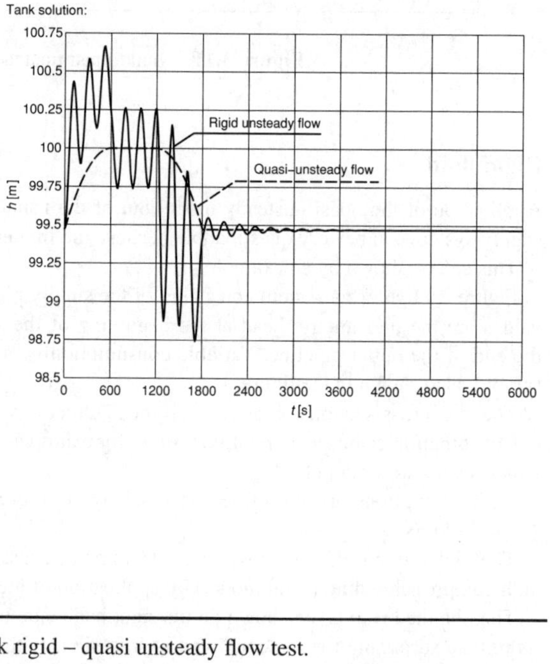

Figure 3.12   Tank rigid – quasi unsteady flow test.

3.2 Storage

3.2.1 Storage equation

The storage tank is a large tank that differs from a simple tank through its shape and size.

In terms of hydraulics, there are no differences whatsoever. The continuity equation for a storage tank is equal to the continuity equation of a simple tank

$$\frac{dV}{dt} + Q_p = Q_k, \tag{3.30}$$

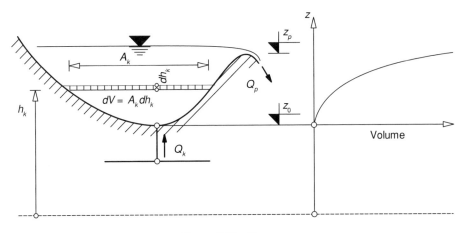

Figure 3.13 Storage.

where Q_k is the filling discharge, Q_p is the overflowing discharge, and V is the water volume. The main difference is in the variable area of horizontal cross-sections; thus the storage volume is set by a volumetric curve, see Figure 3.13.

3.2.2 Fundamental system vector and matrix updating

Updating of the nodal equation by storage as a boundary condition is completely the same as for the simple tank, see expressions (3.16) and (3.17) for steady flow and expressions (3.26) and (3.27) for non-steady flow.

The area of horizontal cross-section A_k is calculated by the numeric derivative of the storage volume graph at the point $z = h_k^+$

$$A_k = \left. \frac{\Delta V}{\Delta h} \right|_{z=h_k^+}. \tag{3.31}$$

Interval $\Delta h = 0.02$ m is implemented in program solution SimpipCore.

3.3 Surge tank

3.3.1 Surge tank role in the hydropower plant

Figure 3.14 shows a scheme of a typical high-pressure hydropower plant, which consists of:

- dam and reservoir,
- intake structure and headrace tunnel,
- surge tank,
- penstock,
- power house with turbines and generators,
- tailrace mains.

Although, in general, each high-pressure hydropower plant consists of the aforementioned system components, each one is unique. Thus, there are different solutions for each of the components. In this

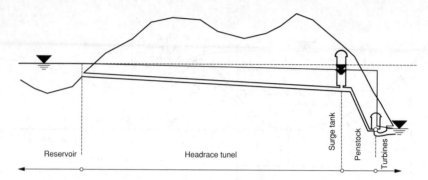

Figure 3.14   Scheme of the spatial disposition and longitudinal section of the hydropower plant.

text we will not deal with the design principles, but will only analyze the hydraulic role of different components in the system and encountered hydraulic problems.

The dam accumulates backwater and increases the hydropower potential, the reservoir balances variable inflow, the supply system (intake structure, headrace tunnel, penstock) transports water to turbines, which use the water power for turbine rotation and electric energy generation, while the outlet mains return used water to the river.

A surge tank plays a particular role in the supply system. It secures inflow to turbines at the moment of the operation's start; for example from stand-by to full operation. The starting time depends on the power supply system that the hydroelectric power plant is connected to and it is relatively short. A supply system is, in general, relatively long and water cannot be quickly accelerated, that is the full turbine flow cannot be achieved in a short time. Thus, directly upstream from the turbine a surge tank is installed, from which water is pulled quickly to turbines via a relatively short pipeline (penstock), see Figure 3.15. The difference between water levels in the storage and surge tank, which is a consequence of the surge tank emptying, gradually accelerates the flow in the headrace tunnel and causes water mass oscillations in the system of the headrace tunnel–surge tank.

Similarly to the fast start of a high-pressure hydroelectric power plant operation within the hydropower system, there is also a fast shutdown. Shutdown can be either a regular or an emergency one. Emergency shutdown occurs when the generator or turbine remain without the load (breakdown or thunderbolt impact to the switchyard). The turbine engine speeds up and there is the possibility of breakdown due to

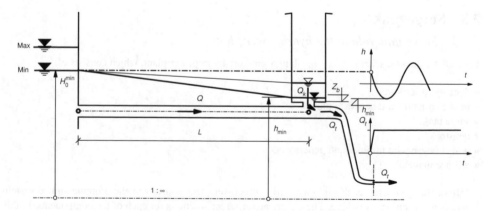

Figure 3.15   Hydropower plant operation start.

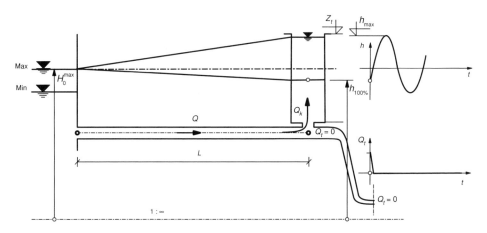

Figure 3.16   Hydropower plant shutdown.

high centrifugal forces. Then the turbine regulator closes the stator blades and stops the flow. Flow stop
is so sudden that it can be considered instantaneous.

The pipeline connecting the surge tank and turbines is called the penstock. Immediately after the
turbine shutdown, the discharge in penstock becomes zero, the surge tank starts to fill due to water
inertia in the headrace tunnel, see Figure 3.16, and there are oscillations of water mass in the headrace
tunnel–surge tank system.

Unlike the headrace tunnel, in which time-dependent flow changes are slow enough that water can
be considered incompressible and the problem can be analyzed as non-steady flow of rigid fluid, flow
changes in the penstock are very fast; thus, at the very beginning of a sudden closing the flow stops at
the turbine cross-section while the upstream flow is still undisturbed, see Figure 3.17. In this case, water
behaves as a compressible fluid, which, due to the upstream inertia, continues to flow. Kinetic energy of

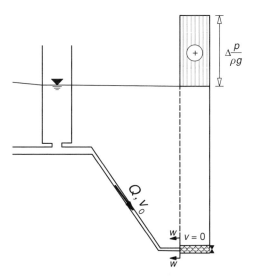

Figure 3.17   Pressure rise in penstock due to stop of the inflow to the turbines.

the undisturbed flow is transformed into potential energy of a slowed flow, thus causing the pressure rise and pipeline expansion.

The pressure change, water compression, and pipeline expansion propagate upstream at a sound velocity. After water is slowed down in the entire penstock, there is a release of the accumulated potential energy. Following the pressure increase phase there is the pressure decrease phase, until all fluctuations due to friction are amortized. Each discharge change causes pressure change; thus, there are similar effects at the start of the turbine operation.

A hydraulic problem of this type is called a *water hammer*, and will be discussed in subsequent chapters.

3.3.2 Surge tank types

The range of water level oscillations in the surge tank at sudden turbine loading or unloading is great and requires a large surge tank size. Since the surge tank is, in general, an underground and relatively expensive structure, its size should be the minimum possible without reduction in its functionality. As will be shown later, the surge tank cross-section area should be large enough to always to amortize the oscillations. Several surge tank types will be described hereinafter.

(a) The cylindrical surge tank shown in Figure 3.18 requires large dimensions. It is usually the subject of theoretical analyses; and cylindrical surge tank parameters are used during the elaboration of preliminary designs for the purpose of assessment of different alternatives.

(b) Figure 3.19 shows a cylindrical surge tank with an asymmetric throttle at the connection with the headrace tunnel. The throttle is designed to provide increased resistance in the surge tank filling phase and a small resistance in the emptying phase. In comparison with the cylindrical surge tank without the throttle, the maximum water rise is reduced. A throttle is an optimum one if it does not cause significant pressure rise in the headrace tunnel.

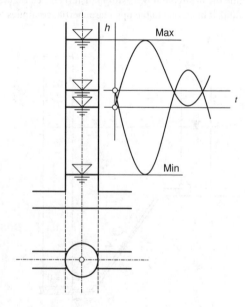

Figure 3.18   Cylindrical surge tank.

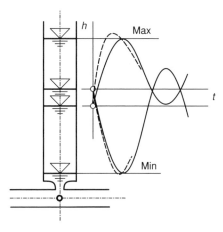

Figure 3.19   Cylindrical surge tank with the asymmetric throttle.

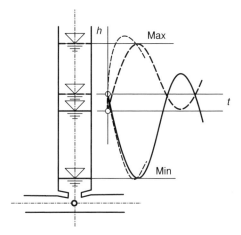

Figure 3.20   Cylindrical surge tank with the Ventouri transition.

(c) Figure 3.20 shows a cylindrical surge tank with a Ventouri transition at the connection with the headrace tunnel. It is used on small headrace tunnels with small longitudinal heads where the benefits of increased velocity in the transition on the surge tank stability are intended to be used. It is recommended to test this type of a surge tank on a hydraulic physical model.

(d) The cylindrical surge tank with air throttle, shown in Figure 3.21, is based on the fact that compressed air at a rising water level slows down tank level rising, acting thus as the surge tank cross-section area is larger. A detailed description of this hydraulic system is given in the PhD theses by Professor Josip Grčić.[1]

(e) Figure 3.22 shows a cylindrical surge tank with a gallery. Its primary role is to reduce the maximum water level rise. Thus, the headrace tunnel is subjected to lower pressure loads than in a case of a simple cylindrical surge tank.

---

[1] Josip Grčić (1918–1977), Professor of Hydraulics in the Faculty of Civil Engineering, University of Zagreb.

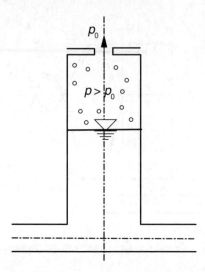

Figure 3.21   Cylindrical surge tank with the air throttle.

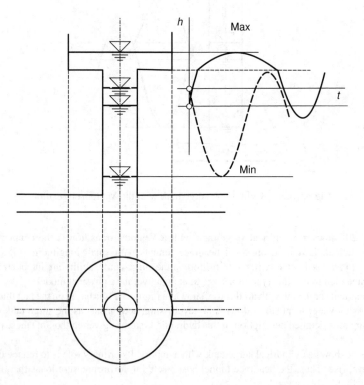

Figure 3.22   Cylindrical surge tank with the gallery.

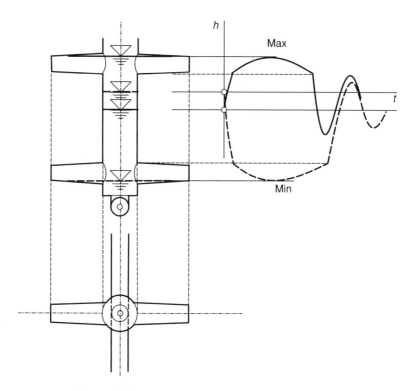

Figure 3.23   Surge tank with upper and lower chamber.

(f) The surge tank with an upper and lower chamber, shown in Figure 3.23, consists of a central cylindrical surge tank and two extended tanks (upper and lower surge tanks), usually constructed as side shafts although they can be differently shaped depending on local geological conditions. The top surge tank reduces the maximum water level rise while the lower surge tank protects the system against too low water levels and prevents air suction. This surge tank type is constructed most frequently.

(g) A differential or Johnson's surge tank is shown in Figure 3.24: the flow into and from the tank branches, with the faster level rising and falling in the riser (inner shaft) while the main chamber level lags behind. This can be seen in practice in different modified forms.

(h) A differential surge tank with an upper and lower chamber is shown in Figure 3.25. For water levels lower than the upper chamber level, it operates as a simple surge tank while for higher levels it behaves as a differential surge tank. An asymmetric throttle can be installed at the connection with the headrace tunnel.

(i) A double surge tank is either constructed to increase the overall tank cross-section area or to achieve the effect of a differential surge tank by alternative cross-section of the two chambers.

There are also surge tank systems that are distributed along a long headrace tunnel; namely when there is a very long oscillation period. However, such a system faces the danger of system resonance. It is rare in practice.

Hydropower plants with long tailrace tunnels that are always under pressure are provided with a tailrace surge tank.

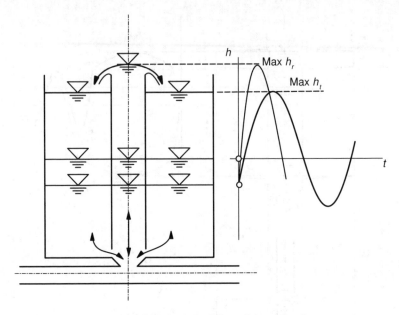

Figure 3.24 Johnson's differential surge tank.

Hydrodynamic conditions are very variable in a system with two surge tanks; thus, the surge tanks should be very carefully sized to avoid amplification of water level due to the turbine speed governor.

In very short headrace tunnels, where water mass acceleration time is relatively short in comparison with the turbine opening velocity, the surge tank can be omitted. In this case, the reservoir takes over the role of the surge tank.

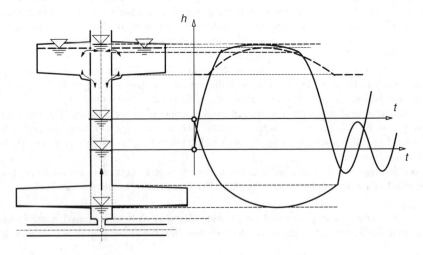

Figure 3.25 Differential surge tank with upper and lower chamber.

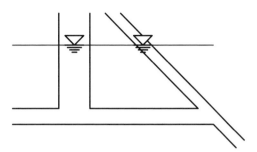

Figure 3.26   Double surge tank.

3.3.3   Equations of oscillations in the supply system

Dynamic equations of the headrace tunnel and surge chamber

Figure 3.27 shows a supply system that consists of a storage tank, intake structure, headrace tunnel, and surge tank with an asymmetric throttle. Time-dependent flow variations are relatively slow, thus water can be considered incompressible (rigid).

The intake structure is a short structure in which the flow inertia can be disregarded. Thus, the dynamic equation is reduced to

$$h_0 = h_1 + \beta_u^{\pm} |Q|\, Q, \tag{3.32}$$

where $\beta_u^{\pm}$ is the asymmetric resistance coefficient at the intake structure.

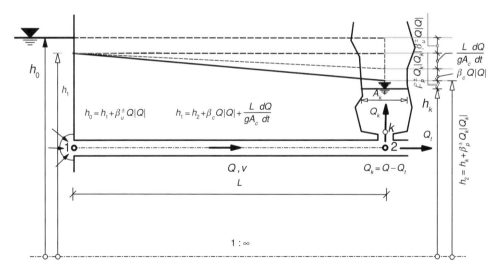

Figure 3.27   Scheme of hydrodynamic relations in the surge tank of general cross-section.

Figure 3.27 shows a scheme of the intake structure position, point 1. The headrace tunnel (pipeline) extends from point 1 to point 2. The equation for non-steady flow of a rigid fluid for the headrace tunnel is

$$h_1 = h_2 + \lambda \frac{L}{D} \frac{v|v|}{2g} + \frac{L}{g} \frac{dv}{dt} \tag{3.33}$$

or expressed by discharge

$$h_1 = h_2 + \beta_c |Q| Q + \frac{L}{g A_c} \frac{dQ}{dt}. \tag{3.34}$$

Between the headrace tunnel and the surge tank – namely, between the points 2 and k – there is a short asymmetric throttle. Its dynamic equation is

$$h_2 = h_k + \beta_p^{\pm} |Q_k| Q_k \tag{3.35}$$

in which, because of the short throttle, the flow inertia is negligible. Taking into account all elements in the inflow pipe that enter the surge tank, a common dynamic equation can be written as

$$h_0 = h_k + \beta_p^{\pm} |Q_k| Q_k + \left(\beta_u^{\pm} + \beta_c\right) |Q| Q + \frac{L}{g A_c} \frac{dQ}{dt}, \tag{3.36}$$

where inflow into the surge tank is equal to

$$Q_k = Q - Q_t. \tag{3.37}$$

Surge tank continuity equation

The surge tank continuity equation can be written as

$$\frac{dV}{dt} = Q - Q_t, \tag{3.38}$$

where V is the volume of water in the surge tank, Q is the discharge in the headrace tunnel, and Q_t is the discharge to the turbines, namely

$$A_k \frac{dh_k}{dt} = Q - Q_t, \tag{3.39}$$

where A_k is the variable area of the surge tank horizontal cross-section, which is calculated from the volumetric curve

$$A_k = \frac{dV}{dh_k}. \tag{3.40}$$

3.3.4 Cylindrical surge tank

Equation of small oscillations

In the analysis of small oscillations in the cylindrical surge tank it is assumed that the inlet resistances are negligible in comparison with the linear resistances in the headrace tunnel. Also, if a cylindrical surge tank without a throttle at the connection with the headrace tunnel is observed, then

$$\beta_u^\pm = \beta_p^\pm = 0. \tag{3.41}$$

Writing $\beta = \beta_c$ and $h = h_k$, the dynamic equation of the cylindrical surge tank becomes

$$\frac{L}{gA_c}\frac{dQ}{dt} + \beta Q|Q| + h = h_0. \tag{3.42}$$

If $z = h - h_0$ is introduced into dynamic equation, then

$$\frac{L}{gA_c}\frac{dQ}{dt} + \beta Q|Q| + z = 0. \tag{3.43}$$

Differentiation of the continuity equation (3.39) in time gives

$$A_k\frac{d^2h}{dt^2} = \frac{dQ}{dt} - \frac{dQ_t}{dt}$$

or

$$\frac{dQ}{dt} = A_k\frac{d^2z}{dt^2} + \frac{dQ_t}{dt}, \tag{3.44}$$

where $z = h - h_0$. If expression (3.44) is introduced into dynamic equation (3.43), a differential equation of the surge tank oscillations is obtained

$$\frac{LA_k}{gA_c}\frac{d^2z}{dt^2} + \beta Q|Q| + z = -\frac{L}{gA_c}\frac{d}{dt}Q_t(t). \tag{3.45}$$

If a homogeneous part is written for Eq. (3.45), and all non-linear terms are disregarded in it, a linear equation is obtained

$$\frac{LA_k}{gA_c}\frac{d^2z}{dt^2} + z = 0. \tag{3.46}$$

or, written in the form

$$\frac{d^2z}{dt^2} + \frac{gA_c}{LA_k}z = 0. \tag{3.47}$$

If the factor in front of z in the second term in the previous equation is written as

$$\omega^2 = \frac{gA_c}{LA_k} \tag{3.48}$$

then a classical equation of oscillations $\ddot{z} + \omega^2 z = 0$ can be recognized, where ω is the angular frequency, while the oscillation period is equal to

$$T = \frac{2\pi}{\omega} = 2\pi \sqrt{\frac{L A_k}{g A_c}}. \tag{3.49}$$

The solution of Eq. (3.47) is

$$z = Z^* \sin \omega t, \tag{3.50}$$

where Z^* is the constant (maximum amplitude), which can be calculated from the initial condition $t = 0$, Q_0. The surge tank discharge is equal to

$$Q = A_k v_k = A_k \frac{dz}{dt} = A_k Z^* \omega \cos \omega t \tag{3.51}$$

from which Z^* is calculated for $t = 0$ as

$$Z^* = \frac{Q_0}{\omega A_k} = \frac{Q_0}{A_k} \sqrt{\frac{L A_k}{g A_c}}. \tag{3.52}$$

Also

$$Z^* = \frac{Q_0}{A_c} \sqrt{\frac{L A_c}{g A_k}}. \tag{3.53}$$

The final form of small oscillations in the surge tank is

$$z = \frac{Q_0}{A_k} \sqrt{\frac{L A_k}{g A_c}} \sin \sqrt{\frac{g A_c}{L A_k}} t \tag{3.54}$$

or

$$z = \frac{Q_0}{A_c} \sqrt{\frac{L A_c}{g A_k}} \sin \sqrt{\frac{g A_c}{L A_k}} t. \tag{3.55}$$

A very useful relation can be derived from the solution of small oscillations in the surge tank

$$T = 2\pi \frac{A_k}{A_d} \frac{Z^*}{v_o} = 2\pi \frac{A_k}{Q_0} Z^*. \tag{3.56}$$

An approximate solution of the maximum oscillation

The maximum amplitude of oscillations in the surge tank for a sudden stop of power plant operation occurs in a time equal to one fourth of the oscillation period, as shown in Figure 3.28. The loss curve $\Delta h = \beta Q^2$ is drawn along the surge tank level curve h_k. If the dynamic equation is written in the form

$$\frac{L}{g A_c} \frac{dQ}{dt} + (h_k - h_0) + \beta Q^2 = 0 \tag{3.57}$$

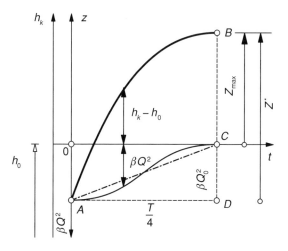

Figure 3.28

and integrated in the time of the interval equal to one fourth of the period $[0, T/4]$

$$\frac{L}{g A_c} \int_{Q_0}^{0} dQ + \int_{0}^{\frac{T}{4}} \left[(h_k - h_0) + \beta Q^2\right] dt = 0 \tag{3.58}$$

$$\underbrace{\qquad\qquad\qquad}_{Area(ABC)}$$

and if the geometric property of a particular integral is applied, then

$$\frac{L}{g} \frac{Q_0}{A_c} = Area(ABC). \tag{3.59}$$

A. Franković[2] assumed that the water level curve, measured from the initial water level, can be approximated by a sinusoid, while the resistance curve can be approximated by the line; namely, it can be linearized. Thus, the area will be equal to

$$Area(ABC) \approx \int_{0}^{T/4} Z^* \sin \frac{2\pi}{T} t \cdot dt - \frac{1}{2} \frac{T}{4} \beta Q_0^2 \tag{3.60}$$

which, after calculations, gives

$$\frac{L}{g} \frac{Q_0}{A_c} = Z^* \frac{T}{2\pi} - \frac{1}{2} \frac{T}{4} \beta Q_0^2. \tag{3.61}$$

---

[2] Ante Franković (1889–1976), Professor of Hydraulic Engineering at the Technical College of the University of Zagreb.

If a relation (3.56) is introduced into Eq. (3.61), a quadratic equation is obtained

$$\frac{L}{g}\frac{Q_0}{A_c} = \frac{A_k}{Q_0}Z^{*2} - \frac{\pi}{4}\beta Q_0 A_k Z^*. \tag{3.62}$$

If expressions for resistances $\Delta h_0 = \beta Q_0^2$ and discharge $Q_0 = A_c v_0$ are introduced, the equation can be written in the form

$$Z^{*2} - \frac{\pi}{4}\Delta h_0 Z^* - \frac{L A_c}{g A_k}v_0^2 = 0 \tag{3.63}$$

with the logical solution of the equation

$$Z^* = \frac{\pi}{8}\Delta h_0 + \sqrt{\left(\frac{\pi}{8}\Delta h_0\right)^2 + \frac{L A_c}{g A_k}v_0^2}. \tag{3.64}$$

Since $Z^* = Z_{max} + \Delta h_0$ the maximum amplitude of oscillations is obtained as

$$Z_{max} = \left(\frac{\pi}{8} - 1\right)\Delta h_0 + \sqrt{\left(\frac{\pi}{8}\Delta h_0\right)^2 + \frac{L A_c}{g A_k}v_0^2}. \tag{3.65}$$

Surge tank stability

A hydropower plant connected to a hydropower system should produce electric power of constant frequency, for example 50 Hz, with the allowable variation of $\approx \pm 0.2\%$ of the normal value. The frequency of the electric current produced by a generator depends on the rotation speed of the rotor and is defined by the expression

$$f = \frac{pn}{60}, \tag{3.66}$$

where f is the current frequency in Hz, n is the number of revolutions in one minute, and p is the number of generator poles. Motion of the rotor (turbine and generator) is defined by the dynamic equation of the engine

$$I\frac{d\omega}{dt} = M_t - M_g, \tag{3.67}$$

where I is the polar moment of inertia of rotating parts, ω is the angular velocity of the generator, M_t is the turbine moment, and M_g is the moment of the generator at the engine axis (loading moment).

A requirement for constant frequency refers to the constant angular velocity of rotation; namely, $d\omega/dt = 0$, which is achieved by equality of the turbine torque and the loading torque

$$M_t = M_g. \tag{3.68}$$

Constant turbine torque is required to maintain the constant load, that is generator power

$$M_t = \frac{P}{\omega} = \eta\frac{\rho g Q_t H_t}{\omega},$$

that is the turbine power shall be constant

$$P = \eta \rho g Q_t H_t, \tag{3.69}$$

where Q_t is the turbine discharge, H_t is the turbine head, and η is the turbine efficiency coefficient.

Turbine operation is controlled by a special device called the turbine speed governor. The turbine moves the blades at the inlet device, thus letting in the discharge in the range from 0 to the installed discharge of a turbine. The start of the turbine operation begins with turbine rotation to the synchronal speed, after which it is gradually loaded to the rated power. The turbine speed governor takes further control. A procedure of constant power maintenance, that is regulation of the rotating speed, is the following:

- if the turbine is accelerating, the speed governor closes the blades (decreasing the discharge);
- if the turbine is decelerating, the speed governor opens the blades (increasing the discharge).

Turbine or generator power is defined by discharge Q_t and energy head H_t, as well as the turbine efficiency coefficient η. They are variable during operation at any discharge or piezometric head change. From the expression for power it is not hard to conclude that a decrease in the piezometric head increases the discharge and vice versa to maintain constant power.

Figure 3.29 shows the transition of hydropower station operation from 50 to 100% power. Power curves with marked changes of duty points $NP_{50\%}$ and $NP_{100\%}$. are given in coordinate system Q_t, H_t.

Hydropower plant transition to the new duty point increases the turbine discharge that is emptying the surge tank, the water level decreases thus decreasing the turbine head H_t in order to maintain the constant power of a turbine, the speed governor increases the discharge Q_t thus additionally influencing the increase in surge tank emptying. There is a similar, although reversed, trend in the phase of surge tank level rise. The speed governor decreases the discharge and increases oscillation, that is the turbine speed governor maintains oscillations; specifically, there is a tendency to oscillation increase (amplification). The turbine speed governor moves the duty point along the power curve, which oscillates around the duty point for the rated power, see points a and b at the constant power curve 100%.

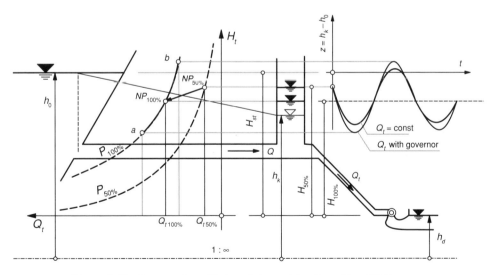

Figure 3.29   Surge tank oscillation after transition from 50 to 100% power.

The surge tank, where oscillations are amortized due to the turbine speed governor, is called the *stable surge tank*. If a surge tank is not sized properly, the impact of the speed governor can be greater than oscillation dumping due to flow resistance and progressive oscillations occur. This surge tank is called an unstable surge tank.

The problem of surge tank stability is reduced to the solution of differential equations of a cylindrical surge tank with the constant power as a boundary condition. The following is available for problem solving

Equation of continuity:

$$A_c v - Q_t = A_k \frac{dz}{dt} \tag{3.70}$$

from which

$$v = u + \frac{A_k}{A_c} \frac{dz}{dt}, \tag{3.71}$$

where

$$u = \frac{Q_t}{A_c}. \tag{3.72}$$

After differentiation in time, the following is obtained

$$\frac{dv}{dt} = \frac{du}{dt} + \frac{A_k}{A_c} \frac{d^2 z}{dt^2}. \tag{3.73}$$

Dynamic equation:

$$\frac{L}{g} \frac{dv}{dt} \pm \beta v^2 + z = 0. \tag{3.74}$$

The algebraic sign of the resistance term is positive for positive and negative for negative velocity since resistances always resist the flow.

Requirement of constant turbine power $P = P_0 = const$

$$\rho g Q_t H_t \eta = \rho g Q_{t0} H_{t0} \eta_0, \tag{3.75}$$

where P is the turbine power during oscillations, P_0 is the rated power, which is expressed by respective turbine hydraulic values of discharge Q_t, energy head H_t, and performance coefficient η. If coefficient $\eta = const$ and head loss at the turbine are expressed by the static loss H_{st} and oscillation z, then it will be

$$Q_t (H_{st} + z) = Q_{t0} H_{t0}. \tag{3.76}$$

When the above expression is divided by A_c, the following is obtained

$$\frac{Q_t}{A_c} (H_{st} + z) = \frac{Q_{t0}}{A_c} H_{t0}. \tag{3.77}$$

If Eq. (3.72) is applied, then it is written as

$$u \left(H_{st} + z \right) = u_0 H_{t0} \tag{3.78}$$

from which

$$u = \frac{u_0 H_{t0}}{H_{st} + z}. \tag{3.79}$$

After differentiation of the obtained expression in time

$$\frac{du}{dt} = -\frac{u_0 H_0}{\left(H_{st} + z \right)^2} \frac{dz}{dt} \tag{3.80}$$

and introduction into Eq. (3.73), the following is obtained

$$\frac{dv}{dt} = -\frac{u_0 H_0}{\left(H_{st} + z \right)^2} \frac{dz}{dt} + \frac{A_k}{A_c} \frac{d^2 z}{dt^2}. \tag{3.81}$$

Furthermore, if Eq. (3.81) is introduced into dynamic equation (3.74) a differential equation of water oscillation in a cylindrical surge tank for constant power is obtained

$$\frac{L}{g} \frac{A_k}{A_c} \frac{d^2 z}{dt^2} - \frac{L}{g} \frac{u_0 H_0}{\left(H_{st} + z \right)^2} \frac{dz}{dt} + z \pm \beta v^2 = 0. \tag{3.82}$$

From the continuity equation (3.71) it is written

$$v = \frac{u_0 H_{t0}}{H_{st} + z} + \frac{A_k}{A_c} \frac{dz}{dt} \tag{3.83}$$

thus v^2 can be expressed in the form

$$v^2 = 2 \frac{u_0 H_{t0}}{H_{st} + z} \frac{A_k}{A_c} \frac{dz}{dt} + \left(\frac{A_k}{A_c} \right)^2 \left(\frac{dz}{dt} \right)^2 + \left(\frac{u_0 H_{t0}}{H_{st} + z} \right)^2. \tag{3 84}$$

After introducing Eq. (3.84) into Eq. (3.82) and arranging, the following is obtained

$$\frac{L}{g} \frac{A_k}{A_c} \frac{d^2 z}{dt^2} \pm \beta \left(\frac{A_k}{A_c} \right)^2 \left(\frac{dz}{dt} \right)^2 + \left(\pm 2\beta \frac{u_0 H_{t0}}{H_{st} + z} \frac{A_k}{A_c} - \frac{L}{g} \frac{u_0 H_0}{\left(H_{st} + z \right)^2} \right) \frac{dz}{dt} + z \pm$$
$$\pm \beta \left(\frac{u_0 H_{t0}}{H_{st} + z} \right)^2 = 0 \tag{3.85}$$

In the general case there is no exact integration of this differential equation. In the case of small amplitudes z, Eq. (3.82) can be linearized around the operating level $z_0 = -\beta v_0^2$ by introduction of

$$z = z_0 + s. \tag{3.86}$$

(For more details see Jaeger, 1949.) Products of small values of a higher order can be omitted in respect to other terms; thus a linearized equation is obtained with the homogeneous part in the form:

$$\frac{d^2 s}{dt^2} + 2a\frac{ds}{dt} + bs = 0. \tag{3.87}$$

D. Thoma[3] was the first who, in this manner, analyzed particular solutions from which he derived the criterion for surge tank stability. In order to be stable, the minimum area of the cylindrical surge tank should be

$$A_k \geq A_{Th} = \frac{v_0^2}{2g}\frac{LA_c}{\Delta h_0(H_{st} - \Delta h_0)}. \tag{3.88}$$

Later on, works by different hydraulic engineers (J. Frank, Ch. Jaeger, A. Franković)[4] showed that Thoma's criterion could not be applied in general to all hydroelectric power plant situations. The limit of Thoma's criterion is $\varepsilon > 40$ where parameter ε is called the Vogt[5] parameter

$$\varepsilon = \frac{LA_c}{gA_k}\frac{v_0^2}{\Delta h_0^2} = \frac{Z^{*2}}{\Delta h_0^2}. \tag{3.89}$$

If the system of headrace tunnel–surge tank–penstock has a size such that the Vogt parameter is within the range $20 < \varepsilon < 40$; them according to Jaeger (1949)

$$A_J = n^* A_{Th}, \tag{3.90}$$

where Jaeger's coefficient is

$$n^* = 1 + 0.482\frac{Z^*}{(H_{st} - \Delta h_0)}. \tag{3.91}$$

For lower ε parameter values, the theory has not yet been elaborated; thus, stability is defined by numerical integration or a hydraulic model.

For complex surge tank types, such as surge tanks with upper and lower chambers, the central cylindrical part is sized to satisfy the criterion of cylindrical surge tank stability. In the general case, surge tank stability can be tested on numerical or physical models in hydraulic labs.

3.3.5 Model of a simple surge tank with upper and lower chamber

Equation of the surge tank with upper and lower chamber

Figure 3.30 shows a scheme of the surge tank upper and lower chamber, with the working volume defined by the volumetric curve $V(z)$.

The surge tank is equipped with a safety weir at elevation Z_p, which overflows the discharge Q_p when surge tank level h_k exceeds the weir elevation. At the surge tank entrance there is an asymmetric throttle $\Delta h = \beta^{\pm}Q_k$; thus, in the general case, the surge tank level differs from the piezometric head at

---

[3]D. Thoma, German hydraulic engineer.
[4]J. Frank, Germany, Charles Jaeger (1901–1989), Swiss, Ante Franković, Croatian hydraulic engineer.
[5]F. Vogt, German hydraulic engineer.

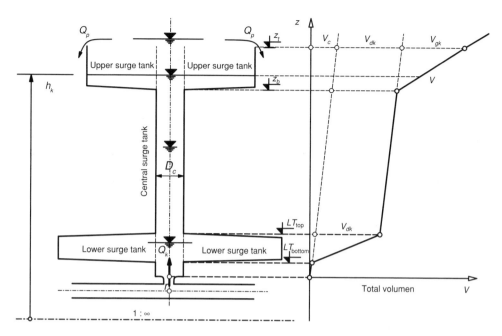

Figure 3.30 Simple surge tank with upper and lower chamber.

the connecting node. The algebraic sign $+$ denotes the loss coefficient for positive and $-$ for negative Q_k. The continuity equation for the surge tank with the weir is

$$\frac{dV}{dt} + Q_p = Q_k,\tag{3.92}$$

where the volumetric change in the time unit is equal to the net difference between the inlet–outlet discharge and the overflow discharge.

Discharge from the node r towards the surge tank is equal to the unbalanced sum of discharges of all connecting elements

$$Q_k = \sum_p{}^r Q^e,\tag{3.93}$$

namely, the equation of the node that includes the surge tank has the form

$$\sum_p{}^r Q^e = \frac{dV}{dt} + Q_p,\tag{3.94}$$

where the overflowing discharge is equal to

$$\begin{array}{ll} h_k > Z_p: & Q_p = a(h_k - Z_p)^b \\ h_k \le Z_p: & Q_p = 0 \end{array},\tag{3.95}$$

where a and b are the weir constants and Z_p is weir elevation. Due to the losses at the inlet throttle, there is a difference between piezometric heads at the inlet node and the surge tank level

$$h_k = h_r - \beta^{\pm} Q_k. \tag{3.96}$$

Secondary variables of the surge tank with upper and lower chamber

For objects such as the surge tank with an upper and lower chamber with a weir and throttle, additional variables are introduced such as the water level in the surge tank h_k, surge tank discharge Q_k, and the overflowing discharge Q_p. Secondary variables are entirely defined by the state of the fundamental primary variables. The iterative condition of the secondary variables is calculated in the phase of calculation of the primary variable increments, see subroutine subroutine IncVar. Calculation of secondary variables is implemented in a subroutine

subroutine CalcSurgeTankVars(inode),

which calls the subroutine

subroutine CalcSimpleTankVars(tank).

Fundamental system vector and matrix updating

If Eq. (3.92) is written in the form

$$\sum_p {}^r Q^e - \underbrace{\left(\frac{dV}{dt} + Q_p \right)}_{Q_k} = 0 \tag{3.97}$$

note that the discharge from the node to the surge tank Q_k is equal to the storage discharge

$$Q_k = \sum_p {}^r Q^e. \tag{3.98}$$

The storage volume that the fundamental vector shall update with is equal to

$$\Delta V_k = \int_{\Delta t} Q_k dt = \int_{\Delta t} \left(\frac{dV}{dt} + Q_p \right) dt. \tag{3.99}$$

Following integration in the time step, the following is obtained

$$\Delta V_k = (V^+ - V) + (1 - \vartheta) \Delta t Q_p + \vartheta \Delta t Q_p^+ \tag{3.100}$$

and added to the global system vector (discharge Q_k is negative, water is withdrawn from the fundamental system)

$$F_r^{new} = F_r^{old} + \Delta V_k. \tag{3.101}$$

Values of the overflowing discharge and the surge tank volume at the beginning and the end of the time step are calculated using the water level in h_k in the surge tank. Overflow in the upper surge tank is active for a water level above weir level and is calculated as

$$
\begin{aligned}
h_k > Z_p : & \quad Q_p^+ = a(h_k^+ - Z_p)^b \\
h_k \leq Z_p : & \quad Q_p^+ = 0
\end{aligned}
\tag{3.102}
$$

The terms that are updating the fundamental system vector are the functions of fundamental variables; thus, the fundamental system matrix shall also be updated by respective derivatives

$$
G_{r,r}^{new} = G_{r,r}^{old} - \frac{\partial \Delta V_k}{\partial h_r^+}
\tag{3.103}
$$

that are calculated as

$$
\frac{\partial \Delta V_k}{\partial h_r^+} = \frac{\partial \Delta V_k}{\partial h_k^+} \frac{dh_k^+}{dh_r^+} = \left(A_k + \vartheta \Delta t \frac{\partial Q_p^+}{\partial h_k^+} \right) \frac{dh_k^+}{dh_r^+},
\tag{3.104}
$$

where

$$
\frac{dh_k^+}{dh_r^+} \cong 1
\tag{3.105}
$$

that do not significantly impact the stability of iteration.

The area of horizontal cross-section A_k is calculated by the numeric derivation of the tank volume graph at the point $z = h_k^+$

$$
A_k = \left. \frac{\Delta V}{\Delta h} \right|_{z=h_k^+}.
\tag{3.106}
$$

Interval $\Delta h = 0.02$ m is implemented. Depending on the water level in tank, that is the weir operation,

$$
\begin{aligned}
h_k^+ > Z_p : & \quad \frac{\partial Q_p^+}{\partial h_k^+} = Q_p^+ \frac{b}{h_k^+ - Z_p} \\
h_k^+ \leq Z_p : & \quad \frac{\partial Q_p^+}{\partial h_k^+} = 0
\end{aligned}
\tag{3.107}
$$

Updating the initial fundamental system with a simple surge tank with an upper and lower chamber as a boundary condition is called in the frontal procedure `FrontU` before elimination of the nodal equation, see program module `Front.f90`. It is implemented in a subroutine

    subroutine SurgeTankNode(inode,iactiv,nactiv),

which calls the subroutine

    subroutine SimpleTankNode(tank, addtovec,addtomtx).

These subroutines are contained in the program module `SurgeTanks.f90`.

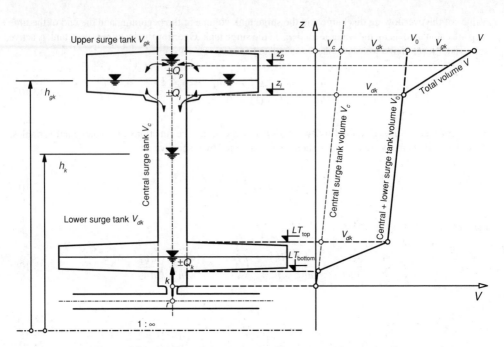

Figure 3.31   Differential surge tank with upper and lower chamber.

3.3.6   Differential surge tank model

Description

Figure 3.31 shows a scheme of the differential surge tank with an upper and lower chamber of the working volume defined by the volumetric curve $V(z)$. It differs from the simple surge tank with an upper and lower chamber described in the previous section in the upper surge tank part. Its purpose is the fast rise of level h_k in the central cylindrical tank and fast deceleration of discharge Q_k from the headrace tunnel. The upper side surge tanks serve to accumulate the surplus water. Something similar occurs in the surge tank emptying phase. Due to the fast water withdrawal from the central surge tank, the upper side tanks are emptied too. Emptying of the upper side tanks may occur simultaneously with the emptying of the lower side tanks.

The upper surge tank is filled through the outlet orifices at the bottom by the discharge $+Q_i$ and the overflowing discharge $+Q_p$ at the top of the central cylindrical tank. Respectively, it is emptied by the discharge $-Q_i$ through the outlet orifices and by reversed overflowing by discharge $-Q_p$ when the level h_b in the side tanks exceeds the weir elevation z_p.

In the surge tank filling phase, when the water level in the central cylindrical tanks exceeds the outlet level $h_k > z_i$, the outflow starts and the side tanks are filled. The water level in the central tank rises faster than in the side parts due to significantly smaller volume in the cylindrical part. When the weir elevation is reached $h_k > z_p$, the side tanks are additionally filled by overflow.

In the surge tank emptying phase, the water level is decreasing faster in the central cylindrical part than in the side tanks. The upper side tanks are emptied when their water level exceeds the level in the central cylindrical part. If the water level in the side parts is above the weir level, the side tank is also emptied by reverse overflow. Note there is no safety overflow outside the differential surge tank.

Equation of the differential surge tank with an upper and lower chamber

Discharge from the node r towards the surge tank is equal to

$$Q_k = \sum_p {}^r Q^e. \tag{3.108}$$

Water level in the central surge tank is

$$h_k = h_r - \beta^{\pm} Q_k^2. \tag{3.109}$$

As long as the level in the central surge tank is lower than the outlet level $h_k \leq z_i$ and the upper surge tank is empty $V_{gk} = 0$ the same continuity equation as for the simple differential surge tank (without external weir) can be applied

$$\frac{dV}{dt} = Q_k. \tag{3.110}$$

Otherwise, apart from the volume change V_0 in the central cylindrical surge tank, the upper surge tank is either filled or emptied

$$\frac{dV_0}{dt} + \frac{dV_{gk}}{dt} = Q_k. \tag{3.111}$$

Filling–emptying of the upper surge tank is governed by the equation

$$\frac{dV_{gk}}{dt} = Q_i + Q_p, \tag{3.112}$$

that is after introducing Eq. (3.112) into Eq. (3.111), the continuity equation for the differential surge tank is obtained in the form

$$\frac{dV_0}{dt} + Q_i + Q_p = Q_k. \tag{3.113}$$

Secondary variables of differential surge tank

For objects such as the surge tank with an upper and lower chamber with a throttle additional variables are introduced such as the water level in the central cylindrical surge tank h_k, the water level in the upper surge tank h_{gk}, tank discharge Q_k, the overflow discharge Q_p, and the outflow discharge through connections is Q_i. Secondary variables are entirely defined by the state of fundamental primary variables. The iterative condition of secondary variables is calculated in the phase of calculation of the primary variable increments, see subroutine subroutine IncVar, where the unbalanced nodal discharge is calculated

$$Q_k = \sum_p {}^r Q^e. \tag{3.114}$$

Calculation of other secondary variables is implemented in the subroutine

subroutine CalcSurgeTankVars(inode),

where the water level in the central cylindrical tank is calculated first

$$h_k = h_r - \beta^{\pm} |Q_k| Q_k. \tag{3.115}$$

Then the subroutine is called

subroutine CalcDifferTankVars(tank),

which calculates the water level h_{gk} in the upper side surge tank, the overflow discharge Q_p, and the outflow discharge Q_i.

Following the integration of Eq. (3.112) in the time step, the volume of the upper side surge tank is obtained

$$V_{gk}^+ = V_{gk} + (1 - \vartheta)\Delta t(Q_i + Q_p) + \vartheta \Delta t(Q_i^+ + Q_p^+). \tag{3.116}$$

Water level h_{gk}, which corresponds to the volume V_{gk}^+, is calculated from the volumetric curve of the upper surge tank, see Figure 3.31. The calculation is implicit (iterative) because the overflow and outflow parameters also depend on the level h_{gk}. Computations of the overflow Q_p and the outflow discharge Q_i as well as their derivative with respect to variable h_k are given hereinafter. These subroutines are comprised in the program module SurgeTanks.f90.

Computation of overflow

Depending on the water level in the central and side surge tank, overflow can be:

(a) No overflow $h_k^+ \le z_p \ge h_{gk}^+$ and $h_k^+ = h_{gk}^+ > z_p$

$$Q_p^+ = 0, \tag{3.117}$$

$$\frac{\partial Q_p^+}{\partial h_k^+} = 0. \tag{3.118}$$

(b) Non-submerged overflow from the central shaft into the upper surge tank $h_k^+ > z_p > h_{gk}^+$

$$Q_p^+ = mB\sqrt{2g} \cdot (h_k^+ - z_p)^{\frac{3}{2}} = a_p(h_k^+ - z_p)^{b_p}, \tag{3.119}$$

$$\frac{\partial Q_p^+}{\partial h_k^+} = Q_p^+ \frac{b_p}{h_k^+ - z_p}, \tag{3.120}$$

where $a_p = mB\sqrt{2g}$ and $b_p = 1.5$.

(c) Submerged overflow from the central shaft into the upper surge tank $h_k^+ > h_{gk}^+ > z_p$

$$Q_p^+ = mB\sqrt{2g} \cdot (h_k^+ - z_p)\sqrt{h_k^+ - h_{gk}^+}, \tag{3.121}$$

$$\frac{\partial Q_p^+}{\partial h_k^+} = \frac{a(3h_k^+ - 2h_{gk}^+ - z_p)}{2\sqrt{h_k^+ - h_{gk}^+}}. \tag{3.122}$$

(d) Non-submerged overflow from the upper surge tank into the central shaft $h_{gk}^+ > z_p > h_k^+$

$$Q_p^+ = -mB\sqrt{2g} \cdot (h_{gk}^+ - z_p)^{\frac{3}{2}} = -a(h_{gk}^+ - z_p)^b,$$ (3.123)

$$\frac{\partial Q_p^+}{\partial h_k^+} = 0.$$ (3.124)

(e) Submerged overflow from the upper surge tank into the central shaft $h_{gk}^+ > h_k^+ > z_p$

$$Q_p^+ = -mB\sqrt{2g} \cdot (h_{gk}^+ - z_p)\sqrt{h_{gk}^+ - h_k^+},$$ (3.125)

$$\frac{\partial Q_p^+}{\partial h_k^+} = \frac{a_p(h_{gk}^+ - z_p)}{2\sqrt{h_{gk}^+ - h_k^+}}.$$ (3.126)

Computation of outflow

Discharge through the orifices between the central and the upper surge tank can be expressed as

$$Q_i = \pm c_i^\pm (h_1 - h_2)^{d_i^\pm},$$ (3.127)

where h_1 is the higher and h_2 is the lower water level, while $c_i^\pm$ and $d_i^\pm$ are the overflow constants for the positive/negative flow direction. It would be the best to define them experimentally in the lab, although they can be calculated approximately depending on the outflow form. The outflow form is usually asymmetric while outflow can be with either a free surface or submerged. Thus, the following can be distinguished:

(f) No outflow $h_k^+ = h_{gk}^+$ and $h_k^+ < h_{gk}^+ = z_i$

$$Q_i^+ = 0,$$ (3.128)

$$\frac{\partial Q_i^+}{\partial h_k^+} - 0.$$ (3.129)

(g) Non-submerged outflow from the central shaft into the upper surge tank $h_k^+ > z_i = h_{gk}^!$

$$Q_i^+ = c_i^+ \left(h_k^+ - z_i \right)^{d_i^+},$$ (3.130)

$$\frac{\partial Q_i^+}{\partial h_k^+} = \frac{d_i^+ Q_i^+}{h_k^+ - z_i}.$$ (3.131)

(h) Submerged outflow from the central shaft into the upper surge tank $h_k^+ > h_{gk}^+ > z_i$

$$Q_i^+ = c_i^+ \left(h_k^+ - h_{gk}^+ \right)^{d_i^+},$$ (3.132)

$$\frac{\partial Q_i^+}{\partial h_k^+} = \frac{d_i^+ Q_i^+}{h_k^+ - h_{gk}^+}.$$ (3.133)

(i) Non-submerged outflow from the upper surge tank into the central shaft $h_{gk}^+ > z_i > h_k^+$

$$Q_i^+ = -c_i^- \left(h_{gk}^+ - z_i \right)^{d_i^-},$$

(3.134)

$$\frac{\partial Q_i^+}{\partial h_k^+} = 0.$$

(3.135)

(j) Submerged outflow from the upper surge tank into the central shaft $h_{gk}^+ > h_k^+ > z_i$

$$Q_i^+ = -c_i^- \left(h_{gk}^+ - h_k^+ \right)^{d_i^-},$$

(3.136)

$$\frac{\partial Q_i^+}{\partial h_k^+} = Q_i^+ \frac{d_i^-}{h_{gk}^+ - h_k^+}.$$

(3.137)

Fundamental system vector and matrix updating

The storage volume that the fundamental vector shall be updated with is equal to

$$\Delta V_k = \int_{\Delta t} Q_k dt.$$

(3.138)

Following the integration of Eq. (3.113) into the time step, the surge tank total volume increment is obtained

$$\Delta V_k = V_0^+ - V_0 + (1 - \vartheta)\Delta t(Q_i + Q_p) + \vartheta \Delta t(Q_i^+ + Q_p^+)$$

(3.139)

and added to the global system vector (discharge Q_k is negative, water is withdrawn from the fundamental system)

$$F_r^{new} = F_r^{old} + \Delta V_k.$$

(3.140)

The terms that are updating the fundamental system vector are the functions of the fundamental variables, thus the fundamental system matrix shall also be updated by respective derivatives

$$G_{r,r}^{new} = G_{r,r}^{old} - \frac{\partial \Delta V_k}{\partial h_r^+}$$

(3.141)

that are calculated as

$$\frac{\partial \Delta V_k}{\partial h_r^+} = \frac{\partial \Delta V_k}{\partial h_k^+} \frac{dh_k^+}{dh_r^+} = A_r + \vartheta \Delta t \frac{\partial Q_i^+}{\partial h_k^+} \frac{dh_k^+}{dh_r^+} + \vartheta \Delta t \frac{\partial Q_p^+}{\partial h_k^+} \frac{dh_k^+}{dh_r^+},$$

(3.142)

where the area of the central cylindrical tank cross-section is equal

$$A_r = \frac{dV_0^+}{dh_k^+} \frac{dh_k^+}{dh_r^+}.$$

(3.143)

In expressions (3.142) and (3.143) it can be adopted that

$$\frac{dh_k^+}{dh_r^+} \cong 1 \qquad (3.144)$$

which does not impact significantly on the stability of iteration, then

$$\frac{\partial \Delta V_k}{\partial h_r^+} = A_r + \vartheta \Delta t \frac{\partial Q_i^+}{\partial h_k^+} + \vartheta \Delta t \frac{\partial Q_p^+}{\partial h_k^+}. \qquad (3.145)$$

Updating the initial fundamental system by a differential surge tank with an upper and lower chamber as a particular boundary condition is called in the frontal procedure `FrontU` before elimination of the nodal equation, see program module `Front.f90`. It is implemented in a subroutine

```
subroutine SurgeTankNode(inode,iactiv,nactiv),
```

which calls the subroutine

```
subroutine DifferTankNode(tank, addtovec,addtomtx).
```

These subroutines are found in the program module `SurgeTanks.f90`.

3.3.7 Example

The headrace tunnel has diameter $D = 5$ m, length $L = 2000$ m, initial discharge $Q_0 = 92$ m^3/s, and a Darcy–Weissbach friction coefficient $\lambda = 0.0137786$, obtained for the headrace tunnel roughness $\varepsilon = 1$ mm. The area of the cylindrical surge tank cross-section is $A_k = 100$ m^2.

The task is to determine the period of oscillations and the maximum water level in the cylindrical surge tank for instantaneous hydroelectric power plant shut down:

(a) according to the formula (3.65) by A. Franković,

(b) by numerical integration by the Runge–Kutta of fourth order method with the integration step of 1 s,

(c) by `SimpipCore` numerical model with the integration step of 1 s,

(d) by determining the surge tank stability to maintain constant power for the lower water level of 25 m.

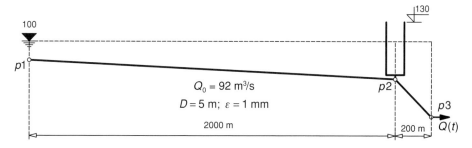

Figure 3.32 Cylindrical surge tank.

ad (a)

$$T = 2\pi\sqrt{\frac{LA_k}{gA_c}} = 2\pi\sqrt{\frac{2000 \cdot 100}{g \cdot 19.635}} = 202.46 \ [s],$$

$$Z_{max} = \left(\frac{\pi}{8} - 1\right)\Delta h_0 + \sqrt{\left(\frac{\pi}{8}\Delta h_0\right)^2 + \frac{LA_c}{gA_k}v_0^2},$$

$$Z_{max} = \left(\frac{\pi}{8} - 1\right) \cdot 6.167 + \sqrt{\left(\frac{\pi}{8} \cdot 6.167\right)^2 + \frac{200 \cdot 19.635}{100g}4.686^2} = 25.999,$$

$$h_{max} = 100 + 25.999 = 125.999.$$

ad (b)

A program source SurgeTank for numerical integration of the surge tank by the Runge–Kutta method is given at www.wiley.com/go/jovic Appendix A and refers to the particular solution of the dynamic equation

$$h_1 = h_2 + \lambda\frac{L}{2gDA_c^2}|Q|Q + \frac{L}{gA_c}\frac{dQ}{dt} \tag{3.146}$$

and the continuity equation:

$$A_k\frac{dh_2}{dt} = Q - Q_t, \tag{3.147}$$

which describes the problem from this example for the sudden shut down Q_t from 100% $\rightarrow$ 0 in the interval of the integration time step. It is assumed that the losses at the inlet of the surge tank connection are negligible. The surge tank water level is equal to the piezometric head in the connection node *p2*.

If approximate solutions are compared, the Runge–Kutta method is considered the most accurate, which gives the maximum water level rise

$$h_{max} = 125.6799 \ m$$

ad (c)

Figure 3.33, on the left hand side, shows the SimpipCore input data for the cylindrical surge tank test given in this example. The right hand side of the same figure shows the comparison between the results obtained by integration according to the Runge–Kutta method and the results obtained by the SimpipCore model, with the same time step. Note the correspondence of the results since the piezometric head graphs completely overlap. The resulting graph also shows the surge tank filling/emptying discharge.

An approximate solution according to the formula by A. Franković gives a somewhat greater solution that is on the safe side.

Approximate solutions under 6.3.7.2 and 6.2.7.3 use the variable resistance coefficient λ unlike an approximate solution under 6.3.7.1.

Input file:	Output file:
VodnaKomora-ispad.simpip	VodnaKomora-ispad.prt

```
; Surge Tank - power plant shut down
; Comparison of the results

Options
     input short
     run rigid
     digits 15
Parameters
        Tz = 1
        Qo = 92
        dt = 1
        D = 5
        eps = 1e-3
        Ak = 100
        Ha = 100
Points
        p1 0 0 0
        p2 2000 0 0
        p3 2200 0 10
Pipes
    c1    p1 p2    D    eps
    c2    p2 p3    D    eps
Graph Q(t)
        0 1
        Tz 0
Charge p3 0 -Qo Q(t)
Piezo  p1 Ha
SurgeTank Komora p2
    0 0
    150 Ak*150
Steady 0
Unsteady 500 dt
Print
        SolPoint p2
```

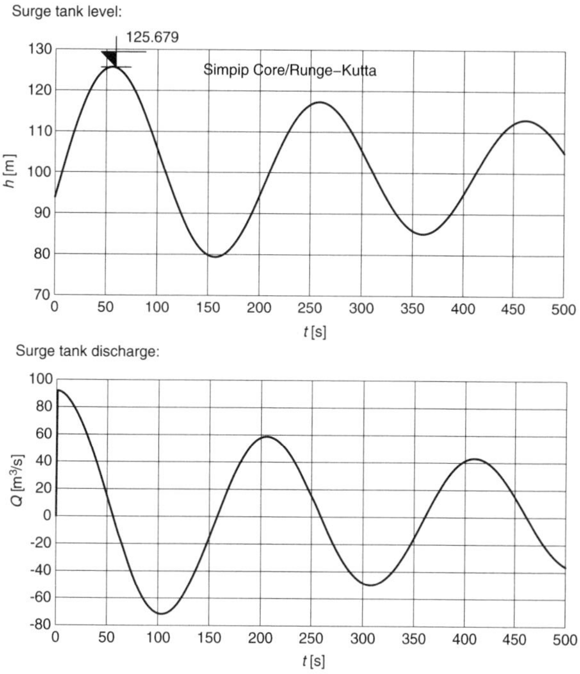

Surge tank level:

Surge tank discharge:

Figure 3.33   Surge tank test.

ad (d)

First, check if the surge tank satisfies Thoma's criterion. The minimum area of a stable surge tank is calculated from the expression (3.88)

$$A_{Th} = \frac{v_0^2}{2g} \frac{L A_c}{\Delta h_0 (H_{st} - \Delta h_0)}$$

and is equal to

$$A_{Th} = \frac{4.685^2}{2g} \frac{2000 \cdot 19.6345}{6.167(75 - 6.167)} = 103.51 \text{ m}^2$$

which is bigger than the area $A_k = 100$. Vogt's parameter is equal to

$$\varepsilon = \frac{L A_c}{g A_k} \frac{v_0^2}{\Delta h_0^2} = \frac{2000 \cdot 19.635}{100g} \frac{4.6856^2}{6.167^2} = 23.11,$$

since $20 < \varepsilon < 40$, Thoma's area shall be corrected according to Jaeger with the correction factor calculated according to the expression (3.91) as $n^* \cong 1.21$.

Thus, the minimum stable surge tank area is equal to

$$A_J = n^* A_{Th} = 1,21 \cdot 103.51 \cong 125 \text{ m}^2. \tag{3.148}$$

That the analyzed surge tank is unstable for the regulation of constant power will be shown by an adequate program solution.

A program `PowerRegulation` solution for the analyzed surge tank, using the Runge–Kutta method, for the transition of the hydropower plant operation from 50% power to 100% power with the turbine speed governor, is given at www.wiley.com/go/jovic Appendix A. The results are shown in Figure 3.34.

It can be observed that the surge tank is unstable because of the oscillation rise instead of dumping (the operational head and the turbine discharge).

Figure 3.35 shows the results for an increased surge tank area; namely, a surge tank area equal to the minimum according to Jaeger (3.148). Note that the surge tank is stable during power regulation.

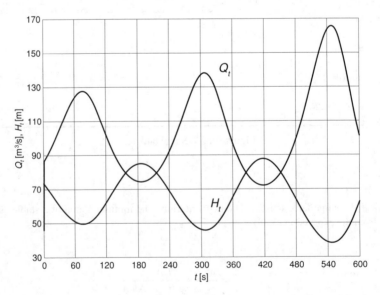

Figure 3.34 Constant power regulation 50%→100%, A_k=100 m², amplification.

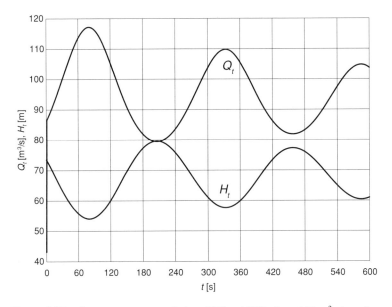

Figure 3.35   Constant power regulation, 50%→100%, $A_k = 125$ m^2, dumping.

3.4   Vessel

3.4.1   Simple vessel

Figure 3.36 shows the scheme of a classical vessel used for water hammer compensation. There is asymmetric dumping $\Delta h = \beta^{\pm} Q_k^2$ at the vessel inlet; thus, in general, the piezometric head in the vessel h_k differs from the piezometric head h at the connection node

$$h_k = h - \beta^{\pm} |Q_k| Q_k. \qquad (3.149)$$

Discharge from the node to the vessel is equal to the unbalanced sum of discharges of all connected elements

$$Q_k = \sum_p {}^r Q^e. \qquad (3.150)$$

The vessel equation of continuity is

$$\frac{dV}{dt} = Q_k, \qquad (3.151)$$

where V is the volume of a liquid; namely

$$\frac{dV}{dt} = A_k \frac{dz}{dt}. \qquad (3.152)$$

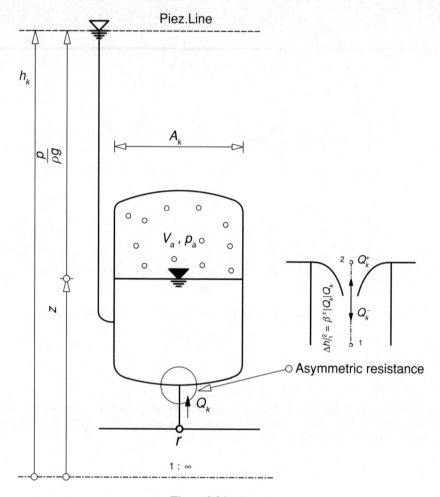

Figure 3.36   Vessel.

Furthermore, since the liquid level in the vessel depends on the pressure, that is the piezometric head, then

$$\frac{dV}{dt} = A_k \underbrace{\frac{dz}{dh_k} \frac{dh_k}{dt}}_{A_k^*(h_k)}.$$

(3.153)

The value $A_k^*(h_k)$ is the area of the respective equivalent storage or tank, variable in terms of the piezometric head; thus the continuity equation for the vessel can be applied

$$\frac{dV}{dt} = A_k^*(h_k)\frac{dh_k}{dt}.$$

(3.154)

The air mass in the vessel is defined by the initial filling, and remains constant during the oscillating process. Thus, the equation for the gas in the form of a polytrope can be applied to this closed thermodynamic system

$$p_{abs}\, V_a^n = const,$$ (3.155)

where p_{abs} is the absolute air pressure, equal to the sum of the relative p and atmospheric pressure p_0, V_a is the air volume and n is the exponent of a polytrope ($n = 1.25 - experimentally!$).

If Eq. (3.155) is applied to the state at any time, and the referent state, then it is written

$$V_a(t)^n \{p_0 + \rho g \,[h_k(t) - z(t)]\} = V_0^n \{p_0 + \rho g \,[h_0 - z_0]\} = const$$ (3.156)

or expressed by the liquid column height

$$V_a(t)^n \left(\frac{p_0}{\rho g} + h_k(t) - z(t) \right) = V_0^n \left(\frac{p_0}{\rho g} + h_0 - z_0 \right)$$ (3.157)

from which the air volume is

$$V_a = V_0 \left(\frac{h_a + h_0 - z_0}{h_a + h_k - z} \right)^{1/n},$$ (3.158)

where $h_a = p_0/\rho g$ denotes a respective atmospheric pressure head (corresponding to 10.1325 m of water column for the nominal value of the mean atmospheric pressure $p_0 = 101325\ [Pa]$).

The air volume V_a at any time, that is any level z, is defined by the reference volume V_0, reference level z_0, and vessel cross-section area A_k; thus the following can be derived from geometric relations, see Figure 3.36

$$V_a = V_0 - A_k(z - z_0).$$ (3.159)

An expression for water level in the vessel depending on the piezometric head is obtained from Eqs (3.159) and (3.158)

$$z = z_0 + \frac{V_0}{A_k} \left[1 - \left(\frac{h_a + h_0 - z_0}{h_a + h_k - z} \right)^{1/n} \right].$$ (3.160)

A derivative of Eq. (3.160) by h_k gives

$$\frac{dz}{dh_k} = \frac{V_0}{n A_k} \frac{\left(\dfrac{h_a + h_0 - z_0}{h_a + h_k - z} \right)^{1/n}}{h_a + h_k - z}.$$ (3.161)

When Eq. (3.158) is introduced into Eq. (3.161) the following can be written

$$\frac{dz}{dh_k} = \frac{V_a}{n A_k \,(h_a + h_k - z)},$$ (3.162)

thus, the equivalent vessel area is

$$A_k^* = \frac{V_a}{n\,(h_a + h_k - z)},\tag{3.163}$$

It is appropriate to adopt the hydrostatic state with the known air volume, piezometric head, and the level in the vessel V_0, h_s, z_s as the reference state. However, any three of these parameters that are set at the time of vessel filling by air can also be adopted.

The vessel nodal equation is an ordinary differential equation with the prescribed initial conditions. Initial conditions are the steady state with all time derivatives equal to zero. Thus, the accumulation discharge is $Q_k = 0$ and the piezometric head in the vessel is equal to the piezometric head at the connecting node.

3.4.2   Vessel with air valves

Figure 3.37 shows a scheme of the vessel with air valves. When the water level is above the level of air valve opening, the vessel functions as a simple vessel. The opening level of the air valves defines the air reference state, namely the volume V_0 and pressure $p_a = p_0$ that is equal to atmospheric pressure. Thus, the reference piezometric head is also equal to the opening level $h_0 = z_0$. When the water level in the

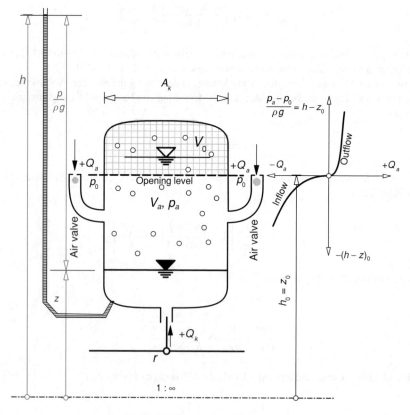

Figure 3.37   Vessel with air valves.

vessel is below the opening level z_0, pressure above the water is equal to the atmospheric pressure and the vessel functions as a simple storage tank. A vessel with air valves is applied to prevent underpressure at high pipeline elevations.

Valve characteristics have the form of a discharge curve

$$h - z_0 = \pm\beta^\pm Q_a^2,$$

(3.164)

where h is piezometric head in the node, z_0 is the level of air valves orifice, and $\beta^\pm$ is the asymmetric resistance coefficient for the air flow Q_a which is positive for release from the pipeline. Figure 3.37 shows the qualitative form of the air valve discharge curve.

The unbalanced sum of discharges Q_k in a node causes the change of water V, that is air volume V_a

$$Q_k = \frac{dV}{dt} = -\frac{dV_a}{dt}.$$

(3.165)

The volume of sucked air is defined by the water level of the volumetric curve. The air mass increment $M_a = \rho_a V_a$ is equal to the mass discharge of emptying

$$\frac{dM_a}{dt} = -\rho_a Q_a.$$

The derivative of the complex term gives

$$V_a \frac{d\rho_a}{dt} + \rho_a \frac{dV_a}{dt} = -\rho_a Q_a$$

from which

$$\frac{V_a}{\rho_a} \frac{d\rho_a}{dt} + \frac{dV_a}{dt} = -Q_a.$$

When Eq. (3.165) is introduced in the previous equation, an air density change equation is obtained in the following form

$$\frac{V_a}{\rho_a} \frac{d\rho_a}{dt} = Q_k - Q_a.$$

(3.166)

Air density change can be triggered by pressure change. Let us assume that a thermodynamic process in the form of a polytrope can be applied to air. Thus the relation between the absolute pressure and density can be written as

$$p_{abs} = cons \cdot \rho_a^n,$$

where n is the exponent of a polytrope. When elementary operations of differentiation are applied and the terms arranged, then

$$\frac{1}{\rho_a} \frac{d\rho_a}{dt} = \frac{1}{np_{abs}} \frac{dp_{abs}}{dt}.$$

(3.167)

After introduction into Eq. (3.166), the following is obtained

$$\frac{V_a}{n p_{abs}} \frac{d p_{abs}}{dt} = Q_k - Q_a.$$

(3.168)

Note that the air pressure change is equal to zero if the water and air discharges on the right hand side are equal; namely, there is atmospheric pressure above the water.[6]

3.4.3   Vessel model

Secondary variables of the vessel

Secondary variables, which are calculated in the phase of primary variable increment calculations, are used for describing the vessel function. Secondary variables are the piezometric head inside the vessel

$$h_k = h - \beta^{\pm} |Q_k| Q_k$$

(3.169)

and the liquid level

$$z = z_0 + \frac{V_0}{A_k} \left[1 - \left(\frac{h_a + h_{k0} - z_0}{h_a + h_k - z} \right)^{1/n} \right].$$

(3.170)

The iterative condition of secondary variables is calculated in the phase of calculation of primary variable increments, see subroutine IncVar. Computation of secondary variables is implemented in a:

subroutine CalcVesselVars(inode)

which calculates secondary variables for the simple vessel and the vessel with air valves. These subroutines are located in the program module Vessels.f90.

Fundamental system vector and matrix updating

Updating the initial fundamental system with a vessel is again based on the volume balance. For the node that vessel is connected to, an equation containing the accumulation discharge can be applied

$$\sum_p {}^r Q^e - Q_k = 0.$$

(3.171)

Then, the accumulation volume will be equal to

$$\Delta V_k = \int_{\Delta t} Q_k dt = \int_{\Delta t} \frac{dV}{dt} dt.$$

(3.172)

Following integration in the time step, the following is obtained

$$\Delta V_k = (V^+ - V) = A_k(z^+ - z)$$

(3.173)

---

[6]The pressure is almost atmospheric, although air relief valves have a large capacity.

and added to the global system vector (discharge Q_k is negative, water is withdrawn from the system)

$$F_r^{new} = F_r^{old} + \Delta V_k. \tag{3.174}$$

The terms updating the fundamental system vector are the functions of fundamental variables; thus, the fundamental system matrix will also be updated by respective derivatives

$$G_{r,r}^{new} = G_{r,r}^{old} - \frac{\partial \Delta V_k}{\partial h_r^+}, \tag{3.175}$$

$$\frac{\partial \Delta V_k}{\partial h_r^+} = \frac{\partial \Delta V_k}{\partial z^+} \frac{dz^+}{h_r^+} = A_k \frac{dz^+}{dh_r^+} = A_k \frac{dz^+}{dh_k^+} \frac{dh_k}{dh_r^+}. \tag{3.176}$$

Since

$$\frac{dh_k^+}{dh_r^+} \cong 1 \tag{3.177}$$

then the term that the matrix is updated by is equal to

$$\frac{\partial \Delta V_k}{\partial h_r^+} = A_k^*. \tag{3.178}$$

Updating of the initial fundamental system by a vessel as a boundary condition is called in the frontal procedure `FrontU` before elimination of the nodal equation, see program module `Front.f90`. It is implemented in a subroutine

                    subroutine VesselNode(inode,iactiv).

These subroutines are contained in the program module `Vessels.f90` and can be applied both for the simple vessel and the vessel with the air valves.

3.4.4   *Example*

Figure 3.38 shows the pumping station pressure pipeline. In point p_0 the pump delivers discharge Q_0 that is elevated to the point p_2 by a pipeline of length L. Due to the sudden power shortage, the pumping discharge becomes zero in a very short time interval. A vessel is connected in the point p_1, which compensates the pumping discharge in the pressure drop phase. In the pressure rise phase; namely, the phase when water is returning, the vessel is filling. Then there are water mass oscillations, that is pressure and discharge oscillations in the pipeline.

  The following should be done:

  (a) Determine the piezometric head and discharge oscillations in the analyzed system shown in the figure, first by numerical integration by the Runge–Kutta fourth order method; then by the `SimpipCore` numerical model and compare the results.

  (b) Pressure variations are shown as envelopes of the maximum and minimum piezometric level along the pipeline. The pipeline is planned for a pressure load of 10 bars. Thus, the vessel should be sized to the maximum piezometric head of 120 m. Similarly, due to minimum piezometric levels the lowest piezometric head shall be above 65 m (see Figure 3.38). For the analyzed vessel, a throttle shall be found to satisfy the requirements.

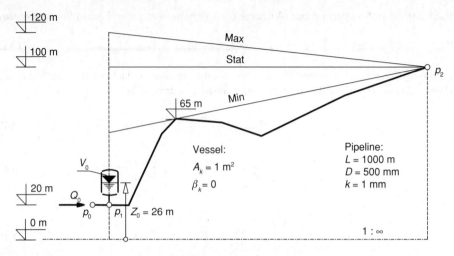

Figure 3.38　Vessel.

ad (a)

A program source `Vessel.f90` for numerical integration by the Runge–Kutta method is given in Appendix A and refers to the solution of the oscillations in the system of the vessel–pressure pipeline shown in Figure 3.38. A sudden stop of pump operation Q_0 from 100% → 0 in the interval of the integration time step will be observed.

The left hand side of Figure 3.39 shows the `SimpipCore` input data. The selected vessel, of the $A_k = 1$ m^2 cross-section area, and the following reference data (filling at hydrostatic condition)

$$V_0 = 1.5 \, \text{m}^3, \ z_0 = 26 \, \text{m}, \ h_{k0} = 100 \, \text{m}.$$

There are no resistances at the connection of the vessel and the pipeline.

A graph on the upper right hand side of the same figure shows the comparison between the numerical results obtained by the Runge–Kutta method and the results obtained by the `SimpipCore` model, for the piezometric head h and level z in the vessel. Note the excellent correspondence between the results; even more so because of the same time step in both cases, being equal to 0.2 s. The figure also shows the graph discharge oscillations in the pipeline.

A graph on the lower right hand side of Figure 3.39 shows the comparison between the numerical results obtained by the Runge–Kutta method and the results obtained by the `SimpipCore` model for the level in the vessel. Note the very good correspondence of both results.

ad (b)

Figure 3.40 shows the influence of the asymmetric throttle on the development of piezometric head oscillations in the vessel.

It can be concluded that the analyzed vessel with the asymmetric throttle of the above given characteristics decreases the upper oscillation to the allowable value.

3.5　Air valves

3.5.1　Air valve positioning

The function of an air valve is to remove air during the filling of an empty pipeline. The filling velocity shall enable free air removal, which is achieved by a small filling discharge, a lot smaller than the duty

Input file: Vessel-test.simpip	Output file: Vessel-test.prt

```
;
; vessel test
;
Options
   input short
   Run rigid
Points
   p0  -20  0  0
   p1   0   0 20
   p2  1000 0 20
Pipes
   c0    p0 p1    0.5  1e-3
   c1    p1 p2    0.5  1e-3
Graph off
   0 1
   0.01 0
Charge p0 0 0.200 off
Piezo  p2 100
Vessel
   vsl p1 1.0 1.5 26 100
;650 5
Steady 0
Unsteady 250 0.2
Print
   SolVessel vsl
```

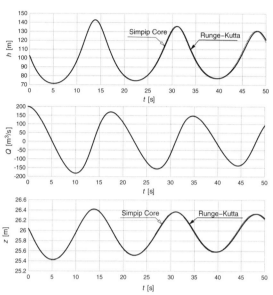

Figure 3.39   Vessel test.

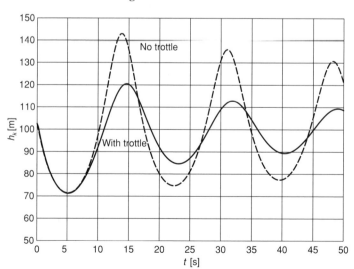

Figure 3.40   Influence of the asymmetric throttle.

one. During normal operation of the pipeline, the air valves are closed. Small quantities of dissolved gas are released automatically through outlets with small hole air valves.

In a pipeline with almost horizontal alignment, air valves play a key role in the system protection against pressure excesses. During breaks in pump unit operation, regardless of whether these are due to a power break or normal shutdown, there is a negative pressure wave propagating downstream. Pressure falls and opens air valves in the direction of wave propagation. A large quantity of air enters the system. Opening the air valves breaks the hydraulic system into several freely oscillating parts. The system oscillates between the adjacent opened air valves while the remaining energy of motion is not spent to overcome motion resistances. At the start of system operation, a positive pressure wave empties the trapped air.

Water supply pipelines are fitted with air valves in all inflection points (convex points) of the vertical alignment. A slight inclination of horizontal pipeline sections towards the air valve is recommended in order to remove all remaining air bubbles by uplift when the discharge is smaller than the critical. If the discharge is greater than the critical, air bubbles will move with the water to the next open air valve. A normal air valve prevents further motion of the air bubble by inclination of the upstream and downstream pipe and gradually removes it. The recommended distance between air valves is approximately 500 to 800 m.

Horizontal or almost horizontal pipeline sections define the critical discharge that will set the air bubble in motion in the direction of the flow.

Critical discharge can be calculated using the formula developed the research center *HR Wallington 2005* (adapted by V. Jović 2006):

$$\frac{v_c}{\sqrt{gD}} = 0.56\sqrt{\sin \beta_c} + a,$$

where β_c is the angle towards the horizontal, see Figure 3.41 – positive for the pipe inclined downstream, D is the pipe diameter, V_z is the volume of the air bubble, while term a is calculated according to the formula

$$a = -0.0755687 \log^2 n + 0.0380147 \log n + 0.606072,$$

where

$$n = 4\frac{V_z}{\pi D^3}.$$

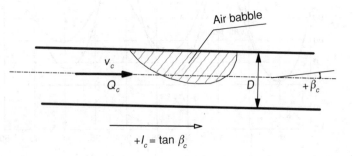

Figure 3.41   An air bubble in the pipe.

The critical discharge to set the air bubble of volume V_z in motion is

$$Q_c = v_c \frac{D^2 \pi}{4}.$$

The calculation of the *critical discharge* to set an air bubble of 0.5 m^3 in motion in a *horizontal pipe* for different pipe diameters is given in below:

Diameter D mm	Q_{min} l/s
200	18
500	260
800	858
1000	1465

The data shows that it is difficult to move air in horizontal pipeline sections of pipelines with a larger diameter because the air is only moved by very large discharges. Thus, particular care should be paid to air valve positioning.

Air removal through air valves is extremely complex. The main mechanisms are shown in Figure 3.42, Figure 3.43, and Figure 3.44.

The velocity of air outflow through the air valve depends on instantaneous hydraulic values of discharge and pressure. If the air emptying through the air valves is normal, water speeds up towards the valve from the two sides; thus at some moment there will be a water mass collision and a surge occurs.

If the emptying velocity is too high, balls rise and trap the air, thus permanently decreasing the discharge due to the obstruction of the flow profile.

Unfortunately, air valve manufacturers do not provide this critical data that will be estimated by the next approximate analysis.

Let the diameter of the ball closing the large opening be equal to the rated diameter of the air valve D. The ball is lighter than water. Its density ρ_k is equal to 25% of water density. The ball weight is equal to

$$G = \rho_k g \frac{4}{3} \pi R^3.$$

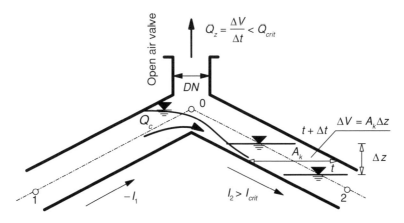

Figure 3.42   Free level overflow.

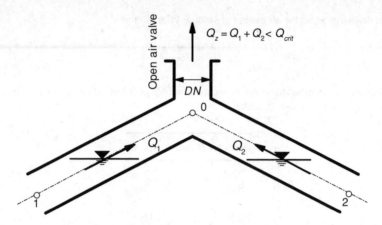

Figure 3.43   Double-sided rising.

The ball is situated in the outflow air current. Thus the hydrodynamic force acting on it is equal to

$$F = C_o R^2 \pi \rho_a \frac{v_z^2}{2},$$

where C_o is the drag coefficient of the acting force, ρ_a is the air density, and v_a is the air velocity in front of a ball. Equilibrium of the acting force and the ball weight

$$\rho_k g \frac{4}{3} \pi R^3 = C_o R^2 \pi \rho_a \frac{v_a^2}{2}$$

gives the air velocity that lifts the ball

$$v_a = \sqrt{\frac{8 g \rho_k}{3 C_o \rho_a}} R,$$

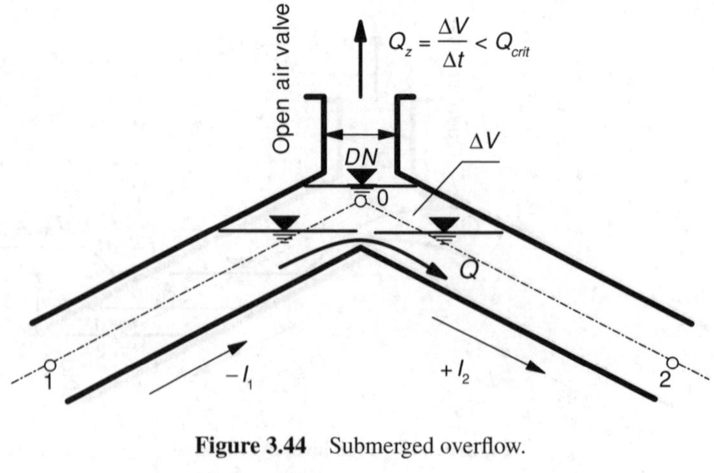

Figure 3.44   Submerged overflow.

that is critical air discharge

$$Q_a = v_a R^2 \pi.$$

The air valve DN 200, drag coefficient of the acting force 0.6, and air density $\rho_a = 1.26$ kg/m^3 give the critical air valve closing velocity $v_a = 29$ m/s and the critical discharge of captured air $Q_a=913$ l/s. Air valve closing due to high velocities is not allowed because damage occurs due to the ball jamming in the hole.

3.5.2   Air valve model

In terms of hydraulics, the air valve acts as a vessel with air valves without initial air volume V_0. In general, there are two types of air valves depending on the air release, that is pressure rise types:

- the air is completely released,
- air is captured and released with delay.

The first type is a simple air valve, with a ball and hole. The second type is the air valve with a special construction of air nozzles. One nozzle with large orifices, which lets in large air quantities, is closed by a pressure rise and captures air. At that time small nozzles remain open until all air is released. Air valves with double nozzles are used as surge dumpers, especially for sewerage pressure pipelines. The valve characteristic has the form of the discharge curve

$$h - z_0 = \pm \beta^\pm Q_a^2, \tag{3.179}$$

where h is the piezometric head in the valve, z_0 is the elevation of the valve opening, and $\beta^\pm$ is the asymmetric coefficient of air flow resistance Q_a, positive in the direction of release from the pipeline. Figure 3.45 shows a quantitative shape of the air valve discharge and volumetric curves.

(a) Air valve is open $h - z_0 < 0$:
     When pressure in the node becomes smaller than the atmospheric pressure, the piezometric head is calculated using the valve characteristic, expression (3.179), where the air flow is calculated from the continuity equation of the node; namely, discharge of the sucked air is equal to the water discharge deficit

$$\sum_p^r Q^e = Q_k. \tag{3.180}$$

Updating the initial fundamental system by an air valve is again based on the volume balance. The volume of water accumulation in a node that the air valve is connected to is equal to

$$\Delta V_k = \int_{\Delta t} Q_k dt = \int_{\Delta t} \frac{dV}{dt} dt. \tag{3.181}$$

After integration in the time step, the following is obtained

$$\Delta V_k = (V^+ - V) \tag{3.182}$$

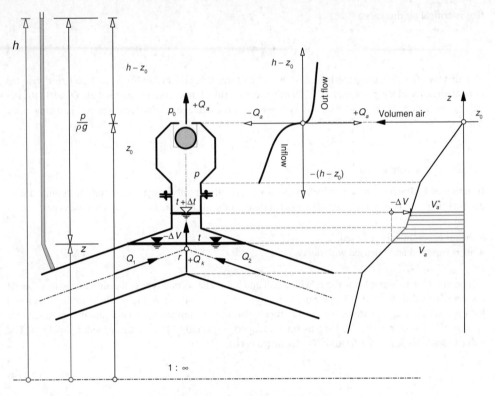

Figure 3.45   Air valve characteristic.

and added to the global system vector

$$F_r^{new} = F_r^{old} + \Delta V_k. \tag{3.183}$$

Terms updating the fundamental system vector are functions of the fundamental variables; thus, the fundamental system matrix will also be updated by respective derivatives. In general, the derivative of the volumetric curve will be prescribed, see Figure 3.45. However, an acceptably rough approximation of the form $\Delta V_k = A_{equ}(z^+ - z)$, will be applied. Thus the matrix updating will be equal to

$$G_{r,r}^{new} = G_{r,r}^{old} - A_{equ}. \tag{3.184}$$

(b) Air valve is closed $h - z_0 \geq 0$:

At the time of air valve closing there are two possibilities:

- $V_0 = 0$, there is no trapped air in the node, standard equations can be applied,
- $V_0 > 0$, there is air trapped in the node, the air valve acts as a simple vessel as described in Section 3.4.2 Vessel with air valves.

Secondary variables and updating of the fundamental system vector and matrix

Updating the initial fundamental system by an air valve as a natural boundary condition is called in the frontal procedure `FrontU` before elimination of the nodal equation, see program module `Front.f90`. It is implemented in a subroutine

> `subroutine AirValveNode(inode,iactiv,nactive),`

while secondary variables are implemented in a subroutine

> `subroutine CalcAirValveVars(node).`

Because of the complexity of the air ration in the valves and pipeline, some simplifications are necessary in the implementation of the air valve. For more information refer to subroutine sources that can be found in the program module `AirValves.f90`.

3.6 Outlets

3.6.1 Discharge curves

Outlets are hydraulic nodal objects discharging liquid from a hydraulic network. There are different outlet types, from the simplest valves to very sophisticated regulation valves. Liquid is discharged through the opening of the cross-section area A at the outflow velocity v. The outflow cross-section area can also depend on the pressure (piezometric head).

The following outlets types will be analyzed.

Gate valve

These are outlets such as gate valves, flat slide valves, and similar. The abridged form of the discharge curve depending on the piezometric head is defined as

$$Q = a\sqrt{(h - z_0)},\tag{3.185}$$

which is obtained from the respective outflow formula:

$$Q = \varphi\varepsilon A_z\sqrt{2g(h - z_0)},\tag{3.186}$$

where A_z is the area of the outflow cross-section under the gate blade, φ is the coefficient of velocity variation at the cross-section of contracted outflow jet, ε is the coefficient of the contraction of the outflow jet, g is the gravity acceleration, and z_0 is the gate elevation. Parameter a is obtained by grouping of the terms:

$$a = \varphi\varepsilon A_z\sqrt{2g}.\tag{3.187}$$

A derivative of the discharge curve with respect to the piezometric head is equal to

$$\frac{\partial Q}{\partial h} = \frac{a}{2\sqrt{(h - z_0)}} = \frac{Q}{2(h - z_0)}.\tag{3.188}$$

Relief valve

A relief discharges a certain water quantity into the atmosphere when pressure in the connection (i.e. piezometric head) exceeds reference one h_{ref}. The cross-section area increases linearly with the pressure rise above the reference, while the outflow velocity depends on the piezometric head rise above the relief elevation z_0.

The discharge curve in abridged form is

$$Q = a(h - h_{ref})\sqrt{h - z_0}. \tag{3.189}$$

A derivative of the discharge curve with respect to the piezometric head is equal to

$$\frac{\partial Q}{\partial h} = \frac{\partial Av}{\partial h} = A\frac{\partial v}{\partial h} + v\frac{\partial A}{\partial h}. \tag{3.190}$$

Namely

$$\frac{\partial Q}{\partial h} = \frac{a(h - h_{ref})}{2\sqrt{h - z_0}} + ah\sqrt{h - z_0}. \tag{3.191}$$

Overflow

The overflow discharge curve in abridged form is defined as

$$Q = a(h - z_0)^b, \tag{3.192}$$

where z_0 is the overflow elevation. A derivative of the discharge curve with respect to the piezometric head is equal to

$$\frac{\partial Q}{\partial h} = Q\frac{b}{h - z_0}. \tag{3.193}$$

General outlet

This is an outlet type with non-linear variation of the outflow cross-section area when the piezometric head h exceeds the reference h_{ref}, and the outflow velocity exceeds the elevation z_0 of the object. An abridged form of the discharge curve depending on the piezometric head is defined by the expression for the area and velocity

$$Q = Av = a(h - h_{ref})^b(h - z_0)^c. \tag{3.194}$$

Most of the regulation valves, which cannot be defined as the aforementioned standard ones, can be modeled by the general outlet discharge curve.

A derivative of the flow curve with respect to the piezometric head is equal to

$$\frac{\partial Q}{\partial h} = \frac{\partial Av}{\partial h} = A\frac{\partial v}{\partial h} + v\frac{\partial A}{\partial h}, \tag{3.195}$$

namely

$$\frac{\partial Q}{\partial h} = Q\left(\frac{b}{h - h_{ref}} + \frac{c}{h - z_0}\right). \tag{3.196}$$

3.6.2   Outlet model

Outlet secondary variables

A secondary variable of the outlet is the outlet discharge Q^{outlet}, which is calculated in the phase computation of the primary variable increments, see subroutine subroutine IncVar. Computation of a secondary variable is implemented in subroutine

subroutine CalcOutletQ(node),

where, depending on the outlet type, the respective subroutine for discharge computation is called:

general outlet

subroutine GeneralOutletQ(Gate,tK,Hk),

relief

subroutine ReliefOutletQ(Gate,tK,Hk),

gate

subroutine GateOutletQ(Gate,tK,Hk),

overflow

subroutine PreljevOutletQ(Gate,tK,Hk).

The aforementioned subroutines can be found in the program module Outlets.f90.

Fundamental system vector and matrix updating

The hydraulic function of an outlet is a natural boundary condition; namely, a discharge that shall be added to the r-th nodal equation, where discharge that updates the nodal equation is negative (extracting from the hydraulic network)

$$\sum_p {}^r Q^e = Q_r^{outlet}. \tag{3.197}$$

Thus, for the steady state

$$\text{vector updating}: \quad F_r^{new} = F_r^{old} + Q_r^{outlet}, \tag{3.198}$$

$$\text{matrix updating}: \quad G_{r,r}^{new} = G_{r,r}^{old} - \frac{\partial Q_r^{outlet}}{\partial h_r}. \tag{3.199}$$

Similarly, for the non-steady state it is

$$\Delta V_r = \int_{\Delta t} Q_r dt = (1 - \vartheta)\Delta t\, Q_r^{outlet} + \vartheta\, \Delta t\, Q_r^{outlet+} \tag{3.200}$$

which is added to the global system vector:

$$F_r^{new} = F_r^{old} + \Delta V_r. \tag{3.201}$$

Respective updating of the global matrix has the following form:

$$G_{r,r}^{new} = G_{r,r}^{old} - \frac{\partial \Delta V_r}{\partial h_r} = G_{r,r}^{old} - \vartheta\, \Delta t\, \frac{\partial Q_r^{outlet+}}{\partial h_r}. \tag{3.202}$$

Updating of the initial fundamental system by an outlet as an essential boundary condition is called in the frontal procedure `FrontU` before elimination of the nodal equation, see program module `Front.f90`. It is implemented in a subroutine

```
subroutine OutletNode(inode,iactiv)
```

which, depending on the outlet type, calls the subroutines

> *general outlet:*
>
> ```
> subroutine GeneralOutlet(Gate,tP,tK,Hp,Hk,Qp,Qk,dQdH),
> ```
>
> *relief:*
>
> ```
> subroutine ReliefOutlet(Gate,tP,tK,Hp,Hk,Qp,Qk,dQdH),
> ```
>
> *gate:*
>
> ```
> subroutine GateOutlet(Gate,tP,tK,Hp,Hk,Qp,Qk,dQdH),
> ```
>
> *overflow:*
>
> ```
> subroutine PreljevOutlet(Gate,tP,tK,Hp,Hk,Qp,Qk,dQdH).
> ```

These subroutines can be found in the program module `Outlets.f90`.

Reference

Jaeger, Ch. (1949) *Technische Hydraulik*. Verlag, Basel.

Further reading

Agroskin, I.I., Dmitrijev, G.T., and Pikalov, F.I. (1969) *Hidraulika (in Croatian)*. Tehnička knjiga, Zagreb, 1969.
Bogomolov, A.I. and Mihajlov, K.A. (1972) Gidravlika, *Stroiizdat* (in Russian). Moskva.
Chow, V.T. (1959) *Open-Channel Hydraulics*. McGraw-Hill Kogakusha Ltd, Tokyo.

Daily, J.W. and Harleman, D.R.F. (1996) *Fluid Dynamics*. Addison-Wesley Pub. Co., Massachusetts.

Davis, C.V. and Sorenson, K.E. (1969) *Handbook of Applied Hydraulics*. 3th edn, McGraw-Hill Co., New York.

Jović, V. (2006) *Fundamentals of Hydromechanics* (in Croatian: Osnove hidromehanike). Element, Zagreb.

Rouse, H. (1969) *Hydraulics* (translation in Serbian: Tehnička hidraulika). Građevinska knjiga, Beograd.

Sever, Z., Franković, B., and Pavlin, Ž., Stanković, V. (2000) *Hydroelectric Power Plants in Croatia*. Hrvatska elektroprivreda, Zagreb.

4

Water Hammer – Classic Theory

4.1 Description of the phenomenon

4.1.1 Travel of a surge wave following the sudden halt of a locomotive

Imagine a train, consisting of railroad cars hauled by a locomotive, moving at a velocity v_0, see Figure 4.1. Locomotive and railroad cars are interconnected by elastic bumpers.

Let the locomotive come to a sudden stop. Its velocity becomes zero. However, the railroad cars behind it are still moving at a velocity v_0 until the first bumper is completely compressed. Bumper compression will last a short time, after which the first car will to a stop while the second one and all the others still keep moving.

The other railway cars will stop in the same way. An observer standing aside will see the railway car stopping as a wave moving from the locomotive towards the end of a train at a velocity c. When the entire train stops, there is a backward relaxation of the bumpers. The last bumper is relaxed first and its accumulated potential energy transforms into kinetic energy of the last car, thus pushing it back. After that, potential energy of the bumper next to the last one is released, and so on, which appears to the observer as a return wave at a velocity c.

In general, wave celerity c can be calculated from the elastic properties of the bumper and the kinetic energy of a train.

4.1.2 Pressure wave propagation after sudden valve closure

A sudden arrest of the fluid flow through the pipe can be compared to the sudden train stopping, see Figure 4.2. Immediately after the sudden valve closure, flow velocity at the valve cross-section becomes zero. Due to inertia, the fluid mass is still moving upstream. In a small time increment Δt, the fluid stops at the length Δl. Kinetic energy of the flow transforms into potential energy; namely, a pressure increase Δp builds up, thus causing the fluid to compress and the pipe to expand.

The pressure surge due to a sudden change in flow velocity is termed a *water hammer* or *hydraulic shock*. The velocity of propagation of the water hammer front, which is the boundary between the disturbed flow $v = 0$ and the undisturbed flow $v = v_0$, will be

$$w = \lim_{\Delta t \to 0} \frac{\Delta l}{\Delta t} = \frac{dl}{dt} \ [\mathrm{m/s}] \tag{4.1}$$

Analysis and Modelling of Non-Steady Flow in Pipe and Channel Networks, First Edition. Vinko Jović.
© 2013 John Wiley & Sons, Ltd. Published 2013 by John Wiley & Sons, Ltd.

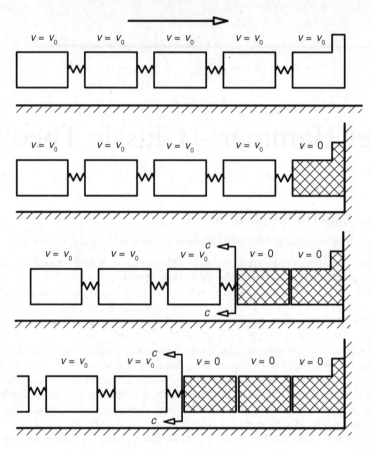

Figure 4.1   Sudden stopping of a locomotive.

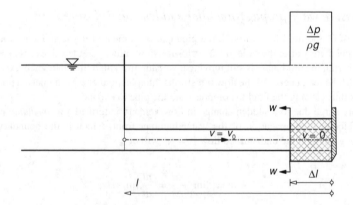

Figure 4.2   Sudden valve closure.

In front of the wave front the flow is undisturbed while behind the flow is arrested, fluid is compressed and the pipeline is expanded.

4.1.3   Pressure increase due to a sudden flow arrest – the Joukowsky water hammer

Pressure increase can be calculated from the Newton's second law

$$m\frac{dv}{dt} = dF \tag{4.2}$$

applied to the constant mass arrested in the time increment Δt. Let us observe the mass $m = \rho A (\Delta l - v_0 \Delta t)$, where ρ is the density, A is the pipe cross-section area, and $v_0 \Delta t$ is the decrease due to mass inflow from upstream. Change in the momentum is caused by the force $\Delta F = \Delta p A$, where p is the pressure; thus

$$\rho A (\Delta l - v_0 \Delta t) \frac{\Delta v}{\Delta t} = \Delta p A. \tag{4.3}$$

From the form of the obtained expression, a pressure rise due to a change in flow velocity $\Delta v = v_0$ can be written

$$\Delta p = \rho \left(\frac{\Delta l}{\Delta t} - v_0 \right) \Delta v = \rho (w - v_0) v_0. \tag{4.4}$$

An absolute velocity w reduced by the flow velocity v_0 is equal to the water hammer relative velocity (celerity), thus it can be written

$$\Delta p = \rho c v_0, \tag{4.5}$$

that is expressed as the piezometric head change

$$\Delta h = \frac{c}{g} v_0. \tag{4.6}$$

Pressure surge was first discovered by N. Joukowsky;[1] thus, in his honor, it is termed the *Joukowsky water hammer.*

4.2   Water hammer celerity

4.2.1   Relative movement of the coordinate system

An observer standing at the side sees the propagation of a pressure wave (water hammer) of a finite amplitude in a pipe, as shown in Figure 4.3, as a wave front propagation at an absolute velocity w. If we imagine relative motion along the pipe at a velocity of the coordinate system w, then an observer sees a steady flow of relative velocity in front of him and behind him.

---

[1] N. Joukowsky, Imperial Technical School, St. Petersburg 1898 and 1900.

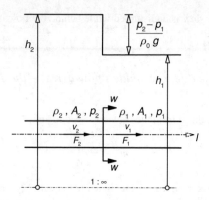

Figure 4.3   Motion of a water hammer of a finite amplitude.

Mass flow towards a movable observer (wave front) is equal to the mass flow from the observer. Figure 4.3 shows the following hydraulic variables: density $\rho(p)$, velocity v, pressure p, area $A(p)$, and pressure force at the cross-section $F = pA(p)$ in front and behind the water hammer front. Relative wave celerity in front and behind the wave front, marked with indexes, will be

$$c_1 = w - v_1$$
$$c_2 = w - v_2 \qquad (4.7)$$

The continuity equation of relative motion is

$$\rho_2 A_2 c_2 = \rho_1 A_1 c_1$$
$$\rho_2 A_2 (w - v_2) = \rho_1 A_1 (w - v_1) \qquad (4.8)$$

from which the absolute velocity of a water hammer of a finite value can be obtained

$$w = \frac{\rho_2 A_2 v_2 - \rho_1 A_1 v_1}{\rho_2 A_2 - \rho_1 A_1} = \frac{(\rho Q)_2 - (\rho Q)_1}{(\rho A)_2 - (\rho A)_1} = \frac{\Delta (\rho Q)}{\Delta (\rho A)}. \qquad (4.9)$$

The momentum equation and the pressure force in relative motion are in equilibrium

$$F_2 + \rho_2 A_2 c_2^2 = F_1 + \rho_1 A_1 c_1^2 \qquad (4.10)$$

from which

$$F_2 + \rho_2 A_2 (w - v_2)^2 = F_1 + \rho_1 A_1 (w - v_1)^2 \qquad (4.11)$$

an additional equation for calculation of a water hammer of a finite value is obtained. If Eq. (4.10) is written in the form

$$F_2 - F_1 = \rho_1 A_1 c_1^2 - \rho_2 A_2 c_2^2 \qquad (4.12)$$

and the following is obtained from Eq. (4.8)

$$c_2 = \frac{\rho_1 A_1}{\rho_2 A_2} c_1 \qquad (4.13)$$

then

$$F_2 - F_1 = \rho_1 A_1 c_1^2 - \rho_2 A_2 \left(\frac{\rho_1 A_1}{\rho_2 A_2}\right)^2 c_1^2 = \frac{\rho_1 A_1}{\rho_2 A_2}(\rho_2 A_2 - \rho_1 A_1) c_1^2 \qquad (4.14)$$

and relative celerity can be written as

$$c_1 = w - v_1 = \pm \sqrt{\frac{F_2 - F_1}{\frac{\rho_1 A_1}{\rho_2 A_2}(\rho_2 A_2 - \rho_1 A_1)}}. \qquad (4.15)$$

Similarly

$$c_2 = w - v_2 = \pm \sqrt{\frac{F_2 - F_1}{\frac{\rho_2 A_2}{\rho_1 A_1}(\rho_2 A_2 - \rho_1 A_1)}}. \qquad (4.16)$$

Celerities c_1 and c_2 are the relative celerities of the pressure wave of a finite magnitude, that is relative *water hammer* celerities. Absolute celerity of the water hammer of a finite magnitude will be

$$w = v_1 \pm c_1 = v_2 \pm c_2, \qquad (4.17)$$

namely

$$w = \frac{v_1 + v_2}{2} \pm \frac{c_1 + c_2}{2}. \qquad (4.18)$$

4.2.2 *Differential pressure and velocity changes at the water hammer front*

For a *differentially small disturbance* the following is valid in limits $v_2 \rightarrow v_1 = v$, $A_2 \rightarrow A_1 = A$, $\rho_2 \rightarrow \rho_1 = \rho$, $c_2 \rightarrow c_1 = c$; namely, finite differences become differentials

$$\rho_2 A_2 - \rho_1 A_1 = d(\rho A),$$

$$F_2 - F_1 = dF.$$

Absolute water hammer celerity is written in the following form

$$w = v \pm c, \qquad (4.19)$$

where relative water hammer celerity in differential form is

$$c = \sqrt{\frac{dF}{d(\rho A)}}. \qquad (4.20)$$

Starting from Eqs (4.19) and (4.9) it can be written

$$w = v \pm c = \frac{d(\rho Q)}{d(\rho A)}. \tag{4.21}$$

Partial differentiation of the numerator gives

$$v \pm c = \frac{v d(\rho A) + \rho A dv}{d(\rho A)} \tag{4.22}$$

from which

$$v \pm c = v + \frac{\rho A}{d(\rho A)} dv \tag{4.23}$$

and finally

$$\pm c = \frac{\rho A}{d(\rho A)} dv, \tag{4.24}$$

that is

$$\pm c d(\rho A) = \rho A dv. \tag{4.25}$$

Furthermore, expression (4.14) in limits becomes a differentially small force in the form

$$dF = d(\rho A) c^2. \tag{4.26}$$

Introducing Eq. (4.25) into Eq. (4.26), it becomes

$$dF = \pm \rho A c dv. \tag{4.27}$$

On the other hand, if the force $F = pA$ is differentiated, the following is obtained

$$dF = d(pA) = dpA + pdA, \tag{4.28}$$

where the increment of the pipe cross-section area can be neglected, then the unknown differential expression at the wave front of a differentially small water hammer has the following form

$$dp = \pm \rho c dv. \tag{4.29}$$

This expression, in general, can be applied to the motion of a differentially small pressure and velocity disturbance in a fluid flow, whether it is a gas or a liquid.

Assuming that the term ρc is constant, which is valid if a fluid is a liquid, then the expression (4.29) can be integrated between two points in front and behind the water hammer front in the form

$$\int_{p_1}^{p_2} dp = \pm \rho c \int_{v_1}^{v_2} dv \tag{4.30}$$

which gives the relation between the pressure and velocity in front and behind the wave front

$$p_2 - p_1 = \pm \rho c \cdot (v_2 - v_1). \tag{4.31}$$

In the liquid flow, pressures can be expressed by the liquid column height, namely, pressure head; thus, expressions (4.29) and (4.31) can be expressed as

$$dh = \pm \frac{c}{g} dv, \tag{4.32}$$

that is

$$h_2 - h_1 = \pm \frac{c}{g}(v_2 - v_1). \tag{4.33}$$

4.2.3   Water hammer celerity in circular pipes

In an abstract water hammer model, it is assumed that only the pipe cross-section is expanding while longitudinal deformation is negligible. Furthermore, deformation of the pipe cross-section has almost no impact on the force. Thus, force change in front and behind the water hammer front can be written as

$$dF = d(pA) \doteq dp \cdot A \tag{4.34}$$

while the water hammer celerity, expression (4.20), is

$$c = \sqrt{\frac{dpA}{d(\rho A)}}. \tag{4.35}$$

Differentiation of the previous expression gives

$$d(\rho A) = \rho dA + A d\rho = \rho A \left(\frac{d\rho}{\rho} + \frac{dA}{A} \right) \tag{4.36}$$

and the water hammer celerity has the following form

$$c = \sqrt{\frac{dp}{\rho \left(\dfrac{d\rho}{\rho} + \dfrac{dA}{A} \right)}}. \tag{4.37}$$

Note that water hammer celerity depends on the ratio between the pressure change and relative changes in the density and pipe cross-section area. Relative changes in the density and pipe cross-section area can be expressed by the pressure increment and the elastic properties of the liquid and pipe. For liquids, relative density change is

$$\frac{d\rho}{\rho} = \frac{dp}{E_v}, \tag{4.38}$$

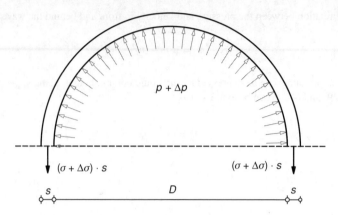

Figure 4.4   Stress change in a pipe due to pressure change.

where E_v is the bulk modulus of elasticity of a liquid. Relative deformation of a cross-section is defined by a relative deformation of the pipe diameter

$$\frac{dA}{A} = \frac{d\left(D^2\frac{\pi}{4}\right)}{D^2\frac{\pi}{4}} = 2\frac{dD}{D}. \tag{4.39}$$

Furthermore, if pipe stress increment due to the pressure increment is observed in the pipe as shown in Figure 4.4

$$dp \cdot D = 2 \cdot d\sigma \cdot s, \tag{4.40}$$

is obtained, from which

$$d\sigma = \frac{D}{2\,s}dp, \tag{4.41}$$

where s is the pipe wall thickness. If Hook's Law is applied, a relationship between the pipe strain and stress is obtained

$$d\sigma = \varepsilon \cdot E_c, \tag{4.42}$$

where $\varepsilon = dD/D$ is the strain and E_c is the pipe modulus of elasticity; thus

$$d\sigma = \frac{dD}{D}E_c. \tag{4.43}$$

By equalizing the stresses, that is Eqs (4.41) and (4.43), the following is obtained

$$\frac{dD}{D}E_c = \frac{D}{2\,s}dp \tag{4.44}$$

and

$$2\frac{dD}{D} = \frac{D}{s\,E_c}dp. \tag{4.45}$$

Finally, relative deformation of the circular pipe cross-section area, expressed by pipe parameters and pressure increment, is

$$\frac{dA}{A} = 2\frac{dD}{D} = \frac{D}{s\,E_c}dp. \tag{4.46}$$

If relative deformations (4.38) and (4.46) are introduced into Eq. (4.36)

$$d\,(\rho A) = \rho A\,\left(\frac{1}{E_v} + \frac{D}{sE_c}\right) \cdot dp \tag{4.47}$$

and the water hammer celerity in a pipe of a circular cross-section becomes

$$c = \sqrt{\frac{dpA}{d\,(\rho A)}} = \sqrt{\frac{1}{\rho\left(\dfrac{1}{E_v} + \dfrac{D}{sE_c}\right)}}, \tag{4.48}$$

where D is the pipe diameter, s is the pipe wall thickness, E_v is the bulk modulus of elasticity of liquid, E_c is the modulus of elasticity of the pipe, and ρ is the liquid density. Water hammer celerity can be expressed by the equivalent *bulk* modulus of elasticity of liquid and pipe

$$c = \frac{1}{\sqrt{\rho\left(\dfrac{1}{E_v} + \dfrac{D}{sE_c}\right)}} = \frac{1}{\sqrt{\dfrac{\rho}{E_e}}} = \sqrt{\frac{E_e}{\rho}}, \tag{4.49}$$

where

$$\frac{1}{E_e} = \frac{1}{E_v} + \frac{D}{sE_c}. \tag{4.50}$$

The following is also valid

$$\frac{1}{c^2} = \frac{1}{\dfrac{E_v}{\rho}} + \frac{1}{\dfrac{sE_c}{\rho D}} = \frac{1}{c_v^2} + \frac{1}{c_c^2}. \tag{4.51}$$

Table 4.1 lists some other expressions for the speed of the water hammer in pipes of different cross-sections.

4.3 Water hammer phases

Water hammer development in a simple pipeline will be analyzed in the following examples. It is assumed that all non-linear terms such as the velocity head and resistances are negligible. It is also assumed that the value of the water hammer celerity c significantly exceeds the value of the velocity v; thus the absolute water hammer speed is $w = v \pm c \approx c$.

Table 4.1   Water hammer velocities.

Water hammer celerity in an infinite space:

$$c = \sqrt{\frac{E_v}{\rho}} \tag{4.52}$$

Water hammer celerity in a thick-walled pipeline:

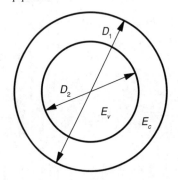

$$c = \cfrac{1}{\sqrt{\rho \left[\cfrac{1}{E_v} + \cfrac{1}{8 E_c} \left(D_1^4 - D_2^4 \right) \right]}} \tag{4.53}$$

Water hammer celerity in a tunnel:

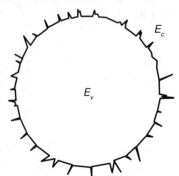

$$c = \cfrac{1}{\sqrt{\rho \left[\cfrac{1}{E_v} + \cfrac{2}{E_s} \right]}} \tag{4.54}$$

Water hammer celerity in a reinforced concrete tunnel:

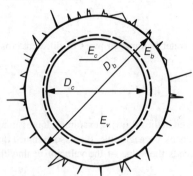

$$\lambda = \left[\cfrac{\cfrac{D_c^2}{4 E_c s}}{\cfrac{D_c^2}{4 s E_c} + \cfrac{D_b^2 - D_c^2}{4 D_b E_b} + \cfrac{(m+1)}{2 m E_s} D_c} \right]$$

where 1/m is the Poisson number of the tunnel wall:

$$c = \cfrac{1}{\sqrt{\rho \left[\cfrac{1}{E_v} + \cfrac{D_c}{E_c s} (1 - \lambda) \right]}} \tag{4.55}$$

4.3.1   Sudden flow stop, velocity change $v_0 \rightarrow 0$

(a) **Initial condition:**

At the moment $t = 0$ water flows out of the tank at a velocity v_0. The piezometric head is constant along the pipeline and equal to the water level in the tank. At that moment, the valve at the end of the pipeline suddenly closes thus causing the water hammer. Due to the outflow velocity change $v_0 \rightarrow 0$ a positive pressure rises in an amount equal to the Joukowsky water hammer; namely, the piezometric head increment: (Figure 4.5)

$$\Delta h = \frac{c}{g} v_0.$$

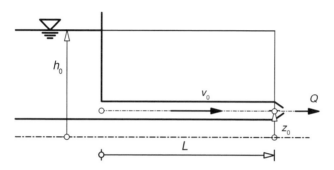

Figure 4.5

(b) **Positive phase, compression:**

Immediately in front of the valve, water is compressing and the pipeline is expanding, propagating upstream at a water hammer celerity w. This phase will last until the water hammer front reaches the tank; namely $0 < t < L/w$. (Figure 4.6)

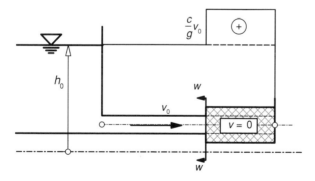

Figure 4.6

(c) **Positive phase, relaxation:**

At the moment when the water hammer front reaches the tank, water velocity in the pipeline is zero, water is compressed and the pipeline expanded. Due to the difference between the pressures in the pipeline and the tank, there is pipeline relaxation. Thus, in this phase, water flows from the pipeline towards the tank. The water hammer reflects from the tank and the water hammer front starts to move towards the valve at a speed w. Duration of this phase is $L/w < t < 2L/w$ (Figure 4.7).

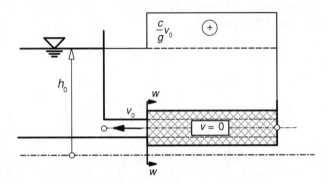

Figure 4.7

(d) **Negative phase, suction:**

At the moment when the water hammer front reaches the valve, water velocity in the pipeline is equal to $-v_0$, while at the valve cross-section it is still zero. Thus, at the valve cross-section, the water hammer is generated again, this time with a negative sign. Here the negative phase of the water hammer starts when water is sucked out and the pipeline contracts. This phase will last until the water hammer front reaches the tank cross-section, namely, in time $2L/w < t < 3L/w$ (Figure 4.8).

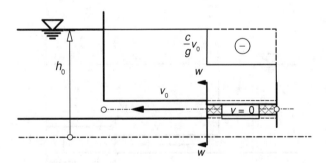

Figure 4.8

(e) **Negative phase, relaxation:**

At the moment when the water hammer front of the negative phase reaches the tank, water velocity in the pipeline is zero and pressure is lower than tank pressure. Thus, water starts to flow again from the tank into the pipeline and there is again pipeline relaxation. The reflected water hammer moves at a speed w towards the valve cross-section. The duration of this phase is $3L/w < t < 4L/w$ (Figure 4.9).

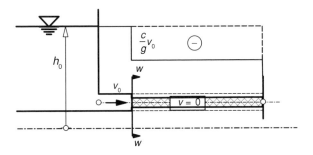

Figure 4.9

(f) **Return to initial state:**

In the return phase, when the negative phase wave front reaches the valve cross-section, the initial condition is established again, i.e. the water flows towards the valve that is completely closed and the cycle is repeated again (Figure 4.10).

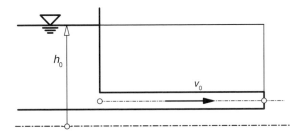

Figure 4.10

The time necessary for the positive or negative phase, generated at the valve cross-section, to travel from the valve to the tank and back is called the *water hammer cycle*

$$\tau_0 = \frac{2L}{w}. \tag{4.56}$$

Figure 4.11 shows pressure development at the valve cross-section and velocity change at the tank cross-section.

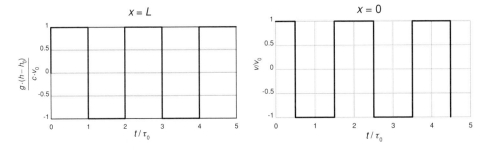

Figure 4.11   Sudden stop, velocity change $v_0 \rightarrow 0$.

4.3.2   Sudden pipe filling, velocity change $0 \to v_0$

(a) Initial condition:

The initial condition is a hydrostatic state. At the left end there is a tank, which defines the initial pressure. There is a closed valve on the right end. After sudden valve opening, let the pipeline fill with water at a velocity $-v_0$. Then, a water hammer will occur: (Figure 4.12)

$$\Delta h = \frac{c}{g} v_0.$$

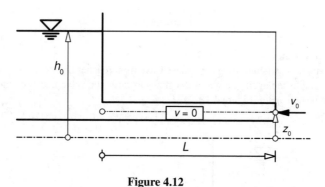

Figure 4.12

(b) Positive phase, compression:

Immediately in front of the valve, water is compressing and the pipeline is expanding, propagating upstream at a water hammer celerity w. This phase will last until the water hammer front reaches the tank; namely, $0 < t < L/w$ (Figure 4.13).

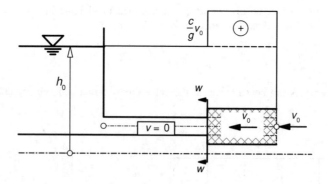

Figure 4.13

(c) Positive phase, relaxation:

At the moment when the water hammer front reaches the tank, water velocity in the pipeline is $-v_0$, water is compressed and the pipeline expanded, which causes relaxation and reflection of the positive phase wave front. This phase occurs in time $L/w < t < 2L/w$ (Figure 4.14).

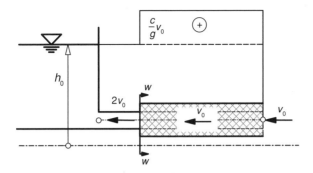

Figure 4.14

(d) **Negative phase, suction:**

At the moment when the water hammer front reaches the tank, water velocity in the pipeline is $-2v_0$, and water still flows into the valve cross-section at a velocity $-v_0$. The negative difference between the velocities generates a negative water hammer phase and water suction from the pipeline. This phase occurs in time $2L/w < t < 3L/w$ (Figure 4.15).

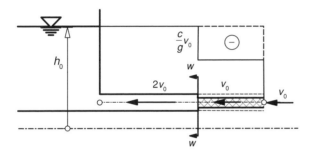

Figure 4.15

(e) **Negative phase, relaxation:**

At the moment when the water hammer front reaches the tank, water velocity in the pipeline is v_0, water is diluted and the pipeline contracted, thus causing relaxation and the reflection of the negative phase wave front. This phase duration is $3L/w < t < 4L/w$ (Figure 4.16).

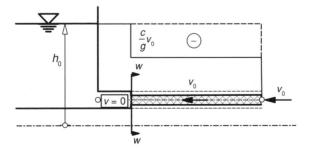

Figure 4.16

(f) **Return to initial condition:**

In the return, when the negative phase wave front reaches the valve cross-section, the initial condition is established again, i.e. water velocity is zero although water flows through the valve at a velocity $-v_0$ and the cycle is repeated again (Figure 4.17).

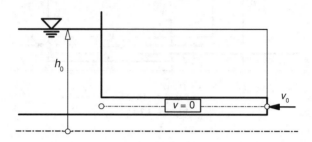

Figure 4.17

Figure 4.18 shows the pressure development at the valve cross-section and the velocity change at the tank cross-section.

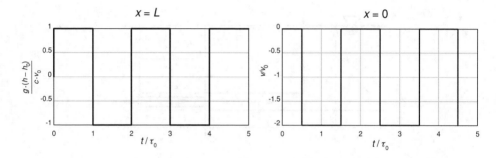

Figure 4.18   Sudden filling, velocity change $0 \to v_0$.

4.3.3   Sudden filling of blind pipe, velocity change $0 \to v_0$

(a) **Initial condition:**

The initial condition is the state at rest under pressure. On the left pipeline end there is a valve, while the right end is completely closed. At the moment $t = 0$ there is initial pressure in the pipeline defined by the initial pressure head h_0. By sudden valve opening water can flow in the pipeline at a velocity v_0 (Figure 4.19).

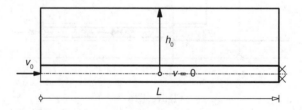

Figure 4.19

(b) **Positive phase, compression:**
      Sudden pipe-filling with water at a velocity v_0 generates the positive water hammer phase and water compression. This phase occurs in time $0 < t < L/w$ (Figure 4.20).

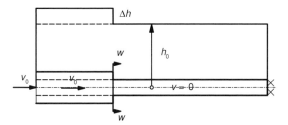

Figure 4.20

(c) **Positive phase, reflection, compression increase:**
      When the water hammer front reaches the blind end of the pipeline, flow velocity is v_0 in the pipeline and zero at the blind end cross-section. This difference generates a new water hammer that increases the existing pressure and the wave front moves as a reflected water hammer of a value equal to the incoming one. Water is even more expanded and the pipeline is additionally expanded. This phase occurs in time $L/w < t < 2L/w$ (Figure 4.21).

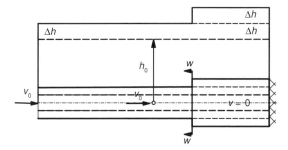

Figure 4.21

(d) **Positive phase, repeated compression:**
      When the water hammer front reaches the free end of a pipe, flow velocity in the pipe is zero. Since water still flows into the pipe at a velocity v_0, a new water hammer is generated, which compresses the water again. The water hammer front again starts to move towards the blind end with increased pressure. This phase occurs in time $2L/w < t < 3L/w$ (Figure 4.22).

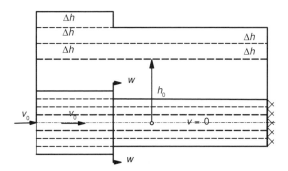

Figure 4.22

(e) **Positive phase, repeated reflection:**

When the wave front reaches the blind end, the wave is reflected again and the pressure increases. This phase occurs in time $3L/w < t < 4L/w$ (Figure 4.23).

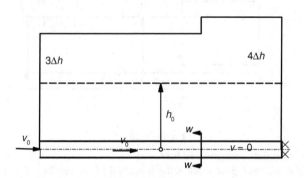

Figure 4.23

Through constant water inflow into a blind pipeline, pressure constantly rises in a step-like form. Figure 4.24 shows pressure development at the beginning and velocity in the middle of blind pipeline for a velocity change $0 \to v_0$.

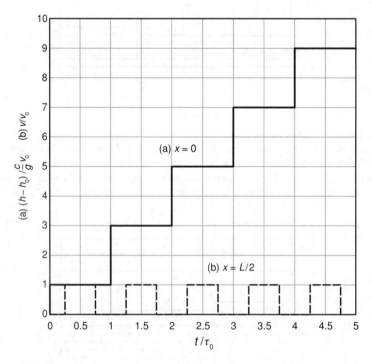

Figure 4.24   Blind pipeline filling; pressure and velocity change.

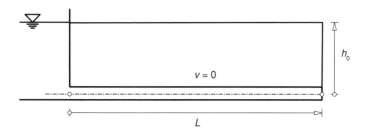

Figure 4.25 Initial condition.

4.3.4 *Sudden valve opening*

Initial condition $t = 0$

Figure 4.25 shows the pipeline in a hydrostatic state. At the right hand end of the pipeline there is a valve that opens suddenly.

First cycle of the water hammer

Immediately after valve opening, pressure at the valve suddenly drops from the initial value $p = \rho g h_0$ to the atmospheric pressure. The sudden change to the piezometric head is followed by a sudden change of velocity from zero to the value equal to

$$v_1 = \frac{g}{c} h_0. \tag{4.57}$$

The generated pressure wave is shown in Figure 4.26a and occurs in time $0 < t < L/c$. When the wave front reaches the tank, the pressure wave is reflected as shown in Figure 4.26b.

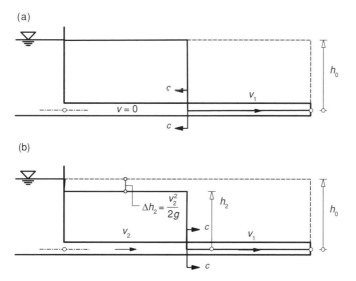

Figure 4.26 Wave propagation in time $0 < t < 2L/c$.

Following the wave reflection from the tank, the difference between the tank pressure and pipe pressure generates the velocity v_2 and a piezometric head h_2 that can be obtained from the velocity head

$$\Delta h_2 = \frac{v_2^2}{2g} \tag{4.58}$$

and water hammer

$$h_2 = h_0 - \Delta h_2 = \frac{c}{g}(v_2 - v_1) \tag{4.59}$$

from which

$$v_2 = \sqrt{c(c + 2v_1) + 2gh_0} - c. \tag{4.60}$$

The outflow velocity v_1 remains unchanged during the water hammer propagation to the tank and back.

Second cycle of the water hammer

At the beginning of the second cycle of the water hammer, the outflow velocity changes from v_1 to v_3, see Figure 4.27c. Thus the stage at the wave head is defined as

$$\frac{c}{g}(v_3 - v_2) = h_2 \tag{4.61}$$

(c)

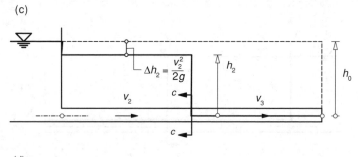

(d)

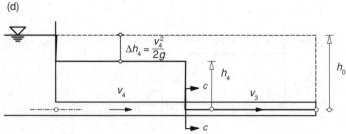

Figure 4.27   Wave propagation in time $2L/c < t < 4L/c$.

from which the outflow velocity can be calculated

$$v_3 = \frac{g}{c}h_2 + v_2. \tag{4.62}$$

Following the repeated wave reflection from the tank, see Figure 4.27d, the difference between the tank pressure and pipe pressure generates the velocity v_4 and a piezometric head h_4 that can be obtained from the velocity head

$$\Delta h_4 = \frac{v_4^2}{2g} \tag{4.63}$$

and water hammer

$$h_4 = h_0 - \Delta h_4 = \frac{c}{g}(v_4 - v_3) \tag{4.64}$$

from which

$$v_4 = \sqrt{c(c + 2v_3) + 2gh_0} - c. \tag{4.65}$$

The outflow velocity v_3 remains unchanged during the water hammer propagation to the tank and back.

In the third, and further, water hammer cycles, the outflow, such as the velocity, at the beginning of the pipe keeps increasing while the piezometric head at the beginning of the pipe keeps decreasing.

Example

Let us imagine a pipeline of $L = 1000$ m length, water hammer celerity is $c = 100$ m/s and the piezometric head is $h_0 = 100$ m.

The water hammer cycle is $\tau_0 = 2$ s. Steady outflow velocity is $v_0 = \sqrt{2gh_0} = 44.2$ m/s.

Velocities at both pipe ends as well as the piezometric head at the beginning of the pipeline were calculated using the previously developed expressions by a spreadsheet program (MS Excel) and are shown in Figure 4.28.

Note that both velocities $v(0)$ and $v(L)$ are asymptotically approaching the steady velocity v_0 in time, while piezometric head $h(0)$ at the beginning of the pipe asymptotically approaches zero. A velocity graph v_r, according to the rigid fluid acceleration problem solution, see Section 2.3.3 Incompressible fluid acceleration, is also shown in the figure

$$v_r = v_0 \tanh \frac{v_0}{2L}t. \tag{4.66}$$

The solution for the sudden valve opening for a rigid fluid is a good approximation of the sudden acceleration of a compressible fluid, see the detail in Figure 4.29.

4.3.5   Sudden forced inflow

Figure 4.30a shows a pipeline in a hydrostatic state. At its left end, immediately after $t = 0$, there is a forced inflow so as to maintain the steady piezometric head h_c, thus generating a water hammer, which propagates towards the tank at a velocity w.

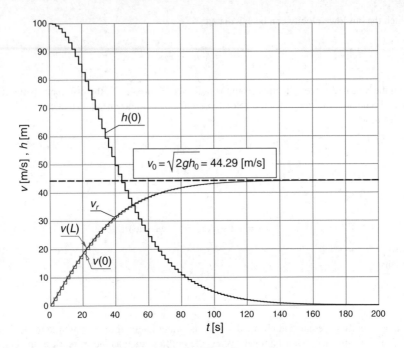

Figure 4.28   Sudden valve opening.

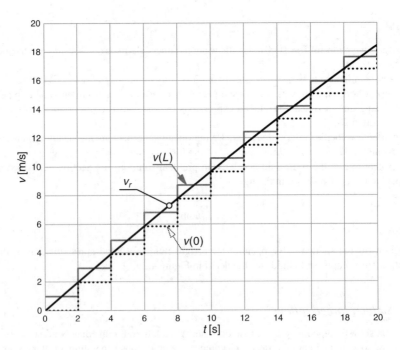

Figure 4.29   Detail, the first 20 s.

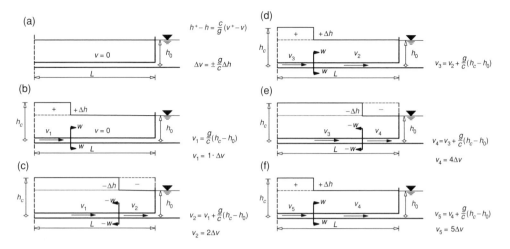

Figure 4.30 Water hammer phases at forced inflow.

The velocity behind the water hammer front is calculated from the equation for the water hammer front (4.33) as follows

$$v_1 = 0 + \frac{g}{c}(h_c - h_0) = \Delta v.$$

It is a positive expansion phase lasting from $0 \leq t \leq L/w$, see Figure 4.30b. When the pressure wave is reflected from the tank, there is a change in velocity behind the wave front

$$v_2 = v_1 + \frac{g}{c}(h_c - h_0) = 2\Delta v.$$

This positive reflection phase lasts from $L/w \leq t \leq 2L/w$, see Figure 4.30c. Again, there is the same situation as at the beginning with the difference that the velocity is no longer zero but the velocity of the reflected wave

$$v_3 = v_2 + \frac{g}{c}(h_c - h_0) = 3\Delta v.$$

This phase is shown in Figure 4.30d and lasts from $2L/w \leq t \leq 3L/w$. Figures 4.30e and f show the next two phases

$$3L/w \leq t \leq 4L/w: \quad v_4 = v_3 + \frac{g}{c}(h_c - h_0) = 4\Delta v,$$

$$4L/w \leq t \leq 5L/w: \quad v_5 = v_4 + \frac{g}{c}(h_c - h_0) = 5\Delta v.$$

Note that constant forced inflow accelerates the water by small velocity increments Δv after each propagation $\Delta t = L/w$, as shown in a few steps in Figure 4.31a for velocities $v(0)$ and $v(L)$ at the beginning and the end of the pipe. In this model of the phenomenon, the water hammer is being kept infinitely, and acceleration increases to the infinite value because there are no resistances.

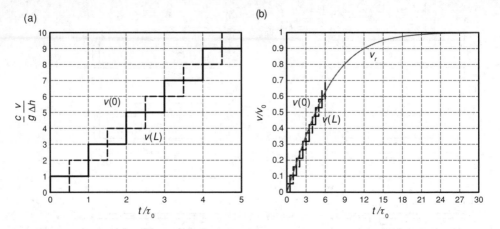

Figure 4.31 Water acceleration at forced inflow (a) detail (b) comparison with the acceleration of a rigid fluid (water).

In a model of rigid fluid motion, for constant forced inflow acceleration is limited and shown as curve v_r in Figure 4.31b.

The model of water hammer acceleration shows a good correspondence with the rigid fluid model immediately after the beginning of acceleration.

Figure 4.32 shows a steady flow after acceleration at constant forced inflow. In real circumstances the flow stills, unlike in the simplified water hammer model where it is infinite.

4.4   Under-pressure and column separation

Sudden flow changes in time cause water hammers with excess pressure increases (positive phases) and decreases (negative phases). Let us observe again a sudden flow stop according to the aforementioned phase, see Section 4.3.1. The maximum and the minimum piezometric head values occur in positive and negative phases as follows

$$\left.\begin{array}{c} h_{max} \\ h_{min} \end{array}\right\} = h_0 \pm \frac{c}{g} v_0 > 0.$$

In the present example, minimum pressures are higher than the atmospheric pressure. When the above expression is analyzed, it becomes obvious that for higher velocities v_0 in negative phase, pressures

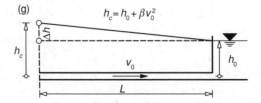

Figure 4.32   Stilling of the forced inflow.

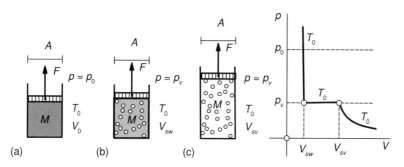

Figure 4.33 Water transition from liquid into vapor.

a lot lower than the atmospheric pressure, that is under-pressures, can occur. As long as the absolute pressure is above the pressure of saturated vapors, water is in a liquid phase and there will be no column separation.

What is happening to water, or any other liquid, in the interval of small pressures, will be explained by an imaginary experimental device shown in Figure 4.33. Figure 4.33a shows the device where water mass M occupies volume V_0 at absolute temperature T_0 and atmospheric pressure p_0. At that moment, the piston force is equal to zero. When piston is pulled back, water is expanding, that is its volume is increasing. When the pressure drops to $p = p_v$, that is volume V_{sw}, vapor bubbles start to form in the water. After that, the piston should not be pulled by force in order to increase the volume, see Figure 4.33b. It seems that the water gives no resistance to the piston, that the water column is separating and transforming into vapor. The volume increases, although there is no pressure change, until the volume V_{sv} is reached, Figure 4.33c. Then, force shall be increased again in order to increase the volume, and the entire water mass has transformed to vapor. The picture on the right shows the pressure $p = p_0 - F/A$ and volume V for a constant absolute temperature T_0. It is one of the water isotherms in the water state equation, which describes the transition from the liquid phase, over the liquid and vapor equilibrium, to the water vapor gas. Thermodynamic coordinates T, p, V of saturated vapor are showed in detail in Section 5.1 Equation of state.

Thus, when the absolute pressure drops to the value of saturated vapor pressure, the thermodynamic state is altered and there is water transition into saturated vapor. The water body transforms into a mixture of water vapor and water drops, that is cavitation and water column separation occur.

The equilibrium state in front and behind the water hammer front can also be analyzed in the case of *two water phases by relative motion*. If, in front and behind the water hammer front, a relative balance of forces and momentum is set as well as the relative balance of the mass flow, as shown in Section 4.2.1 Relative movement of the coordinate system, two equations are obtained for calculation of the absolute velocity of motion w and piezometric head h_2, or velocity v_2 behind the wave front depending on the hydraulic conditions at the pipe ends. Therefore, for a horizontal pipeline at an elevation z_0 relevant equations for determining the unknowns are written

$$w = \frac{\rho_2 A_2 v_2 - \rho_1 A_1 v_1}{\rho_2 A_2 - \rho_1 A_1}, \tag{4.67}$$

$$\rho_2 g (h_2 - z_0) A_2 + \rho_2 A_2 (w - v_2)^2 = \rho_1 g (h_1 - z_0) A_1 + \rho_1 A_1 (w - v_1)^2. \tag{4.68}$$

The unknowns w and h_2, that is w and v_2 are calculated by an iterative procedure. The density and deformation of an elastic pipe $\rho(p)$, $A(p)$ are functions of pressure, while pressure is obtained from the piezometric head as $p = \rho g (h - z_0)$.

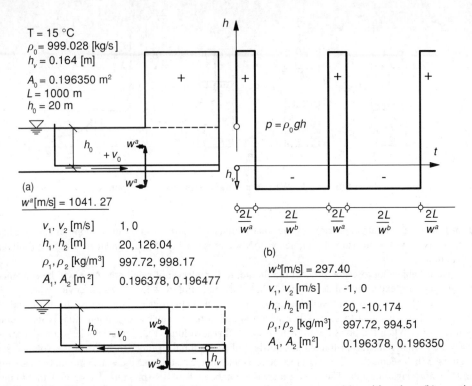

Figure 4.34 Water column separation below the saturated vapor pressure (a) positive phase (b) negative phase.

Figure 4.34 shows the sudden closing of a pipeline of length L. The initial velocity and initial piezometric head are v_0 and h_0, respectively. These values are marked in the figure, together with other data.

(a) **Positive phase:**

At the moment $t = 0$ the valve closes suddenly, a positive phase of the water hammer is generated, the pressure wave propagates towards the tank, water is compressed and the pipe expands. In front of the wave front, marked with index 1, water is in a liquid state with prescribed values: $v_1, h_1, \rho_1(p_1), A_1(p_1)$. Behind the wave front, marked with index 2, fluid is in a liquid state with prescribed value $v_2 = 0$, from which the values $w, v_2, h_2, \rho_2(p_2), A_1(p_2)$ are calculated. The calculation results are shown in Figure 4.34a. The water hammer speed is $-w^a$. At the moment when the positive phase reaches the tank, the wave reflects, and the relaxation propagates towards the tank at a velocity $-v_0$ due to a pressure head h_0 as a boundary condition. The wave returns at a speed $+w^a$.

(b) **Negative phase:**

In the return, when the reflected wave head reaches the right hand end, water is moving at a velocity $-v_0$ in the entire pipeline, except in the valve cross-section. At that moment, a water hammer negative phase is generated. In the front of the wave head there is water in a liquid phase while behind the wave front there water is in a saturated water vapor phase. Namely, balance in the liquid phase is not possible because pressures below the absolute zero, that is below the saturated vapor pressure, are obtained.

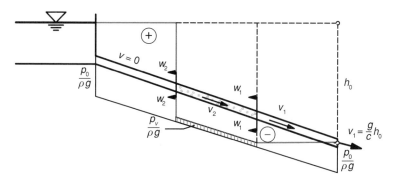

Figure 4.35 Column separation.

The calculation results are shown in Figure 4.34b. The absolute water hammer speed is $-w^b$. Note that the wave velocity is a lot smaller than the previously obtained water hammer speed because sound velocity in a liquid phase is higher than sound velocity in a gas phase. Figure 4.34 shows a graph of piezometric head development from which it can be observed that the negative phase is longer than the positive one because of the respective slower wave propagation.

An instructive example of water column separation will be shown as the flow generated after a sudden opening of an inclined pipeline, as shown in Figure 4.35. Immediately after the valve is suddenly opened at the pipe end, the water hammer front starts to propagate upstream towards the tank at a speed w_1, increasing the underpressure (value of the absolute pressure is dropping). When the absolute pressure drops to the value of saturated water vapor pressure p_v pressure remains constant while velocity changes from the value of outflow velocity v_1 to the value v_2. The water column starts to separate; namely it transforms into saturated vapor, which moves upstream with the new speed of the front w_2. Velocities v_2 and w_2 and density ρ_2 are obtained from the equilibrium of two water phases, in front and behind the water hammer front.

The saturated vapor zone, that is the separated column, moves towards the tank and then the front is reflected. Then wave reflection and refilling are expected, that is the saturated vapor zone vanishes.

From an engineering aspect, column separation is unacceptable and the problem should be solved by mitigation measures to prevent generation of extremely low pressures. In a real flow, the problem is solved by engineering solutions; namely, by extension of the valve opening or installation of a surge tank or a device in an appropriate location to eliminate column separation.

4.5 Influence of extreme friction

(a) **Initial state:**

Unlike previously analyzed water hammer situations with negligible resistances to flow, a water hammer generated by sudden closure of the flow from the tank through the pipe with extreme friction will be analyzed here.

Figure 4.36 shows the outflow at a velocity v_0 through the pipeline where all available energy is used to overcome friction resistance. Velocity heads are also negligible and the water hammer speed is $w \approx c$.

(b) **Positive phase of compression and friction resistance release:**

Immediately after sudden closing of the valve at the moment t_0 the Joukowsky surge is generated based on the change of the outflow velocity v_0. The water hammer front is propagating upstream, moving the positive phase over small to greater pressures prevailing in front of the wave front, as shown in Figure 4.37.

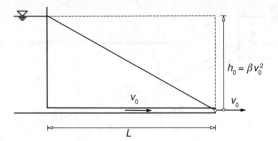

Figure 4.36   Extreme friction (a) Initial state.

Since there is a head difference towards the valve due to the motion of the water hammer front, the flow relaxes in a manner such that behind the wave front a flow of velocity v_x is established towards the valve. The piezometric head behind the front is decreasing while at the valve it is increasing. Velocity v_x is dropping towards the valve where it is zero. Characteristic pressures and velocities are shown in Figure 4.37 at several characteristic moments with the marked water hammer front at cross-section $x = L/2$ in time t_2.

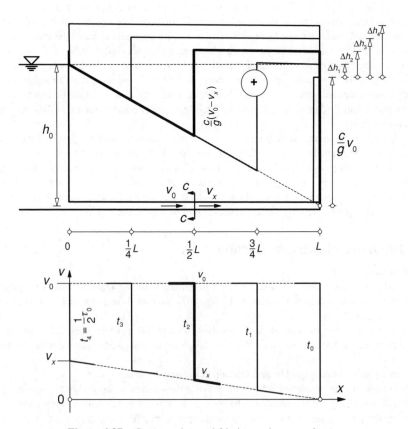

Figure 4.37   Compression and friction resistance release.

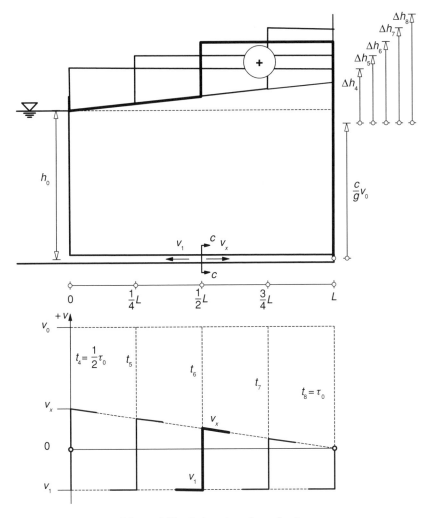

Figure 4.38   Relaxation after reflection.

The top figure shows the piezometric heads in front and behind the water hammer front, while the bottom figure shows the respective velocity changes related to wave fronts at respective times.

(c) **Positive phase, relaxation, and overcoming friction resistance:**

At the moment when the water hammer reaches the tank cross-section, pressure in the pipeline is greater than pressure in the tank. Then, the flow is established towards the tank at a velocity v_1. Due to a velocity difference at the water hammer front $v_x - v_1$ and flow towards the tank, the piezometric head is still rising. Figure 4.38 shows the piezometric heads and velocities along the pipe for characteristic time stages as well as for the wave front position in the middle of the pipeline.

(d) **Negative phase, suction:**

The pipeline relaxation phase ends when the water hammer front reaches the valve cross-section again. Since water flows at a velocity v_1 in front of the valve, while velocity at the valve is zero, there is a pressure drop. The water hammer negative phase moves towards the tank and behind the front

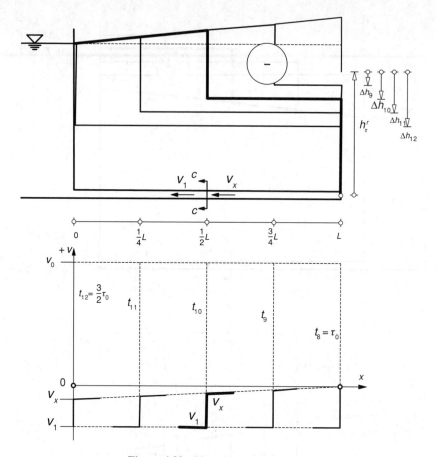

Figure 4.39 Negative phase, suction.

there is a new relaxation velocity v_x towards the tank due to the previously established piezometric head inclination. Figure 4.39 shows the piezometric heads and velocities in front and behind the water hammer front for several characteristic time stages as well as for the wave front position in the middle of the pipeline.

(e) **Negative phase, relaxation:**

When the negative phase front reaches the pipeline end, due to the difference in pressure between the tank and the pipeline, in the pipeline there is reflection of the negative phase and relaxation by the flow velocity v_2. The water hammer front moves at a velocity $v_2 - v_x$ and propagates towards the valve. Figure 4.40 shows piezometric heads and velocities in front and behind the water hammer front for several characteristic time stages as well as for the wave front position in the middle of the pipeline.

When the wave front reaches the closed valve, a flow is established similar to that at the moment of sudden valve closure, although at a velocity $v_2 < v_0$. Again, there is a positive phase of contraction and the previously described procedure is repeated.

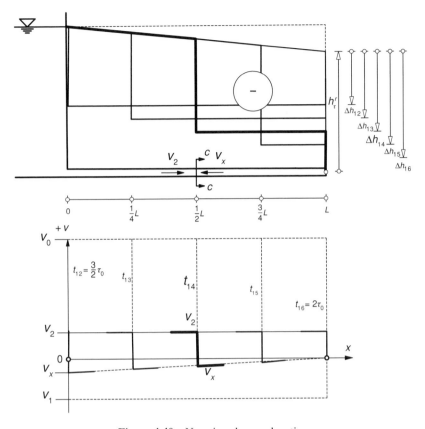

Figure 4.40   Negative phase, relaxation.

Piezometric head and velocity development in time at characteristic pipeline cross-sections are shown in Figure 4.41 and Figure 4.42.

Note that, at the moment of flow stop in the pipeline with extreme friction, positive phase pressures are increasing while negative phase pressures are decreasing in comparison to the Joukowsky surge. Reflection of the positive phase h_r is somewhat above, while reflection of the negative phase is below, the tank water level. The maximum pressure increase Δh_{τ_0} occurs at the end of the first cycle while the maximum pressure drop $\Delta h_{2\tau_0}$ occurs at the end of the second water hammer cycle.

Levels of reflection of individual phases and the increase or decrease in pressure cannot be correctly determined by simple analysis, but by modelling a flow which includes friction.

4.6   Gradual velocity changes

4.6.1   Gradual valve closing

Let us observe a water hammer at the valve cross-section caused by a flow arrest in two subsequent sudden velocity changes $v_0(t_1) \rightarrow v_2(t_2) \rightarrow 0$ according to the diagram shown in Figure 4.43. Each

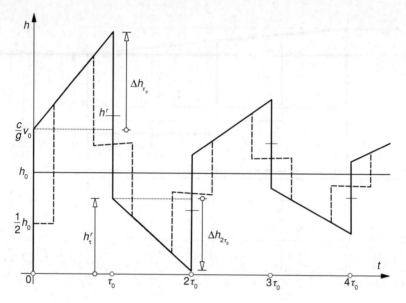

Figure 4.41 Piezometric heads at the valve cross-section and in the middle of the pipeline.

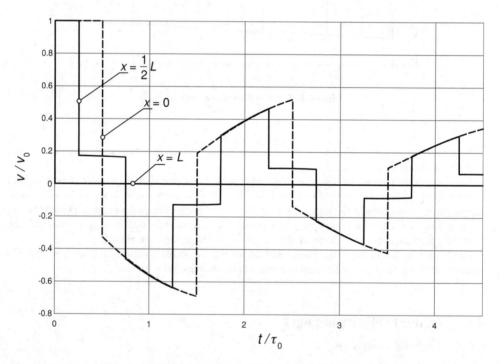

Figure 4.42 Velocities at the pipe end and in the middle of the pipeline.

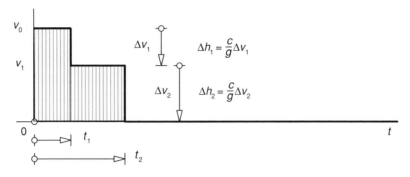

Figure 4.43 Gradual velocity change law.

gradual velocity change causes pressure, that is piezometric head, change which value is given by the Joukowsky equation.

$$\Delta h = \frac{c}{g} \Delta v.$$

A diagram of pressure changes at the valve cross-section is shown in Figure 4.44, where positive and negative phases are drawn separately for each of the velocity changes.

Note that the negative phase, which alternates with the positive in water hammer cycles, can be "packed" in between the positive phases in a way such that the negative phase diagram overlaps the positive phase diagram. The resulting "packed" overlapped phases are shown in Figure 4.45.

The resulting superposed piezometric head changes at moment t are obtained by simple addition of all positive and negative phases, see Figure 4.46.

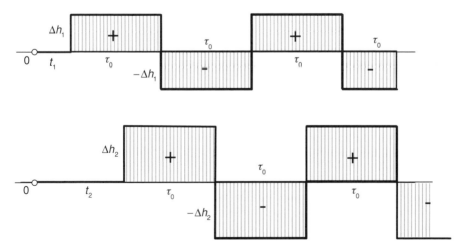

Figure 4.44 Positive and negative pressure change phases.

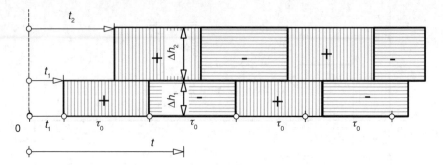

Figure 4.45   Overlapping of positive and negative phases.

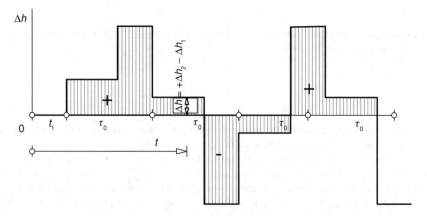

Figure 4.46   Resulting change at the valve cross-section.

4.6.2   Linear flow arrest

Continuous velocity changes can be observed as a series of infinitely small gradual changes. In this case, construction of pressure changes at the valve cross-section as described in the previous section can be applied.

Let the flow be gradually arrested in time T_z. Figure 4.47a shows a velocity graph $v(t) \to 0$, Figure 4.47b shows a diagram of velocity changes $\Delta v(t)$, while Figure 4.47c shows an affine diagram of Δh changes.

The shape of the overlapped positive and negative phases of the water hammer is obtained by the sliding of the affine diagram by the water hammer cycle τ_0. Figure 4.48d shows the overlapped positive and negative phases.

The resulting diagram of pressure changes at the valve cross-section is obtained by superposition of positive and negative changes as shown in the Figure 4.48e.

The described procedure for linear change construction can also be applied to other laws of velocity changes at the valve cross-section.[2]

---

[2]R.W. Angus, Simple graphical solution for pressure rise in pipe and pump discharge lines, *Engineering Journal of the Engineering Institute of Canada*, **February**, 1935.

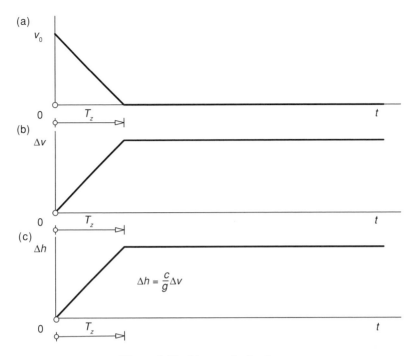

Figure 4.47   Linear velocity change.

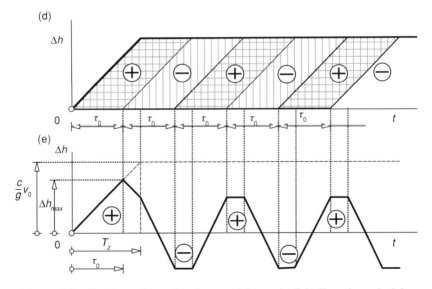

Figure 4.48   Positive and negative phases and the result of the linear law principle.

The maximum value of the water hammer occurs at the end of the water hammer cycle τ_0. If the valve closure time T_z is shorter than the water hammer cycle τ_0, then the maximum Joukowsky water hammer will occur. Otherwise, the maximum water hammer value decreases and can be calculated from linear relations that are valid in the closure interval T_z

$$\Delta h_{max} = \frac{\tau_0}{T_z} \frac{c}{g} v_0, \tag{4.69}$$

where τ_0 / T_z is the water hammer reduction factor. *Thus, the extended valve closure time decreases the water hammer, which is the key to understanding all procedures applied to prevent a water hammer.*

Example

Water flows through an $L = 5000$ m long pipeline at a velocity $v_0 = 1.6$ m/s. Gradual velocity closure time shall be defined to achieve the maximum pressure rise of 30 m of the water column. Water hammer propagation velocity is 1050 m/s.

The water hammer cycle is calculated first

$$\tau_0 = \frac{2L}{c} = 9.524\,\text{s},$$

then, the Joukowsky surge

$$\frac{c}{g} v_0 = 171.254\,\text{m v. s.}$$

The necessary closure time is calculated from Eq. (4.69)

$$T_z = \frac{\tau_0}{\Delta h_{max}} \frac{c}{g} v_0 = 54.36\,\text{s}.$$

4.7 Influence of outflow area change

Water hammer analysis, described in the previous section, assumes the prescribed law of velocity change at the valve cross-section. However, it is rare in practice. Instead of prescribed velocity change law, the more common case is the prescribed valve closure (opening) law; namely, how the outflow area A_i changes in time while closure (or opening) lasts several water hammer cycles τ_0.

Pressure changes are related to the outflow through the valve and velocity changes in front of the valve, see the scheme in Figure 4.49. A flow with negligible resistance and velocity heads is observed, which complies with the classical water hammer analysis; thus, the outflow must be dumped. The initial area of the dumped outflow is defined by a steady state before a discharge change $Q_0 = v_0 A_0$

$$A_i^0 = \frac{v_0 A_0}{\mu \sqrt{2g(h_0 - z_0)}}, \tag{4.70}$$

where $\mu \approx \varepsilon$ is the outflow coefficient, which is approximately equal to the coefficient of the outflow jet contraction.

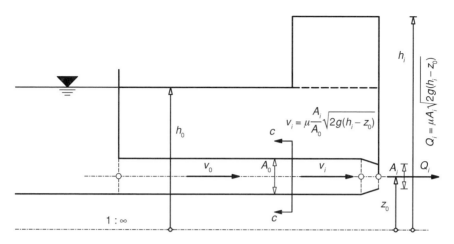

Figure 4.49 Pressure rise at the outflow.

Gate lowering generates a positive phase of the water hammer that moves upstream at a speed $w \approx c$. In the first cycle of the water hammer $t = \tau_0$ velocity and pressure at the valve cross-section are related by the following expressions:

$$h_1 - h_0 = -\frac{c}{g}(v_1 - v_0), \tag{4.71}$$

$$v_1 = \mu \frac{A_1}{A_0}\sqrt{2g(h_1 - z_0)} \tag{4.72}$$

that can be solved for the unknowns v_1, h_1.

In the second cycle there is a negative phase, which decreases the size of a water hammer and is equal to the double positive phase from the first cycle. Thus, at the end of the water hammer second cycle $t = 2\tau_0$ velocity and pressure are related by the following expressions

$$h_2 - h_1 = -\frac{c}{g}(v_2 - v_1) - 2(h_1 - h_0), \tag{4.73}$$

$$v_2 = \mu \frac{A_2}{A_0}\sqrt{2g(h_2 - z_0)} \tag{4.74}$$

from which the unknowns v_2, h_2 can be obtained. In each subsequent cycle, negative phases from the previous cycle shall be included; thus, the following expressions can be written for the n-th cycle

$$h_n - h_{n-1} = -\frac{c}{g}(v_n - v_{n-1}) - 2(h_{n-1} - h_0), \tag{4.75}$$

$$v_n = \mu \frac{A_n}{A_0}\sqrt{2g(h_n - z_0)} \tag{4.76}$$

from which new values v_n, h_n are calculated. Note that the surge wave is always reflected to the static water level in the tank h_0, which complies with the initial assumptions of the classic analyses of the water hammer where the velocity head and resistances can be neglected.

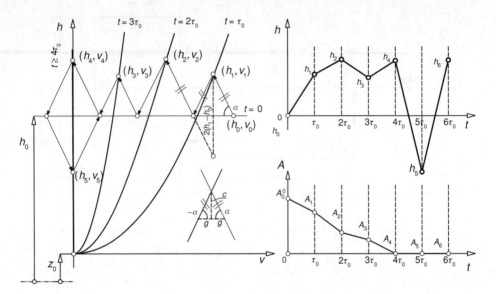

Figure 4.50   Graphic solution.

4.7.1   Graphic solution

The graphic solution of pressure changes for real valve closure and opening laws is based on the graphical solution of Eqs (4.75) and (4.76), better known as the Schnyder–Bergeron method. The procedure is as follows. Since velocity at the end of each interval $t = n\tau_0$ is defined by the outflow formula

$$v_n = \mu \frac{A_n}{A_0}\sqrt{2g(h_n - z_0)} \tag{4.77}$$

at times $t = \tau_0, 2\tau_0, 3\tau_0, \ldots$, this relationship will be shown as a family of parabolas, as shown in Figure 4.50.

At the moment $t = 0$, h_0, v_0 are the prescribed values and represent the intersecting point between the parabola for $t = 0$ and the line $h = h_0$. At the end of the first interval, equation

$$v_1 = \mu \frac{A_1}{A_0}\sqrt{2g(h_1 - z_0)} \tag{4.78}$$

is presented as a parabola at $t = \tau_0$. It is not hard to observe that the equation

$$h_1 - h_0 = -\frac{c}{g}(v_1 - v_0) \tag{4.79}$$

in the same coordinate system is part of the line drawn from h_0, v_0 at the end angle α whose tangent is c/g. The intersecting point of this line and the parabola is the point h_1, v_1, which is a solution of Eqs (4.78) and (4.79).

In the second cycle $t = 2\tau_0$ there is a negative phase of the first cycle which decreases the water hammer and is equal to the double positive phase. Thus, at the end of the second water hammer cycle $t = 2\tau_0$, value $2(h_1 - h_0)$ shall be reflected. Reflection is carried out according to the presented construction in Figure 4.50, in the way that the line is drawn through the point h_1, v_1 at an angle $-\alpha$ to the water hammer reflection level h_0. The solution of the water hammer in the second cycle h_2, v_2 is obtained as an intersection between the curve $t = 2\tau_0$ and the line at an angle α drawn through that point. Arrows drawn on the line segment inclined at $\pm c/g$ vividly show the water hammer propagation and reflection.

The procedure is repeated for the remaining water hammer cycles. The results obtained by this construction are show in Figure 4.50 on the right. After complete closure, the pressure fluctuates around the equilibrium level h_0, as can be concluded from the graphical solution. An amortization of the fluctuations is not expected due to disregarded resistances. It is interesting to note that at partial closure, pressure fluctuations shall be always damped by themselves.

4.7.2 Modified graphical procedure

For a slow pipe closure and an outflow with friction resistances, the procedure described in the previous section can be modified. Again, the starting point is the initial dumped outflow where the initial area is defined by the steady discharge $Q_0 = v_0 A_0$

$$A_i^0 = \frac{v_0 A_0}{\mu \sqrt{2g(h_0 - z_0)}}. \tag{4.80}$$

The difference is in the level of reflection that is not equal to the tank water level in every water hammer cycle. Namely, for a slow closure, it can be assumed that the reflection level h_r will be equal to the steady equilibrium level in each water hammer cycle. Figure 4.51 shows a modified graphical procedure.

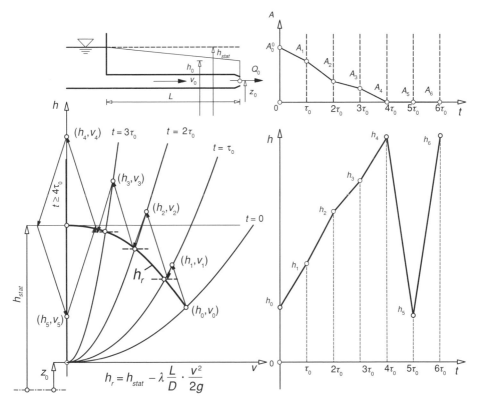

Figure 4.51 Modified graphical procedure.

The reflection level is a curve obtained by subtraction of the resistance curve from the hydrostatic level

$$h_r = h_{stat} - \lambda \frac{L}{D} \frac{v^2}{2g}. \tag{4.81}$$

Although the procedure is based on an estimate of the reflection level, it gives satisfactory results for a slow closure with friction.

4.8   Real closure laws

The most efficient water hammer control method is the control of valve closure speed (it also applies to valve opening); namely, that gate manipulation is slow enough.

It is not uncommon to combine standard gates with additional ones, so-called surge arresters or controllers, which will, in the case of fast gate closure, redirect the entire discharge and gradually arrest during a closure time that is long enough. An example is shown in Figure 4.52 where the surge control valve is connected to the turbine spiral that closes fast enough. The best solution for pressure regulation is mechanical connection of the turbine blades ring with the opening of the surge arrester valve, so there will be no change in discharge in penstock during that time.

If the extended linear valve closure, that is the decrease of the outflow area in time $T_z \gg \tau_0$, is analyzed, note that there is a non-linear discharge decrease, see Figure 4.53. At the beginning of the closure, the discharge changes more slowly than the outflow area. The reason is the velocity increase due to the pressure rise in front of the valve. This phenomenon is more emphasized for the outflow with greater friction and smaller outflow opening reduction. Due to the slow initial discharge decrease, complex closure laws can be applied, that is faster closing at the beginning and then slower later.

The aim is to achieve an almost linear outflow arrest, which can be realized by the change in velocity of the regulation device closing or by a combination of two or more regulation valves of different sizes.

In practice, it is common to use auxiliary non-regulated valves, such as ball valves, butterfly valves etc., as regulation valves. These solutions sometimes lead to pipeline ruptures; and are thus more expensive solutions in the end. Valve regulation should be reliable, and thus valve selection becomes a very sensitive issue in terms of either hydraulic properties or the price.

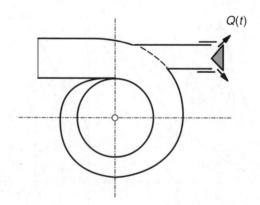

Figure 4.52   Pressure regulator at the turbine spiral.

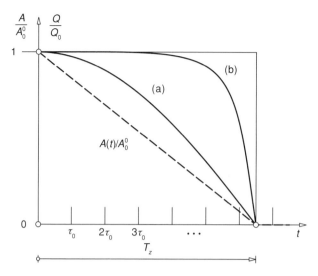

Figure 4.53   Linear law of outflow area reduction: (a) great outflow reduction, small friction (b) small outflow reduction, high friction.

4.9   Water hammer propagation through branches

When a water hammer propagates through the branches there is a surge transformation and reflection. A branch is a junction of $n \geq 2$ pipelines as shown in Figure 4.55. Surface area A_j and water hammer celerity c_j are prescribed for each j-th hand.

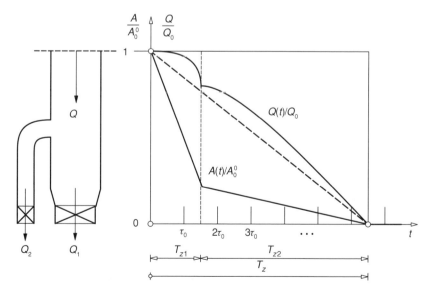

Figure 4.54   Complex valve closure law.

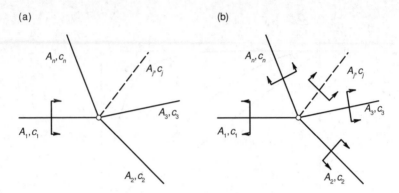

Figure 4.55 Water hammer propagation through the branch.

Let the incident water hammer reach the i-th branch hand, then the transmission coefficient will be

$$t = \frac{\dfrac{2A_i}{c_i}}{\displaystyle\sum_{j=1}^{n} \dfrac{A_j}{c_j}} \qquad (4.82)$$

while the coefficient of reflection is $r = t - 1$. Addition by the index j in the denominator includes all hands and therefore also the incident one.

Note that wave transmission and reflection do not depend on flow velocity. Details can be found in (Rouse, 1969).

It would be interesting to observe water hammer transformation in the junction of two pipes as a simple branch, first when the surge enters the wide pipe from the narrow one, see Figure 4.56. For the junction, the coefficient of transmission is

$$t = \frac{\dfrac{2A_1}{c_1}}{\dfrac{A_1}{c_1} + \dfrac{A_2}{c_2}} \qquad (4.83)$$

and the coefficient of reflection $r = t - 1$.

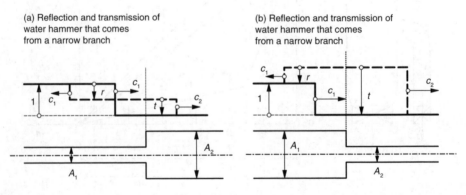

(a) Reflection and transmission of water hammer that comes from a narrow branch

(b) Reflection and transmission of water hammer that comes from a narrow branch

Figure 4.56 Two branched junctions. Ths dashed line marks the water hammer after passing through the branch.

Note that when the water hammer reaches the wider pipe the surge decreases and in the smaller diameter pipe it increases. This is similar to other waves such as waves in channels or sea waves.

In the particular case when an area $A_2 \to \infty$, as it is when the pipeline is connected to the tank, the coefficient of transmission tends to 0 while the coefficient of reflection is equal to -1. Similarly, when $A_2 \to 0$, as in the case of the blind pipe branch, the coefficient of transmission tends to 2, while the coefficient of reflection is equal to 1. Thus, the wave reflected off the closed pipeline reflects in a double amount. The analysis of the water hammer phase, see Section 4.3, gave the same results.

4.10   Complex pipelines

Classical analysis of the water hammer becomes very complex if, for example, the pipeline changes its properties along the alignment, whether this is pipe diameter, pipe material, or pipe branching, in particular if friction resistances cannot be disregarded. Nowadays, numerical procedures for non-steady flow of a compressible fluid,that is liquid, have been developed, which can provide adequate answers relating to the water hammer in very complex pipelines and channels, just as this book deals with. Thus, classical analyses of complex pipelines will be regarded as one episode in the history of development of hydraulic calculations, and will not be elaborated upon here. Anyone interested in the topic can refer to the classical textbooks and books given in the further reading for this chapter.

4.11   Wave kinematics

4.11.1   Wave functions

An elemental water hammer is generated by the disturbance of pressure and velocity, which are interconnected by the expression $dp = \pm \rho c d v$. It is often expressed by the piezometric head

$$dh = \pm \frac{c}{g} dv. \tag{4.84}$$

A disturbance is propagating along a pipe at a speed

$$w = v \pm c \tag{4.85}$$

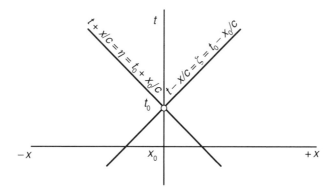

Figure 4.57   Trajectories of positive and negative wave fronts.

in the form of two waves, one being a positive wave moving in the direction $+x$ at a speed $w = v + c$, the other being a negative wave moving in the direction $-x$ at a speed $w = v - c$. In the classical theory of the water hammer, a flow where $v \ll c$ is observed; thus the speed of positive and negative waves is equal to celerity $w^{\pm} = \pm c$.

Let us observe the motion of positive and negative waves in an infinitely long pipe where initial conditions are prescribed by the steady state, that is the piezometric head and velocity h_0, v_0. In elementary time, the wave front moves in the value

$$dx = \pm c dt. \tag{4.86}$$

The motion of the wave front in the plane x, t is the line obtained by integration of the expression

$$x - x_0 = c(t - t_0) \tag{4.87}$$

that can be written in the following form

$$t - \frac{x}{c} = t_0 - \frac{x_0}{c} = const = \zeta. \tag{4.88}$$

This line characterizes the positive wave front that is in position x_0 at time t_0, that is the constant $\zeta = t_0 - x_0/c$ is valid, see Figure 4.57. A positive wave with the trajectory described by the characteristic ζ has a wave height that is, in comparison to the undisturbed state, equal to

$$F^+(\zeta) = h - h_0 = \frac{c}{g}(v - v_0) \tag{4.89}$$

and is obtained by integration of an elemental wave (4.84). The value of the wave function F^+ is constant along the characteristic line $\zeta = const$

$$F^+(\zeta) = const. \tag{4.90}$$

Similarly, for the negative wave, see Figure 4.57

$$x - x_0 = -c(t - t_0), \tag{4.91}$$

$$t + \frac{x}{c} = t_0 + \frac{x_0}{c} = const = \eta. \tag{4.92}$$

The wave function of the negative wave is equal to

$$F^-(\eta) = h - h_0 = -\frac{c}{g}(v - v_0) \tag{4.93}$$

and is constant along the characteristic line $\eta = const$

$$F^-(\eta) = const. \tag{4.94}$$

A law of superposition can be applied to the water hammer because the expressions are linear. Thus, for example, if positive and negative waves are propagating through the pipe before the collision, the piezometric heads and velocities given in Figure 4.58 are valid.

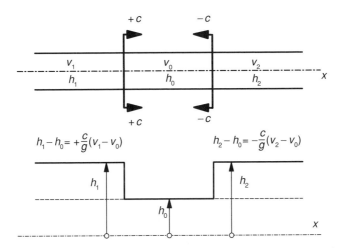

Figure 4.58 Positive and negative waves in the pipe (before collision).

Thus, the wave function of the positive wave will be

$$F_1^+ = (h_1 - h_0) = +\frac{c}{g}(v_1 - v_0). \tag{4.95}$$

Similarly, the wave function of the negative wave

$$F_2^- = (h_2 - h_0) = -\frac{c}{g}(v_2 - v_0). \tag{4.96}$$

Following the collision, positive and negative waves are summed as follows

$$h_3 - h_0 = (h_1 - h_0) + (h_2 - h_0), \tag{4.97}$$

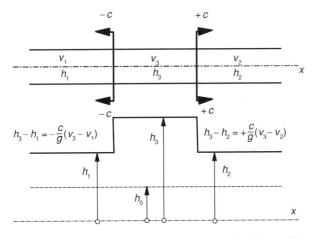

Figure 4.59 Positive and negative waves in the pipe (after collision).

that is expressed by velocities

$$h_3 - h_0 = \frac{c}{g}(v_1 - v_0) - \frac{c}{g}(v_2 - v_0).$$ (4.98)

By adding Eqs (4.97) and (4.98), it is obtained

$$2(h_3 - h_0) = \underbrace{\left[(h_1 - h_0) + \frac{c}{g}(v_1 - v_0)\right]}_{2F_1^+} + \underbrace{\left[(h_2 - h_0) - \frac{c}{g}(v_2 - v_0)\right]}_{2F_2^-},$$ (4.99)

from which

$$h_3 - h_0 = F_1^+ + F_2^-.$$ (4.100)

Thus, the resulting water hammer is defined by the summing of the wave functions of the positive and negative waves

$$h - h_0 = F^+(\zeta) + F^-(\eta).$$ (4.101)

Furthermore, following the wave summation, there will be a new stage of the variables in front and behind the wave front. Thus, the following will be valid for the positive wave front

$$h_3 - h_2 = \frac{c}{g}(v_3 - v_2).$$ (4.102)

Similarly, for the negative wave front

$$h_3 - h_1 = -\frac{c}{g}(v_3 - v_1).$$ (4.103)

If expression (4.102) and (4.96) are added together, after arranging it is obtained

$$h_3 - h_0 = \frac{c}{g}(v_3 - v_0) - 2\frac{c}{g}(v_2 - v_0).$$ (4.104)

Adding the expressions (4.103) and (4.95) together by their arranging, the following is obtained

$$h_3 - h_0 = -\frac{c}{g}(v_3 - v_0) + 2\frac{c}{g}(v_1 - v_0)$$ (4.105)

and when the obtained equation is deducted from Eq. (4.104), it is written

$$\frac{c}{g}(v_3 - v_0) = \frac{c}{g}(v_1 - v_0) + \frac{c}{g}(v_2 - v_0).$$ (4.106)

The obtained expression for the velocity change can be written as

$$\frac{c}{g}(v_3 - v_0) = F_1^+ - F_2^+.$$ (4.107)

Therefore, the resulting state of velocities is also defined by the wave functions of the positive wave and written as

$$\frac{c}{g}(v - v_0) = F^+(\zeta) - F^-(\eta). \tag{4.108}$$

4.11.2   General solution

The obtained result is a general solution for water hammers in pipes. It is, in general, Rieman's solution of linearized partial differential equations of the water hammer, as will be shown in Chapter 5. Thus, for water hammers in pipes it is valid that

$$h - h_0 = F^+(\zeta) + F^-(\eta), \tag{4.109}$$

$$\frac{c}{g}(v - v_0) = F^+(\zeta) - F^-(\eta). \tag{4.110}$$

A wave function in the following form can be obtained from the previous two equations

$$F^+(\zeta) = \frac{1}{2}\left[(h - h_0) + \frac{c}{g}(v - v_0)\right], \tag{4.111}$$

$$F^-(\eta) = \frac{1}{2}\left[(h - h_0) - \frac{c}{g}(v - v_0)\right]. \tag{4.112}$$

Characteristic lines that describe the wave trajectory are defined by the expressions

$$\zeta = t - \frac{x}{c}, \tag{4.113}$$

$$\eta = t + \frac{x}{c}. \tag{4.114}$$

Wave functions that are constant on lines are defined by constant values $\zeta = const$ and $\eta = const$

$$F^+(\zeta) = const, \ F^-(\eta) = const. \tag{4.115}$$

Reference

Rouse, H. (1969) Hydraulics, *Tehnička Hidraulika*, (a translation into Serbian). Građevinska knjiga, Beograd.

Further reading

Abbot, M.B. (1970) *Computational Hydraulics – Elements of the Theory of Surface Flow*. Pitman.
Agroskin, I.I., Dmitrijev, G.T., and F.I. Pikalov (1969) *Hidraulika*. Tehnička knjiga, Zagreb.
Allievi, L. (1925) *Theory of Water Hammer.* translated by E.E. Halmos, ASME, New York.
Angus, R.W. (1937) Water hammer in pipes, including those supplied by centrifugal pumps: graphical treatment. *Proceedings Institute of Mechanical Engineers*, 136: 245.
Angus, R.W. (1939) Water-hammer pressures in compound and branched pipes. *Transactions A.S.C.E.*, 104: 340.

Bergeron, L. (1935) Etude des variations de régime dans les conduites d'eau: solution graphique générale. *Revue générale de l'hydraulique*, 1: 12.

Bogomolov, A.I. and Mihajlov, K.A. (1972) *Gidravlika*, Stroiizdat, Moskva.

Budak, B.M., Samarskii, A.A., and Tikhonov, A.N. (1980) *Collection of Problems on Mathematical Physics* [in Russian]. Nauka, Moscow.

Cunge, J.A., Holly, F.M., and Verwey, A. (1980) *Practical Aspects of Computational River Hydraulics*. Pitman Advanced Publishing Program, Boston.

Davis, C.V. and Sorenson, K.E. (1969) *Handbook of Applied Hydraulics*. 3th edn, McGraw-Hill Co., New York.

Dracos, Th. (1970) Die Berechnung istatationärer Abfüsse in offenen Gerinnen beliebiger Geometrie, *Schweizerische Bauzeitung*, 88. Jahrgang Heft 19.

Fox, J.A. (1977) *Hydraulic Analysis of Unsteady Flow in Pipe Networks*. Macmillan Press Ltd, London, UK; Wiley, New York, USA.

Godunov, S.K. (1971) *Equations of Mathematical Physics* (in Russian Uravnjenija matematičjeskoj fizici). Izdateljstvo Nauka, Moskva.

Irons, B.M. (1970) A frontal solution program, *Int. J. Num. Meth.* 2: 5–32.

Jaeger, Ch. (1949) *Technische Hydraulik*. Verlag, Basel.

Jeffrey, A. (1976) *Quasilinear Hyperbolic System and Waves*. Pitman, Boston.

Johnson, R.D. (1915) The differential surge tank. *Transactions A.S.C.E.*, 78.

Joukowsky, N., (1904) Water hammer (translated by O. Simin), *Proceedings American Water Works Association*. 24.

Jović, V. (1987) Modelling of non-steady flow in pipe networks, *Proc. 2nd Int. Conf. NUMETA '87*. Martinus Nijhoff Pub, Swansea.

Jović, V. (2006) *Fundamentals of Hydromechanics* (in Croatian: Osnove hidromehanike). Element, Zagreb.

Jović, V. (1995) Finite elements and the method of characteristics applied to water hammer modelling. *Engineering Modelling*, 8: 51–58.

Polyanin, A.D. (2002) *Handbook of Linear Partial Differential Equations for Engineers and Scientists*. Chapman & Hall/CRC, London.

Rouse, H. (1946) *Elementary Mechanics of Fluids*. John Wiley Sons, London.

Rouse, H. (1961) *Fluid Mechanics for Hydraulic Engineers*. Dover Pub. Inc, New York.

Smirnov, D.N., Zubob, L.B. (1975) *Hammer Water in Pressure Pipelines*. Stroiizdat, Moskva (Gidravličjeskij udar v napornjih vodovodah in Russian).

Streeter, V.L., Wylie, E.B. (1967) *Hydraulic Transients*. McGraw Hill Book Co., New York, London, Sydney.

Streeter, V.L., Wylie, E.B. (1993) *Fluid Transients*. FEB Press, Ann Arbor, Mich.

Watters, G.Z. (1984) *Analysis and Control of Unsteady Flow in Pipe Networks*. Butterworths, Boston.

Wylie, E.B., and Streeter, V.L. (1993) *Fluid Transients in Systems*. Prentice Hall, Englewood Cliffs, New Jersey, USA.

5

Equations of Non-steady Flow in Pipes

5.1   Equation of state

A liquid equation of state is defined by a general p, V, T surface of matter

$$F(p, V, T) = 0, \tag{5.1}$$

where p is the absolute pressure, V is the volume, and T is the absolute temperature of a liquid. The p, V, T surface is complex, even for the simplest substance. Figure 5.1a shows a p, V, T surface for a simple substance that expands on freezing while Figure 5.1b shows a substance that shrinks when frozen. On the p, V, T surface there is critical point T_c and a triple line where the solid, liquid, and gas phase are in equilibrium. Water is a substance characterized by a series of anomalies; thus, its p, V, T surface is extremely complex.

At temperatures above the critical one, gas cannot transform into liquid no matter how much pressure is applied. The critical temperature T_c of a substance is the maximum temperature at which the substance can exist in a liquid phase.

For practical purposes, phase projections of the equation of state are used.

5.1.1   p,T *phase diagram*

Figure 5.2 shows a p, T phase diagram of a substance (water). In this projection, characteristic phase transitions can be observed. A curve drawn from the origin to the triple point T_{tr}, (which is a projection of a the *triple line*), is a set of points where solid and gas (vapor) coexist in thermodynamic equilibrium. A curve drawn from the triple point to the critical point T_c shows the conditions for equilibrium between the liquid (water) and gas (vapor). A line left of the vaporization curve is the *melting curve*, where solid (ice) and liquid (water) are in equilibrium.

Unlike the melting curve, which has no end, the condensation curve ends at the critical point. The thermodynamic coordinates of the critical point are the critical temperature T_c, the critical pressure p_c, and the critical specific volume v_{sc}, that is the molar volume v_{mc}. Above the critical temperature, liquid, that is drops of liquid, cannot be formed by isothermal compression of gas, that is there is no phase boundary between the liquid and the gaseous phase. If pressure is increased, in the region $p > p_c$ and

---

Analysis and Modelling of Non-Steady Flow in Pipe and Channel Networks, First Edition. Vinko Jović.
© 2013 John Wiley & Sons, Ltd. Published 2013 by John Wiley & Sons, Ltd.

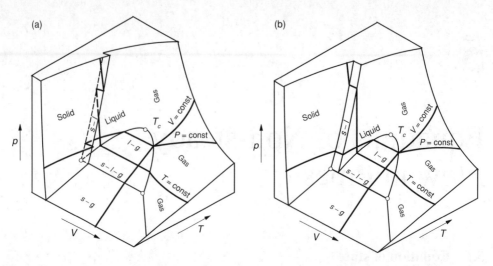

Figure 5.1 Spatial presentation of the equation of state.

$T > T_c$ there is a continuous transition from thin to very dense gases, which, by their properties, do not differ much from a liquid.

At the critical point, the volume of a substance in the gaseous and liquid phase is the same. In the triple point all three phases are in equilibrium. For water, equilibrium of all three phases is achieved at $p_t = 6.112$ mbar and $T_t = 273.16$ K (0.01 °C).

5.1.2 p,V *phase diagram*

A typical p, V phase diagram for water is shown in Figure 5.3. Isotherms passing through the equilibrium region between the liquid and vapor have a very interesting shape. If an isothermal compression at a prescribed temperature is observed, starting from a vapor, then volume is decreasing and pressure is

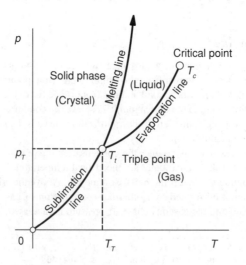

Figure 5.2 Phase diagram, p,T projection.

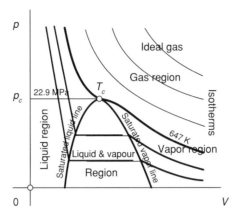

Figure 5.3 Phase p, V diagram for water $T_c = 374.15\,°C$, $p_c = 647.30\,$Mpa, $\rho = 315.46\,$kg/m^3.

increasing, until an equilibrium area of liquid and vapor is achieved. There, pressure remains constant, although the volume is decreasing, and the vapor starts to condense into liquid.

That pressure is called the saturated vapor pressure for a respective temperature. When the entire gas transforms into liquid (*left edge of the boundary region of liquid and vapor*), by further compression of the liquid the volume decreases again, although significantly less since liquid is much less compressible than gas.

In stages where the liquid and vapor are in equilibrium, the vapor is called *saturated vapor* while the liquid is called a *saturated liquid*.

The vapor region is approximated by an ideal gas in the form

$$\frac{pV}{T} = \frac{p_0 V_0}{T_0} = const, \tag{5.2}$$

where V is the volume of gas and T is the absolute temperature. A constant is dependent on mass and gas, and is called *the individual gas constant*

$$R_p = \frac{p_0 V_0}{T_0}, \tag{5.3}$$

where V_0 is the volume of gas at the temperature $T_0 = 273.15\,$K $(0\,°C)$ and atmospheric pressure $p_0 = 101325\,$Pa. An equation for pressure can be written from Eq. (5.2), in which the pressure depends upon R_p and the specific volume, that is density, in the form

$$p = \rho R_p T. \tag{5.4}$$

Since the equation of state for gas includes the temperature, gas behavior depends on the type of thermodynamic process. Otherwise, the liquid's temperature is omitted from the hydraulic calculations.

In order to determine the state of the gas, a polytrophic equation is introduced. A polytrophic process is a thermodynamic process which occurs with an interchange of both heat δQ and work δW between the system and its surroundings. The polytrophic thermodynamic process for a gas can be expressed by an equation of state in the form of a hyperbole

$$p \cdot V^n = const, \tag{5.5}$$

that is

$$\frac{p}{\rho^n} = \frac{p_0}{\rho_0^n} = const, \tag{5.6}$$

where n is the polytrophic index.

Note that all thermodynamic processes may be expressed as polytrophic with the respective value of the polytrophic index

- *isochoric process (V = const)* $\to n = \infty$,
- *isobaric process (p = const)* $\to n = 0$,
- *isothermal process (T = const)* $\to n = 1$,
- *adiabatic process ($\delta Q = 0$)* $\to n = \kappa$.

Of particular technical importance are the polytrophic indexes within the range $1 < n < \kappa$. *A region of water in the form of liquid and vapor* is situated on the right from the sublimation and ice melting point shown on the *p–T* projection. An isotherm for water as the liquid can be obtained in that area using the compressibility data. The compressibility of a liquid (water) is described by the compressibility modulus (reciprocal value of the bulk modulus) E_V

$$dp = -\varepsilon_V E_V = -E_V \frac{dv}{V}, \tag{5.7}$$

where ε_V is the volume dilatation. Due to mass conservativity $m = \rho V$ at the level of a particle of the liquid of volume V the following relationship is valid

$$V d\rho + \rho dv = 0, \tag{5.8}$$

from which

$$\varepsilon_V = -\frac{d\rho}{\rho}. \tag{5.9}$$

By introducing Eq. (5.9) into Eq. (5.7), a differential form of the equation of the state of an elastic liquid is obtained

$$d\rho = \rho \frac{dp}{E_V}. \tag{5.10}$$

For the pressure above the saturated vapor pressure, water is in the liquid phase and it can be integrated from the density of saturated water ρ_v and pressure p_v, that is $\rho > \rho_v$ and $p > p_v$

$$\int_{\rho_v}^{\rho} \frac{d\rho}{\rho} = \int_{p_v}^{p} \frac{dp}{E_V}. \tag{5.11}$$

The lower integration boundary for a prescribed isotherm $T = const$ is obtained from expressions (5.22) and (5.23). The density dependence on pressure is obtained by calculation of the integral

$$\rho = \rho_v e^{\frac{p-p_v}{E_V}}. \tag{5.12}$$

By definition, the speed of sound is equal to

$$c^2 = \frac{dp}{d\rho} = \sqrt{\frac{E_V}{\rho}}. \tag{5.13}$$

In the flow of an elastic liquid (water), changes of density and the speed of sound, due to usual pressure changes, are negligible (*within the pressure range from 1 to 100 bar changes are smaller than 0.5%*); thus, due to $\rho = cons = \rho_0$, the following can be used

$$c_0 = \sqrt{\frac{E_V}{\rho_0}} = const. \tag{5.14}$$

At the saturated vapor pressure there is a step-like transition from the saturated water density to the saturated vapor density. The thermodynamic behavior of vapor can be described by a polytrophe in the form

$$\frac{p}{\rho^n} = \frac{p_v}{\rho_{sv}^n} = const, \tag{5.15}$$

where ρ_{sv} is the saturated vapor density and p_v is the saturated vapor pressure, which are calculated from expressions (5.22) and (5.24). For a long term pressure state below the saturated vapor pressure $p < p_v$ an isothermal process ($n = 1$) can be expected, while a short-term state is closer to an adiabatic process ($n = 1.4$). Thus, for $p < p_v$

$$\rho = \rho_{sv} \left(\frac{p}{p_v} \right)^{1/n}. \tag{5.16}$$

If Eq. (5.15) is differentiated, then

$$\frac{dp}{d\rho} = n \frac{p}{\rho} \tag{5.17}$$

from which the speed of sound in the vapor region will be

$$c = \sqrt{\frac{dp}{d\rho}} = \sqrt{n \frac{p}{\rho}}. \tag{5.18}$$

Thus, the state of water as a fluid is presented by isotherms in the respective phase projection. Reconstruction of an isotherm for a prescribed temperature can be done according to polynomial approximations of physical properties of water.[1]

Water density at atmospheric pressure:

$$0 \leq T°C \leq 4 \quad \rho[\text{kg/m}^3]$$
$$\rho = 1000 - 0.00735675 \cdot (T - 4)^2 + 0.00138764 \cdot (T - 4), \tag{5.19}$$

$$4 \leq T°C \leq 100$$
$$\rho = 1000 + 1.5573 \cdot 10^{-5} \cdot (T - 4)^3 - $$
$$-0.0057198 \cdot (T - 4)^2 - 0.027281 \cdot (T - 4). \tag{5.20}$$

---

[1] An approximation by V. Jović based on different data tables.

Bulk modulus of elasticity at atmospheric pressure:

$$E_V = 2.0307 + 0.013556 \cdot T + 1.7302 \cdot 10^{-4} \cdot T^2 + 4.7285 \cdot 10^{-7} \cdot T^3$$
$$0 \le T[^{\circ}C] \le 100 \quad E_V[\text{GPa}].$$

(5.21)

Saturated vapor pressure:

$$h_v = \frac{p_v}{\rho_0 g} =$$
$$8.9770 \cdot 10^{-8} T^4 - 2.5105 \cdot 10^{-6} T^3 + 3.6507 \cdot 10^{-4} T^2 +$$
$$+ 1.3975 \cdot 10^{-3} T + 0.064841.$$
$$0 \le T[^{\circ}C] \le 100 \quad h_v[\text{m v.s.}]$$

(5.22)

Saturated water density:

$$0 \le T^{\circ}C \le 250 \quad \rho[\text{kg/m}^3]$$
$$\rho = 1000 - 0.0025 \cdot (T-4)^2 - 0.1914 \cdot (T-4).$$

(5.23)

Saturated vapor density:

$$0 \le T^{\circ}C \le 100 \quad \rho[\text{kg/m}^3]$$
$$\rho = 4.24052 \cdot 10^{-9} T^4 - 1.38234 \cdot 10^{-8} T^3 + 1.51691 \cdot 10^{-5} T^2 +$$
$$+ 3.09045 \cdot 10^{-4} T + 4.79846 \cdot 10^{-3}.$$

(5.24)

Kinematic viscosity – Poiseuille's formula:

$$\nu = \frac{0.0178}{1 + 0.0337 \cdot T + 0.000221 \cdot T^2}$$
$$T[^{\circ}C] \quad \nu[\text{cm}^2/\text{s}].$$

(5.25)

The calculation of physical properties of water for a prescribed temperature according to polynomial approximations (5.19) to (5.25) is implemented in the module `Fluid.F90`.

Figure 5.4 shows the water isotherm at $T = 15\,^{\circ}C$, reconstructed using the previously described principles and formulas from Eqs (5.19) to (5.24).

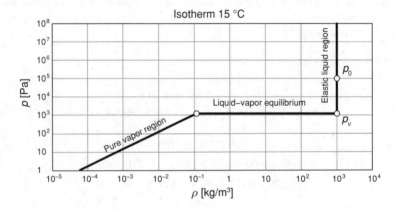

Figure 5.4   Isothermal state of water.

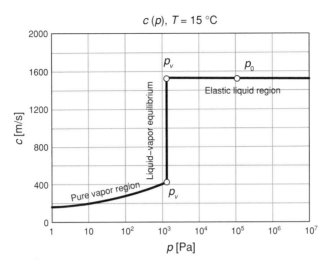

Figure 5.5 Dependence of the speed of sound on pressure in water at 15 °C.

Atmospheric pressure and saturated vapor pressure are marked in the diagram. Three characteristic regions on the isotherm are also marked:

(a) the elastic liquid region,
(b) the equilibrium region between the liquid and vapor, where pressure is constant and,
(c) the pure vapor region.

Figure 5.5 shows the dependence of the speed of sound in water at 15 °C temperature at absolute pressure. The following can be observed at the curve for the speed of sound:

(a) in the region above the saturated vapor pressure (elastic liquid), the speed of sound is almost constant,
(b) at the saturated vapor pressure, the speed is changing from the values corresponding to vapor to the speed corresponding to liquid,
(c) below the saturated vapor pressure, the speed of sound corresponds to the velocities in gas.

5.2 Flow of an ideal fluid in a streamtube

5.2.1 Flow kinematics along a streamtube

Mass of the fluid particle

Fluid flow along a streamtube is observed as the motion of a series of fluid particles of constant mass as a *logical extension of the mechanics of material points*. It is the *Lagrangean*[2] approach to the description of fluid flow. In the following text, all differential values related to the motion of fluid particles will be marked by notation δ. The mass of a particle has a differentially small value δM. It is defined as the mass

---

[2] J.L. Lagrange, Italian mathematician and astronomer (1736–1813).

flowing through the streamtube cross-section of a finitely small cross-section area A at uniform velocity v in a differentially small time increment δt

$$\delta M = \rho \delta V = \rho A \delta l = \rho A v \delta t = \rho Q \delta t, \tag{5.26}$$

where ρ is the fluid density, δV is the differentially small particle volume, δl is differentially small particle length, and Q is the volumetric flow rate. Mass flow rate in the streamtube will be

$$\dot{M} = \frac{\delta M}{\delta t} = \rho \frac{\delta V}{\delta t} = \rho A \frac{\delta l}{\delta t} = \rho A v = \rho Q \tag{5.27}$$

and remains constant in the entire streamtube

$$\dot{M} = \rho Q = const. \tag{5.28}$$

The volumetric flow rate is the volume of fluid δV which passes through a given surface in time δt

$$Q = \dot{V} = \frac{\delta V}{\delta t}. \tag{5.29}$$

Equation of continuity

Because, according to the streamtube definition, there is no flow through the pipe perimeter, the mass flow rate remains unchanged, that is the total change of the mass flow rate is equal to zero as can be expressed in differential form

$$\delta(\dot{M}) = \frac{\partial \dot{M}}{\partial t} \delta t + \frac{\partial \dot{M}}{\partial l} \delta l = 0, \tag{5.30}$$

where the first term in the total differential denotes the change of mass flow rate in time, while the second denotes the change along the stream axis, that is the streamline.

If Eq. (5.27) is applied to the mass flow rate

$$\delta(\dot{M}) = \frac{\partial \rho A \dfrac{\delta l}{\delta t}}{\partial t} \delta t + \frac{\partial \rho Q}{\partial l} \delta l = \frac{\partial \rho A}{\partial t} \delta l + \frac{\partial \rho Q}{\partial l} \delta l = 0, \tag{5.31}$$

from which the equation of continuity for the streamtube is obtained in the form

$$\frac{\delta(\dot{M})}{\delta l} = \frac{\partial \rho A}{\partial t} + \frac{\partial \rho Q}{\partial l} = 0. \tag{5.32}$$

Volumetric changes of a particle

Let us observe a compressible liquid particle of constant mass $\delta M = \rho \delta V$, moving through the stream-tube. Then, the particle changes its shape. A change in the particle's shape is caused by adaptation to the streamtube, when the volume of the particle can change if the liquid is compressible and the pipe expandable.

Figure 5.6 shows the volumetric change of a compressible particle of constant mass in a given position under the impact of internal pressure, where the volume of the particle is equal to

$$\delta V = A \delta l = A v \delta t = Q \delta t. \tag{5.33}$$

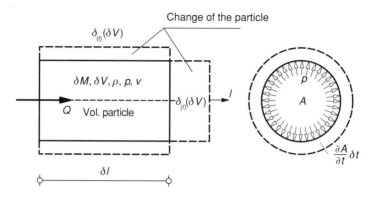

Figure 5.6 Volumetric strain of the fluid particle.

Volumetric change consists of the time-dependent change (expansion of the pipe cross-section), and the change along the flow due to expansion of the particle's length is:

$$\delta\left(\delta V\right) = \underbrace{\frac{\partial \delta V}{\partial t}\delta t}_{\delta_{(t)}\delta V} + \underbrace{\frac{\partial \delta V}{\partial l}\delta l}_{\delta_{(l)}\delta V}. \tag{5.34}$$

If Eq. (5.33) is applied to the volume of a particle, then

$$\delta\left(\delta V\right) = \frac{\partial A}{\partial t}\delta l \delta t + \frac{\partial Q}{\partial l}\delta l \delta t, \tag{5.35}$$

from which the rate of volume change is obtained

$$\delta\left(\frac{\delta V}{\delta t}\right) = \delta\dot{V} = \frac{\partial A}{\partial t}\delta l + \frac{\partial Q}{\partial l}\delta l. \tag{5.36}$$

The rate of volume change per unit of length, namely the relative rate of particle volume change – the rate of volumetric strain will be

$$\dot{\varepsilon}_{V_{\delta M}} = \frac{\delta\dot{V}}{\delta l} = \frac{\partial A}{\partial t} + \frac{\partial Q}{\partial l}. \tag{5.37}$$

Application of the material derivative to the equation of continuity

It is more appropriate, except in specific cases, to observe the fluid flow as a flow-through mechanical system; namely, the flow through the control volume, extracted from the streamtube by two cross-sections, and fixed in space and time. Instead of monitoring the particle motion along the flow, the values at a point in time are observed. This monitoring approach is called the *Euler*[3] approach to fluid flow description. The connection between the Lagrangean and the Euler approach is defined by the material derivative. The material derivative can be defined as the application of the Reynolds transport theorem or use of the differential calculus. The differential calculus will be used here.

---

[3]Leonard Euler, Swiss mathematician and physicist (1707–1783).

Let some variable of the flow along the streamtube be equal to $e(l, t)$; then the total derivative can be applied to it

$$de = \frac{\partial e}{\partial t}dt + \frac{\partial e}{\partial l}dl,$$

that is

$$\frac{de}{dt} = \frac{\partial e}{\partial t} + \frac{dl}{dt}\frac{\partial e}{\partial l}.$$

Since the flow velocity is equal to

$$v = \frac{dl}{dt},$$

a relation between the total derivative and partial derivatives is obtained in the form

$$\frac{de}{dt} = \frac{\partial e}{\partial t} + v\frac{\partial e}{\partial l} \tag{5.38}$$

which is called the material derivative. The first term on the right hand side is called the local derivative. It is the change with respect to time at a certain point. The second term on the right hand side is called the convective derivative.

If, in the equation of continuity (5.32), complex terms are partially derived, then it is written

$$A\left(\frac{\partial \rho}{\partial t} + v\frac{\partial \rho}{\partial l}\right) + \rho\left(\frac{\partial A}{\partial t} + \frac{\partial Q}{\partial l}\right) = 0. \tag{5.39}$$

Since the term in the first parenthesis is the material derivative of density, and the term in the second parenthesis is the rate of volumetric strain, the equation of continuity is obtained in the following form

$$A\frac{d\rho}{dt} + \rho\dot{\varepsilon}_{V_{\delta M}} = 0. \tag{5.40}$$

5.2.2   Flow dynamics along a streamtube

Total energy of a particle

If the flow through the streamtube is observed as the motion of a series of particles of constant mass, then each particle has the total energy

$$E = U + E_p + E_k \tag{5.41}$$

that consists of the internal U, potential E_p, and kinetic energy E_k. A change of total energy in a mechanical-thermodynamic system is equal to the change of work carried out over the system; namely, the law of total energy conservation expressed in the form of the rate of change in the unit of time δt can be applied to every particle

$$\frac{\delta E}{\delta t} = \frac{\delta Q^0}{\delta t} + \frac{\delta W_n}{\delta t}, \tag{5.42}$$

where δE is a differentially small change of total energy, δQ^0 is differentially small heat added to the system, while δW_n is differentially small work of normal (pressure) forces.

The flow of an ideal fluid in a streamtube is a mechanical system with no heat transfer with the environment; thus, the first term on the right hand side of Eq. (5.42) is equal to zero. By introducing Eq. (5.41) into Eq. (5.42), the energy equation of fluid particle motion is written in the following form

$$\frac{\delta U}{\delta t} + \frac{\delta E_p}{\delta t} + \frac{\delta E_k}{\delta t} = \frac{\delta W_n}{\delta t}. \tag{5.43}$$

Potential energy of a particle

In the homogeneous field of the gravity force, the potential energy of a fluid particle of mass δM will be

$$E_p = \delta M \cdot gz = (\rho A \delta l) \, gz \tag{5.44}$$

for which the total derivative will be

$$\delta E_p = \frac{\partial E_p}{\partial t} \delta t + \frac{\partial E_p}{\partial l} \delta l,$$

that is it follows that

$$\frac{\delta E_p}{\delta t} = \underbrace{\frac{\partial E_p}{\partial t}}_{=0} + v \frac{\partial E_p}{\partial l}.$$

Since potential energy is not time-dependent, the first term on the right hand side is equal to zero, and the rate of change of the potential energy of a liquid particle is equal to

$$\frac{\delta E_p}{\delta t} = \rho A g v \frac{\partial z}{\partial l} \delta l = \rho Q g \frac{\partial z}{\partial l} \delta l, \tag{5.45}$$

that is it follows that

$$\frac{\delta E_p}{\delta t} = \dot{M} g \frac{\partial z}{\partial l} \delta l. \tag{5.46}$$

Kinetic energy of a particle

The kinetic energy of a liquid particle of mass δM is equal to

$$E_k = \delta M \frac{v^2}{2} = (\rho A \delta l) \frac{v^2}{2}. \tag{5.47}$$

The rate of change of kinetic energy will be determined from the total derivative

$$\delta E_k = \frac{\partial E_k}{\partial t} \delta t + \frac{\partial E_k}{\partial l} \delta l,$$

that is it follows that

$$\frac{\delta E_k}{\delta t} = \frac{\partial E_k}{\partial t} + v\frac{\partial E_k}{\partial l}. \tag{5.48}$$

By introducing Eq. (5.47) into Eq. (5.48), it is written as

$$\frac{\delta E_k}{\delta t} = \frac{\partial(\delta M \frac{v^2}{2})}{\partial t} + v\frac{\partial\left(\delta M \frac{v^2}{2}\right)}{\partial l}.$$

Since the particle mass is constant, then

$$\frac{\delta E_k}{\delta t} = \delta M \left(\frac{\partial \frac{v^2}{2}}{\partial t} + v\frac{\partial \frac{v^2}{2}}{\partial l}\right) = \delta M \left(v\frac{\partial v}{\partial t} + v^2\frac{\partial v}{\partial l}\right) = \delta M v \left(\frac{\partial v}{\partial t} + v\frac{\partial v}{\partial l}\right).$$

If we introduce the expression for the particle mass from Eq. (5.26)

$$\frac{\delta E_k}{\delta t} = \rho Q \delta t v \left(\frac{\partial v}{\partial t} + v\frac{\partial v}{\partial l}\right),$$

the rate of change of the kinetic energy of a particle is obtained in the form

$$\frac{\delta E_k}{\delta t} = \rho Q \left(\frac{\partial v}{\partial t} + v\frac{\partial v}{\partial l}\right)\delta l, \tag{5.49}$$

that is, when a material derivative for the velocity v is used, then

$$\frac{\delta E_k}{\delta t} = \rho Q \frac{dv}{dt}\delta l = \dot{M}\frac{dv}{dt}\delta l. \tag{5.50}$$

Internal energy of a particle

The change of the internal energy of a particle, due to volumetric expansion, is a reversible thermodynamic process. The decrease of internal energy is equal to the work spent on the particle expansion and vice versa, that is the increase in internal energy is equal to the work spent on the compression of a particle. Thus, for the particle observed as a thermodynamic system, the following can be applied

$$\delta U = -p\delta(\delta V), \tag{5.51}$$

where $\delta(\delta V)$ is the differential change of a liquid particle volume. When the expression (5.35) is used, then

$$\delta U = -p\delta(\delta V) = -p\left(\frac{\partial A}{\partial t} + \frac{\partial Q}{\partial l}\right)\delta l\delta t, \tag{5.52}$$

that is the rate of change of internal energy of a particle is equal to

$$\frac{\delta U}{\delta t} = -p\left(\frac{\partial A}{\partial t} + \frac{\partial Q}{\partial l}\right)\delta l = -p\dot{\varepsilon}_{V_{\delta M}}\delta l. \tag{5.53}$$

Note that internal energy decreases at expansion and increases at compression of a particle.

Work of normal (pressure) forces of a particle

Fluid particles moving through a streamtube have the imaginary form of an elementary cylinder, according to Figure 5.7, with normal (pressure) forces acting on its perimeter.

A change in the work of the pressure forces consists of a change generated by pressure acting on the cylinder perimeter and pipe compression in time δt (work on volume change)

$$-p\delta l \frac{\partial A}{\partial t} \delta t \tag{5.54}$$

and the change generated by particle displacement in time δt over the length $\delta l = v\delta t$. The work performed by pressure forces pA acting on the cylinder's top and bottom sides A is $pAv\delta t = pQ\delta t$. A change in work due to particle displacement in time δt will be equal to

$$\left(pQ - \frac{\partial pQ}{\partial l}\frac{\delta l}{2}\right)\delta t - \left(pQ + \frac{\partial pQ}{\partial l}\frac{\delta l}{2}\right)\delta t = -\frac{\partial pQ}{\partial l}\delta l. \tag{5.55}$$

Thus, the total change of normal (pressure) forces' work is equal to

$$\delta W_n = -p\frac{\partial A}{\partial t}\delta l\delta t - \frac{\partial pQ}{\partial l}\delta l\delta t, \tag{5.56}$$

that is the rate of change of normal forces' work is equal to

$$\frac{\delta W_n}{\delta t} = -p\frac{\partial A}{\partial t}\delta l - \frac{\partial pQ}{\partial l}\delta l. \tag{5.57}$$

If the partial derivative is applied to the last term, and after grouping of the terms, it can be written as

$$\frac{\delta W_n}{\delta t} = -p\underbrace{\left(\frac{\partial A}{\partial t} + \frac{\partial Q}{\partial l}\right)\delta l}_{\frac{\delta U}{\delta t}} - Q\frac{\partial p}{\partial l}\delta l, \tag{5.58}$$

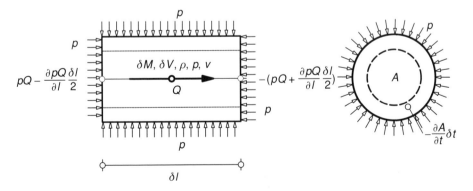

Figure 5.7 Work of normal forces.

where the first term on the right hand side is equal to the rate of change of internal energy. If, in the second term, the volumetric flow rate is expressed as $Q = \dot{M}/\rho$, the following is obtained

$$\frac{\delta W_n}{\delta t} = \frac{\delta U}{\delta t} - Q \frac{\partial p}{\partial l} \delta l = \frac{\delta U}{\delta t} - \dot{M} \frac{1}{\rho} \frac{\partial p}{\partial l}. \tag{5.59}$$

Note that the work of the normal forces is spent on the increase of internal energy and in overcoming the pressure gradient along the flow, that is the change of pressure energy.

Dynamic equation for a streamtube

If the expressions for the rate of change of the potential energy (5.46), kinetic energy (5.50), and the work of internal forces (5.59) are introduced into energy equation (5.43), it can be written as

$$\frac{\delta U}{\delta t} + \dot{M}g \frac{\partial z}{\partial l} \delta l + \dot{M} \frac{dv}{dt} \delta l = \frac{\delta U}{\delta t} - \dot{M} \frac{1}{\rho} \frac{\partial p}{\partial l}, \tag{5.60}$$

that is following arrangement

$$\dot{M} \left(\frac{dv}{dt} + g \frac{\partial z}{\partial l} + \frac{1}{\rho} \frac{\partial p}{\partial l} \right) = 0. \tag{5.61}$$

The obtained equation is the dynamic equation of motion of an ideal compressible fluid particle in the streamtube. If the total acceleration term (the first term in parenthesis) is expressed by the material derivative of velocity v, then

$$\dot{M} \left(\frac{\partial v}{\partial t} + v \frac{\partial v}{\partial l} + g \frac{\partial z}{\partial l} + \frac{1}{\rho} \frac{\partial p}{\partial l} \right) = 0. \tag{5.62}$$

If Eq. (5.62) is abridged by the mass flow rate $\dot{M}$ a dynamic equation of the non-steady flow of an ideal compressible fluid along the streamtube is obtained

$$\frac{\partial v}{\partial t} + v \frac{\partial v}{\partial l} + g \frac{\partial z}{\partial l} + \frac{1}{\rho} \frac{\partial p}{\partial l} = 0. \tag{5.63}$$

5.3   The real flow velocity profile

5.3.1   Reynolds number, flow regimes

Flow of a real fluid can be *laminar*, *transitional*, or *turbulent*, depending on the Reynolds number – a dimensionless criteria to determine the flow regime. For the flow of a real fluid in circular pipes, the Reynolds number is

$$R_e = \frac{vD}{\nu}, \tag{5.64}$$

where $v = Q/A$ [m/s] is the mean flow velocity, D [m] is the pipe diameter, and ν [m^2/s] is the coefficient of kinematic viscosity. The critical Reynolds number is $R_e = 2320$. Below that number, the flow is laminar. Laminar flow can exist even for larger Reynolds numbers than the critical one; however, it is highly unstable. Even the smallest disturbance to the upstream flow will transform the flow into a

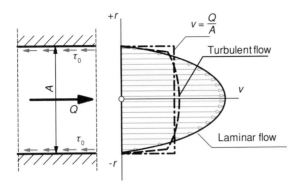

Figure 5.8   Developed velocity profile.

turbulent one. A narrow region above the critical Reynolds number is unstable and is called the transient region, above which a turbulent flow is developed.

Depending on the relative roughness of the pipe, turbulent flow can be divided into:

- turbulent rough flow, where resistances depend only on relative roughness,
- turbulent smooth flow,[4] where resistances depend only on the Reynolds number,
- turbulent transient flow, where resistances depend both on relative roughness and the Reynolds number.

5.3.2   Velocity profile in the developed boundary layer

The flow regime in pipes depends on the developed boundary layer. The boundary layer develops from the beginning of the pipe and increases in thickness along the flow. At a certain length, measured from the beginning of the pipe – when the boundary layer reaches the pipe axis, that is it connects on both sides – further flow is characterized by a fully developed boundary layer and the respective developed velocity profile. That length is called the transitional section of boundary layer development. At the beginning of the pipeline, the laminar boundary layer is developed first, which, depending on the Reynolds number, transfers to the transient and turbulent boundary layer. The flow regime in a pipe with a developed boundary layer is determined by the developed boundary layer type, that is the type of the boundary layer at the connection point.

The length of the transitional section depends on the flow type and the intensity of turbulence in the incoming stream and moves, according to the measurements, from about 40 to 300 pipe diameters. Since it is relatively short in comparison to the pipeline length, the transitional section is disregarded in engineering calculations. Figure 5.8 shows the comparison between the developed laminar and turbulent velocity profile for the same mean flow velocity, which represents the velocity profile of an ideal undisturbed flow.

The developed laminar velocity profile is a rotational paraboloid, defined by the Hagen–Poisseuille law

$$v = J\frac{\rho g}{4\mu}\left(R_0^2 - r^2\right), \tag{5.65}$$

---

[4]One should bear in mind that physically rough conduits may behave like smooth ones if there is a thick viscous boundary sublayer.

where J is the piezometric head line gradient, equal to the gradient of the energy line, R_0 is the pipe radius, ρ is the fluid density, and μ is the coefficient of dynamic viscosity. According to the *Hagen–Poisseuille law*, the gradient of the piezometric head line is equal to

$$J = -\frac{dh}{dl} = \frac{8\mu}{\rho g R_0^2} \bar{v}. \tag{5.66}$$

After Eq. (5.66) is introduced into Eq. (5.65), a dimensionless expression for the laminar velocity profile is obtained in the form

$$\frac{v}{\bar{v}} = 2\left[1 - \left(\frac{r}{R_0}\right)^2\right]. \tag{5.67}$$

With the development of turbulence the velocity profile becomes more and more smoothed in comparison to the laminar one, due to the transversal transfer of the mass and momentum by turbulent vortices. Due to the complexity of the turbulent boundary layer and the influence of roughness, it is not possible to establish an analytical expression for the velocity profile; thus numerous approximations are used such as, for example logarithmic, the "one-seventh" etc. Apart from this, note that there are turbulent fluctuations around the mean values, which additionally increase the complexity of the velocity profile analyses.

For the purposes of necessary further calculations, a simple approximation of the developed velocity profile will be proposed here

$$\frac{v}{\bar{v}} = \frac{n+2}{n}\left(1 - \left|\frac{r}{R_0}\right|^n\right) \tag{5.68}$$

which approximates the laminar flow by setting the value $n = 2$, an ideal flow by $n = \infty$, and the turbulent flow by $2 < n < \infty$.

5.3.3   Calculations at the cross-section

Mass flow

$$Q_m = \rho \int_A v dA = \rho A \bar{v} = \rho Q, \tag{5.69}$$

where

$$\bar{v} = \frac{\int_A v dA}{A} = \frac{Q}{A}. \tag{5.70}$$

Momentum flow: Boussinesq coefficient

$$Q_K = \int_A \rho v dQ = \rho \int_A v^2 dA = \rho \beta \bar{v}^2 A = \rho \beta Q \bar{v} = \beta \dot{M} \bar{v}, \tag{5.71}$$

where

$$\beta = \frac{\int\limits_A v^2 \, dA}{\bar{v}^2 A}. \tag{5.72}$$

A correction number β is called the Boussinesq[5] coefficient. It reflects the variability of the momentum flow across the cross-section. If a velocity profile is approximated by expression (5.68) the following is obtained

$$\beta = \frac{n+2}{n+1}. \tag{5.73}$$

Thus, for example, for an ideal flow the Boussinesq coefficient number is one, for a laminar flow it is $4/3$, while for turbulent flow with $n = 20$ it is equal to $24/21 \approx 1.043$.

Kinetic energy flow: Coriolis coefficient

$$Q_{ke} = \int\limits_A \rho v \frac{v^2}{2} \, dA = \frac{1}{2} \rho \alpha \bar{v}^3 A = \rho \alpha \frac{\bar{v}^2}{2} Q = \alpha \dot{M} \frac{\bar{v}^2}{2}, \tag{5.74}$$

where

$$\alpha = \frac{\int\limits_A v^3 \, dA}{\bar{v}^3 A}. \tag{5.75}$$

α is called the Coriolis[6] coefficient and it used for correction of non-uniformity of the kinetic energy flow across the cross-section. If a velocity profile is approximated by expression (5.68) the following is obtained

$$\alpha = \frac{3(n+2)^2}{(n+1)(3n+2)}. \tag{5.76}$$

Thus, for example, for an ideal flow the Coriolis coefficient is 1, for a laminar flow it is 2, while for turbulent flow with $n = 20$ it is equal to $242/217 \approx 1.12$.

5.4   Control volume

Equations of non-steady flow in pipes can be derived from the analysis of flow in a finite pipe segment – which is the control volume. The control volume is a pipe segment between two cross-sections perpendicular to the pipe flow axis as shown in Figure 5.9. The flow axis is a centroidal axis of a cross-section, also being the line that connects maximum velocities. It is slightly curved in space. There is a hydrostatic equilibrium normal to the flow; thus the compressive force at the cross-section can be calculated from pressure at the cross-section centriod, that is pressures along the flow axis.

---

[5] Joseph Valentin Boussinesque, French mathematician and physicist (1842–1929).
[6] Gaspard Gustave Coriolis, French scientist (1792–1843).

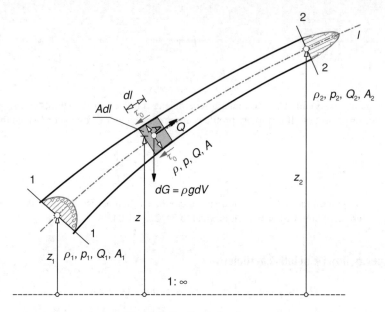

Figure 5.9 Control volume.

Unlike an ideal fluid flow in a streamtube, velocity distribution in a cross-section is not uniform. It is assumed that in all cross-sections there is a velocity profile the same as in the case of a developed steady boundary layer.

The control volume is fixed in space and time and can be observed as a mechanical-thermodynamic flow system.

The required equations of non-steady flow in pipes will be obtained by application of the mass and energy conservation laws on the control volume.

5.5 Mass conservation, equation of continuity

5.5.1 Integral form

The rate of change of a mass contained in a control volume + net flow of mass inflow in a unit of time through boundary cross-sections is equal to zero

$$\frac{dM}{dt} = \frac{\partial}{\partial t} \int_{l_1}^{l_2} \rho A dl + (\rho Q)_2 - (\rho Q)_1 = 0. \tag{5.77}$$

that is the mass flow $\dot{M}_a$ of accumulation in a control volume + net flow of mass inflow into control volume $(\dot{M}_2 - \dot{M}_1)$ is equal to zero

$$\frac{dM}{dt} = \dot{M}_a + \dot{M}_2 - \dot{M}_1 = 0. \tag{5.78}$$

5.5.2 Differential form

The last two members in Eq. (5.77) can be written as an integral of a partial gradient while a partial derivative in time can be put under the integral, as integration boundaries are not time-dependent, so we have

$$\int_{l_1}^{l_2} \frac{\partial \rho A}{\partial t} dl + \int_{l_1}^{l_2} \frac{\partial \rho Q}{\partial l} dl = 0, \tag{5.79}$$

that is the following is valid

$$\int_{l_1}^{l_2} \left(\frac{\partial \rho A}{\partial t} + \frac{\partial \rho Q}{\partial l} \right) dl = 0. \tag{5.80}$$

Since the obtained definite integral is equal to zero for any of the integration boundaries, the integrand function vanishes. Thus, a differential form of the conservation law is obtained

$$\frac{\partial \rho A}{\partial t} + \frac{\partial \rho Q}{\partial l} = 0. \tag{5.81}$$

The obtained equation is the equation of the mass continuity for pipe flow, which is completely identical to the previously obtained Eq. (5.32) for a streamtube or, when dismembered,

$$A \left(\frac{\partial \rho}{\partial t} + v \frac{\partial \rho}{\partial l} \right) + \rho \left(\frac{\partial A}{\partial t} + \frac{\partial Q}{\partial l} \right) = 0. \tag{5.82}$$

Since the expression in the first parentheses is the material derivation of density, and the expression in the second one is equal to the rate of volume change, the continuity equation is obtained in the form

$$A \frac{d\rho}{dt} + \rho \dot{\varepsilon}_{V_{\delta M}} = 0. \tag{5.83}$$

5.5.3 Elastic liquid

Depending on the problem being analyzed, different forms of the continuity equation can be used. For example, if a non-steady flow of a liquid is observed where, due to great pressure changes, density change cannot be disregarded, that liquid can be considered elastic. Density changes along the flow can be disregarded; thus, the equation of continuity will be

$$\frac{\partial \rho A}{\partial t} + \rho_0 \frac{\partial Q}{\partial l} = 0 \tag{5.84}$$

where the fist term depends on pressure changes, that is the mass flow of accumulation depends on pressure. In this case, the chain rule in the environment of atmospheric pressure p_0 and density ρ_0, can be applied to the first term as

$$\frac{\partial \rho A}{\partial t} = \frac{d(\rho A)}{dp} \bigg|_0 \frac{\partial p}{\partial t}, \tag{5.85}$$

where the mark "$|_0$" emphasizes the state in atmospheric pressure conditions. If a complex term is dismembered

$$\left.\frac{d\left(\rho A\right)}{dp}\right|_0 = A_0\left.\frac{d\rho}{dp}\right|_0 + \rho_0\frac{dA}{dp},$$ (5.86)

where A_0 is the pipe cross-section in atmospheric pressure conditions. From the equation of state for an elastic liquid (5.10) written as

$$\left.\frac{d\rho}{dp}\right|_0 = \frac{\rho_0}{E_V}$$ (5.87)

and from the elastic properties of a circular pipe

$$\frac{dA}{A_0} = 2\frac{dD}{D} = \frac{D}{s\,E_c}dp$$ (5.88)

the following is obtained

$$\frac{dA}{dp} = A_0\frac{D}{sE_c}.$$ (5.89)

When expressions (5.87) and (5.89) are introduced into Eq. (5.86),

$$\left.\frac{d\left(\rho A\right)}{dp}\right|_0 = A_0\left(\frac{\rho_0}{E_v} + \frac{\rho_0 D}{s E_c}\right) = A_0\left.\left(\frac{1}{c_v^2} + \frac{1}{c_c^2}\right)\right|_0 = \frac{A_0}{c_0^2}$$ (5.90)

is obtained, where c_0, the water hammer celerity, is constant, that is it does not depend on pressure. Expression (5.85) becomes equal to

$$\frac{\partial\rho A}{\partial t} = \frac{A_0}{c_0^2}\frac{\partial p}{\partial t}$$ (5.91)

while the equation of continuity for the elastic liquid and elastic pipeline, expressed by variables p, Q, becomes

$$\frac{A_0}{\rho_0 c_0^2}\frac{\partial p}{\partial t} + \frac{\partial Q}{\partial l} = 0.$$ (5.92)

For a liquid with small density changes it can be written $p = \rho_0 g\left(h - z\right)$; then, the equation of continuity for the flow of the elastic liquid in elastic pipeline, after density is annulled, expressed by variables h, Q, will be

$$\frac{gA_0}{c_0^2}\frac{\partial h}{\partial t} + \frac{\partial Q}{\partial l} = 0.$$ (5.93)

The equation of continuity (5.150) can be integrated between two points on the flow axis

$$g\frac{A_0}{c_0^2}\int_{l_1}^{l_2}\frac{\partial h}{\partial t}dl + Q_2 - Q_1 = 0$$ (5.94)

where the first term denotes the accumulation flux

$$Q_a = \frac{\partial V}{\partial t} = \int_{l_1}^{l_2} g \frac{A_0}{c_0^2} \frac{\partial h}{\partial t} dl. \tag{5.95}$$

5.5.4   Compressible liquid

If a non-steady flow of a compressible fluid is analyzed, a functional relation of density with pressure shall be known. Then, the equation of continuity will be expressed by *primitive variables of pressure and velocity*. If a discharge is expressed as $Q = Av$, the equation of continuity (5.81) assumes the dismembered form

$$\frac{\partial \rho A}{\partial t} + \frac{\partial \rho A v}{\partial l} = A \frac{\partial \rho}{\partial t} + \rho \frac{\partial A}{\partial t} + \rho A \frac{\partial v}{\partial l} + v \frac{\partial \rho A}{\partial l} = 0. \tag{5.96}$$

Since the liquid and pipe are compressible, that is density $\rho(p)$ and the cross-section area $A(p)$ are pressure dependent, when a chain rule is applied the previous equation is written as

$$A \frac{d\rho}{dp} \frac{\partial p}{\partial t} + \rho \frac{dA}{dp} \frac{\partial p}{\partial t} + \rho A \frac{\partial v}{\partial l} + v\rho \frac{dA}{dp} \frac{\partial p}{\partial l} + vA \frac{d\rho}{dp} \frac{\partial p}{\partial l} = 0. \tag{5.97}$$

Following grouping of members in the form

$$\left(A \frac{d\rho}{dp} + \rho \frac{dA}{dp} \right) \frac{\partial p}{\partial t} + \rho A \frac{\partial v}{\partial l} + v \left(A \frac{d\rho}{dp} + \rho \frac{dA}{dp} \right) \frac{\partial p}{\partial l} = 0, \tag{5.98}$$

and extraction from the parentheses, it is written

$$\left(A \frac{d\rho}{dp} + \rho \frac{dA}{dp} \right) \left(\frac{\partial p}{\partial t} + v \frac{\partial p}{\partial l} \right) + \rho A \frac{\partial v}{\partial l} = 0. \tag{5.99}$$

The term in the first parenthesis will be written using the speed of sound in a liquid and pipe as

$$A \frac{d\rho}{dp} + \rho \frac{dA}{dp} = A \left(\frac{\rho}{E_v} + \frac{\rho D}{sE_c} \right) = A \left(\frac{1}{c_v^2} + \frac{1}{c_c^2} \right) = \frac{A}{c^2}.$$

After introduction into Eq. (5.99) and arrangement, the equation of continuity is obtained in the form

$$\frac{1}{\rho c^2} \left(\frac{\partial p}{\partial t} + v \frac{\partial p}{\partial l} \right) + \frac{\partial v}{\partial l} = 0. \tag{5.100}$$

5.6   Energy conservation law, the dynamic equation

5.6.1   Total energy of the control volume

The rate of change of total energy (internal + potential + kinetic energy) of a control volume is equal to the power of surface forces acting on a control volume

$$\frac{dU}{dt} + \frac{dE_p}{dt} + \frac{dE_k}{dt} = \frac{d(W_n + W_o)}{dt}, \tag{5.101}$$

where W_n is the work of normal forces (pressure) and W_o is the work of resistance forces (friction along the pipe mantle).

5.6.2   Rate of change of internal energy

The rate of change of internal energy or the power of internal forces is obtained by integration of the work of internal forces on expansion of a thermodynamic system

$$
\frac{dU}{dt} = -\int_{l_1}^{l_2} p \dot{\varepsilon}_V \, dl = -\int_{l_1}^{l_2} p \left(\frac{\partial A}{\partial t} + \frac{\partial Q}{\partial l} \right) dl,
\tag{5.102}
$$

where $\dot{\varepsilon}_V$ is the volumetric strain per unit of length, defined by Eq. (5.37).

5.6.3   Rate of change of potential energy

The rate of change of potential energy or the power of gravity forces is obtained by integration along the control volume

$$
\frac{dE_p}{dt} = \int_{l_1}^{l_2} \rho g Q \frac{\partial z}{\partial l} dl,
\tag{5.103}
$$

where the rate of change of potential energy per unit of length is defined by Eq. (5.45).

5.6.4   Rate of change of kinetic energy

The rate of change of kinetic energy is obtained by integration of the change of kinetic energy of a control volume, which consists of the rate of change of kinetic energy in a control volume and the difference of the kinetic energy flow through control cross-sections

$$
\frac{dE_k}{dt} = \frac{\partial}{\partial t} \int_{l_1}^{l_2} \int_A \rho \frac{v^2}{2} dA \, dl + \int_{A_2} \rho \frac{v^2}{2} v dA - \int_{A_1} \rho \frac{v^2}{2} v dA.
\tag{5.104}
$$

Integrals per cross-section, according to Eqs (5.71) and (5.74), can be written using the Boussinesq and Coriolis coefficients

$$
\frac{dE_k}{dt} = \frac{\partial}{\partial t} \int_{l_1}^{l_2} \rho \beta \frac{\bar{v}^2}{2} A dl + \left(\rho \alpha \frac{\bar{v}^2}{2} Q \right)_2 - \left(\rho \alpha \frac{\bar{v}^2}{2} Q \right)_1.
\tag{5.105}
$$

The last two members can be written as an integral of a partial gradient

$$
\frac{dE_k}{dt} = \frac{\partial}{\partial t} \int_{l_1}^{l_2} \rho \beta \frac{\bar{v}^2}{2} A dl + \int_{l_1}^{l_2} \frac{\partial}{\partial l} \left(\rho \alpha \frac{\bar{v}^2}{2} Q \right) dl.
\tag{5.106}
$$

The partial derivative in time can be put under the integral, since integration limits are not time-dependent

$$
\frac{dE_k}{dt} = \int_{l_1}^{l_2} \frac{\partial}{\partial t} \left(\rho \beta \frac{\bar{v}^2}{2} A \right) dl + \int_{l_1}^{l_2} \frac{\partial}{\partial l} \left(\rho \alpha \frac{\bar{v}^2}{2} Q \right) dl,
\tag{5.107}
$$

that is following the grouping

$$\frac{dE_k}{dt} = \int_{l_1}^{l_2} \left\{ \frac{\partial}{\partial t} \left(\rho \beta \frac{\bar{v}^2}{2} A \right) + \frac{\partial}{\partial l} \left(\rho \alpha \frac{\bar{v}^2}{2} Q \right) \right\} dl. \tag{5.108}$$

An integrand function in expression (5.108) will be dismembered by partial derivation of complex terms

$$\frac{\partial}{\partial t} \left(\rho \beta \frac{\bar{v}^2}{2} A \right) + \frac{\partial}{\partial l} \left(\rho \alpha \frac{\bar{v}^2}{2} Q \right) =$$

$$\rho A \frac{\partial}{\partial t} \left(\beta \frac{\bar{v}^2}{2} \right) + \beta \frac{\bar{v}^2}{2} \frac{\partial (\rho A)}{\partial t} + \alpha \frac{\bar{v}^2}{2} \frac{\partial (\rho Q)}{\partial l} + \rho Q \frac{\partial}{\partial l} \left(\alpha \frac{\bar{v}^2}{2} \right) = \tag{5.109}$$

$$\rho A \bar{v} \frac{\partial \beta \bar{v}}{\partial t} + \rho Q \bar{v} \frac{\partial \alpha \bar{v}}{\partial l} + \frac{\bar{v}^2}{2} \frac{\partial (\rho Q)}{\partial l} (\alpha - \beta).$$

Following the grouping of members and application of the equation of continuity (5.81), it is written

$$\rho Q \frac{\partial \beta \bar{v}}{\partial t} + \rho Q \bar{v} \frac{\partial \alpha \bar{v}}{\partial l} + \rho Q \frac{\bar{v}^2}{2} \underbrace{\frac{1}{\rho Q} \frac{\partial (\rho Q)}{\partial l} (\alpha - \beta)}_{\zeta^*} =$$

$$\rho Q \left(\frac{\partial \beta \bar{v}}{\partial t} + \bar{v} \frac{\partial \alpha \bar{v}}{\partial l} + \zeta^* \frac{\bar{v}^2}{2} \right), \tag{5.110}$$

where

$$\zeta^* = \frac{1}{\rho Q} \frac{\partial (\rho Q)}{\partial l} (\alpha - \beta). \tag{5.111}$$

The rate of change of kinetic energy in the final form will be

$$\frac{dE_k}{dt} = \int_{l_1}^{l_2} \rho Q \left(\frac{\partial \beta \bar{v}}{\partial t} + \bar{v} \frac{\partial \alpha \bar{v}}{\partial l} + \zeta^* \frac{\bar{v}^2}{2} \right) dl. \tag{5.112}$$

5.6.5 Power of normal forces

The power of normal (pressure) forces consists of the power of pressure forces at the inflow and outflow cross-sections

$$(pQ)_1 - (pQ)_2 = - \int_{l_1}^{l_2} \frac{\partial pQ}{\partial l} dl \tag{5.113}$$

and the power of pressure forces along the pipe mantle (deformable pipe)

$$- \int_{l_1}^{l_2} p \frac{\partial A}{\partial t} dl. \tag{5.114}$$

The total power of normal forces through the pipe control volume is equal to

$$\frac{dW_n}{dt} = -\int_{l_1}^{l_2} \left(\frac{\partial p\,Q}{\partial l} + p\frac{\partial A}{\partial t} \right) dl. \tag{5.115}$$

By partial derivation of the first member of the integrand and grouping, it is obtained that:

$$\frac{dW_n}{dt} = -\int_{l_1}^{l_2} \left(Q\frac{\partial p}{\partial l} + p\frac{\partial Q}{\partial l} + p\frac{\partial A}{\partial t} \right) dl =$$

$$-\int_{l_1}^{l_2} p\left(\frac{\partial A}{\partial t} + \frac{\partial Q}{\partial l} \right) dl - \int_{l_1}^{l_2} Q\frac{\partial p}{\partial l} dl. \tag{5.116}$$

The first integral on the right hand side is the rate of change of internal energy of a control volume; thus it is written

$$\frac{dW_n}{dt} = \frac{dU}{dt} - \int_{l_1}^{l_2} Q\frac{\partial p}{\partial l} dl. \tag{5.117}$$

The power of normal forces through the control volume of the streamtube is used to increase the internal energy and overcome the pressure gradient along the flow, that is a change of pressure energy.

5.6.6 Power of resistance forces

Friction along the pipe mantle, namely the shear stress τ_0 along the pipe perimeter, resists the flow through the pipe control volume. The work of resistance forces is always negative; thus, the power of resistance forces is equal to

$$\frac{dW_o}{dt} = -\int_{l_1}^{l_2} \tau_0 O\bar{v}\,dl = -\int_{l_1}^{l_2} \tau_0 O\frac{Q}{A}dl = -\int_{l_1}^{l_2} Q\frac{\tau_0}{R}dl, \tag{5.118}$$

where O is the wetted perimeter and R is the hydraulic radius.

5.6.7 Dynamic equation

After introducing the rate of change of kinetic energy (5.112), the rate of change of potential energy (5.103), the power of normal forces (5.117), and the power of resistance forces (5.118) into the energy equation (5.101), the following is obtained

$$\frac{dU}{dt} + \int_{l_1}^{l_2} \rho g Q\frac{\partial z}{\partial l}dl + \int_{l_1}^{l_2} \rho Q\left(\frac{\partial \beta \bar{v}}{\partial t} + \bar{v}\frac{\partial \alpha \bar{v}}{\partial l} + \zeta^*\frac{\bar{v}^2}{2} \right)dl =$$

$$\frac{dU}{dt} - \int_{l_1}^{l_2} Q\frac{\partial p}{\partial l}dl - \int_{l_1}^{l_2} Q\frac{\tau_0}{R}dl. \tag{5.119}$$

In the obtained energy equation, after canceling the power of internal forces and extracting the mass discharge ρQ, an integral form of the power change over the observed pipe control volume is obtained

$$\int_{l_1}^{l_2} \rho Q \left(\frac{\partial \beta \bar{v}}{\partial t} + \bar{v} \frac{\partial \alpha \bar{v}}{\partial l} + g \frac{\partial z}{\partial l} + \frac{1}{\rho} \frac{\partial p}{\partial l} + \frac{\tau_0}{\rho R} + \zeta^* \frac{\bar{v}^2}{2} \right) dl = 0. \tag{5.120}$$

Since the obtained definite integral is equal to zero for any of the integration limits, the integrand function vanishes. Thus, a differential form of the power change is obtained:

$$\rho Q \left(\frac{\partial \beta \bar{v}}{\partial t} + \bar{v} \frac{\partial \alpha \bar{v}}{\partial l} + g \frac{\partial z}{\partial l} + \frac{1}{\rho} \frac{\partial p}{\partial l} + \frac{\tau_0}{\rho R} + \zeta^* \frac{\bar{v}^2}{2} \right) = 0. \tag{5.121}$$

After reduction with the mass discharge ρQ, a differential equation of non-steady flow in pipes is obtained:

$$\frac{\partial \beta \bar{v}}{\partial t} + \bar{v} \frac{\partial \alpha \bar{v}}{\partial l} + g \frac{\partial z}{\partial l} + \frac{1}{\rho} \frac{\partial p}{\partial l} + \frac{\tau_0}{\rho R} + \zeta^* \frac{\bar{v}^2}{2} = 0. \tag{5.122}$$

5.6.8 Flow resistances, the dynamic equation discussion

Note that the dynamic equation (5.122) is a one-dimensional model of a complex spatial non-steady flow in pipes, derived using several assumptions due to the purposes of a simple analysis; primarily for monitoring energy and other stream-related values using the mean flow velocity $\bar{v} = Q/A$. In terms of engineering, it is an acceptable description of real flow. However, for practical purposes, further simplifications are required. Although by using the Boussinesq coefficient β and the Coriolis coefficient α it is possible to express the kinetic members related to the variation of the flow velocity profile by mean velocity, the problem of their real values still remains. The reason for this is the unknown flow velocity profile of the developed boundary layer.

From an approximate analysis of values of those correction coefficients for different types of developed steady boundary layers, described in Section 5.3.3, it can be observed that for the turbulent flow values approach 1, which corresponds to the uniform velocity distribution. Thus, for practical purposes, for relatively small specific kinetic energy in comparison with the potential and pressure energy, it is adopted that $\alpha = \beta = 1$.

When comparing the obtained dynamic equation of a real flow in a pipe (5.122) and the dynamic equation of flow of an ideal fluid in a streamtube (5.63) there is a difference in the form of the two new terms

$$\frac{\tau_0}{\rho R} + \zeta^* \frac{\bar{v}^2}{2} = 0. \tag{5.123}$$

The first term in Eq. (5.123) represents the influence of resistance forces, that is it is related to the friction in the developed boundary layer, which is well researched in the case of a steady flow. It is expressed as

$$\tau_0 = c_f \rho \frac{\bar{v}^2}{2}, \tag{5.124}$$

where c_f is the friction coefficient dependent on the development of the boundary layer, ρ is the density, and $\bar{v}$ is the velocity of undisturbed flow; for pipes it is equal to the mean flow velocity.

If Eq. (5.124) is used, the first term in Eq. (5.123) becomes

$$\frac{c_f}{R}\frac{\bar{v}^2}{2}. \qquad (5.125)$$

For circular cross-section pipes $R = D/4$, where D is the pipe diameter, it can be written as

$$\frac{4c_f}{D}\frac{\bar{v}^2}{2} = \frac{\lambda}{D}\frac{\bar{v}^2}{2}, \qquad (5.126)$$

where $\lambda = 4c_f$ is Darcy–Weissbach friction coefficient in circular cross-section pipes. Coefficient λ depends on the Reynolds number and relative roughness. It is determined from the Moody chart, see Chapter 2, which represents the synthesis of the tests carried out by Nikuradze and the Colebrook–White analyses of measurements of the resistance to flow in technical pipes.

The second term in Eq. (5.123) is the result of the integration of a non-uniform velocity profile, where ζ^* is defined by Eq. (5.111). The term is cancelled for a uniform velocity profile, that is when $\alpha = \beta$, namely for the steady flow. In a general case of non-steady flow of a compressible fluid it changes kinetic energy, depending on the gradient of power of pressure forces in cross-section pQ.

It can be thought of as the change of friction in non-steady flow. Namely, research carried out so far has shown the differences in energy dissipation between steady and non-steady flow. Due to flow complexity, generalization is not possible; thus, resistances in non-steady flow are determined by the same procedure used in a steady flow.

Taking into account all that is presented in the discussion, the relevant dynamic equation of a non-steady flow of a compressible fluid will be

$$\frac{\partial v}{\partial t} + v\frac{\partial v}{\partial l} + g\frac{\partial z}{\partial l} + \frac{1}{\rho}\frac{\partial p}{\partial l} + \frac{\lambda}{D}\frac{v^2}{2} = 0, \qquad (5.127)$$

where $v = Q/A$ is the mean flow velocity. The average symbol, denoted by putting a line above the velocity symbol, will be omitted from the following text.

Using the Darcy–Weissbach expression for the gradient of energy line

$$J_e = \frac{\lambda}{D}\frac{v^2}{2g} \qquad (5.128)$$

and the gradient of the pipe elevation along the stream axis

$$J_0 = -\frac{\partial z}{\partial l}, \qquad (5.129)$$

the dynamic equation obtains the following form

$$\frac{\partial v}{\partial t} + v\frac{\partial v}{\partial l} + \frac{1}{\rho}\frac{\partial p}{\partial l} + g(J_e - J_0) = 0. \qquad (5.130)$$

Similarly, the dynamic equation can be written in the form of gradients of members in the head form

$$\frac{1}{g}\frac{\partial v}{\partial t} + \frac{\partial}{\partial l}\frac{v^2}{2g} + \frac{\partial z}{\partial l} + \frac{1}{\rho g}\frac{\partial p}{\partial l} + J_e = 0. \qquad (5.131)$$

5.7   Flow models

5.7.1   Steady flow

Steady flow of incompressible fluid

Equations of steady flow are obtained from non-steady ones, simply by omitting all partial derivatives in time

$$\frac{\partial \dots}{\partial t} = 0. \tag{5.132}$$

The density of an *incompressible* fluid is constant; thus, the equation of continuity (5.81), after Eq. (5.132) is applied, is reduced to constant discharge along the flow

$$Q(l) = const. \tag{5.133}$$

The dynamic equation for an *incompressible* fluid is obtained when Eq. (5.132) is applied to Eq. (5.131)

$$\frac{\partial}{\partial l}\left(z + \frac{p}{\rho g} + \alpha\frac{v^2}{2g}\right) + J_e = 0 \tag{5.134}$$

or is expressed by averaged specific mechanic energy

$$\frac{\partial H}{\partial l} + J_e = 0 \tag{5.135}$$

where

$$H = z + \frac{p}{\rho g} + \alpha\frac{v^2}{2g}. \tag{5.136}$$

The dynamic equation (5.135) can be integrated between two points on the flow axis as follows

$$H_2 - H_1 + \int_{l_1}^{l_2} J_e dl = 0. \tag{5.137}$$

Figure 5.10 shows the integrated dynamic equation in the head form.

Steady flow of compressible fluid

If Eq. (5.132) is applied to the continuity equation (5.81), the equation of continuity for the steady flow of *compressible* fluid in a pipe is obtained

$$\frac{\partial \rho Q}{\partial l} = 0 \quad \Rightarrow \quad \dot{M} = \rho Q = const. \tag{5.138}$$

In steady flow, mass discharge along the flow is constant. If Eq. (5.132) is applied to the dynamic equation (5.131), then

$$\frac{\partial}{\partial l}\left(\frac{v^2}{2g}\right) + \frac{\partial z}{\partial l} + \frac{1}{\rho g}\frac{\partial p}{\partial l} + J_e = 0. \tag{5.139}$$

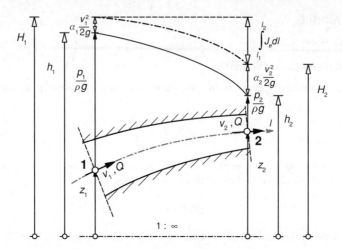

Figure 5.10 Head form of the dynamic equation for steady flow of an incompressible fluid.

If the equation is multiplied by elemental displacement along the flow axis, the following is obtained

$$\frac{\partial}{\partial l}\left(\frac{v^2}{2g}\right)dl + \frac{\partial z}{\partial l}dl + \frac{1}{\rho g}\frac{\partial p}{\partial l}dl + J_e dl = 0, \tag{5.140}$$

that is spatial total differentials are obtained, namely, the differential equation of steady flow of a compressible fluid

$$d\left(\frac{v^2}{2g}\right) + dz + \frac{dp}{\rho g} + J_e dl = 0. \tag{5.141}$$

This dynamic equation can be integrated between two points in the form

$$\left(z + \frac{v^2}{2g}\right)_2 - \left(z + \frac{v^2}{2g}\right)_1 + \frac{1}{g}\int_{l_1}^{l_2}\frac{dp}{\rho} + \int_{l_1}^{l_2} J_e dl = 0. \tag{5.142}$$

The marked pressure integral can be calculated if the change of density $\rho(p)$ is known for the prescribed thermodynamic process. For a polytrophic thermodynamic process $p = c \cdot \rho^n$ where c is constant, developed integration gives

$$\left(z + \frac{v^2}{2g} + \frac{n}{n-1}\frac{p}{\rho g}\right)_2 - \left(z + \frac{v^2}{2g} + \frac{n}{n-1}\frac{p}{\rho g}\right)_1 + \int_{l_1}^{l_2} J_e dl = 0. \tag{5.143}$$

Density changes can be disregarded for a flow with the small Mach number $M_a = v/c < 0.25$, where v is the flow velocity and c is the speed of sound. Then, the steady flow of compressible fluid can be observed as an incompressible fluid flow with constant volume discharge, and the dynamic equation (5.135) can be applied.

5.7.2 Non-steady flow

Non-steady flow of compressible fluid

Equation of continuity. Non-steady flow in pipes is generally described by the equation of continuity (5.81) that has the following developed form for a compressible fluid (5.100)

$$\frac{1}{\rho c^2}\left(\frac{\partial p}{\partial t} + v\frac{\partial p}{\partial l}\right) + \frac{\partial v}{\partial l} = 0, \tag{5.144}$$

that is the form

$$\frac{\partial p}{\partial t} + v\frac{\partial p}{\partial l} + \rho c^2\frac{\partial v}{\partial l} = 0. \tag{5.145}$$

Dynamic equation. The dynamic equation in the following form is used for a compressible fluid flow modelling

$$\frac{\partial v}{\partial t} + v\frac{\partial v}{\partial l} + \frac{1}{\rho}\frac{\partial p}{\partial l} + g\,(J_e - J_0) = 0, \tag{5.146}$$

where J_e the gradient of the energy line and J_0 the gradient of the pipe axis are equal to

$$J_e = \frac{\lambda}{2gD}v\,|v|, \quad J_0 = -\frac{\partial z}{\partial l}. \tag{5.147}$$

The equation of continuity and the dynamic equation contain three unknowns: p, v, ρ pressure, velocity, and density. The system cannot be solved without a third equation that connects pressure and density. The third equation is the equation of state

$$F(p, V, T) = 0, \tag{5.148}$$

which, in general, introduces the new unknown – the temperature T. If an isothermal process is observed, which is the most common case in a fluid flow, then these three equations completely describe non-steady flow in pipes. Then, density $\rho(p)$ and the speed of sound $c(p)$ are functionally dependent on pressure p. These functional connections are determined from the equation of state, see Section 5.1, Figure 5.4 and Figure 5.5.

Otherwise, the system of equations shall be expanded by thermodynamic equations.

Non-steady flow of an elastic liquid

Equation of continuity. The equation of continuity for an elastic liquid and elastic pipe, expressed by variables p, Q, according to Eq. (5.92), will be

$$\frac{A_0}{\rho_0 c_0^2}\frac{\partial p}{\partial t} + \frac{\partial Q}{\partial l} = 0, \tag{5.149}$$

that is expressed by variables h, Q

$$\frac{g A_0}{c_0^2}\frac{\partial h}{\partial t} + \frac{\partial Q}{\partial l} = 0, \tag{5.150}$$

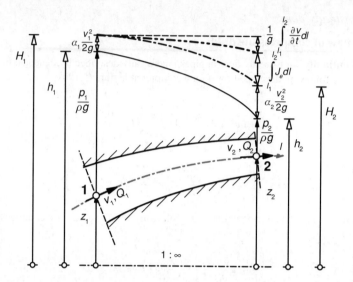

Figure 5.11   Head form of the dynamic equation for non-steady flow of an elastic liquid.

where the water hammer celerity c_0, liquid density ρ_0, and the pipe cross-section area A_0 are constant and equal to the values at normal atmospheric pressure.

Dynamic equation. In elastic liquid flow, it can be assumed that density is constant along the pipe length; thus, the dynamic equation (5.146) can be written in the form

$$\frac{1}{g}\frac{\partial v}{\partial t} + \frac{\partial H}{\partial l} + J_e = 0, \tag{5.151}$$

where $H = z + \frac{p}{\rho g} + \alpha\frac{v^2}{2g} = h + \alpha\frac{v^2}{2g}$ is the energy head and h is the piezometric head. The dynamic equation can be integrated between two points on the flow axis

$$\left(z + \frac{p}{\rho_0 g} + \alpha\frac{v^2}{2g}\right)_2 - \left(z + \frac{p}{\rho_0 g} + \alpha\frac{v^2}{2g}\right)_1 + \int_{l_1}^{l_2} J_e dl + \frac{1}{g}\int_{l_1}^{l_2}\frac{\partial v}{\partial t}dl = 0 \tag{5.152}$$

Figure 5.11 shows the integrated dynamic equation in the head form.

Non-steady flow of liquid and saturated vapor

When, in elastic fluid flow, pressures at some point drop below the saturated vapor pressure, density and the speed of sound change drastically, see Section 5.1.2. The flow becomes a two-phase flow. Although the same equations, and thus the solution methods, as for a compressible flow can be applied, for practical reasons it would be appropriate to differentiate the liquid region from the saturated liquid/vapor region. Namely, the liquid phase region is far larger than the saturated liquid/vapor phase region; thus, in that part simpler algorithms for the analysis of elastic liquid can be used.

Non-steady flow of incompressible fluid in rigid pipes

For an incompressible (rigid) fluid $\rho = \rho_0$ and rigid pipe $\partial A/\partial t = 0$, the equation of continuity will be

$$Q(l) = Q(t),\tag{5.153}$$

that is the discharge is constant along the flow at each moment t and equal to $Q(t)$ in time. The dynamic equation is

$$\frac{1}{g}\frac{\partial v}{\partial t} + \frac{\partial}{\partial l}\left(z + \frac{p}{\rho_0 g} + \frac{v^2}{2g}\right) + J_e = 0\tag{5.154}$$

or, expressed by averaged specific mechanic energy

$$\frac{1}{g}\frac{\partial v}{\partial t} + \frac{\partial H}{\partial l} + J_e = 0.\tag{5.155}$$

Integration between two points along the flow at the distance $L = l_2 - l_1$ gives

$$\int_{l_1}^{l_2}\left(\frac{1}{g}\frac{\partial v}{\partial t} + \frac{\partial H}{\partial l} + J_e\right) dl = 0.\tag{5.156}$$

Since integration limits are not time-dependent, an ordinary differential equation is obtained

$$\frac{L}{g}\frac{dv}{dt} + H_2 - H_1 + \int_{l_1}^{l_2} J_e dl = 0,\tag{5.157}$$

which is used in rigid fluid flow modelling.

Non-steady flow of an incompressible fluid in compressible pipes

For an incompressible fluid $\rho = const$, and a compressible pipe $\partial A/\partial t \neq 0$, the equation of continuity will be

$$\frac{\partial A}{\partial t} + \frac{\partial Q}{\partial l} = 0\tag{5.158}$$

thus, the discharge is not constant along the flow. Since the equation of continuity expresses volumetric change of a particle, see expression (5.40)

$$\dot{\varepsilon}_{V_{\delta M}} = \frac{\delta \dot{V}}{\delta l} = \frac{\partial A}{\partial t} + \frac{\partial Q}{\partial l} = 0,\tag{5.159}$$

then, the volumetric strain of a particle is equal to zero.

The dynamic equation for $\rho = \rho_0$ has the form

$$\frac{1}{g}\frac{\partial v}{\partial t} + \frac{\partial}{\partial l}\left(z + \frac{p}{\rho_0 g} + \frac{v^2}{2g}\right) + J_e = 0,\tag{5.160}$$

or, expressed by the averaged specific mechanic energy

$$\frac{1}{g}\frac{\partial v}{\partial t} + \frac{\partial H}{\partial l} + J_e = 0. \tag{5.161}$$

5.8　Characteristic equations

5.8.1　Elastic liquid

The dynamic equation and equation of continuity can be transformed into ordinary differential equations along characteristic curves that coincide with the trajectories of positive and negative elementary waves.[7] For an elastic liquid and pipe, density $\rho = \rho_0$ and the speed of sound $c = c_0$ are constant; thus, the equation of continuity (5.145) can be applied

$$\frac{\partial p}{\partial t} + v\frac{\partial p}{\partial l} + \rho_0 c_0^2\frac{\partial v}{\partial l} = 0. \tag{5.162}$$

After division by $\rho_0 c_0$, the equation can be written in the form

$$\frac{\partial}{\partial t}\left(\frac{p}{\rho_0 c_0}\right) + v\frac{\partial}{\partial l}\left(\frac{p}{\rho_0 c_0}\right) + c_0\frac{\partial v}{\partial l} = 0. \tag{5.163}$$

The dynamic equation (1.130), divided by density, is

$$\frac{\partial v}{\partial t} + v\frac{\partial v}{\partial l} + \frac{\partial}{\partial l}\left(\frac{p}{\rho_0}\right) + g(J_e - J_0) = 0.$$

For an elastic liquid and pipe, a direct transformation of characteristics is possible, and the procedure is the following.

Positive characteristic

If the equation of continuity is added to the dynamic equation, the following is obtained

$$\frac{\partial}{\partial t}\left(\frac{p}{\rho_0 c_0}\right) + \frac{\partial v}{\partial t} + v\frac{\partial v}{\partial l} + v\frac{\partial}{\partial l}\left(\frac{p}{\rho_0 c_0}\right) + c_0\frac{\partial v}{\partial l} + \frac{\partial}{\partial l}\left(\frac{p}{\rho_0}\right) + g(J_e - J_0) = 0 \tag{5.164}$$

and members can be grouped as follows

$$\frac{\partial}{\partial t}\left(v + \frac{p}{\rho_0 c_0}\right) + v\left(\frac{\partial v}{\partial l} + \frac{\partial}{\partial l}\frac{p}{\rho_0 c_0}\right) + c_0\left(\frac{\partial v}{\partial l} + \frac{\partial}{\partial l}\frac{p}{\rho_0 c_0}\right) + g(J_e - J_0) = 0. \tag{5.165}$$

Following the extraction of the common factor, it is written as

$$\frac{\partial}{\partial t}\left(v + \frac{p}{\rho_0 c_0}\right) + (v + c_0)\frac{\partial}{\partial l}\left(v + \frac{p}{\rho_0 c_0}\right) + g(J_e - J_0) = 0. \tag{5.166}$$

---

[7]These are the elementary waves propagating in the direction of the positive or negative coordinate axis.

If the symbol

$$\Gamma^+ = v + \frac{p}{\rho_0 c_0} \tag{5.167}$$

is introduced into Eq. (5.166), it can be written as

$$\frac{\partial \Gamma^+}{\partial t} + (v + c_0) \frac{\partial \Gamma^+}{\partial l} + g\,(J_e - J_0) = 0. \tag{5.168}$$

The total differential of the function Γ^+ is

$$d\Gamma^+ = \frac{\partial \Gamma^+}{\partial t} dt + \frac{\partial \Gamma^+}{\partial l} dl$$

which, after division by dt gives

$$\frac{d\Gamma^+}{dt} = \frac{\partial \Gamma^+}{\partial t} + \frac{\partial \Gamma^+}{\partial l} \frac{dl}{dt},$$

where

$$\frac{dl}{dt} = v + c_0 = w^+ \tag{5.169}$$

is the velocity of the positive wave in the absolute coordinate system, thus

$$\frac{d\Gamma^+}{dt} = \frac{\partial \Gamma^+}{\partial t} + (v + c_0) \frac{\partial \Gamma^+}{\partial x}. \tag{5.170}$$

Introducing Eq. (5.170) into Eq. (5.168), the following is obtained

$$\frac{d\Gamma^+}{dt} + g\,(J_e - J_0) = 0, \tag{5.171}$$

that is an ordinary differential equation is valid along the positive wave trajectory $w^+(l, t)$:

$$\frac{d}{dt}\left(v + \frac{p}{\rho_0 c_0}\right) + g\,(J_e - J_0) = 0. \tag{5.172}$$

Negative characteristic

Similarly, an ordinary differential equation is obtained along the negative wave trajectory. If the equation of continuity is deducted from the dynamic equation and a grouping similar to the previous one is carried out, then

$$\frac{\partial}{\partial t}\left(v - \frac{p}{\rho_0 c_0}\right) + (v - c_0) \frac{\partial}{\partial l}\left(v - \frac{p}{\rho_0 c_0}\right) + g\,(J_e - J_0) = 0. \tag{5.173}$$

If the symbol

$$\Gamma^- = v - \frac{p}{\rho_0 c_0} \tag{5.174}$$

is introduced into Eq. (5.173), it can be written as

$$\frac{\partial \Gamma^-}{\partial t} + (v - c_0) \frac{\partial \Gamma^-}{\partial l} + g\,(J_e - J_0) = 0. \tag{5.175}$$

The total differential of the function Γ^+ is

$$d\Gamma^- = \frac{\partial \Gamma^-}{\partial t} dt + \frac{\partial \Gamma^-}{\partial l} dl$$

which, after division by dt gives

$$\frac{d\Gamma^-}{dt} = \frac{\partial \Gamma^-}{\partial t} + \frac{\partial \Gamma^-}{\partial l}\frac{dl}{dt},$$

where

$$\frac{dl}{dt} = v - c_0 = w^- \tag{5.176}$$

is the velocity of the negative wave in the absolute coordinate system, thus

$$\frac{d\Gamma^-}{dt} = \frac{\partial \Gamma^-}{\partial t} + (v - c_0)\frac{\partial \Gamma^-}{\partial x}. \tag{5.177}$$

Introducing Eq. (5.177) into Eq. (5.175), the following is obtained

$$\frac{d\Gamma^-}{dt} + g(J_e - J_0) = 0, \tag{5.178}$$

that is an ordinary differential equation is valid along the negative wave trajectory $w^-(l, t)$

$$\frac{d}{dt}\left(v - \frac{p}{\rho_0 c_0}\right) + g(J_e - J_0) = 0. \tag{5.179}$$

Characteristic equations, variables p, v

Elementary wave trajectories in the direction and count direction of flow are called characteristics $\gamma^\pm$

$$\gamma^\pm: \quad \frac{dl}{dt} = v \pm c_0 \tag{5.180}$$

along which the equations for respective wave functions are valid

$$\frac{d\Gamma^\pm}{dt} + g(J_e - J_0) = 0 \tag{5.181}$$

where

$$\Gamma^\pm = v \pm \frac{p}{\rho_0 c_0}, \tag{5.182}$$

that is

$$\Gamma^\pm: \quad \frac{d}{dt}\left(v \pm \frac{p}{\rho_0 c_0}\right) + g(J_e - J_0) = 0. \tag{5.183}$$

Characteristic equations, variables *h*, *v*

Characteristic equations, expressed by variables h, v, are obtained by a similar procedure

$$\gamma^\pm: \quad \frac{dl}{dt} = v \pm c_0, \tag{5.184}$$

$$\frac{d\Gamma^\pm}{dt} + g\left(J_e \pm \frac{v}{c_0} J_0\right) = 0, \tag{5.185}$$

where

$$\Gamma^\pm = v \pm \frac{g}{c_0} h, \tag{5.186}$$

that is

$$\Gamma^\pm: \quad \frac{d}{dt}\left(v \pm \frac{g}{c_0} h\right) + g\left(J_e \pm \frac{v}{c_0} J_0\right) = 0. \tag{5.187}$$

5.8.2   *Compressible fluid*

General procedure of transformation of hyperbolic equations

If two partial differential equations for U, V with linear coefficients are observed

$$
\begin{aligned}
a_1 \frac{\partial U}{\partial t} + b_1 \frac{\partial U}{\partial l} + c_1 \frac{\partial V}{\partial t} + d_1 \frac{\partial V}{\partial l} + e_1 &= 0 \\
a_2 \frac{\partial U}{\partial t} + b_2 \frac{\partial U}{\partial l} + c_2 \frac{\partial V}{\partial t} + d_2 \frac{\partial V}{\partial l} + e_2 &= 0,
\end{aligned}
\tag{5.188}
$$

then, depending on coefficients $a_1, b_1, \ldots$ and $a_2, b_2, \ldots$, the equations can be hyperbolic, parabolic, or elliptic. Hyperbolic equations can be transformed into characteristics in several ways. A procedure developed by R. Courant[8] and K. O. Friedrichs,[9] described in the article by Th. Dracos (1970),[10] will be used here, without discussion of mathematical particularities. The procedure of the system transformation into characteristic form is carried out by several second order determinants

$$A = [a\,c], \quad 2B = [a\,d] + [b\,c]$$
$$C = [b\,d], \quad D = [a\,b], \quad E = [b\,c],$$
$$F = [a\,e], \quad G = [b\,e],$$

that are expressed by the system coefficients, where the symbol [**pq**] denotes the determinant value

$$\begin{vmatrix} p_1 & q_1 \\ p_2 & q_2 \end{vmatrix} = p_1 q_2 - p_2 q_1.$$

---

[8]Richard Courant, German mathematician (1888–1972).
[9]Kurt Otto Friedrichs, German mathematician (1901–1982).
[10]Themistocles Dracos, Swiss Professor Emeritus at the Department of Civil, Environmental and Geomatic Engineering.

The characteristic solution, that is the inclination of the curve of characteristics $w = dl/dt$ in the plane l, t is defined in the form of the quadratic equation $Aw^2 - 2Bw + C = 0$ with the solution

$$w^{\pm} = \frac{B \pm \sqrt{B^2 - AC}}{A}. \tag{5.189}$$

The term under the square root defines the character of the system of differential equations; thus:

- $B^2 - AC < 0$ - elliptic system without real solutions,
- $B^2 - AC = 0$ - parabolic system with one real solution,
- $B^2 - AC > 0$ - hyperbolic system with two real solutions.

If there is a hyperbolic system; then there are two real systems of characteristics with the curves defined by differential equations

$$\begin{aligned} \gamma^+ : \quad dl - w^+ dt &= 0 \\ \gamma^- : \quad dl - w^- dt &= 0. \end{aligned} \tag{5.190}$$

Positive and negative characteristics are the trajectories of positive and negative waves that are propagating in positive or negative directions of the axis l at an absolute velocity w^+ or w^-.

Respective differential equations on characteristics are obtained by other previously prepared determinants in the expressions

$$\begin{aligned} \Gamma^+ : \quad DdU + \left(Aw^+ - E\right) dv + \left(Fw^+ - G\right) dt &= 0 \\ \Gamma^- : \quad DdU + \left(Aw^- - E\right) dv + \left(Fw^- - G\right) dt &= 0. \end{aligned} \tag{5.191}$$

Transformation into characteristic equations

Non-steady flow of a compressible fluid is described by the equation of continuity (5.145)

$$\frac{\partial p}{\partial t} + v \frac{\partial p}{\partial l} + \rho c^2 \frac{\partial v}{\partial l} = 0$$

and the dynamic equation (5.146)

$$\frac{\partial v}{\partial t} + v \frac{\partial v}{\partial l} + \frac{1}{\rho} \frac{\partial p}{\partial l} + g \left(J_e - J_0\right) = 0.$$

Comparing members besides partial derivations of these equations, with the members beside the general system (5.188), the values of the coefficients are written:

	$U = p$		$V = v$		
	a	b	c	d	e
(1)	1	v	0	ρc^2	0
(2)	0	ρ^{-1}	1	v	$g \left(J_e - J_0\right)$

that is the determinants

$$A = \begin{vmatrix} 1 & 0 \\ 0 & 1 \end{vmatrix} = 1, \; 2B = \begin{vmatrix} 1 & \rho c^2 \\ 0 & v \end{vmatrix} + \begin{vmatrix} v & 0 \\ \rho^{-1} & 1 \end{vmatrix} = 2v,$$

$$C = \begin{vmatrix} v & \rho c^2 \\ \rho^{-1} & v \end{vmatrix} = v^2 - c^2, \; D = \begin{vmatrix} 1 & v \\ 0 & \rho^{-1} \end{vmatrix} = \rho^{-1},$$

$$E = \begin{vmatrix} v & 0 \\ \rho^{-1} & 1 \end{vmatrix} [b \; c] = v, \; F = \begin{vmatrix} 1 & 0 \\ 0 & g(J_e - J_0) \end{vmatrix} = g(J_e - J_0),$$

$$G = \begin{vmatrix} v & 0 \\ g & g(J_e - J_0) \end{vmatrix} = g v(J_e - J_0).$$

Calculation of the inclination (5.189) of the characteristics defines the velocity of propagation of the absolute wave speed

$$w^\pm = v \pm c \tag{5.192}$$

after which the positive and negative characteristic equations $\gamma^\pm$ are calculated:

$$dl - (v \pm c)\,dt = 0. \tag{5.193}$$

When $dU = dp$, $dY = dv$ and other required values are introduced into expression (5.191), the respective equations of wave functions on characteristic $\Gamma^\pm$ are obtained

$$\frac{dp}{\rho} + [(v \pm c) - v]\,dv + [g(J_e - J_0)(v \pm c) - vg(J_e - J_0)]\,dt = 0.$$

After arranging, the equations of two wave functions $\Gamma^\pm$ along characteristic $\gamma^\pm$ are obtained in the form

$$\frac{dv}{dt} \pm \frac{1}{\rho c}\frac{dp}{dt} + g(J_e - J_0) = 0, \tag{5.194}$$

where density and water hammer celerity are pressure-dependent functions $\rho(p)$, $c(p)$.

5.9 Analytical solutions

5.9.1 Linearization of equations – wave equations

If we start from the equation of continuity (5.149), the following is obtained for a pipe with a constant cross-section area

$$\frac{1}{\rho c^2}\frac{\partial p}{\partial t} + \frac{\partial v}{\partial x} = 0. \tag{5.195}$$

The pipe has no inclination, and if velocity heads are omitted and the resistance term is linearized, the dynamic equation is obtained

$$\rho\frac{\partial v}{\partial t} + \frac{\partial p}{\partial x} + rv = 0. \tag{5.196}$$

Equations (5.195) and (5.196) are wave equations with the pressure wave $p(x, t)$ and the velocity wave $v(x, t)$ functions as the unknowns.

5.9.2 Riemann general solution

For the flow without resistances the following equations are applied

$$\frac{1}{\rho c^2} \frac{\partial p}{\partial t} + \frac{\partial v}{\partial x} = 0, \tag{5.197}$$

$$\rho \frac{\partial v}{\partial t} + \frac{\partial p}{\partial x} = 0, \tag{5.198}$$

with the general solution in the form

$$p(x, t) = p_0 + \varphi(x + ct) + \psi(x - ct), \tag{5.199}$$

$$v(x, t) = v_0 - \frac{1}{\rho c} [\varphi(x + ct) - \psi(x - ct)], \tag{5.200}$$

where φ is the wave function of the positive wave (wave propagating at velocity $+c$) and ψ is the wave function of the negative wave (wave propagating at velocity $-c$). Values p_0, v_0 are the solutions of the steady flow. It is a Riemann[11] general solution of the wave equation. Wave functions φ, ψ can be determined from the prescribed values of the pressure and velocity waves

$$\varphi(x + ct) = \frac{1}{2} [(p - p_0) - \rho c (v - v_0)], \tag{5.201}$$

$$\psi(x + ct) = \frac{1}{2} [(p - p_0) + \rho c (v - v_0)]. \tag{5.202}$$

It will be shown that expressions (5.199) and (5.200) are general solution of Eqs (5.197) and (5.198). If written $\zeta = x + ct$ and $\eta = x - ct$, then

$$p(x, t) = p_0 + \varphi(\zeta) + \psi(\eta), \tag{5.203}$$

$$v(x, t) = v_0 - \frac{1}{\rho c} [\varphi(\zeta) - \psi(\eta)]. \tag{5.204}$$

This statement will be proved if Eqs (5.203) and (5.204) are introduced into Eqs (5.197) and (5.198); thus, the partial derivatives are calculated using the chain rule

$$\frac{\partial p}{\partial x} = \frac{d\varphi}{d\zeta} \frac{\partial \zeta}{\partial x} + \frac{d\psi}{d\eta} \frac{\partial \eta}{\partial x} = \frac{d\varphi}{d\zeta} + \frac{d\psi}{d\eta}, \tag{5.205}$$

$$\frac{\partial p}{\partial t} = \frac{d\varphi}{d\zeta} \frac{\partial \zeta}{\partial t} + \frac{d\psi}{d\eta} \frac{\partial \eta}{\partial t} = c \left(\frac{d\varphi}{d\zeta} - \frac{d\psi}{d\eta} \right), \tag{5.206}$$

---

[11] Georg Friedrich Bernhard Riemann, German mathematician (1826–1866).

$$\frac{\partial v}{\partial x} = \frac{1}{\rho c}\left(\frac{d\varphi}{d\zeta}\frac{\partial \zeta}{\partial x} - \frac{d\psi}{d\eta}\frac{\partial \eta}{\partial x}\right) = -\frac{1}{\rho c}\left(\frac{d\varphi}{d\zeta} - \frac{d\psi}{d\eta}\right), \tag{5.207}$$

$$\frac{\partial v}{\partial t} = \frac{1}{\rho c}\left(\frac{d\varphi}{d\zeta}\frac{\partial \zeta}{\partial t} - \frac{d\psi}{d\eta}\frac{\partial \eta}{\partial t}\right) = -\frac{1}{\rho}\left(\frac{d\varphi}{d\zeta} + \frac{d\psi}{d\eta}\right). \tag{5.208}$$

After introduction, the following is obtained

$$\frac{1}{\rho c}\left(\frac{d\varphi}{d\zeta} - \frac{d\psi}{d\eta}\right) - \frac{1}{\rho c}\left(\frac{d\varphi}{d\zeta} - \frac{d\psi}{d\eta}\right) = 0, \tag{5.209}$$

$$0 = 0,$$

$$-\left(\frac{d\varphi}{d\zeta} + \frac{d\psi}{d\eta}\right) + \frac{d\varphi}{d\zeta} + \frac{d\psi}{d\eta} = 0, \tag{5.210}$$

$$0 = 0.$$

thus, expressions (5.199) and (5.200) are the general Riemann solution.

The general Riemann solution describes the water hammer kinematics as shown in Section 6.1 Solutions by the method of characteristics. If the solutions are written

$$\Gamma^+ = \varphi, \Gamma^- = \psi, \tag{5.211}$$

$$\gamma^+ = \zeta, \gamma^- = \eta, \tag{5.212}$$

and compared, note that the Riemann solution is the subset from the solution of non-steady flow using the characteristics.

5.9.3   Some analytical solutions of water hammer

Figure 5.12 shows several examples with known analytical solutions (V. Jović, R. Lucić, 1999) obtained by the Laplace[12] transforms. Solutions as inverse transforms without possible rationalization of marked infinite sums will be given hereinafter.

The Heaviside[13] step function $H(u)$ is used:

(a) Sudden closing $v_0 \to 0$

$$\frac{p - p_0}{\rho c v_0} = \sum_{m=0}^{\infty}(-1)^m\left\{H\left[t - \frac{(2m + 1)L - x}{c}\right] - H\left[t - \frac{(2m + 1)L + x}{c}\right]\right\}, \tag{5.213}$$

$$\frac{v - v_0}{v_0} = -\sum_{m=0}^{\infty}(-1)^m\left\{H\left[t - \frac{(2m + 1)L - x}{c}\right] + H\left[t - \frac{(2m + 1)L + x}{c}\right]\right\}. \tag{5.214}$$

---

[12] Pierre-Simone Laplace, French mathematician and astronomer (1749–1827).

[13] Oliver Heaviside, English electrical engineer (1850–1925). Between 1880 and 1887 Heaviside created the notation for the differential operator in calculations and invented the method of solution of differential equations so as to transform them into ordinary algebraic equations; which, initially, caused numerous polemics in scientific circles due to the lack of mathematically strict proof for his procedure. His famous quotation: "Mathematics is an experimental science, and definitions do not come first, but later on" was his answer to the criticism that use of his operators was not strictly defined. On some other occasion, he said: "Shall I refuse my dinner because I do not fully understand the process of digestion?"

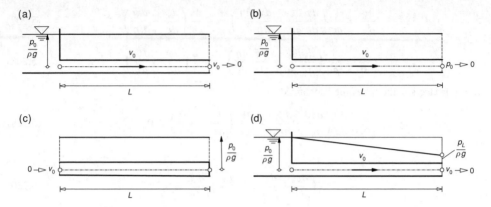

Figure 5.12   Initial and boundary conditions of analytical solutions.

(b) Sudden closing $p_0 \rightarrow 0$

$$\frac{p - p_0}{p_0} = -\sum_{m=0}^{\infty} \left\{ H \left[t - \frac{(2m+1)L - x}{c} \right] - H \left[t - \frac{(2m+1)L + x}{c} \right] \right\}, \qquad (5.215)$$

$$\frac{\rho c v}{p_0} = \sum_{m=0}^{\infty} \left\{ H \left[t - \frac{(2m+1)L - x}{c} \right] + H \left[t - \frac{(2m+1)L + x}{c} \right] \right\}. \qquad (5.216)$$

(c) Sudden filling of a blind pipe $0 \rightarrow v_0$

$$\frac{p - p_0}{\rho c v_0} = \sum_{m=0}^{\infty} \left\{ H \left[t - \frac{2mL + x}{c} \right] + H \left[t - \frac{2(m+1)L - x}{c} \right] \right\}, \qquad (5.217)$$

$$\frac{v}{v_0} = \sum_{m=0}^{\infty} \left\{ H \left[t - \frac{2mL + x}{c} \right] - H \left[t - \frac{2(m+1)L - x}{c} \right] \right\}. \qquad (5.218)$$

(d) Sudden closing, linearized friction resistances $v_0 \rightarrow 0$.
    The resistance constant is calculated from the steady state

$$r = \frac{p_0 - p_L}{v_0 L}. \qquad (5.219)$$

The solution for pressure will be

$$\frac{p - p_0}{\rho c v_0} = U(x, t), \qquad (5.220)$$

where

$$U(x, t) = -ce^{-\delta t} S(x, t),$$
(5.221)

$$S(x, t) = \frac{2c^2}{L} \sum_{k=0}^{\infty} \frac{b_k^2}{\omega_k \left(\delta^2 + \omega_k^2\right)^2} \frac{\sin b_k x}{\sin b_k L} \left[\left(\delta^2 - \omega_k^2\right) \sin \omega_k t + 2\delta \omega_k \cos \omega_k t\right],$$
(5.222)

$$\delta = \frac{r}{2\rho}$$

$$b_k = \frac{(2k + 1)\pi}{2L}$$

$$\omega_k = \sqrt{c^2 b_k^2 - \delta^2}.$$

The solution for velocity will be

$$\frac{v}{v_0} = V(x, t),$$
(5.223)

where

$$V(x, t) = \frac{4}{\pi} e^{-\delta t} \sum_{k=0}^{\infty} \frac{\cos b_k x}{(2k + 1) \sin b_k L} \left(\cos \omega_k t + \frac{\delta}{\omega_k} \sin \omega_k t\right).$$
(5.224)

Reference

Dracos, Th. (1970) Die Berechnung istatationärer Abfüsse in offenen Gerinnen beliebiger Geometrie, *Schweizerische Bauzeitung*, 88. Jahrgang Heft 19.

Further reading

Abbot, M.B. (1979) *Computational Hydraulics – Elements of the Theory of Surface Flow*. Pitman, Boston.

Budak, B.M., Samarskii, A.A., and Tikhonov, A.N. (1980) *Collection of Problems on Mathematical Physics* [in Russian]. Nauka, Moscow.

Cunge, J.A., Holly, F.M., and Verwey, A. (1980) *Practical Aspects of Computational River Hydraulics*. Pitman Advanced Publishing Program, Boston.

Davis, C.V. and Sorenson, K.E. (1969) *Handbook of Applied Hydraulics*. 3th edn, McGraw-Hill Co., New York.

Fox, J.A. (1977) *Hydraulic Analysis of Unsteady Flow in Pipe Networks*. Macmillan Press Ltd, London, UK; Wiley, New York, USA.

Godunov, S.K. (1971) *Equations of Mathematical Physics* (in Russian: Uravnjenija matematičjeskoj fizici). Izdateljstvo Nauka, Moskva.

Jeffrey, A. (1976) *Quasilinear Hyperbolic System and Waves*, Pitman, Boston.

Jović, V. (1987) Modelling of non-steady flow in pipe networks, *Proc. 2nd Int. Conf. NUMETA '87*, Martinus Nijhoff Pub., Swansea.

Jović, V. (2006) *Fundamentals of Hydromechanics* (in Croatian: Osnove hidromehanike). Element, Zagreb.

Jović, V. (1995) Finite elements and the method of characteristics applied to water hammer modelling, *Engineering Modelling*, 8: 51–58.

Polyanin, A.D. (2002) *Handbook of Linear Partial Differential Equations for Engineers and Scientists.* Chapman & Hall/CRC, London.

Sears, F.W. (1953) *An Introduction To Thermodynamics, The Kinetic Theory Of Gases And Statistical Mechanics.* Addison-Wesley Pub. Co, New York.

Smirnov, D.N. and Zubob, L.B. (1975) *Water Hammer in Pressurised Pipelines* (in Russian: Gidravličjeskij udar v napornjih vodovodah). Stroiizdat, Moskva.

Streeter, V.L. and Wylie, E.B. (1993) *Fluid Transients.* FEB Press, Ann Arbor, Mich.

Walton, A.J. (1976) *Three Phases of Matter.* McGraw Hill Co, New York.

Watters, G.Z. (1984) *Analysis and Control of Unsteady Flow in Pipe Networks.* Butterworths, Boston.

Wylie, E.B. and Streeter, V.L. (1993) *Fluid Transients in Systems.* Prentice Hall, Englewood Cliffs, New Jersey, USA.

6

Modelling of Non-steady Flow of Compressible Liquid in Pipes

6.1 Solution by the method of characteristics

6.1.1 Characteristic equations

Equations of non-steady flow in pipes can be written as ordinary differential equations along characteristic curves (characteristics) $\gamma^\pm$ defined by equations

$$\gamma(l, t)^\pm : \quad \frac{dl}{dt} = v \pm c, \tag{6.1}$$

where $w^\pm = v \pm c$ is the absolute velocity of elementary wave propagation. The characteristics are trajectories of positive and negative elementary waves. Equations that describe wave functions $\Gamma^\pm$ can be applied along the characteristics

$$\frac{d\Gamma^\pm}{dt} + g\left(J_e \pm \frac{v}{c}J_0\right) = 0, \tag{6.2}$$

where

$$\Gamma^\pm = v \pm \frac{g}{c}h, \tag{6.3}$$

that is written in the form

$$\Gamma(l, t)^\pm : \quad \frac{d}{dt}\left(v \pm \frac{g}{c}h\right) + g\left(J_e \pm \frac{v}{c}J_0\right) = 0. \tag{6.4}$$

Thus, in non-steady flow of an incompressible liquid $v \ll c$, the term with the gradient J_0 of the pipe axis can be omitted. Also, due to relatively small velocities v, $w^\pm \approx \pm c$, the characteristics become lines, and it can be written

$$\frac{dl}{dt} = \pm c, \tag{6.5}$$

$$\frac{d\Gamma^\pm}{dt} + g J_e = 0. \tag{6.6}$$

Analysis and Modelling of Non-Steady Flow in Pipe and Channel Networks, First Edition. Vinko Jović.
© 2013 John Wiley & Sons, Ltd. Published 2013 by John Wiley & Sons, Ltd.

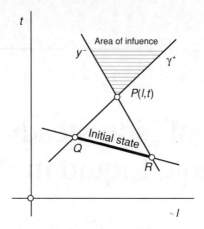

Figure 6.1   Dependence and influence area for the point within the mesh of characteristics.

The gradient of the energy line is calculated from the Darcy–Weisbach equation for the steady flow

$$J_e = \frac{\lambda}{2gD} |v| \, v. \tag{6.7}$$

Flow variables are calculated from

$$v = \frac{1}{2} \left(\Gamma^+ + \Gamma^- \right), \tag{6.8}$$

$$h = \frac{c}{2g} \left(\Gamma^+ - \Gamma^- \right). \tag{6.9}$$

6.1.2   Integration of characteristic equations, wave functions

Integration of ordinary differential equations is possible for prescribed initial conditions, that is the starting point is the prescribed stage. If in the plane l, t two points Q and R are selected with the prescribed values of the solution, then characteristic equations (6.5) will define the stage in the point P at the intersection of positive and negative characteristics, see Figure 6.1.

Integration of Eqs (6.5) and (6.6) from the initial prescribed stage to the point P, once along the positive, then along the negative characteristic, is written as

$$\int_Q^P \frac{dl}{dt} dt = + \int_Q^P c \, dt, \quad \int_R^P \frac{dl}{dt} dt = - \int_R^P c \, dt, \tag{6.10}$$

$$\int_Q^P \left(\frac{d\Gamma^+}{dt} + g J_e \right) dt = 0, \quad \int_R^P \left(\frac{d\Gamma^-}{dt} + g J_e \right) dt = 0. \tag{6.11}$$

Integration of Eq. (6.10) defines the coordinates l_P, t_P of the point P

$$l_P - l_Q = c(t_P - t_Q),\tag{6.12}$$

$$l_P - l_R = c(t_P - t_R),\tag{6.13}$$

while integration of Eq. (6.11) will give two algebraic equations to determine the unknown values $\Gamma_P^{\pm}$ in the point P

$$\Gamma_P^+ - \Gamma_Q^+ + g \bar{J}_e\Big|_Q^P (t_P - t_Q) = 0,\tag{6.14}$$

$$\Gamma_P^- - \Gamma_R^- + g \bar{J}_e\Big|_R^P (t_P - t_R) = 0.\tag{6.15}$$

Integration of the term with the resistances is carried out based on the mean value of the integral theorem

$$\bar{J}_e\Big|_Q^P = \frac{1}{2}\left(J_{eQ} + J_{eP}\right),\tag{6.16}$$

$$\bar{J}_e\Big|_R^P = \frac{1}{2}\left(J_{eR} + J_{eP}\right),\tag{6.17}$$

where the values of the gradient are calculated from the corresponding velocities, according to the formula

$$J_e = \frac{\lambda}{2gD}|v|\,v.\tag{6.18}$$

If the expression (6.8) is applied to the gradient of the energy line, then

$$J_e = \frac{\lambda}{8gD}\left|(\Gamma^+ + \Gamma^-)\right|(\Gamma^+ + \Gamma^-):\tag{6.19}$$

$$\bar{J}_e\Big|_Q^P = \frac{\lambda_Q}{16gD}\left|(\Gamma_Q^+ + \Gamma_Q^-)\right|(\Gamma_Q^+ + \Gamma_Q^-) + \frac{\lambda_P}{16gD}\left|(\Gamma_P^+ + \Gamma_P^-)\right|(\Gamma_P^+ + \Gamma_P^-),\tag{6.20}$$

$$\bar{J}_e\Big|_R^P = \frac{\lambda_R}{16gD}\left|(\Gamma_R^+ + \Gamma_R^-)\right|(\Gamma_R^+ + \Gamma_R^-) + \frac{\lambda_P}{16gD}\left|(\Gamma_P^+ + \Gamma_P^-)\right|(\Gamma_P^+ + \Gamma_P^-).\tag{6.21}$$

The unknown values Γ_P^+, Γ_P^- can be calculated by iteration; thus, according to Eq. (6.14) and Eq. (6.15), the iterative form can be written as

$$\Gamma_P^+ = f_1\left(\Gamma_P^+, \Gamma_P^-\right) = \Gamma_Q^+ - g \bar{J}_e\Big|_Q^P (t_P - t_Q),\tag{6.22}$$

$$\Gamma_P^- = f_1\left(\Gamma_P^+, \Gamma_P^-\right) = \Gamma_R^- - g \bar{J}_e\Big|_R^P (t_P - t_R)\tag{6.23}$$

which converges quickly for the prescribed initial iterative values

$$\Gamma_P^+ = \Gamma_Q^+,\tag{6.24}$$

$$\Gamma_P^- = \Gamma_R^-.\tag{6.25}$$

6.1.3   Integration of characteristic equations, variables h, v

Integration of Eqs (6.5) and (6.4) from the initial prescribed stage to the point P, once along the positive, then along the negative characteristic, is written as

$$\int_Q^P \frac{dl}{dt} dt = + \int_Q^P c\, dt, \quad \int_R^P \frac{dl}{dt} dt = - \int_R^P c\, dt, \tag{6.26}$$

$$\int_Q^P \left[\frac{d}{dt}\left(v + \frac{g}{c} h \right) + g J_e \right] dt = 0, \quad \int_R^P \left[\frac{d}{dt}\left(v - \frac{g}{c} h \right) + g J_e \right] dt = 0. \tag{6.27}$$

Integration of Eq. (6.26) defines the coordinates l_P, t_P of the point P

$$l_P - l_Q = c(t_P - t_Q), \tag{6.28}$$

$$l_P - l_R = c(t_P - t_R), \tag{6.29}$$

while integration of Eq. (6.27) will give two algebraic equations to determine the unknown values h_P, v_P in the point P

$$v_P - v_Q + \frac{g}{c}(h_P - h_Q) + g\bar{J}_e\big|_Q^P (t_P - t_Q) = 0, \tag{6.30}$$

$$v_P - v_R - \frac{g}{c}(h_P - h_R) + g\bar{J}_e\big|_R^P (t_P - t_R) = 0. \tag{6.31}$$

Integration of the term with the resistances is carried out based on the mean value of the integral theorem

$$\bar{J}_e\big|_Q^P = \frac{1}{2}\left(J_{eQ} + J_{eP} \right), \tag{6.32}$$

$$\bar{J}_e\big|_R^P = \frac{1}{2}\left(J_{eR} + J_{eP} \right), \tag{6.33}$$

where the values of the gradient are calculated from the corresponding velocities, according to the formula

$$J_e = \frac{\lambda}{2gD} |v|\, v. \tag{6.34}$$

The unknown values h_P, v_P can be calculated by iteration. The sum of Eqs (6.30) and (6.31) gives

$$2v_P - v_Q - v_R - \frac{g}{c}(h_Q - h_R) + g\bar{J}_e\big|_Q^P (t_P - t_Q) + g\bar{J}_e\big|_R^P (t_P - t_R) = 0 \tag{6.35}$$

from which

$$v_P = \frac{1}{2}\left[(v_Q + v_R) + \frac{g}{c}(h_Q - h_R) - g\bar{J}_e\big|_Q^P (t_P - t_Q) - g\bar{J}_e\big|_R^P (t_P - t_R) \right], \tag{6.36}$$

while the difference is

$$-v_Q + v_R + 2\frac{g}{c}h_P + \frac{g}{c}(-h_Q - h_R) + g\bar{J}_e\big|_Q^P (t_P - t_Q) - g\bar{J}_e\big|_R^P (t_P - t_R) = 0 \tag{6.37}$$

from which

$$h_P = \frac{c}{2g} \left[(v_Q - v_R) + \frac{g}{c}(h_Q + h_R) - g\bar{J}_e\big|_Q^P (t_P - t_Q) + c\bar{J}_e\big|_R^P (t_P - t_R) \right]. \tag{6.38}$$

Iteration converges quickly for the initial values

$$v_P = \frac{1}{2} \left[(v_Q + v_R) + \frac{g}{c}(h_Q - h_R) \right]. \tag{6.39}$$

6.1.4 The water hammer is the pipe with no resistance

Mesh of the characteristics

The problem is described by characteristic equations in the form

$$\gamma^\pm : \quad \frac{dx}{dt} = \pm c. \tag{6.40}$$

According to the water hammer classical analyses resistances are negligible; thus, the wave functions are described by equation

$$\frac{d\Gamma^\pm}{dt} = 0, \tag{6.41}$$

where the values of the wave functions are

$$\Gamma^\pm = v \pm \frac{g}{c} h. \tag{6.42}$$

Let us observe propagation of the positive and negative wave in an infinite pipe with steady initial conditions, that is a piezometric head and velocity $h(x, 0)$, $v(x, 0)$.

Propagation of the positive wave front in the plane x, t is the line obtained by integration of the equation of the positive characteristic γ^+

$$x - x_0 = c(t - t_0). \tag{6.43}$$

This line characterizes the positive wave front that was at the position x_0 at the time t_0, that is the constant $\gamma^+ = x_0 - ct_0$ can be applied. The value of the wave function Γ^+ is constant along the characteristic line γ^+

$$\Gamma^+(\gamma^+) = \text{const}, \tag{6.44}$$

that is

$$\Gamma^+ = v + \frac{g}{c} h = v_0 + \frac{g}{c} h_0 = \Gamma_0^+. \tag{6.45}$$

Something similar can be applied to the negative wave

$$x - x_0 = -c(t - t_0). \tag{6.46}$$

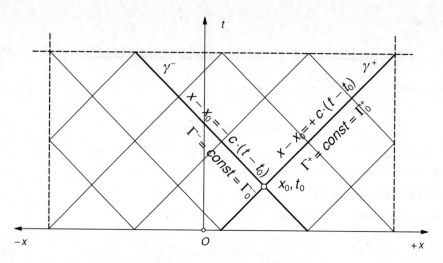

Figure 6.2 Mesh of characteristics.

The wave function of the negative wave Γ^- is constant along the characteristic line γ^-

$$\Gamma^-(\gamma^-) = const, \qquad (6.47)$$

that is

$$\Gamma^- = h - h_0 = -\frac{c}{g}(v - v_0) = \Gamma_0^-. \qquad (6.48)$$

The mesh of characteristics is a set of lines of the form (6.43) and (6.46) complete with wave functions of the form (6.45) and (6.48), as shown in Figure 6.2.

Furthermore, velocities and piezometric heads on any point of the plane x, t can be calculated from the prescribed values of the wave functions

$$v = \frac{1}{2}\left(\Gamma^+ + \Gamma^-\right), \qquad (6.49)$$

$$h = \frac{c}{2g}\left(\Gamma^+ - \Gamma^-\right). \qquad (6.50)$$

Discretization

If it is applied that

$$x = i\Delta x; \quad i = 1, 2, 3, \ldots m$$
$$t = k\Delta t; \quad k = 1, 2, 3, \ldots n, \qquad (6.51)$$

then each pair of the coordinates x, t has the corresponding discrete coordinates i, k, that is values of the wave functions $\Gamma(x, t)^\pm$ have the corresponding discrete values

$$\Gamma_{i,k}^+, \ \Gamma_{i,k}^-. \qquad (6.52)$$

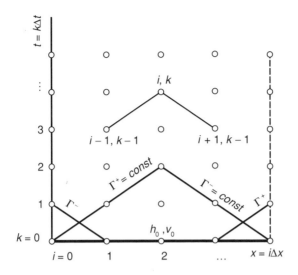

Figure 6.3 Discretization.

Likewise, the solution of the wave problem $h(x, t)$ and $v(x, t)$ is expressed in discrete coordinates

$$h_{i,k}, \quad v_{i,k}. \tag{6.53}$$

Figure 6.3 shows the discretization of the wave problem on a pipe of length L, which forms a regular equidistant rectangular mesh of nodes. In a regular discretization mesh real coordinates x, t are replaced by discrete coordinates i, k, while the discretization step is calculated in a manner such that pipe is divided into m segment $\Delta x = L/m$. The time step is calculated so the points will lie on characteristic lines $\gamma^{\pm}$, that is $\Delta t = \Delta x / c$.

Calculations

Within the regular discretization mesh, according to Eqs (6.44) and (6.47), a recursion is applied

$$\Gamma^+_{i,k} = \Gamma^+_{i-1,k-1}, \tag{6.54}$$

$$\Gamma^-_{i,k} = \Gamma^-_{i+1,k-1}. \tag{6.55}$$

Values of the solution in discrete coordinates can be calculated from the prescribed values of the wave function, according to expressions (6.49) and (6.50)

$$h_{i,k} = \frac{c}{2g} \left(\Gamma^+_{i,k} - \Gamma^-_{i,k} \right), \tag{6.56}$$

$$v_{i,k} = \frac{1}{2} \left(\Gamma^+_{i,k} + \Gamma^-_{i,k} \right). \tag{6.57}$$

Values of the wave functions from the prescribed values of the solution in discrete coordinates can be calculated from expressions (6.45) and (6.48)

$$\Gamma_{i,k}^+ = v_{i,k} + \frac{g}{c} h_{i,k}, \tag{6.58}$$

$$\Gamma_{i,k}^- = v_{i,k} - \frac{g}{c} h_{i,k}. \tag{6.59}$$

Initial and boundary conditions

Initial conditions are the steady ones $h(x, 0) = h_0$ and $v(x, 0) = v_0$, and are written in discrete coordinates

$$\left. \begin{array}{c} h_{i,0} = h_0 \\[2mm] v_{i,0} = v_0 \end{array} \right\} \; i = 0, 1, 2, 3, \ldots m. \tag{6.60}$$

If Eq. (6.42) is applied, initial values of the wave function are obtained

$$\left. \begin{array}{c} \Gamma_{i,0}^+ = v_0 + \dfrac{g}{c} h_0 \\[3mm] \Gamma_{i,0}^- = v_0 - \dfrac{g}{c} h_0 \end{array} \right\} \; i = 0, 1, 2, 3, \ldots m. \tag{6.61}$$

Boundary conditions are the most explicit functions; namely, the piezometric head $h(t)$ and velocity $v(t)$. However, they can also be implicit functions $f(h, v)$. On the boundaries $x = 0$ and $x = L$, the unknown wave function values are calculated from the boundary conditions. Since in the points for the values $i = 0 \; i \; k > 0$, see Figure 6.3, there are only wave function values for the negative wave, wave function values for the positive wave will be calculated from the boundary condition. Similarly, in the points for values $i = m$ and $k > 0$, there are only wave function values for the positive wave, while the wave function for the negative wave is calculated from the boundary condition.

If, the piezometric head $h_{r,k}$ is prescribed on the boundary $r = 0 \; ili \; r = m$, then the unknown value of the wave function is calculated from the expression

$$h_{r,k} = \frac{c}{2g} \left(\Gamma_{r,k}^+ - \Gamma_{r,k}^- \right). \tag{6.62}$$

Similarly, if the velocity $v_{r,k}$ is prescribed on the boundary $r = 0$ or m, then the unknown value of the wave function is calculated from the expression

$$v_{r,k} = \frac{1}{2} \left(\Gamma_{r,k}^+ + \Gamma_{r,k}^- \right). \tag{6.63}$$

Implicit boundary conditions, such as the *real closing law*, are complex boundary conditions where velocity is prescribed implicitly by the outflow formula

$$v(t) = \mu A(t) \sqrt{2g(h - z)} \tag{6.64}$$

where $A(t)$ is the time-dependent area of the outflow cross-section while μ is the discharge coefficient. Written in discrete coordinates

$$v_{r,k} = \mu A_k \sqrt{2g(h_{j,k} - z)}. \tag{6.65}$$

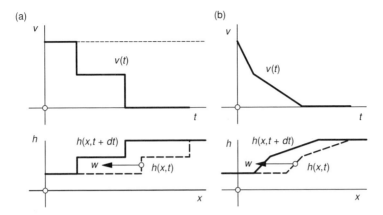

Figure 6.4 Character of boundary conditions and wave propagation (a) sudden and (b) gradual changes.

The unknown values $h_{r,k}$, $v_{r,k}$ can be calculated from the expressions (6.62), (6.63), and (6.65) as well as the value of the wave function $\Gamma_{r,k}$ of the positive or negative wave depending on the boundary $r = 0$ or $r = m$.

Sudden changes, front discontinuity

The character of the boundary condition change reflects the shape of the water hammer wave. Figure 6.4 shows examples of sudden and gradual change of velocity as boundary conditions. The same can be applied to sudden and gradual changes of pressure as a boundary condition.

Sudden changes of velocity or pressure are accompanied by a water hammer with a pronounced front with discontinuity. This discontinuity is kept permanently owing to the wave nature of the water hammer.

Unlike sudden changes, gradual changes of velocity or pressure are accompanied by a water hammer with a similar wave front with the continuity of basic variables $\mathbb{C}^0$.

Example

Figure 6.5 shows the wave function values for sudden and instantaneous change of velocity $v_0 \to 0$ at the end of the pipeline (sudden closing). Initial conditions are $h_0 = 0$ and $v_0 = 1$, while the water hammer ratio is $c/g = 100$. Values of the wave functions for the initial conditions are: $\Gamma_{i,0}^+ = v_0 = 1$ and $\Gamma_{i,0}^- = v_0 = 1$.

Boundary conditions are constant level h_0 on the left pipe end and the prescribed graph of outflow velocity variation on the right pipe end. Discrete values of boundary conditions for wave functions are obtained from the expressions (6.62) and (6.63)

$$x = 0 :\Rightarrow \Gamma_{0,k}^+ = \Gamma_{0,k}^-, \tag{6.66}$$

$$x = L :\Rightarrow \Gamma_{m,k}^- = -\Gamma_{m,k}^+. \tag{6.67}$$

A sudden change of velocity at the end of the pipeline generates the duality at the wave front profile. In front of the wave front the flow is undisturbed while behind it the flow is disturbed. For the disturbed state, the negative wave is defined by the velocity $v_{m,0+} = 0$ and it is written

$$v_{m,0+} = \frac{1}{2}\left(\Gamma_{m,0}^+ + \Gamma_{m,0+}^-\right),$$

$$0 = \frac{1}{2}\left(1 + \Gamma_{m,0+}^-\right) \Rightarrow \Gamma_{m,0+}^- = -1.$$

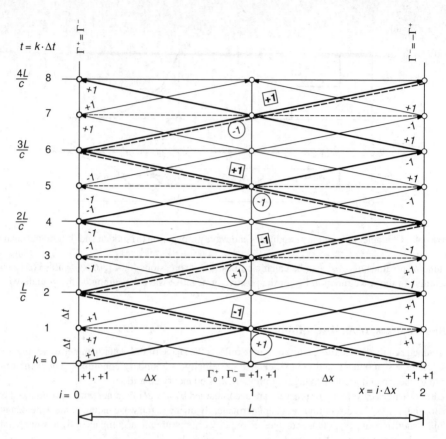

Figure 6.5 Mesh of characteristics for sudden closing.

Thus, both values are written by the wave function for the wave front. The wave function value which corresponds to the state in front of the wave is written in the circle, while the value corresponding to the state behind the wave front is written into square.

In Figure 6.5 the dashed line denotes the trajectory of the point *in front* and the continuous line the trajectory *behind* the water hammer front. Values of all wave functions are written by the corresponding characteristics.

The water hammer solution using the wave functions as shown in the figure clearly depicts the water hammer kinematics and reflection conditions on the boundaries. It can be said that the mesh of characteristics with the values of wave functions represents an accurate solution of the analyzed water hammer problem. It enables water hammer presentation at any moment. Thus, for example, Figure 6.6 shows the water hammer phase by direct reading of wave functions from the mesh of characteristics in Figure 6.5 and application of expressions (6.62) and (6.63).

Recursive calculation

A recursive calculation can be used to obtain the water hammer solution in the point x, t, as shown in Figure 6.7. Characteristic lines – positive and negative wave trajectories – are drawn from the

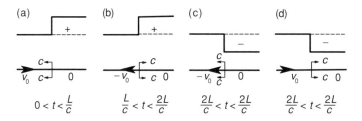

Figure 6.6 Kinematics of water hammer phases for sudden closing.

prescribed point backwards. At the points of the boundary, positive characteristics shift into negative ones and vice versa.

Construction ends when trajectories intersect the line $t = 0$, that is when they reach the initial conditions with the prescribed values h_0, v_0. This is followed by the backward phase, each trajectory is monitored from point to point and boundary values are calculated.

The procedure ends by calculation of the wave function values in the starting point; namely, the variables h, v. Numerical calculation is easily implemented in *Fortran* in the form of two recursive functions Fp(x,t) and Fm(x,t) for the calculation of the values of the positive and negative wave functions. A source for the interactive program solution, called ChtxRekurzija, is shown hereinafter and also in Program solutions & tests/Programs Chtx characteristics from www.wiley.com/go/jovic.

Boundary conditions (left – prescribed piezometric head, right – velocity) are written in the program module as the arrays tbdc, Hbdc, and Vbdc (polygonal graphs). The program module contains these functions as well as the function PgVal, which calculated the value of a polygonal graph for the prescribed argument.

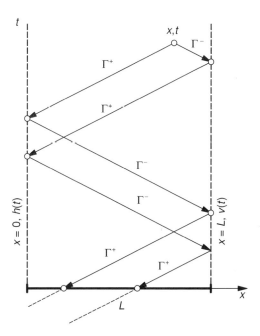

Figure 6.7 Recursion.

```fortran
program ChtxRekurzija
! water hammer in pipe
implicit none
real x,t,h,v
real g/9.81/,c/981/,L/1962/
real fpVal,fmVal,Tau0
real h0/0/,v0/1/
! time bdc
real tbdc(2)/0,0.01/
! right bdc: prescribed v(t)
real Vbdc(2)/1,0/
! left bdc: prescribed h(t)
real Hbdc(2)/0,0/
  Tau0=2*L/c     ! cycle of water hammer
  write(*,*) "Tau0",Tau0
  do while (.True.)
    write(*,'(a\)') "x,t: "
    read(*,*) x,t
    fpVal = Fp(x,t)
    fmVal = Fm(x,t)
    h = c/(2*g)*(fpVal-fmVal)
    v = (fpVal+fmVal)/2
    write(*,*) "x,t,Fp,Fm",x,t,fpVal,fmVal
    write(*,'(a,4f10.3)') "x,t,h,v",x,t,h,v
  enddo
contains
  recursive real function Fp(x,t)
  real x,t,xx,tt,hh,vv
    xx=0
    tt=t-x/c
    if(tt.gt.0) then
      hh=PgVal(size(Hbdc),tbdc,Hbdc,tt)
      Fp=2*g/c*hh+Fm(xx,tt) ! BDC for x=0
    else
      Fp=v0+g/c*h0  ! apply initial conditions
    endif
  end function Fp
  recursive real function Fm(x,t)
  real x,t,value,xx,tt,hh,vv
    xx=L
    tt=t+(x-L)/c
    if(tt.gt.0) then
      vv=PgVal(size(Hbdc),tbdc,Vbdc,tt)
      Fm=2*vv-Fp(xx,tt) ! BDC for x=L
    else
      Fm=v0-g/c*h0  ! apply initial conditions
    endif
  end function Fm
  real Function PgVal(Ndata, x, y, xval)
```

Figure 6.8   Interactive program for recursion.

```
! returns value of polygonal graf
integer Ndata
real x(Ndata),y(ndata),xval
real nagib
integer i
  If (xval .le. x(1)) Then
    PgVal = y(1)
  ElseIf (xval .gt. x(Ndata)) Then
    PgVal = y(Ndata)
  Else
    do i = 2 , Ndata
      If (xval .eq. x(i)) Then
        PgVal = y(i)
        return
      ElseIf (xval .lt. x(i)) Then
        nagib = (y(i) - y(i - 1)) / (x(i) - x(i - 1))
        PgVal = y(i - 1) + nagib * (xval - x(i - 1))
        return
      End If
    enddo
  End If
End Function PgVal
end program
```

Figure 6.8 (*Continued*)

6.1.5   Water hammers in pipes with friction

Discretization

Let us observe a water hammer in a pipe after a sudden arrest of the flow with friction resistances. In the steady flow, water flows through the pipe at the velocity v_0. At the left end there is a water tank with the prescribed constant level h_0. A pipe of length L and diameter D has the absolute hydraulic roughness ε. The flow stops instantaneously.

The problem will be solved by the method of characteristics using numeric integration of wave functions in an equidistant mesh of characteristics, complete with monitoring of the discontinuity at the water hammer front. Discretization of the x axis into m equal segments of the length $\Delta x = L/m$ defines discretization of the time axis with the step $\Delta t = \Delta x/c$. Real coordinates x, t will be replaced by discrete coordinates $i\Delta x, k\Delta t$; thus forming the discretization mesh shown in Figure 6.9.

In order to monitor the discontinuity at the water hammer front, a function shall be defined which determines whether the node is on the front trajectory and its affiliation with the type of discontinuity characteristic

$$tip(i, k) \begin{cases} = +1 \ for \ \gamma^+ \\ = -1 \ for \ \gamma^- \\ = 0 \ not \ on \ front \end{cases} . \tag{6.68}$$

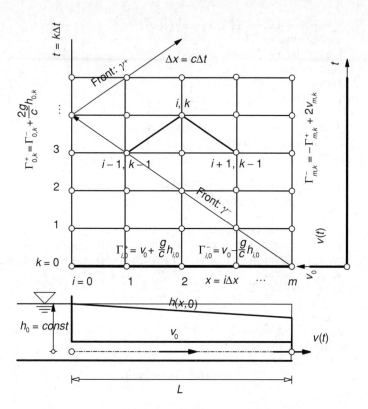

Figure 6.9   Discretization.

Initial conditions

Initial conditions are the steady flow with the prescribed velocity v_0 and piezometric head line $h(x, 0)$. Wave functions of initial conditions in discrete coordinates are

$$\Gamma_{i,0}^+ = v_0 + \frac{g}{c} h_{i,0},\tag{6.69}$$

$$\Gamma_{i,0}^- = v_0 - \frac{g}{c} h_{i,0}.\tag{6.70}$$

At the time $t = 0^+$ the outflow velocity is changing instantaneously $v_0 \to 0$

$$v\left(L, 0^+\right) = v_{m,0^+} = 0,\tag{6.71}$$

thus, the discontinuity occurs at the wave front and the respective value of the negative wave function

$$\hat{\Gamma}_{m,0^+}^- = v_{m,0^+} - \Gamma_{i,0}^+.\tag{6.72}$$

where "^" marks the values on characteristics with discontinuities.

Calculations, iteration

Discrete values of wave functions in points $0 < i < m$, $k > 0$ are calculated according to expressions (6.22) and (6.23), iteratively

$$\Gamma_{i,k}^{+} = \Gamma_{i-1,k-1}^{+} - g \bar{J}_{e}\Big|_{(i-1,k-1)}^{(i,k)} \Delta t, \tag{6.73}$$

$$\Gamma_{i,k}^{-} = \Gamma_{i+1,k-1}^{-} - g \bar{J}_{e}\Big|_{(i+1,k-1)}^{(i,k))} \Delta t, \tag{6.74}$$

where

$$\bar{J}_{e}\Big|_{(i-1,k-1)}^{(i,k)} = \frac{1}{2}\left(J_{(i-1,k-1)} + J_{(i,k)}\right), \tag{6.75}$$

$$\bar{J}_{e}\Big|_{(i+1,k-1)}^{(i,k)} = \frac{1}{2}\left(J_{(i+1,k-1)} + J_{(i,k)}\right). \tag{6.76}$$

The gradient of the energy line in discrete points is calculated based on the expression

$$J_{e} = \frac{\lambda}{2gD}\,|v|\,v. \tag{6.77}$$

The values of the piezometric head and velocity in discrete coordinates are calculated from the wave functions, according to

$$h_{i,k} = \frac{c}{2g}\left(\Gamma_{i,k}^{+} - \Gamma_{i,k}^{-}\right), \tag{6.78}$$

$$v_{i,k} = \frac{1}{2}\left(\Gamma_{i,k}^{+} + \Gamma_{i,k}^{-}\right). \tag{6.79}$$

For the nodes on the positive characteristic $\hat{\gamma}^{+}$, trajectories of the wave front are calculated from the values of the wave function of the discontinuity $\hat{\Gamma}_{i,k}^{+}$ using the stage along the positive characteristic $\hat{\gamma}^{+}$ of discontinuity. The value of the wave function $\hat{\Gamma}_{i,k}^{-}$ is calculated along the negative characteristic γ^{-}.

For the nodes on the negative characteristic $\hat{\gamma}^{-}$ trajectories of the wave front are calculated from the values of the wave function of the discontinuity $\hat{\Gamma}_{i,k}^{-}$ using the stage along the negative characteristic $\hat{\gamma}^{-}$ of discontinuity. The value of the wave function $\hat{\Gamma}_{i,k}^{+}$ is calculated along the positive characteristic γ^{+}.

For the points on the wave front trajectory, the following can be applied

$$\hat{h}_{i,k} = \frac{c}{2g}\left(\hat{\Gamma}_{i,k}^{+} - \hat{\Gamma}_{i,k}^{-}\right), \tag{6.80}$$

$$\hat{v}_{i,k} = \frac{1}{2}\left(\hat{\Gamma}_{i,k}^{+} + \hat{\Gamma}_{i,k}^{-}\right). \tag{6.81}$$

Boundary conditions

The following expressions are used for determining the discrete value of wave functions on the boundaries, that is in points $r = 0$ or $r = m$

$$h_{r,k} = \frac{c}{2g}\left(\Gamma_{r,k}^{+} - \Gamma_{r,k}^{-}\right), \tag{6.82}$$

$$v_{r,k} = \frac{1}{2}\left(\Gamma_{r,k}^{+} + \Gamma_{r,k}^{-}\right). \tag{6.83}$$

On the left side $x = 0$, the piezometric head is prescribed, and the unknown value of the positive wave shall be calculated

$$\Gamma^+_{0,k} = \Gamma^-_{0,k} + \frac{2g}{c} h_{0,k}. \tag{6.84}$$

Value of the positive wave function $\hat{\Gamma}^+_{0,k}$ for the wave front is calculated as

$$\hat{\Gamma}^+_{0,k} = \hat{\Gamma}^-_{0,k} + \frac{2g}{c} h_{0,k}. \tag{6.85}$$

On the right side $x = L$, velocity is prescribed, and the unknown value of the negative wave shall be calculated

$$\Gamma^-_{m,k} = -\Gamma^+_{m,k} + 2v_{m,k}. \tag{6.86}$$

The value of the negative wave function $\hat{\Gamma}^-_{m,k}$. for the wave front is calculated as

$$\hat{\Gamma}^-_{m,k} = -\hat{\Gamma}^+_{m,k} + 2v_{m,k}. \tag{6.87}$$

Sensitivity tests of the solution to discretization

In Program solutions & tests/ Programs Chtx characteristics from www.wiley.com/go/jovic, a source of the program for the solution of the water hammer (with wave head discontinuity) in the pipe with friction resistances is given in

                 Program ChtxFrictionDisc

for sudden valve closing. A solution of wave functions by the method of characteristics is analyzed, together with the monitoring of the water hammer front.

     The data on the pipe and other data are defined within the program and are changed easily. Water hammer calculations were carried out for a pipe of $L = 1000$ m length, $D = 500$ mm in diameter, $\varepsilon = 1$ mm hydraulic roughness, $v_0 = 5$ m/s initial velocity and tank level $h_0 = 100$ m, for pipe discretization into $m = 1, 2, 4$, and 8 equal segments $\Delta x = L/m$. Water hammer celerity is $c = 1000$ m/s, see the test example shown in Figure 6.10.

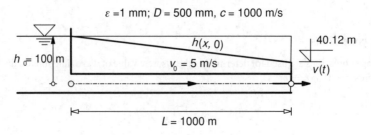

Figure 6.10   Test example.

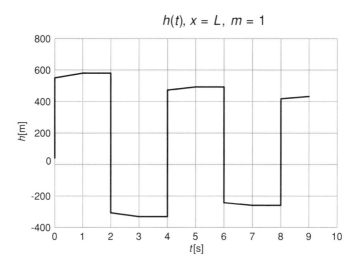

Figure 6.11 Discretization $m = 1$, $\Delta x = L$.

Figure 6.11, Figure 6.12, Figure 6.13, and Figure 6.14 show piezometric heads $h(t)$ at the pipe end; namely, $x = L$, for each discretization m.

Note that integration of the term with resistances is sensitive to discretization size m, as was expected since resistances are non-linear. Thus, on the example $m = 1$, a break point is observed on the curve within every water hammer cycle. The shape of the curve is surprising; namely, it is an inclined line to the breaking point and then constant after the breaking point.

Similar phenomenon can be observed for other discretization sizes m, occurring m times within the water hammer cycle. The curve gradually approaches the line. In limits, when $m \to \infty$ solution approaches an accurate water hammer solution with the friction resistances.

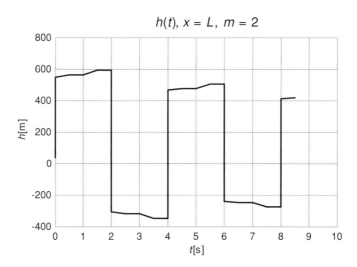

Figure 6.12 Discretization $m = 2$, $\Delta x = L/2$.

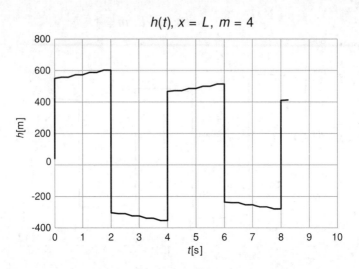

Figure 6.13   Discretization $m = 4$, $\Delta x = L/4$.

Recursive calculation with friction

Similar to the recursive calculation of wave functions for the water hammer without friction resistances, recursive calculation of water hammer with friction resistances can be carried out. Program solutions & tests/Programs Chtx characteristics from www.wiley.com/go/jovic provides a code of fortran recursive program for a water hammer

```
Program ChtxTrenjeRekurzija
```

which uses two recursive functions Fp(x,t) and Fm(x,t) for calculations of wave function values of the positive and negative waves.

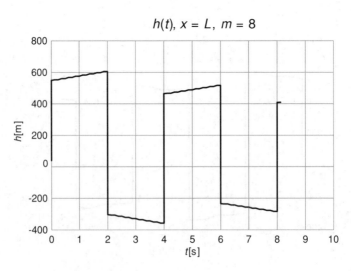

Figure 6.14   Discretization $m = 8$, $\Delta x = L/8$.

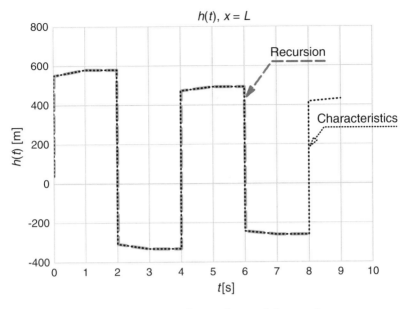

Figure 6.15 Recursion vs. characteristics $m = 1$.

Figure 6.15 shows the comparison between the recursive calculation of the piezometric head at the closing profile and the solution by the method of characteristics for discretization $m = 1$ in the pipe according to Figure 6.10 from the previous examples.

It was shown that recursion with friction resistances is equal to the method of characteristics, as was expected. In both cases, the resistance term is integrated over an interval equal to the pipe length L. Insisting on an accurate recursive calculation leads to the need for insertion of new recursive points on backward trajectories, as was drawn as an idea in Figure 6.16. However, this idea has not been realized yet.

Comparison with analytical solution

Equations of non-steady flow in pipes in the linearized form of the resistance term are written as

$$\frac{1}{\rho c^2}\frac{\partial p}{\partial t} + \frac{\partial v}{\partial x} = 0, \tag{6.88}$$

$$\rho \frac{\partial v}{\partial t} + \frac{\partial p}{\partial x} + rv = 0, \tag{6.89}$$

where the resistance constant is calculated from the steady flow piezometric line

$$r = \frac{p_0 - p_L}{v_0 L}.$$

Values are $p_0 = \rho g h_0$ and $p_L = \rho g h_L$ where

$$h_L = h_0 - \frac{\lambda L}{D}\frac{v_0^2}{2g}$$

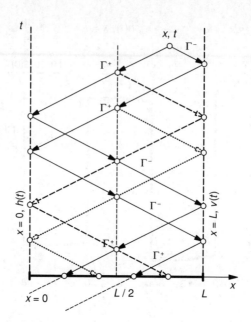

Figure 6.16   Recursive calculation improvement.

is the piezometric head at the end of the pipeline. Test data are shown in Figure 6.10. Comparison between the calculated analytical solution acccording to solution 5.220, 5.223, and numerical solution by the method of characteristics (discretization $m = 8$) is shown in Figure 6.17. If the numerical solution for the adopted discretization is considered "accurate" it can be observed that in the first water hammer cycle solutions correspond to each other. The differences are observed in subsequent cycles because the

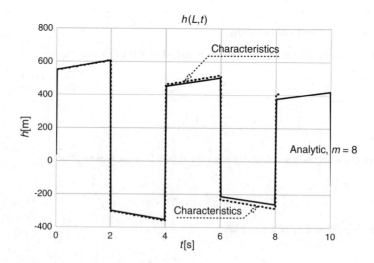

Figure 6.17   Comparison of analytical and numerical solution, $x = L$.

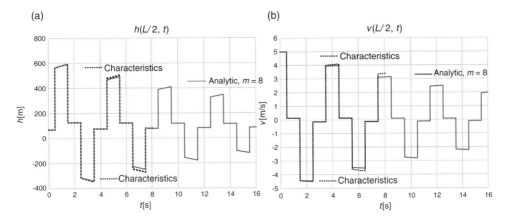

Figure 6.18 Comparison of analytical and numerical solution, $x = L/2$.

constant by the resistance term does not change in analytical solution. These results are expected. In analytical solution, the dumping parameter remains the same as the one from the first cycle, while in numerical solution it changes based on the Reynolds number.

Figure 6.18a and b shows comparison of the results for the pipe cross-section $x = L/2$, where, apart from the change of the piezometric head, velocity change is also presented. There is good correspondence in the first water hammer cycles.

Note that a great number of analytical solution order members shall be taken into account in order to achieve satisfactory accuracy. Thus, in comparison with the method of characteristics, it shall be considered a far "more expensive" solution.

6.2   Subroutine `UnsteadyPipeMtx`

For modelling of non-steady flow of a compressible liquid, a subroutine `UnsteadyPipeMtx` shall be updated by calling the subroutines `FemUnsteadyPipeMtx` and `ChtxUnsteadyPipeMtx` for the finite element matrix and vector computation:

```
. . .
    if runmode = QuasiUnsteady then
      call QuasiUnsteadyPipeMtx(ielem)
    else if runmode = RigidUnsteady then
      call RgdUnsteadyPipeMtx(ielem)
    else if runmode = FullUnsteady then
      if(FemPipes) then
         call FemUnsteadyPipeMtx(ielem)
      else
         call ChtxUnsteadyPipeMtx(ielem)
      endif
    endif
. . .
```

Parameter `FemPipes` is an optional *logical parameter* with the implemented *default* value `.false.`.

6.2.1   Subroutine FemUnsteadyPipeMtx

FEM integration, the conservation law procedure

Calculation starts from the equation of continuity for the non-steady flow of elastic liquid in pipes in the form

$$\frac{gA}{c^2}\frac{\partial h}{\partial t} + \frac{\partial Q}{\partial l} = 0, \tag{6.90}$$

and dynamic equation in the form

$$\frac{1}{gA}\frac{\partial Q}{\partial t} + \frac{\partial H}{\partial l} + \frac{\lambda}{2gDA^2}Q|Q| = 0, \tag{6.91}$$

where specific energy $H(h, Q) = h + \frac{Q^2}{2gA^2}$ is expressed by the piezometric head and discharge. Introducing the symbol for

$$\beta = \frac{\lambda}{2gDA^2} \tag{6.92}$$

The dynamic equation becomes

$$\frac{1}{gA}\frac{\partial Q}{\partial t} + \frac{\partial H}{\partial l} + \beta|Q|Q = 0. \tag{6.93}$$

The equation of continuity can be integrated along the finite element

$$\int_{\Delta l}\left(\frac{gA}{c^2}\frac{\partial h}{\partial t} + \frac{\partial Q}{\partial l}\right)dl = 0 \tag{6.94}$$

which gives the equation of the mass conservation law on a finite element

$$\frac{gA}{c^2}\int_{\Delta l}\frac{\partial h}{\partial t}dl + (Q_2 - Q_1) = 0. \tag{6.95}$$

Using the mean value of the integral theorem, it can be written

$$\frac{gA}{c^2}\frac{\Delta l}{2}\left(\frac{\partial h_1}{\partial t} + \frac{\partial h_2}{\partial t}\right) + (Q_2 - Q_1) = 0. \tag{6.96}$$

After integration with the time step Δt

$$\frac{gA}{c^2}\int_{\Delta t}\frac{\Delta L}{2}\left(\frac{\partial h_1}{\partial t} + \frac{\partial h_2}{\partial t}\right)dt + \int_{\Delta t}(Q_2 - Q_1)\,dt = 0 \tag{6.97}$$

and repetitive reusing of the aforementioned time-dependent integration rules, the first elemental equation is obtained

$$F_1: \qquad \begin{aligned} &\frac{\Delta l}{2}\frac{gA}{c^2}\left[\left(h_1^+ + h_2^+\right) - (h_1 + h_2)\right] + \\ &+ (1 - \vartheta)\,\Delta t\,(Q_2 - Q_1) + \vartheta\,\Delta t\left(Q_2^+ - Q_1^+\right) = 0. \end{aligned} \qquad (6.98)$$

Integration of the dynamic equation along the finite element

$$\int\limits_{\Delta l}\left(\frac{1}{gA}\frac{\partial Q}{\partial t} + \frac{\partial H}{\partial l} + \beta\,|Q|\,Q\right)dl = 0 \qquad (6.99)$$

gives the energy conservation law (in head form) on the finite element

$$\frac{1}{gA}\int\limits_{\Delta l}\frac{\partial Q}{\partial t}\,dl + H_2 - H_1 + \int\limits_{\Delta l}\beta\,|Q|\,Q\,dl = 0. \qquad (6.100)$$

If the mean value of the integral theorem is applied, then

$$\frac{\Delta l}{2gA}\left(\frac{\partial Q_1}{\partial t} + \frac{\partial Q_2}{\partial t}\right) + H_2 - H_1 + \int\limits_{\Delta l}\beta\,|Q|\,Q\,dl = 0. \qquad (6.101)$$

According to the mean value of the integral theorem, the resistance term is written as

$$\int\limits_{\Delta l}\beta\,|Q|\,Q\,dl = \frac{\Delta l}{2}\left(\beta_1\,|Q_1|\,Q_1 + \beta_2\,|Q_2|\,Q_2\right); \qquad (6.102)$$

thus, integration along the element will be

$$\frac{\Delta l}{2gA}\left(\frac{\partial Q_1}{\partial t} + \frac{\partial Q_2}{\partial t}\right) + H_2 - H_1 + \frac{\Delta l}{2}\left(\beta_1\,|Q_1|\,Q_1 + \beta_2\,|Q_2|\,Q_2\right) = 0. \qquad (6.103)$$

After integration with the time step Δt

$$\begin{aligned} &\frac{\Delta l}{2gA}\int\limits_{\Delta t}\left(\frac{\partial Q_1}{\partial t} + \frac{\partial Q_2}{\partial t}\right)dt + \int\limits_{\Delta t}(H_2 - H_1)\,dt + \\ &+ \frac{\Delta l}{2}\int\limits_{\Delta t}\left(\beta_1\,|Q_1|\,Q_1 + \beta_2\,|Q_2|\,Q_2\right)dt = 0 \end{aligned} \qquad (6.104)$$

finally, the second elemental equation is obtained in the form

$$F_2: \quad \begin{aligned} & \frac{\Delta l}{2gA} \left[(Q_1^+ + Q_2^+) - (Q_1 + Q_2) \right] + \\ & + (1 - \vartheta)\Delta t (H_2 - H_1) + \vartheta \Delta t \left(H_2^+ - H_1^+ \right) + \\ & + (1 - \vartheta)\Delta t \left(\beta_1 |Q_1| Q_1 + \beta_2 |Q_2| Q_2 \right) \frac{\Delta l}{2} + \\ & + \vartheta \Delta t \left(\beta_1^+ |Q_1^+| Q_1^+ + \beta_2^+ |Q_2^+| Q_2^+ \right) \frac{\Delta l}{2} = 0. \end{aligned} \tag{6.105}$$

The Newton–Raphson iterative form for the finite element is formally written using the matrix-vector operations

$$[\underline{H}] \cdot [\Delta h^+] + [\underline{Q}] \cdot [\Delta Q^+] = [\underline{F}], \tag{6.106}$$

where matrix

$$[\underline{H}] = \begin{bmatrix} \dfrac{\Delta l}{2} \dfrac{gA}{c^2} & \dfrac{\Delta l}{2} \dfrac{gA}{c^2} \\ -\vartheta \Delta t & +\vartheta \Delta t \end{bmatrix} \tag{6.107}$$

and matrix

$$[\underline{Q}] = \begin{bmatrix} -\vartheta \Delta t & +\vartheta \Delta t \\ \left(\dfrac{\Delta l}{2gA} + \vartheta \Delta t \Delta l \beta_1^+ |Q_1^+| - \dfrac{\vartheta \Delta t}{gA^2} Q_1^+ \right) & \left(\dfrac{\Delta l}{2gA} + \vartheta \Delta t \Delta l \beta_2^+ |Q_2^+| + \dfrac{\vartheta \Delta t}{gA^2} Q_2^+ \right) \end{bmatrix}. \tag{6.108}$$

The right hand side vector is

$$[\underline{F}] = - \begin{bmatrix} F_1 \\ F_2 \end{bmatrix}. \tag{6.109}$$

The resistance term

$$\beta = \frac{\lambda \Delta l}{2gDA^2} \tag{6.110}$$

is calculated using the function aMoody, which returns accurate values for each $|Q| \neq 0$.
   Thus, for $Q = 0$ instead of the term $\beta |Q| Q$, the term $\beta^* Q$ shall be applied into Eq. (6.105), where

$$\beta^* = \frac{64\nu}{2g D^2 A} \tag{6.111}$$

because the flow is laminar. Likewise, the derivative $\vartheta \Delta t \Delta l \beta |Q|$ in matrix (6.108) shall be replaced by $\vartheta \Delta t \Delta l \beta^*/2$.

Finite element matrix and vector

Elemental discharge increments are calculated from the Newton–Raphson form of the elemental equation (6.106) in such a manner that the previous equation is calculated by the inverse matrix $[\underline{Q}]^{-1}$

$$\left[\Delta Q^{+}\right] = \left[\underline{Q}\right]^{-1}\left[\underline{F}\right] - \left[\underline{Q}\right]^{-1}\left[\underline{H}\right] \cdot \left[\Delta h^{+}\right]. \tag{6.112}$$

Introducing the symbols

$$\left[\underline{A}\right] = \left[\underline{Q}\right]^{-1}\left[\underline{H}\right] \tag{6.113}$$

$$\left[\underline{B}\right] = \left[\underline{Q}\right]^{-1}\left[\underline{F}\right] \tag{6.114}$$

an expression for elemental discharge increments is obtained

$$\left[\Delta Q^{+}\right] = \left[\underline{B}\right] - \left[\underline{A}\right] \cdot \left[\Delta h^{+}\right]. \tag{6.115}$$

The finite element matrix A^{e} and vector B^{e} for non-steady modelling have the form

$$A^{e} = \vartheta \, \Delta t \begin{bmatrix} +\underline{A}_{11} & +\underline{A}_{12} \\ -\underline{A}_{21} & -\underline{A}_{22} \end{bmatrix}, \tag{6.116}$$

$$B^{e} = (1 - \vartheta)\Delta t \begin{bmatrix} +Q_{1} \\ -Q_{2} \end{bmatrix} + \vartheta \, \Delta t \begin{bmatrix} +Q_{1}^{+} \\ -Q_{2}^{+} \end{bmatrix} + \vartheta \, \Delta t \begin{bmatrix} +\underline{B}_{1} \\ -\underline{B}_{2} \end{bmatrix}. \tag{6.117}$$

The unknown elemental discharge increments are calculated from Eq. (6.112) following the calculation of the unknown nodal piezometric head increment, subroutine IncVar.

The entire procedure is implemented numerically in the subroutine

subroutine FemUnsteadyPipeMtx(ielem),

which is located in the program module Pipes.f90.

6.2.2 Subroutine ChtxUnsteadyPipeMtx

Interpolation on finite element

Figure 6.19 shows possible positions of characteristics on a finite element, drawn from points R and S backwards. From point R to point P it is a negative characteristic, while from point S to point Q it is a positive characteristic. The known values of the solution (prescribed stage) lie at the level t while the unknown ones (the unknown stage) lie at the level $t + \Delta t$.

If points P and Q lie within the finite element, the state of variables is defined explicitly; or, if characteristics are intersecting the time axis, the state of variables is defined implicitly. Anyhow, variables in points P and Q, that is P' and Q', are defined by linear interpolations. The interpolation parameter for the negative characteristic is

$$\lambda_{P} = \frac{\Delta l_{1}}{\Delta L} \tag{6.118}$$

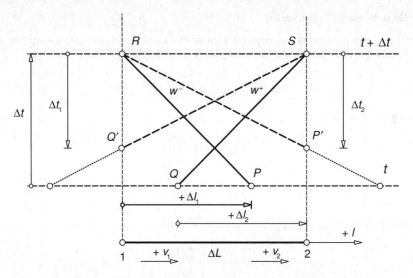

Figure 6.19 Possible positions of characteristics on finite element.

which is used for interpolation of some value f between the known values in finite element nodes

$$f_P = (1 - \lambda_P)f_1 + \lambda_P f_2. \tag{6.119}$$

The interpolation parameter for the positive characteristic is

$$\lambda_Q = \frac{\Delta l_2}{\Delta L} \tag{6.120}$$

which is used for interpolation of some value f between the known values in finite element nodes

$$f_Q = (1 - \lambda_Q)f_2 + \lambda_Q f_1. \tag{6.121}$$

Interpolation parameters define positions of characteristics. If an interpolation parameter is within the interval $0 \le \lambda \le 1$, the position of the characteristic is explicit while for $\lambda > 1$ it is implicit. Values of interpolation parameters are calculated from geometric relations; namely, by integration of two characteristic equations

$$\frac{dl}{dt} = w^\pm = v \pm c. \tag{6.122}$$

Integration is carried out between two points on the characteristic

$$\int_a^b dl = \int_a^b (v \pm c)\, dt \tag{6.123}$$

using the mean value of the integral theorem

$$(l_b - l_a) \doteq \frac{v_a + v_b}{2}(t_b - t_a) \pm c(t_b - t_a).$$ (6.124)

The interpolation parameter for the *positive characteristic*

$$l_S - l_Q = \frac{v_Q + v_S}{2}(t_S - t_Q) + c(t_S - t_Q).$$ (6.125)

Value v_Q is interpolated using expressions (6.120) and (6.121); thus, it is written

$$v_Q = (1 - \lambda_Q)v_2 + \lambda_Q v_1.$$ (6.126)

After introducing this into Eq. (6.125), the following is obtained

$$l_S - l_Q = \frac{1}{2}\left[(1 - \lambda_Q)v_2 + \lambda_Q v_1 + v_S\right](t_S - t_Q) + c(t_S - t_Q).$$ (6.127)

The left side can be expressed by Eq. (6.120) as

$$l_S - l_Q = \lambda_Q \Delta L,$$ (6.128)

$$\lambda_Q 2\Delta L = \left[(1 - \lambda_Q)v_2 + \lambda_Q v_1 + v_2^+\right]\Delta t + 2c\Delta t,$$ (6.129)

from which

$$\lambda_Q = \frac{\left(v_2 + v_2^+ + 2c\right)\Delta t}{2\Delta L + (v_2 - v_1)\Delta t},$$ (6.130)

where v_2^+ is the unknown value of velocity at the end of the time interval.
The interpolation parameter for the *negative characteristic*

$$l_R - l_P = \frac{v_P + v_R}{2}(t_R - t_P) - c(t_R - t_P).$$ (6.131)

Value v_P is interpolated using expressions (6.118) and (6.119); thus, it is written

$$v_P = (1 - \lambda_P)v_1 + \lambda_P v_2.$$ (6.132)

After introducing this into Eq. (6.131) the following is obtained

$$l_R - l_P = \frac{1}{2}[(1 - \lambda_P)v_1 + \lambda_P v_2 + v_R](t_R - t_P) - c(t_R - t_P).$$ (6.133)

The left hand side can be expressed by Eq. (6.118) as

$$l_R - l_P = -\lambda_P \Delta L,$$ (6.134)

$$-\lambda_P 2\Delta L = \left[(1 - \lambda_P)v_1 + \lambda_P v_2 + v_1^+\right]\Delta t - 2c\Delta t,$$ (6.135)

from which

$$\lambda_P = -\frac{\left(v_1 + v_1^+ - 2c\right)\Delta t}{2\Delta L + (v_2 - v_1)\Delta t},$$ (6.136)

where v_1^+ is the unknown value of velocity at the end of the time interval.

Interpolation parameters per time axis for the implicit position of characteristics are obtained from geometric ratios

$$\tau_{P'} = \frac{\Delta t_2}{\Delta t} = \frac{1}{\lambda_P},$$ (6.137)

$$\tau_{Q'} = \frac{\Delta t_1}{\Delta t} = \frac{1}{\lambda_Q},$$ (6.138)

while interpolation of values at intersecting points gives

$$f_{P'} = (1 - \tau_{P'})f_2^+ + \tau_{P'}f_2,$$ (6.139)

$$f_{Q'} = (1 - \tau_{Q'})f_1^+ + \tau_{Q'}f_1,$$ (6.140)

where the sign "+" refers to the state at the end, while the absence of the sign refers to the state at the beginning of the time interval.

Integration along elemental characteristics

Furthermore, wave function $\Gamma^\pm$ is also integrated between two points a and b

$$\int_a^b \frac{dQ}{A} \pm \frac{g}{c}\int_a^b dh + g\int_a^b J_e Q dt = 0;$$ (6.141)

thus, using the mean value of the integral theorem, it is written as

$$\frac{Q_b - Q_a}{A} \pm \frac{g}{c}(h_b - h_a) + g\frac{J_{ea} + J_{eb}}{2}(t_b - t_a) = 0.$$ (6.142)

First node
explicit positions $0 \le \lambda_P \le 1$:

$$h_P = (1 - \lambda_P)h_1 + \lambda_P h_2,$$ (6.143)

$$Q_P = (1 - \lambda_P)Q_1 + \lambda_P Q_2,$$ (6.144)

$$J_{eP} = \frac{\lambda}{2gDA^2}|Q_P|Q_P,$$ (6.145)

$$J_{e1} = \frac{\lambda}{2gDA^2}|Q_1^+|Q_1^+,$$ (6.146)

$$\Gamma^-: \quad \frac{(Q_1^+ - Q_P)}{A} - \frac{g}{c}(h_1^+ - h_P) + \frac{g}{2}(J_{eP} + J_{e1})\,\Delta t = 0,$$ (6.147)

$$\frac{\partial \Gamma^-}{\partial h_1^+} = -\frac{g}{c}, \tag{6.148}$$

$$\frac{\partial \Gamma^-}{\partial h_2^+} = 0, \tag{6.149}$$

$$\frac{\partial \Gamma^-}{\partial Q_1^+} = \frac{1}{A} + \frac{\lambda \Delta t}{2DA^2} \left| Q_1^+ \right|, \tag{6.150}$$

$$\frac{\partial \Gamma^-}{\partial Q_2^+} = 0. \tag{6.151}$$

implicit position $\lambda_P > 1$:

$$\tau_{P'} = \frac{\Delta t_2}{\Delta t} = \frac{1}{\lambda_P}, \tag{6.152}$$

$$\Delta t_2 = \tau_{P'} \Delta t, \tag{6.153}$$

$$h_{P'} = (1 - \tau_{P'}) h_2^+ + \tau_{P'} h_2, \tag{6.154}$$

$$Q_{P'} = (1 - \tau_{P'}) Q_2^+ + \tau_{P'} Q_2, \tag{6.155}$$

$$J_{eP'} = \frac{\lambda}{2gDA^2} \left| Q_{P'} \right| Q_{P'}, \tag{6.156}$$

$$J_{e1} = \frac{\lambda}{2gDA^2} \left| Q_1^+ \right| Q_1^+, \tag{6.157}$$

$$\Gamma^-: \quad \frac{(Q_1^+ - Q_{P'})}{A} - \frac{g}{c}(h_1^+ - h_{P'}) + \frac{g}{2}\left(J_{eP'} + J_{e1}\right) \Delta t_2 = 0, \tag{6.158}$$

$$\frac{\partial \Gamma^-}{\partial h_1^+} = -\frac{g}{c}, \tag{6.159}$$

$$\frac{\partial \Gamma^-}{\partial h_2^+} = \frac{g}{c}(1 - \tau_{P'}), \tag{6.160}$$

$$\frac{\partial \Gamma^-}{\partial Q_1^+} = \frac{1}{A} + \frac{\lambda \Delta t_2}{2DA^2} \left| Q_1^+ \right|, \tag{6.161}$$

$$\frac{\partial \Gamma^-}{\partial Q_2^+} = (1 - \tau_{P'}) \left(-\frac{1}{A} + \frac{\lambda \Delta t_2}{2DA^2} \left| Q_{P'} \right| \right). \tag{6.162}$$

Second node
Explicit position $0 \le \lambda_Q \le 1$:

$$h_Q = (1 - \lambda_Q) h_2 + \lambda_Q h_1, \tag{6.163}$$

$$Q_Q = (1 - \lambda_Q) Q_2 + \lambda_Q Q_1, \tag{6.164}$$

$$J_{eQ} = \frac{\lambda}{2gDA^2} \left| Q_Q \right| Q_Q, \tag{6.165}$$

$$J_{e2} = \frac{\lambda}{2gDA^2} \left|Q_2^+\right| Q_2^+,$$

(6.166)

$$\Gamma^+: \quad \frac{(Q_2^+ - Q_Q)}{A} + \frac{g}{c}(h_2^+ - h_Q) + \frac{g}{2}\left(J_{eQ} + J_{e2}\right)\Delta t = 0,$$

(6.167)

$$\frac{\partial \Gamma^+}{\partial h_1^+} = 0,$$

(6.168)

$$\frac{\partial \Gamma^+}{\partial h_2^+} = +\frac{g}{c},$$

(6.169)

$$\frac{\partial \Gamma^+}{\partial Q_1^+} = 0,$$

(6.170)

$$\frac{\partial \Gamma^+}{\partial Q_2^+} = \frac{1}{A} + \frac{\lambda \Delta t}{2DA^2}\left|Q_2^+\right|.$$

(6.171)

Implicit position $\lambda_Q > 1$:

$$\tau_{Q'} = \frac{\Delta t_1}{\Delta t} = \frac{1}{\lambda_Q},$$

(6.172)

$$\Delta t_1 = \tau_{Q'}\Delta t,$$

(6.173)

$$h_{Q'} = (1 - \tau_{Q'})h_1^+ + \tau_{Q'}h_1,$$

(6.174)

$$Q_{Q'} = (1 - \tau_{Q'})Q_1^+ + \tau_{Q'}Q_1,$$

(6.175)

$$J_{eQ'} = \frac{\lambda}{2gDA^2}\left|Q_{Q'}\right| Q_{Q'},$$

(6.176)

$$J_{e2} = \frac{\lambda}{2gDA^2}\left|Q_2^+\right| Q_2^+,$$

(6.177)

$$\Gamma^+: \quad \frac{(Q_2^+ - Q_{Q'})}{A} + \frac{g}{c}(h_2^+ - h_{Q'}) + \frac{g}{2}\left(J_{eQ'} + J_{e2}\right)\Delta t_1 = 0,$$

(6.178)

$$\frac{\partial \Gamma^+}{\partial h_1^+} = -\frac{g}{c}(1 - \tau_{Q'}),$$

(6.179)

$$\frac{\partial \Gamma^+}{\partial h_2^+} = +\frac{g}{c},$$

(6.180)

$$\frac{\partial \Gamma^+}{\partial Q_1^+} = (1 - \tau_{Q'})\left(-\frac{1}{A} + \frac{\lambda \Delta t_1}{2DA^2}\left|Q_{Q'}\right|\right),$$

(6.181)

$$\frac{\partial \Gamma^+}{\partial Q_2^+} = \frac{1}{A} + \frac{\lambda \Delta t_1}{2DA^2}\left|Q_2^+\right|.$$

(6.182)

The Newton–Raphson iterative form for the finite element is formally written using the matrix-vector operations

$$[\underline{H}] \cdot [\Delta h^+] + [\underline{Q}] \cdot [\Delta Q^+] = [\underline{F}],$$

(6.183)

where the matrix is

$$[\underline{H}] = \begin{bmatrix} \dfrac{\partial \Gamma^-}{\partial h_1^+} & \dfrac{\partial \Gamma^-}{\partial h_2^+} \\[2ex] \dfrac{\partial \Gamma^+}{\partial h_1^+} & \dfrac{\partial \Gamma^+}{\partial h_2^+} \end{bmatrix}, \qquad (6.184)$$

and the matrix

$$[\underline{Q}] = \begin{bmatrix} \dfrac{\partial \Gamma^-}{\partial Q_1^+} & \dfrac{\partial \Gamma^-}{\partial Q_2^+} \\[2ex] \dfrac{\partial \Gamma^+}{\partial Q_1^+} & \dfrac{\partial \Gamma^+}{\partial Q_2^+} \end{bmatrix}. \qquad (6.185)$$

The right hand side vector is

$$[\underline{F}] = -\begin{bmatrix} \Gamma^- \\ \Gamma^+ \end{bmatrix}. \qquad (6.186)$$

Finite element matrix and vector

The subsequent procedure of finite element matrix and vector computation is completely the same as for the FEM integration and implemented in the subroutine

> `subroutine ChtxUnsteadyPipeMtx(ielem),`

located in the program module `Pipes.f90`.

6.3 Comparison tests

6.3.1 Test example

Figure 6.20 shows the data for testing of the numerical solutions of finite element integration using the conservation laws and integration along the characteristics.

The pipe is divided into 10 equal finite elements of $\Delta l = 0.1\,L$ length. The iteration time step is $\Delta t = \Delta l/c$ and is thus equal to the water hammer propagation time for one finite element. The `SimpipCore` program will be used for test example solving. The test file `Test SimpipCore FEM-CHTX.simpip` can be viewed in Program solutions & tests/6 Chapter Non-steady (see Programs Chtx, test example) from www.wiley.com/go/jovic.

Solutions will be compared for the piezometric head at the valve cross-section (pipeline end) in the conditions of water hammer generation due to sudden velocity change in the interval Δt.

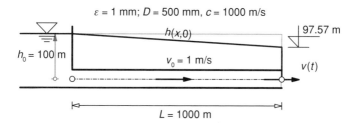

$\varepsilon = 1$ mm; $D = 500$ mm, $c = 1000$ m/s
$h(x,0)$ 97.57 m
$h_0 = 100$ m
$v_0 = 1$ m/s $v(t)$
$L = 1000$ m

Figure 6.20 Pipe in `SimpipCore` tests.

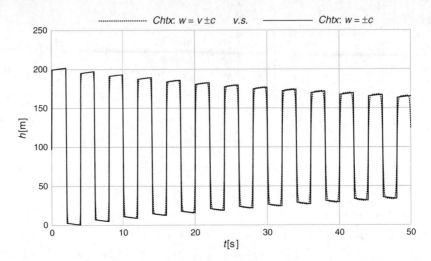

Figure 6.21   `SimpipCore` test, characteristics.

First, solutions were compared that depend on the procedure of defining the positions of characteristics on a finite element for the positions calculated using the velocities $w^\pm = v \pm c$ and $w^\pm = \pm c$. The results are shown in Figure 6.21. The differences are negligible, as was expected because $v \ll c$. Solution by the method of characteristics is an accurate solution within the numerical accuracy range.

(a) Test – comparison of solutions between the characteristics $w^\pm = \pm c$ and FEM integration, $\vartheta = 0.5$. This is shown in Figure 6.22.

    The conservation law method shows the over-shooting at the end of every water hammer cycle, which increases constantly. Without further, more complex analyses, it can be considered that differences are the result of the inadequate accuracy of the approximate integrations of time integrals.

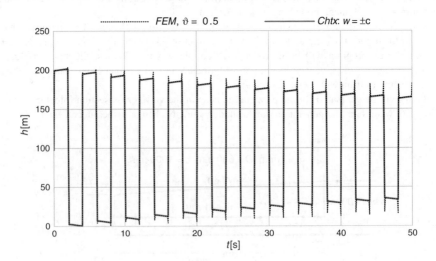

Figure 6.22   (a) Test.

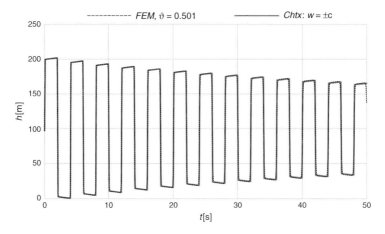

Figure 6.23   (b) Test.

(b) Test – comparison between the characteristics $w^\pm = \pm c$ and FEM integration,$\vartheta = 0.501$. This is shown in Figure 6.23.

Even with the small increase of the parameter ϑ of time integration the over-shooting is removed.

(c) Test – comparison between the characteristics $w^\pm = \pm c$ and FEM integration, $\vartheta = 0.55$.

(d) Test – comparison between the characteristics $w^\pm = \pm c$ and FEM integration $\vartheta = 0.75$.

These are shown in Figure 6.24 and Figure 6.25. By further increase of the time integration parameter ϑ, the solution is smoothed, the period of the wave phenomenon is preserved while the wave sharpness seriously deteriorates.

6.3.2   Conclusion

The tests performed show that high accuracy solutions can be obtained by the SimpipCore software for equidistant finite elements and the water hammer real problem, which is equivalent to the classical procedure by the method of characteristics.

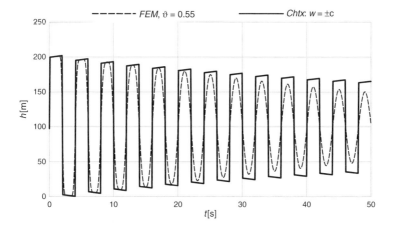

Figure 6.24   (c) Test.

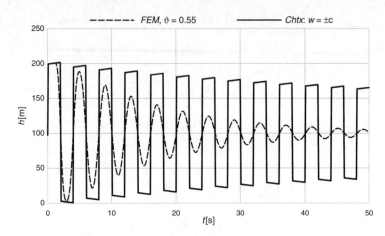

Figure 6.25 (d) Test.

Further reading

Davis, C.V. and Sorenson, K.E. (1969) *Handbook of Applied Hydraulics*. 3th edn, McGraw-Hill Co., New York.

Dracos, Th. (1970) Die Berechnung istatationärer Abfüsse in offenen Gerinnen beliebiger Geometrie, *Schweizerische Bauzeitung*, 88. Jahrgang Heft 19.

Fox, J.A. (1977) *Hydraulic Analysis of Unsteady Flow in Pipe Networks*. Macmillan Press Ltd, London, UK; Wiley, New York, USA.

Godunov, S.K. (1971) *Equations of Mathematical Physics* (in Russian: Uravnjenija matematičjeskoj fizici). Izdateljstvo Nauka, Moskva.

Jeffrey, A. (1976) *Quasilinear Hyperbolic System and Waves*. Pitman, Boston.

Jović, V. (1977) Non–steady Flow in Pipes and Channels by Finite Element Method, *Proceedings of XVII Congress of the IAHR*, **2**, pp 197–204. IAHR, Baden–Baden.

Jović, V. (1987) Modelling of non-steady flow in pipe networks, *Proc. 2nd Int. Conf. NUMETA '87*, Martinus Nijhoff Pub. Swansea.

Jović, V. (1995) Finite elements and the method of characteristics applied to water hammer modeling. *Engineering Modelling*, 8: 51–58.

Jović, V. (2006) *Fundamentals of Hydromechanics* (in Croatian: Osnove hidromehanike). Element, Zagreb.

Streeter, V.L. and Wylie, E.B. (1967) *Hydraulic Transients*. McGraw–Hill Book Co., New York, London, Sydney.

Streeter, V.L. and Wylie, E.B. (1993) *Fluid Transients*. FEB Press, Ann Arbor, Mich.

Watters, G.Z. (1984) *Analysis and Control of Unsteady Flow in Pipe Networks*. Butterworths, Boston.

Wylie, E.B. and Streeter, V.L. (1993) *Fluid Transients in Systems*. Prentice Hall, Englewood Cliffs, New Jersey, USA.

7

Valves and Joints

7.1 Valves

7.1.1 Local energy head losses at valves

Valves are short hydraulic branches that regulate the flow in a hydraulic network. Flow regulation is achieved by different levels of valve opening. Depending on the opening of the valve, there is local resistance to flow. In general, valve resistances are asymmetric, namely they can be different in positive and negative directions of flow. Local loss of the energy line can be written as

$$\Delta H|_1^2 = \xi_v^{\pm}(o)\frac{|v|v}{2g},\tag{7.1}$$

where $\zeta_v^{\pm}$ is the coefficient of local resistances depending on the valve opening o while v is the velocity for the rated valve cross-section.

Energy head losses expressed by discharge are

$$\Delta H|_1^2 = \beta_v^{\pm}(o)|Q|Q,\tag{7.2}$$

where $\beta_v^{\pm}$ is the resistance coefficient depending on the valve opening o. For a completely closed valve, the resistance coefficient is infinite.

Figure 7.1 shows the heads and losses on a finite element for a positive discharge (flow in the direction from the first to the second node).

Some standard valves according to (Jovic, 2006) and shown in Table 7.1 are inbuilt in the $SimpipCore$ program solution.

For calculation of coefficients of local resistances[1] of standard valves (a), (b), and (c) ζ and β respective subroutines are used that are implemented in the program module hydraulics.f90, see www.wiley.com/go/jovic.

---

[1] Calculations of resistances in valves are based on theoretic and experimental data. The respective procedures can be found in the international literature. Non-critical acceptance of the data from the literature should be avoided because of typing and other errors. One of the most reliable and comprehensive reference books is I.E. Idelcik, 1969, *Memento des pertes de charge*, Eyrolles, Paris. An acceptable critical approach to local resistances is given in the book by V. Jović, 2006, *Osnove hidromehanike* (in Croatian), Element, Zagreb.

Analysis and Modelling of Non-Steady Flow in Pipe and Channel Networks, First Edition. Vinko Jović.
© 2013 John Wiley & Sons, Ltd. Published 2013 by John Wiley & Sons, Ltd.

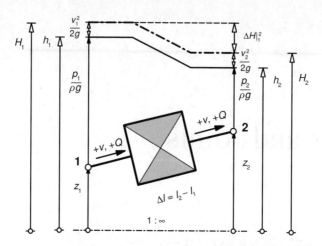

Figure 7.1   Valve finite element, steady flow.

Table 7.1   Valve types and coefficients of local resistances as a function of valve opening.

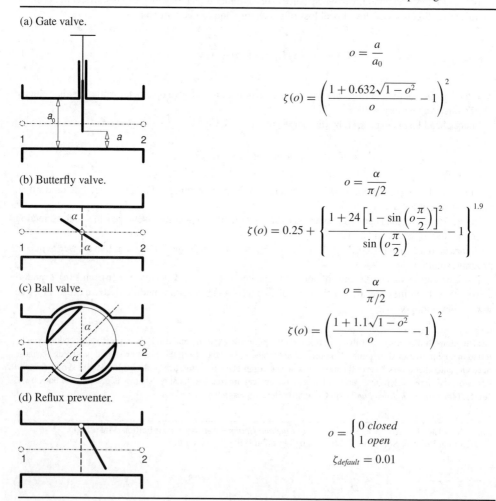

(a) Gate valve.

$$o = \frac{a}{a_0}$$

$$\zeta(o) = \left(\frac{1 + 0.632\sqrt{1 - o^2}}{o} - 1 \right)^2$$

(b) Butterfly valve.

$$o = \frac{\alpha}{\pi/2}$$

$$\zeta(o) = 0.25 + \left\{ \frac{1 + 24\left[1 - \sin\left(o\frac{\pi}{2}\right)\right]^2}{\sin\left(o\frac{\pi}{2}\right)} - 1 \right\}^{1.9}$$

(c) Ball valve.

$$o = \frac{\alpha}{\pi/2}$$

$$\zeta(o) = \left(\frac{1 + 1.1\sqrt{1 - o^2}}{o} - 1 \right)^2$$

(d) Reflux preventer.

$$o = \begin{cases} 0 \ closed \\ 1 \ open \end{cases}$$

$$\zeta_{default} = 0.01$$

7.1.2   *Valve status*

In steady flow, a valve can be partially and completely open (`VALVE_IS_OPEN`) or closed (`VALVE_IS_CLOSED`), which is generally given in the input data. In non-steady flow, apart from the aforementioned statuses, there are also transient statuses during which the valve is opening (`VALVE_OPEN_IT`) or closing (`VALVE_CLOSE_IT`) which are determined in computation time. The status constants are defined as parameters in the program module `GlobalVars.f90`.

The valve status and calculation of coefficients $\beta_v^{\pm}$ between the two time stages is defined by the procedure

```
integer function & CalcValveStatus(Valve,tP,tK,bp_p,bp_k,bm_p,bm_k)
```

that is implemented in the program module `valves.f90`.

The status of the reflux preventer in steady flow is explicitly set by input data (prescribed initial condition) and can be either *open* or *closed*.

The status of the reflux preventer in quasi unsteady and non-steady flow of an incompressible (rigid) liquid, unlike other valve types, is determined dynamically from the current iterative values of the variables. It is implemented as a function in the module `reflux.f90`, which returns the value of the status

```
status=GetRefluxStatus(ielem).
```

The status of the reflux preventer in non-steady flow of a compressible liquid, unlike other valve types, is determined dynamically from the current iterative values of the variables. It is implemented as function in the module `reflux.f90`, which returns the value of the status

```
status= UnsRefluxStatus(ielem).
```

7.1.3   *Steady flow modelling*

In subroutine `Steady` a call for a subroutine for computation of the steady flow matrix and vector for an elemental valve type is added as follows:

```
select case(Elems(ielem).tip)
      case ...

            . . .
      case (VALVE_OBJ)
            call SteadyValveMtx(ielem)
      case ...

            . . .
endselect
```

Subroutine `SteadyValveMtx`

This subroutine calculates the finite element matrix and vector by calling the respective subroutine depending on the valve type:

- for the reflux preventer, the following subroutine is called

```
call SteadyRefluxMtx(ielem),
```

- for all other types call

```
call SteadyGeneralValveMtx(ielem).
```

The procedure `SteadyGeneralValveMtx` tests the valve status that can be either `VALVE_IS_OPEN` or `VALVE_IS_CLOSED`.

If the valve is open, a general procedure `SteadyOpenValveMtx` for valve and matrix computation is called.

Subroutine `SteadyOpenValveMtx`

The following elemental equation can be applied to the valve

$$F^e: \quad h_2 - h_1 + \beta_v^{\pm} |Q| Q = 0 \tag{7.3}$$

which does not differ formally from the elemental equation of the pipe element. The difference can be observed on the coefficient $\beta_v^{\pm}$, which depends on the flow direction. Thus, computation of the valve finite element matrix and vector is formally equal to the computation of the pipe element. A starting point is the Newton–Raphson form, which is formally written using the matrix-vector operations

$$[\underline{H}] \cdot [\Delta h] + [\underline{Q}] \cdot [\Delta Q] = [\underline{F}], \tag{7.4}$$

where the scalar value $[\underline{F}]$ is equal to

$$[\underline{F}] = -\left(h_2 - h_1 + \beta_v^{\pm} |Q| Q\right), \tag{7.5}$$

while vector $[\underline{H}]$ has the form

$$[\underline{H}] = \left[\frac{\partial F_e}{\partial h_1} \quad \frac{\partial F_e}{\partial h_2} \right] = [-1 \quad 1]. \tag{7.6}$$

The scalar value $[\underline{Q}]$ is equal to

$$[\underline{Q}] = \frac{\partial F^e}{\partial Q} = 2\beta_v^{\pm} |Q|. \tag{7.7}$$

When the previous expression is multiplied by the inverse term $[\underline{Q}]^{-1}$ the following is obtained

$$[\underline{A}] \cdot [\Delta h] + [\Delta Q] = [\underline{B}], \tag{7.8}$$

from which the value of the elemental discharge increment can be calculated as

$$[\Delta Q] = [\underline{B}] - [\underline{A}][\Delta h], \tag{7.9}$$

where

$$[\underline{A}] = [\underline{Q}]^{-1} [\underline{H}], \tag{7.10}$$

$$[\underline{B}] = [\underline{Q}]^{-1} [\underline{F}]. \tag{7.11}$$

In the aforementioned expressions $[\underline{A}]$ is a two-term vector while $[\underline{B}]$ is a scalar. A process of elimination of elemental discharges from nodal equations of continuity defines the structure of the finite element matrix

$$A^e = \begin{bmatrix} +\underline{A} \\ -\underline{A} \end{bmatrix} \qquad (7.12)$$

and vector

$$B^e = \begin{bmatrix} +Q \\ -Q \end{bmatrix} + \begin{bmatrix} +\underline{B} \\ -\underline{B} \end{bmatrix}. \qquad (7.13)$$

The procedure is carried out numerically. If the valve is closed, then the element discharge is equal to $Q = 0$, which implies that the vectors $[\underline{A}] = 0$ and $[\underline{B}] = 0$, that is the finite element matrix $[A^e] = 0$ and vector $[B^e] = 0$, are equal to zero.

Subroutine *SteadyRefluxMtx*
This subroutine calculates the steady flow finite element matrix and vector for the preventer depending on the reflux preventer status. The reflux preventer status in steady flow is known from the input data.

If the valve is opened, then the procedure `SteadyOpenRefluxMtx` for the valve matrix and vector computation is called.

Subroutine `SteadyOpenRefluxMtx` differs from subroutine `SteadyOpenGeneralValveMtx` only because there is no negative discharge for the reflux preventer.

If the valve is closed, then the element discharge is equal to $Q = 0$, which implies that the vectors $[\underline{A}] = 0$ and $[\underline{B}] = 0$, that is the finite element matrix $[A^e] = 0$ and vector $[B^e] = 0$ are equal to zero.

7.1.4   Non-steady flow modelling

In subroutine `Unsteady` a call for a subroutine for computation of the non-steady flow matrix and vector for an elemental valve type is added as follows:

```
select case(Elems(ielem).tip)
        case ...

              . . .
        case (VALVE_OBJ)
                call UnSteadyValveMtx(ielem)
        case ...

              . . .
endselect
```

Subroutine **UnsteadyValveMtx**

This subroutine calculates the finite element matrix and vector by calling the respective subroutine depending on the valve type:

• for the reflux preventer, the following subroutine is called

```
                call UnsteadyRefluxMtx(ielem),
```

- for all other types subroutines are called depending on the type of non-steady calculation:

```
if runmode = QuasyUnsteady then
  call QuValveMtx(ielem)
else if runmode = RigidUnsteady then
  call RgdValveMtx(ielem)
else if runmode = FullUnsteady then
    call UnsValveMtx(ielem)
endif.
```

Subroutine `QuValveMtx`

This subroutine calculates the finite element matrix and vector for quasi unsteady flow by branching computation according to the valve status.

(a) If the valve is opened at the end of the time stage then

$$status = VALVE\_IS\_OPEN.$$

The equation of a quasi unsteady flow in a pipe is equal to the steady flow equation

$$h_2 - h_1 + \beta_v^{\pm} |Q| Q = 0. \tag{7.14}$$

Variables of the state h, Q are time-dependent functions. By integration of the elemental equation in time interval Δt the following is obtained

$$F^e = \int_{\Delta t} \left[h_2 - h_1 + \beta_v^{\pm} |Q| Q \right] dt = \tag{7.15}$$

$$(1 - \vartheta)\Delta t \left(h_2 - h_1 + \beta_v^{\pm} |Q| Q \right) + \vartheta \Delta t \left(h_2^+ - h_1^+ + \beta_v^{\pm} |Q^+| Q^+ \right) = 0.$$

The Newton–Raphson iterative form for a finite element is formally written using the matrix-vector operations

$$[\underline{H}] \cdot [\Delta h] + [\underline{Q}] \cdot [\Delta Q] = [\underline{F}], \tag{7.16}$$

where the vector is equal to

$$[\underline{H}] = \left[-\vartheta \Delta t \quad +\vartheta \Delta t \right] = \vartheta \Delta t \left[-1 \quad +1 \right], \tag{7.17}$$

while the scalar terms are

$$[\underline{F}] = -F^e, \tag{7.18}$$

$$[\underline{Q}] = 2\vartheta \Delta t \beta_v^{\pm} |Q^+|. \tag{7.19}$$

When the previous expression is multiplied by the inverse term $[\underline{Q}]^{-1}$ the following is obtained

$$[\underline{A}] \cdot [\Delta h] + [\Delta Q] = [\underline{B}], \tag{7.20}$$

from which the value of the elemental discharge increment can be calculated as

$$[\Delta Q] = [\underline{B}] - [\underline{A}] \cdot [\Delta h],\tag{7.21}$$

where

$$[\underline{A}] = [\underline{Q}]^{-1}[\underline{H}],\tag{7.22}$$

$$[\underline{B}] = [\underline{Q}]^{-1}[\underline{F}].\tag{7.23}$$

In the aforementioned expressions $[\underline{A}]$ is a two-term vector while $[\underline{B}]$ is a scalar. A process of elimination of elemental discharges from nodal equations of continuity defines the structure of the finite element matrix

$$A^e = \vartheta\,\Delta t \begin{bmatrix} +\underline{A} \\ -\underline{A} \end{bmatrix}\tag{7.24}$$

and vector

$$B^e = (1 - \vartheta\,\Delta t) \begin{bmatrix} +Q \\ -Q \end{bmatrix} + \vartheta\,\Delta t \begin{bmatrix} +Q^+ \\ -Q^+ \end{bmatrix} + \vartheta\,\Delta t \begin{bmatrix} +\underline{B} \\ -\underline{B} \end{bmatrix}.\tag{7.25}$$

Calculation of the scalar term $[\underline{Q}]^{-1}$, vector $[\underline{A}]$, scalar term $[\underline{B}]$, the finite element matrix $[A^e]$, and vector $[B^e]$ of the valve is carried out numerically.

(b) If the valve status at the end of the time interval is such that the valve is closed or should be closed

> status = VALVE_IS_CLOSED or VALVE_CLOSE_IT

then the discharge is equal to zero: $Q^+ = 0$, which implies that the vector $[\underline{A}] - 0$ and scalar $[\underline{B}] = 0$ since discharge increment, according to the expression (7.21), is always zero for each increment Δh, that is the matrix $[A^e] = 0$. The finite element vector $[B^e]$ should contain information on the discharge from the beginning of the time interval; thus, according to Eq. (7.25), it is

$$B^e = (1 - \vartheta\,\Delta t) \begin{bmatrix} +Q \\ -Q \end{bmatrix}.\tag{7.26}$$

(c) If the valve status at the end of the time interval is such that the valve should be opened

> status = VALVE_OPEN_IT

then there will be a problem of integration of resistances between the two time stages. Namely, at the beginning of the time interval the valve is completely closed and the resistance coefficient is $\beta_v^{\pm} = \infty$. The problem is solved by full implicit integration; namely, the time integration parameter $\vartheta = 1$. Further procedure is equal to the procedure described for the open valve.

Subroutine RgdValveMtx
This subroutine calculates the finite element matrix and vector for non-steady flow of incompressible (rigid) fluid by branching computation according to the valve status.

(a) If the valve is opened at the end of the time stage then

$$status = VALVE\_IS\_OPEN.$$

The equation of a non-steady flow of a rigid fluid in a valve is

$$h_2 - h_1 + \beta_v^{\pm} |Q| Q + \frac{\Delta l}{gA} \frac{dQ}{dt} = 0. \tag{7.27}$$

Variables of the state h, Q are time-dependent functions. By integration of the elemental equation in time interval Δt the following is obtained

$$F^e = \int\limits_{\Delta t} \left[h_2 - h_1 + \beta_v^{\pm} |Q| Q + \frac{\Delta l}{gA} \frac{dQ}{dt} \right] dt =$$

$$(1 - \vartheta)\Delta t \left(h_2 - h_1 + \beta_v^{\pm} |Q| Q \right) + \tag{7.28}$$

$$+ \vartheta \Delta t \left(h_2^+ - h_1^+ + \left(\beta_v^{\pm} \right)^+ |Q^+| Q^+ \right) +$$

$$+ \frac{\Delta l}{gA} (Q^+ - Q) = 0.$$

The Newton–Raphson iterative form for a finite element is formally written using the matrix-vector operations

$$[\underline{H}] \cdot [\Delta h] + [\underline{Q}] \cdot [\Delta Q] = [\underline{F}], \tag{7.29}$$

where the vector is equal to

$$[\underline{H}] = [-\vartheta \Delta t \quad +\vartheta \Delta t] = \vartheta \Delta t [-1 \quad +1], \tag{7.30}$$

while the scalar terms are

$$[\underline{F}] = -F^e, \tag{7.31}$$

$$[\underline{Q}] = \frac{\Delta l}{gA} + 2\vartheta \Delta t \beta_v^{\pm} |Q^+|. \tag{7.32}$$

When the previous expression is multiplied by the inverse term $[\underline{Q}]^{-1}$ the following is obtained

$$[\underline{A}] \cdot [\Delta h] + [\Delta Q] = [\underline{B}], \tag{7.33}$$

from which the value of the elemental discharge increment can be calculated as

$$[\Delta Q] = [\underline{B}] - [\underline{A}] \cdot [\Delta h], \tag{7.34}$$

where

$$[\underline{A}] = [\underline{Q}]^{-1} [\underline{H}],$$ (7.35)

$$[\underline{B}] = [\underline{Q}]^{-1} [\underline{F}].$$ (7.36)

In the aforementioned expressions $[\underline{A}]$ is a two-term vector while $[\underline{B}]$ is a scalar. A process of elimination of the elemental discharges from the nodal equations of continuity defines the structure of the finite element matrix

$$A^e = \vartheta \, \Delta t \begin{bmatrix} +\underline{A} \\ -\underline{A} \end{bmatrix}$$ (7.37)

and vector

$$B^e = (1 - \vartheta \, \Delta t) \begin{bmatrix} +Q \\ -Q \end{bmatrix} + \vartheta \, \Delta t \begin{bmatrix} +Q^+ \\ -Q^+ \end{bmatrix} + \vartheta \, \Delta t \begin{bmatrix} +\underline{B} \\ -\underline{B} \end{bmatrix}.$$ (7.38)

Calculation of the scalar term $[\underline{Q}]^{-1}$, vector $[\underline{A}]$, scalar term $[\underline{B}]$, the finite element matrix $[A^e]$, and vector $[B^e]$ of the valve is carried out numerically.

(b) If the valve status at the end of the time interval is such that the valve is closed or should be closed

        status = VALVE_IS_CLOSED or VALVE_CLOSE_IT

then the discharge is equal to zero: $Q^+ = 0$, which implies that the vector $[\underline{A}] = 0$ and scalar $[\underline{B}] = 0$ since discharge increment, according to the expression (7.34), is always zero for each increment Δh, that is the matrix $[A^e] = 0$. The finite element vector $[B^e]$ should contain information on the discharge from the beginning of the time interval; thus, according to Eq. (7.38), it is

$$B^e = (1 - \vartheta \, \Delta t) \begin{bmatrix} +Q \\ -Q \end{bmatrix}.$$ (7.39)

(c) If the valve status at the end of the time interval is such that the valve should be opened

        status = VALVE_OPEN_IT

then there will be a problem of integration of resistances between the two time stages. Namely, at the beginning of the time interval the valve is completely closed and the resistance coefficient is $\beta_v^{\pm} = \infty$. The problem is solved by full implicit integration; namely, the time integration parameter $\vartheta = 1$. The further procedure is equal to the procedure described for the open valve.

Subroutine *UnsValveMtx*
This subroutine calculates the finite element matrix and vector by branching computation according to the valve status.

(a) If the valve is opened at the end of the time stage then

        status = VALVE_IS_OPEN.

The valve finite element matrix and vector will be obtained by integration of the mass and the specific mechanical energy conservation law between the initial and end state. Equations are formally equal to the equations for pipes.

The mass conservation law on a finite element is

$$\frac{gA}{c^2} \int_{\Delta l} \frac{\partial h}{\partial t} dl + (Q_2 - Q_1) = 0, \tag{7.40}$$

where the water hammer celerity is large enough for the water hammer to always pass the valve. Since the valves are significantly more rigid than pipes of the respective diameter, a celerity $c = 10\,000$ m/s large enough is adopted for all valve types.

Using the mean value theorem of integral, it is written as

$$\frac{gA}{c^2} \frac{\Delta l}{2} \left(\frac{\partial h_1}{\partial t} + \frac{\partial h_2}{\partial t} \right) + (Q_2 - Q_1) = 0. \tag{7.41}$$

After integration with the time step Δt

$$\frac{gA}{c^2} \int_{\Delta t} \frac{\Delta L}{2} \left(\frac{\partial h_1}{\partial t} + \frac{\partial h_2}{\partial t} \right) dt + \int_{\Delta t} (Q_2 - Q_1) dt = 0 \tag{7.42}$$

and repeatedly reusing the aforementioned time-dependent integration rules, the first elemental equation is obtained

$$F_1: \quad \frac{\Delta l}{2} \frac{gA}{c^2} \left[\left(h_1^+ + h_2^+ \right) - (h_1 + h_2) \right] + $$
$$+ (1 - \vartheta) \Delta t (Q_2 - Q_1) + \vartheta \Delta t \left(Q_2^+ - Q_1^+ \right) = 0. \tag{7.43}$$

The specific mechanic energy conservation law in the head form is

$$\frac{1}{gA} \int_{\Delta l} \frac{\partial Q}{\partial t} dl + H_2 - H_1 + \Delta H|_1^2 = 0 \tag{7.44}$$

and is shown in Figure 7.2. The local loss (7.2) is expressed as

$$\Delta H|_1^2 = \beta_v^{\pm} |\bar{Q}| \bar{Q}, \tag{7.45}$$

where $\bar{Q} = (Q_1 + Q_2)/2$ is the mean discharge on the element.

Using the mean value theorem of the integral, it is written as

$$\frac{\Delta l}{2gA} \left(\frac{\partial Q_1}{\partial t} + \frac{\partial Q_2}{\partial t} \right) + H_2 - H_1 + \beta_v^{\pm} |\bar{Q}| \bar{Q} = 0. \tag{7.46}$$

After integration with the time step Δt

$$\frac{\Delta l}{2gA} \int_{\Delta t} \left(\frac{\partial Q_1}{\partial t} + \frac{\partial Q_2}{\partial t} \right) dt + \int_{\Delta t} (H_2 - H_1) dt + \int_{\Delta t} \beta_v^{\pm} |\bar{Q}| \bar{Q} dt = 0 \tag{7.47}$$

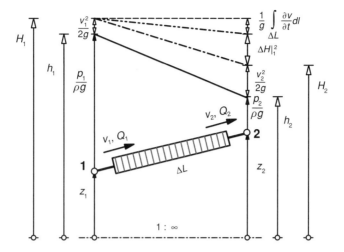

Figure 7.2 Finite element of a general valve, non-steady flow.

finally, the second elemental equation is obtained in the form

$$F_2: \quad \frac{\Delta l}{2gA}\left[(Q_1^+ + Q_2^+) - (Q_1 + Q_2)\right] +$$
$$+ (1-\vartheta)\Delta t\,(H_2 - H_1) + \vartheta\,\Delta t\,\left(H_2^+ - H_1^+\right) +$$
$$+ (1-\vartheta)\Delta t\,\left[\beta_v^\pm\,|\bar{Q}|\,\bar{Q}\right] + \vartheta\,\Delta t\,\left[\beta_v^\pm\,|\bar{Q}|\,\bar{Q}\right]^+ = 0. \tag{7.48}$$

The Newton–Raphson iterative form for a finite element is formally written using the matrix-vector operations

$$\left[\underline{H}\right]\cdot\left[\Delta h^+\right] + \left[\underline{Q}\right]\cdot\left[\Delta Q^+\right] = \left[\underline{F}\right], \tag{7.49}$$

where the matrix

$$\left[\underline{H}\right] = \begin{bmatrix} \dfrac{\Delta l}{2}\dfrac{gA}{c^2} & \dfrac{\Delta l}{2}\dfrac{gA}{c^2} \\[2mm] -\vartheta\,\Delta t & +\vartheta\,\Delta t \end{bmatrix} \tag{7.50}$$

and matrix

$$\left[\underline{Q}\right] = \begin{bmatrix} -\vartheta\,\Delta t & +\vartheta\,\Delta t \\[2mm] \left(\dfrac{\Delta l}{2gA} + \vartheta\,\Delta t\,\Delta l\beta_v^\pm\,|\bar{Q}^+| - \dfrac{\vartheta\,\Delta t}{gA^2}Q_1^+\right) & \left(\dfrac{\Delta l}{2gA} + \vartheta\,\Delta t\,\Delta l\beta_v^\pm\,|\bar{Q}^+| + \dfrac{\vartheta\,\Delta t}{gA^2}Q_2^+\right) \end{bmatrix}. \tag{7.51}$$

The right hand side vector is:

$$\left[\underline{F}\right] = -\begin{bmatrix} F_1 \\ F_2 \end{bmatrix}. \tag{7.52}$$

The elemental discharge increments are calculated from the Newton–Raphson form of the elemental equations (7.49) in a manner such that the equation is multiplied by the inverse matrix $[\underline{Q}]^{-1}$

$$[\Delta Q^+] = [\underline{Q}]^{-1}[\underline{F}] - [\underline{Q}]^{-1}[\underline{H}] \cdot [\Delta h^+]. \tag{7.53}$$

Introducing the symbols

$$[\underline{A}] = [\underline{Q}]^{-1}[\underline{H}], \tag{7.54}$$

$$[\underline{B}] = [\underline{Q}]^{-1}[\underline{F}] \tag{7.55}$$

an expression for elemental discharge increments is obtained

$$[\Delta Q^+] = [\underline{B}] - [\underline{A}] \cdot [\Delta h^+]. \tag{7.56}$$

The finite element matrix A^e and vector B^e for non-steady modelling have the form

$$A^e = \vartheta\,\Delta t \begin{bmatrix} +\underline{A}_{11} & +\underline{A}_{12} \\ -\underline{A}_{21} & -\underline{A}_{22} \end{bmatrix}, \tag{7.57}$$

$$B^e = (1-\vartheta)\Delta t \begin{bmatrix} +Q_1 \\ -Q_2 \end{bmatrix} + \vartheta\,\Delta t \begin{bmatrix} +Q_1^+ \\ -Q_2^+ \end{bmatrix} + \vartheta\,\Delta t \begin{bmatrix} +\underline{B}_1 \\ -\underline{B}_2 \end{bmatrix}. \tag{7.58}$$

Calculation of the matrices: $[\underline{Q}]^{-1}$, $[\underline{A}]$, A^e and vectors: $[\underline{B}]$, B^e of the valve finite element is carried out numerically.

(b) If the valve status at the end of the time interval is such that the valve is closed or should be closed

status = VALVE_IS_CLOSED or VALVE_CLOSE_IT

then the discharges are equal to zero: $Q_1^+ = Q_2^+ = 0$, which implies that the matrix $[\underline{A}] = 0$ and vector $[\underline{B}] = 0$ since discharge increment, according to the expression (7.53), is always zero for each increment Δh, that is the matrix $A^e = 0$. The finite element vector B^e should contain information on the discharge from the beginning of the time interval; thus, according to Eq. (7.58), it is

$$B^e = (1 - \vartheta\,\Delta t) \begin{bmatrix} +Q_1 \\ -Q_2 \end{bmatrix}. \tag{7.59}$$

(c) If the valve status at the end of the time interval is such that the valve should be opened

status = VALVE_OPEN_IT

then there will be a problem of integration of resistances between the two time stages. Namely, at the beginning of the time interval the valve is completely closed and the resistance coefficient is $\beta_v^{\pm} = \infty$. The problem is solved by fully implicit integration: the time integration parameter $\vartheta = 1$. The further procedure is equal to the procedure described for the open valve.

Subroutine `UnsteadyRefluxMtx`

This subroutine calculates the finite element matrix and vector for the reflux preventer depending on the type of non-steady calculation:

```
if runmode = QuasyUnsteady then
  call QuRefluxMtx(ielem)
else if runmode = RigidUnsteady then
  call RgdRefluxMtx(ielem)
else if runmode = FullUnsteady then
    call UnsRefluxMtx(ielem)
endif.
```

Subroutine `QuRefluxMtx`

This subroutine calculates the finite element matrix and vector for quasi unsteady flow by branching computation according to the reflux preventer status.

(a) If the reflux preventer is opened at the end of the time stage

```
status = VALVE_IS_OPEN
```

then the subroutine is called

```
call QuOpenRefluxMtx(ielem).
```

This subroutine is procedurally equal to the procedure described in Section 7.1.4 Subroutine `QuValveMtx`, under (a).

(b) If the reflux preventer status at the end of the time interval is such that the valve is closed or should be closed

```
status = VALVE_IS_CLOSED or VALVE_CLOSE_IT
```

then the subroutine is called

```
call QuCloseRefluxMtx(ielem).
```

This subroutine is procedurally equal to the procedure described in Section 7.1.4 Subroutine `QuValveMtx`, under (b).

(c) If the reflux preventer status at the end of the time interval is such that the valve should be opened

```
status = VALVE_OPEN_IT,
```

then the subroutine is called

```
call QuOpenItRefluxMtx(ielem).
```

This subroutine is procedurally equal to the procedure described in Section 7.1.4 Subroutine `QuValveMtx`, under (c).

Subroutine `RgdRefluxMtx`

This subroutine calculates the finite element matrix and vector for an incompressible fluid flow by branching according to the reflux preventer status.

(a) If the reflux preventer is opened at the end of the time stage

```
status = VALVE_IS_OPEN
```

then the subroutine is called

call RgdOpenRefluxMtx(ielem).

This subroutine is procedurally equal to the procedure described in Section 7.1.4 Subroutine RgdValveMtx, under (a).

(b) If the reflux preventer status at the end of time interval is such that the valve is closed or should be closed

status = VALVE_IS_CLOSED or VALVE_CLOSE_IT

then the subroutine is called

call RgdCloseRefluxMtx(ielem).

This subroutine is procedurally equal to the procedure described in Section 7.1.4 Subroutine RgdValveMtx, under (b).

(c) If the reflux preventer status at the end of the time interval is such that the valve should be opened

status = VALVE_OPEN_IT

then the subroutine is called

call RgdOpenItRefluxMtx(ielem).

This subroutine is procedurally equal to the procedure described in Section 7.1.4 Subroutine RgdValveMtx, under (c).

Subroutine *UnsRefluxMtx*

This subroutine calculates the finite element matrix and vector for an incompressible liquid flow by branching according to the reflux preventer status.

(a) If the reflux preventer is opened at the end of the time stage

status = VALVE_IS_OPEN

then the subroutine is called

call UnsOpenRefluxMtx(ielem).

This subroutine is procedurally equal to the procedure described in Section 7.1.4 Subroutine UnsValveMtx, under (a).

(b) If the reflux preventer status at the end of the time interval is such that the valve is closed or should be closed

status = VALVE_IS_CLOSED or VALVE_CLOSE_IT

then the subroutine is called

call UnsCloseRefluxMtx(ielem).

This subroutine is procedurally equal to the procedure described in Section 7.1.4 Subroutine UnsValveMtx, under (b).

(c) If the reflux preventer status at the end of the time interval is such that the valve should be opened

status = VALVE_OPEN_IT

then the subroutine is called

`call UnsOpenItRefluxMtx(ielem).`

This subroutine is procedurally equal to the procedure described in Section 7.1.4 Subroutine `UnsValveMtx`, under (c).

7.2   Joints

7.2.1   Energy head losses at joints

Hydraulic network joints are relatively short hydraulic branches to connect pipe and other branch types. In general, joint resistances are asymmetric; namely, resistances can be different in positive and negative directions. Local loss of the energy line can be written as

$$\Delta H|_1^2 = \zeta_j^{\pm} \frac{|v|v}{2g}, \tag{7.60}$$

where $\zeta_j^{\pm}$ is the coefficient of local resistances depending on the joint type while v is the velocity for the rated joint cross-section. Energy head losses expressed by discharge are

$$\Delta H|_1^2 = \beta_j^{\pm} Q|Q, \tag{7.61}$$

where $\beta_j^{\pm}$ is the resistance coefficient of the joint. Some standard joints, shown in Table 7.2, are inbuilt in the programming solution `SimpipCore`.

Some joints, such as transitions and bends, apart from flow resistances due to form change also have friction resistances due to relatively long length. The coefficient of local energy loss for transitions is

$$\zeta_j^+ = \zeta_o + f_\lambda \lambda_{D_2}, \tag{7.62}$$

where ζ_o is the form resistance coefficient, and f_λ is the factor that the coefficient of friction resistance is multiplied by λ_{D_2} calculated for the diameter D_2. Calculation of resistances for transitions given in Table 7.2, module `Joints.f90`, is carried out in the procedure

`logical function CalcJointLoss(Joint)`

using the respective procedures from the program module `hydraulics.f90`.

By selection of adequate diameters and lengths of elements, procedures for transitions also model the resistances for sudden changes shown in Figure 7.3.

7.2.2   Steady flow modelling

In a subroutine `Steady` a call for a subroutine for computation of the steady flow matrix and vector for elemental joint type is added as follows:

```
select case(Elems(ielem).tip)
      case ...

            . . .
      case (JOINT_OBJ)
            call SteadyJointMtx(ielem)
      case ...

            . . .
endselect
```

Table 7.2 Built-in joints types and coefficients of local resistances.

(a) Transition.

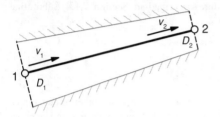

Contraction $D_1 > D_2$, $v = v_2$:

$$\zeta_j^+ = \zeta_o + f_\lambda \lambda_{D_2}; \; \zeta_j^- = \zeta_o + f_\lambda \lambda_{D_2}$$

$$\beta_j^\pm = \frac{\zeta_j^\pm}{2g A_2^2}$$

Expansion $D_1 < D_2$, $v = v_1$:

$$\zeta_j^+ = \zeta_o + f_\lambda \lambda_{D_1}; \; \zeta_j^- = \zeta_o + f_\lambda \lambda_{D_2}$$

$$\beta_j^\pm = \frac{\zeta_j^\pm}{2g A_1^2}$$

(b) Bend.

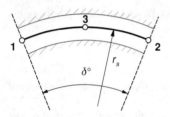

$$\zeta_j^\pm = \zeta_o + \frac{\lambda}{D} r_s \frac{\delta^\circ \pi}{180^\circ}$$

$$\zeta_o = \left[0.131 + 0.163 \left(\frac{D}{r_s} \right)^{3.5} \right] \frac{\delta^\circ}{90^\circ}$$

$$\beta_j^\pm = \frac{\zeta_j^\pm}{2g A^2}$$

(c) Sharp bend.

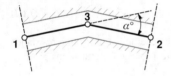

$$\zeta_j^\pm = 3 \sin^2 \frac{\alpha}{2} - \sin^4 \frac{\alpha}{2}$$

$$\beta_j^\pm = \frac{\zeta_j^\pm}{2g A^2}$$

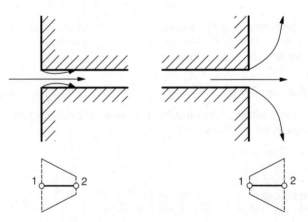

Figure 7.3 Sudden changes.

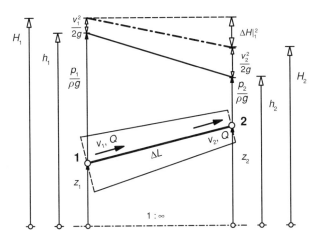

Figure 7.4   General joint, steady flow.

Subroutine `SteadyJointMtx`

Figure 7.4 shows the heads and losses in a steady flow on a finite element for a positive discharge (flow in the direction from the first to the second node). The elemental equation can be applied to the joint in steady flow

$$F^e: \quad h_2 + \frac{Q^2}{2gA_2^2} - h_1 - \frac{Q^2}{2gA_1^2} + \beta_j^{\pm} |Q| \, Q = 0 \tag{7.63}$$

which includes velocity heads. A starting point is the Newton–Raphson form, which is formally written using the matrix-vector operations

$$[\underline{H}] \cdot [\Delta h] + [\underline{Q}] \cdot [\Delta Q] = [\underline{F}], \tag{7.64}$$

where the scalar value $[F]$ is equal to

$$[\underline{F}] = -\left[h_2 - h_1 + \frac{Q^2}{2g}\left(\frac{1}{A_2^2} - \frac{1}{A_1^2}\right) + \beta_j^{\pm} |Q| \, Q \right] \tag{7.65}$$

while vector $[\underline{H}]$ has the form

$$[\underline{H}] = \left[\frac{\partial F^e}{\partial h_1} \quad \frac{\partial F^e}{\partial h_2} \right] = [-1 \quad 1]. \tag{7.66}$$

The scalar value $[\underline{Q}]$ is equal to

$$[\underline{Q}] = \frac{\partial F^e}{\partial Q} = \frac{Q}{g}\left(\frac{1}{A_2^2} - \frac{1}{A_1^2}\right) + 2\beta_j^{\pm} |Q|. \tag{7.67}$$

When the previous expression is multiplied by the inverse term $[\underline{Q}]^{-1}$ the following is obtained

$$[\underline{A}] \cdot [\Delta h] + [\Delta Q] = [\underline{B}], \tag{7.68}$$

from which the value of the elemental discharge increment can be calculated as

$$[\Delta Q] = [\underline{B}] - [\underline{A}] \cdot [\Delta h], \tag{7.69}$$

where

$$[\underline{A}] = [\underline{Q}]^{-1} [\underline{H}], \tag{7.70}$$

$$[\underline{B}] = [\underline{Q}]^{-1} [\underline{F}]. \tag{7.71}$$

In the aforementioned expressions $[\underline{A}]$ is a two-term vector while $[\underline{B}]$ is a scalar. A process of elimination of elemental discharges from nodal equations of continuity defines the structure of the finite element matrix

$$A^e = \begin{bmatrix} +\underline{A} \\ -\underline{A} \end{bmatrix} \tag{7.72}$$

and vector

$$B^e = \begin{bmatrix} +Q \\ -Q \end{bmatrix} + \begin{bmatrix} +\underline{B} \\ -\underline{B} \end{bmatrix}. \tag{7.73}$$

The procedure is carried out numerically.

7.2.3   Non-steady flow modelling

In the subroutine Unsteady a call for a subroutine for computation of the non-steady flow matrix and vector for the elemental joint type is added as follows:

```
select case(Elems(ielem).tip)
      case ...
            ...
      case (JOINT_OBJ)
            call UnsteadyJointMtx(ielem)
      case ...
            ...
endselect
```

Subroutine UnsteadyJointMtx

This subroutine calculates the joint finite element matrix and vector by calling the respective subroutine depending on the type of non-steady calculation:

```
if runmode = QuasyUnsteady then
  call QuUnsteadyJointMtx(ielem)
else if runmode = RigidUnsteady then
  call RgdUnsteadyJointMtx(ielem)
else if runmode = FullUnsteady then
   call NonSteadyJointMtx(ielem)
endif
```

Subroutine `QuUnsteadyJointMtx`
The equation of quasi unsteady flow in a joint is equal to the steady flow equation

$$H_2 - H_1 + \beta_j^\pm |Q| Q = 0 \tag{7.74}$$

where the energy heads are equal to

$$H_1 = h_1 + \frac{Q^2}{2g A_1^2}, \quad H_2 = h_2 + \frac{Q^2}{2g A_2^2}.$$

Variables of the state are time-dependent functions. By integration of the elemental equation in time interval Δt the following is obtained

$$
\begin{aligned}
F^e &= \int_{\Delta t} \left(H_2 - H_1 + \beta_j^\pm |Q| Q\right) dt = \\
&(1 - \vartheta)\Delta t \left(H_2 - H_1 + \beta_j^\pm |Q| Q\right) + \\
&+ \vartheta \Delta t \left(H_2^+ - H_1^+ + \beta_j^{\pm +} |Q^+| Q^+\right) = 0.
\end{aligned}
\tag{7.75}
$$

The Newton–Raphson iterative form for a finite element is formally written using the matrix-vector operations

$$\left[\underline{H}\right] \cdot [\Delta h] + \left[\underline{Q}\right] \cdot [\Delta Q] = \left[\underline{F}\right] \tag{7.76}$$

where the vector is equal to

$$\left[\underline{H}\right] = \left[-\vartheta \Delta t \quad +\vartheta \Delta t\right] = \vartheta \Delta t \left[-1 \quad +1\right] \tag{7.77}$$

while the scalar terms are

$$\left[\underline{F}\right] = -F^e, \tag{7.78}$$

$$\left[\underline{Q}\right] = \vartheta \Delta t \frac{Q^+}{g} \left(\frac{1}{A_2^2} - \frac{1}{A_1^2}\right) + \vartheta \Delta t 2\beta_j^{\pm +} |Q^+|. \tag{7.79}$$

When the expression (7.76) is multiplied by the inverse term $[\underline{Q}]^{-1}$ the following is obtained

$$\left[\underline{A}\right] \cdot [\Delta h] + [\Delta Q] = \left[\underline{B}\right], \tag{7.80}$$

from which the value of the elemental discharge increment can be calculated as

$$[\Delta Q] = \left[\underline{B}\right] - \left[\underline{A}\right] \cdot [\Delta h], \tag{7.81}$$

where

$$\left[\underline{A}\right] = \left[\underline{Q}\right]^{-1} \left[\underline{H}\right], \tag{7.82}$$

$$\left[\underline{B}\right] = \left[\underline{Q}\right]^{-1} \left[\underline{F}\right]. \tag{7.83}$$

In the aforementioned expressions $[\underline{A}]$ is a two-term vector while $[\underline{B}]$ is a scalar. A process of elimination of elemental discharges from nodal equations of continuity defines the structure of the finite element matrix

$$A^e = \vartheta \, \Delta t \begin{bmatrix} +\underline{A} \\ -\underline{A} \end{bmatrix} \qquad (7.84)$$

and vector

$$B^e = (1 - \vartheta \, \Delta t) \begin{bmatrix} +\underline{Q} \\ -\underline{Q} \end{bmatrix} + \vartheta \, \Delta t \begin{bmatrix} +\underline{Q}^+ \\ -\underline{Q}^+ \end{bmatrix} + \vartheta \, \Delta t \begin{bmatrix} +\underline{B} \\ -\underline{B} \end{bmatrix}. \qquad (7.85)$$

Calculation of the scalar term $[\underline{Q}]^{-1}$, vector $[\underline{A}]$, scalar term $[\underline{B}]$, the finite element matrix $[A^e]$, and vector $[B^e]$ of the valve is carried out numerically.

Subroutine `RgdUnsteadyJointMtx`
The equation of non-steady flow of a rigid fluid in a joint is

$$H_2 - H_1 + \beta_j^{\pm} |Q| \, Q + \frac{\Delta l}{gA} \frac{dQ}{dt} = 0. \qquad (7.86)$$

Variables of the state are time-dependent functions. By integration of the elemental equation in time interval Δt the following is obtained:

$$\begin{aligned}
F^e = \int_{\Delta t} \left(H_2 - H_1 + \beta_j^{\pm} |Q| \, Q + \frac{\Delta l}{gA} \frac{dQ}{dt} \right) dt = \\
(1 - \vartheta)\Delta t \left(H_2 - H_1 + \beta_j^{\pm} |Q| \, Q \right) + \\
+ \vartheta \, \Delta t \left(H_2^+ - H_1^+ + \beta_j^{\pm +} |Q^+| \, Q^+ \right) + \\
+ \frac{\Delta l}{gA} (Q^+ - Q) = 0.
\end{aligned} \qquad (7.87)$$

The Newton–Raphson iterative form for a finite element is formally written using the matrix-vector operations

$$[\underline{H}] \cdot [\Delta h] + [\underline{Q}] \cdot [\Delta Q] = [\underline{F}], \qquad (7.88)$$

where the vector is equal to

$$[\underline{H}] = [-\vartheta \, \Delta t \quad +\vartheta \, \Delta t] = \vartheta \, \Delta t [-1 \quad +1], \qquad (7.89)$$

while the scalar terms are

$$[\underline{F}] = -F^e, \qquad (7.90)$$

$$[\underline{Q}] = \frac{\Delta l}{gA} + \vartheta \, \Delta t \frac{Q^+}{g} \left(\frac{1}{A_2^2} - \frac{1}{A_1^2} \right) + \vartheta \, \Delta t 2\beta_v^{\pm} |Q^+|. \qquad (7.91)$$

When the previous expression is multiplied by the inverse term $[\underline{Q}]^{-1}$ the following is obtained

$$[\underline{A}] \cdot [\Delta h] + [\Delta Q] = [\underline{B}],\tag{7.92}$$

from which the value of the elemental discharge increment can be calculated as

$$[\Delta Q] = [\underline{B}] - [\underline{A}] \cdot [\Delta h],\tag{7.93}$$

where

$$[\underline{A}] = [\underline{Q}]^{-1}[\underline{H}],\tag{7.94}$$

$$[\underline{B}] = [\underline{Q}]^{-1}[\underline{F}].\tag{7.95}$$

In the aforementioned expressions $[\underline{A}]$ is a two-term vector while $[\underline{B}]$ is a scalar. A process of elimination of elemental discharges from nodal equations of continuity defines the structure of the finite element matrix

$$A^e = \vartheta\,\Delta t \begin{bmatrix} +\underline{A} \\ -\underline{A} \end{bmatrix}\tag{7.96}$$

and vector

$$B^e = (1 - \vartheta\,\Delta t)\begin{bmatrix} +\underline{Q} \\ -\underline{Q} \end{bmatrix} + \vartheta\,\Delta t\begin{bmatrix} +\underline{Q}^+ \\ -\underline{Q}^+ \end{bmatrix} + \vartheta\,\Delta t\begin{bmatrix} +\underline{B} \\ -\underline{B} \end{bmatrix}.\tag{7.97}$$

Calculation of the scalar term $[\underline{Q}]^{-1}$, vector $[\underline{A}]$, scalar term $[\underline{B}]$, the finite element matrix $[A^e]$, and vector $[B^e]$ of the valve is carried out numerically.

Subroutine *NonSteadyJointMtx*
The joint finite element matrix and vector will be obtained by integration of the mass and specific mechanical energy conservation law between the initial and end state. The mass conservation law on a finite element is

$$\frac{gA}{c^2} \int_{\Delta l} \frac{\partial h}{\partial t}\,dl + (Q_2 - Q_1) = 0,\tag{7.98}$$

where the water hammer celerity is adopted to be equal to the water hammer celerity in a pipe with the diameter equal to the mean joint diameter.

Using the mean value theorem of the integral, it is written as

$$\frac{gA}{c^2}\frac{\Delta l}{2}\left(\frac{\partial h_1}{\partial t} + \frac{\partial h_2}{\partial t}\right) + (Q_2 - Q_1) = 0.\tag{7.99}$$

After integration with the time step Δt

$$\frac{gA}{c^2}\int_{\Delta t}\frac{\Delta L}{2}\left(\frac{\partial h_1}{\partial t} + \frac{\partial h_2}{\partial t}\right)dt + \int_{\Delta t}(Q_2 - Q_1)\,dt = 0\tag{7.100}$$

and repeatedly re-using the aforementioned time-dependent integration rules, the first elemental equation is obtained

$$F_1: \quad \frac{\Delta l}{2} \frac{g A}{c^2} \left[\left(h_1^+ + h_2^+ \right) - \left(h_1 + h_2 \right) \right] +$$
$$+ (1 - \vartheta) \Delta t \left(Q_2 - Q_1 \right) + \vartheta \Delta t \left(Q_2^+ - Q_1^+ \right) = 0. \tag{7.101}$$

The specific mechanic energy conservation law in the head form is

$$\frac{1}{g A} \int\limits_{\Delta l} \frac{\partial Q}{\partial t} dl + H_2 - H_1 + \Delta H \big|_1^2 = 0 \tag{7.102}$$

and this is shown in Figure 7.5. The local loss (7.2) is expressed as

$$\Delta H \big|_1^2 = \beta_j^\pm \left| \bar{Q} \right| \bar{Q} = 0, \tag{7.103}$$

where $\bar{Q} = (Q_1 + Q_2)/2$ is the mean discharge on the element.
   Using the mean value theorem of the integral, it is written as

$$\frac{\Delta l}{2 g A} \left(\frac{\partial Q_1}{\partial t} + \frac{\partial Q_2}{\partial t} \right) + H_2 - H_1 + \beta_j^\pm \left| \bar{Q} \right| \bar{Q} = 0. \tag{7.104}$$

After integration with the time step Δt

$$\frac{\Delta l}{2 g A} \int\limits_{\Delta t} \left(\frac{\partial Q_1}{\partial t} + \frac{\partial Q_2}{\partial t} \right) dt + \int\limits_{\Delta t} (H_2 - H_1) dt + \int\limits_{\Delta t} \beta_j^\pm \left| \bar{Q} \right| \bar{Q} dt = 0 \tag{7.105}$$

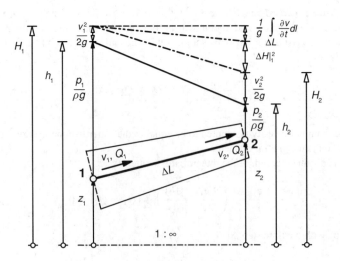

Figure 7.5   General joint, non-steady flow.

finally, the second elemental equation is obtained in the form

$$F_2 : \quad \frac{\Delta l}{2g\,A} \left[(Q_1^+ + Q_2^+) - (Q_1 + Q_2) \right] +$$
$$+ (1 - \vartheta)\Delta t\, (H_2 - H_1) + \vartheta\,\Delta t\, \left(H_2^+ - H_1^+ \right) + \tag{7.106}$$
$$+ (1 - \vartheta)\Delta t \left[\beta_j^\pm \left| \bar{Q} \right| \bar{Q} \right] + \vartheta\,\Delta t \left[\beta_j^\pm \left| \bar{Q} \right| \bar{Q} \right]^+ = 0.$$

The Newton–Raphson iterative form for a finite element is formally written using the matrix-vector operations

$$[\underline{H}] \cdot [\Delta h^+] + [\underline{Q}] \cdot [\Delta Q^+] = [\underline{F}], \tag{7.107}$$

where the matrix

$$[\underline{H}] = \begin{bmatrix} \dfrac{\Delta l}{2}\dfrac{g\,A}{c^2} & \dfrac{\Delta l}{2}\dfrac{g\,A}{c^2} \\[2mm] -\vartheta\,\Delta t & +\vartheta\,\Delta t \end{bmatrix} \tag{7.108}$$

and matrix

$$[\underline{Q}] = \begin{bmatrix} -\vartheta\,\Delta t & +\vartheta\,\Delta t \\[2mm] \left(\dfrac{\Delta l}{2g\,A} + \vartheta\,\Delta t\,\Delta l \beta_j^\pm \left| \bar{Q}^+ \right| - \dfrac{\vartheta\,\Delta t}{g\,A^2} Q_1^+ \right) & \left(\dfrac{\Delta l}{2g\,A} + \vartheta\,\Delta t\,\Delta l \beta_j^\pm \left| \bar{Q}^+ \right| + \dfrac{\vartheta\,\Delta t}{g\,A^2} Q_2^+ \right) \end{bmatrix}. \tag{7.109}$$

The right hand side vector is

$$[\underline{F}] = - \begin{bmatrix} F_1 \\ F_2 \end{bmatrix}. \tag{7.110}$$

The elemental discharge increments are calculated from the Newton–Raphson form of the elemental equations in such a manner that the equation is multiplied by the inverse matrix $[\underline{Q}]^{-1}$

$$[\Delta Q^+] = [\underline{Q}]^{-1} [\underline{F}] - [\underline{Q}]^{-1} [\underline{H}] \cdot [\Delta h^+]. \tag{7.111}$$

Introducing the symbols

$$[\underline{A}] = [\underline{Q}]^{-1} [\underline{H}], \tag{7.112}$$

$$[\underline{B}] = [\underline{Q}]^{-1} [\underline{F}]. \tag{7.113}$$

an expression for elemental discharge increments is obtained

$$[\Delta Q^+] = [\underline{B}] - [\underline{A}] \cdot [\Delta h^+]. \tag{7.114}$$

The finite element matrix A^e and vector B^e for non-steady modelling have the form

$$A^e = \vartheta \, \Delta t \begin{bmatrix} +\underline{A}_{11} & +\underline{A}_{12} \\ -\underline{A}_{21} & -\underline{A}_{22} \end{bmatrix}, \tag{7.115}$$

$$B^e = (1-\vartheta)\Delta t \begin{bmatrix} +Q_1 \\ -Q_2 \end{bmatrix} + \vartheta \, \Delta t \begin{bmatrix} +Q_1^+ \\ -Q_2^+ \end{bmatrix} + \vartheta \, \Delta t \begin{bmatrix} +\underline{B}_1 \\ -\underline{B}_2 \end{bmatrix}. \tag{7.116}$$

Calculation of the scalar term $[Q]^{-1}$, vector $[A]$, scalar term $[B]$, the matrix A^e, and vector B^e of the joint finite element is carried out numerically.

7.3 Test example

Figure 7.6 shows a linear system built of pipes, transitions, and valves, which enable energy and piezometric line modelling. Input data are given in the first column of Table 7.3, while the print file with the results for the state at the tenth second is given on the right hand side.

Figure 7.6 shows the energy and piezometric lines according to the print file for the tenth second.

A transient state, from the hydrostatic state to the steady state after a sudden drop of the piezometric head at the end of the pipeline is being modeled. The transient state lasts for about 10 seconds. Figure 7.7 shows the discharge at the valve obtained by modelling of the non-steady flow with a time step of 1 second.

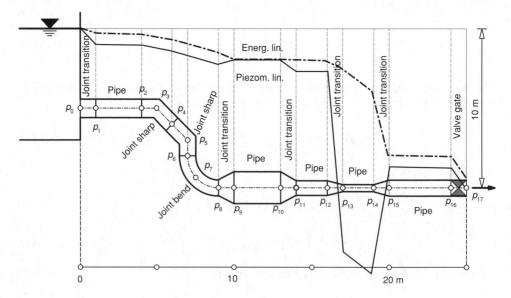

Figure 7.6   Modelling of local losses.

Table 7.3   Test example. File: `LocalLosses.simpip`.

```
;                                              Print file:
;      Local losses modelling
;                                        Stage:    10    Time:     10.0000
;      notice:                           Point     Piez.head      SumQ
; Local resistance can be modeled        p0    10.0000      -0.760558
; for a system with no branches          p1     8.89894      0.00000
;                                         p2     8.83314      0.00000
Parameters                               p3     0.00000      0.00000
    Da = 10                              p4     8.51353      0.00000
    D0 = 0.5                             p5     0.00000      0.00000
    D1 = 1                              p6     8.19392      0.00000
    D2 = 0.25                            p7     0.00000      0.00000
    s = 0.05                             p8     7.60538      0.00000
    eps = 0.1/1000                       p9     7.87251      0.00000
Points                                   p10    7.87053      0.00000
    p0 0 0 5                             p11    7.12973      0.00000
    p1 1 0 5                             p12    7.08586      0.00000
    p2 4 0 5                             p13   -4.88144      0.00000
    p3 5 0 5                             p14   -6.45494      0.00000
    p4 6 0 4                             p15    0.747047     0.00000
    p5 7 0 3                             p16    0.659315     0.00000
    p6 7 0 2                             p17    0.00000      0.760558
    p7 9-2*cos(Pi/4) 0 2-2*sin(Pi/4)     El.    Name     Q1       Q2
    p8 9 0 0
    p9 10 0 0                            1      c1    0.760558      0.760558
    p10 13 0 0                           2      c2    0.760558      0.760558
    p11 14 0 0                           3      c3    0.760558      0.760558
    p12 16 0 0                           4      c4    0.760558      0.760558
    p13 17 0 0                           5      c5    0.760558      0.760558
    p14 19 0 0                           6      vlv   0.760558      0.760558
    p15 20 0 0                           7      Entrance   0.760558      0.760558
    p16 24 0 0                           8      SharpI    0.760558      0.760558
    p17 25 0 0                           9      SharpII   0.760558      0.760558
Pipes                                    10     Bend90    0.760558      0.760558
    c1 p1 p2 D0 eps s Steel              11     ExpanI    0.760558      0.760558
    c2 p9 p10 D1 eps s Steel             12     ContrI    0.760558      0.760558
    c3 p11 p12 D0 eps s Steel            13     ExpanII   0.760558      0.760558
    c4 p13 p14 D2 eps s Steel            14     ContrII   0.760558      0.760558
    c5 p15 p16 D0 eps s Steel
Joints
    TRANSITION Entrance p0 p1 Da D0
    SHARP SharpI p2 p3 p4 D0
    SHARP SharpII p4 p5 p6 D0
    BEND Bend90 p6 p7 p8 D0 s eps
    TRANSITION ExpanI p8 p9 D0 D1
    TRANSITION ContrI p10 p11 D1 D0
    TRANSITION ExpanII p12 p13 D0 D2
    TRANSITION ContrII p14 p15 D2 D0
Valve
    GATE vlv p16 p17 D0 0.5
Graph h(t)
    0 10
    1 0
Piezo p0 10
Piezo p17 0 1 h(t)
Steady 0
Unsteady 10 1
Print
    solStage 0 10
```

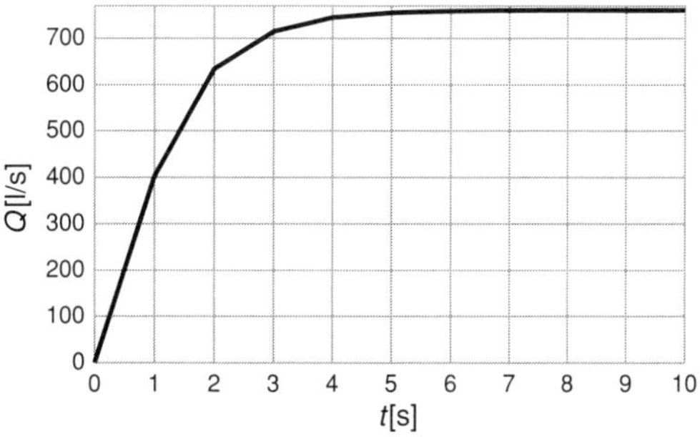

Figure 7.7   Discharge at the valve.

Reference

Jović, V. (2006) *Fundamentals of Hydromechanics* (in Croatian: Osnove hidromehanike). Element, Zagreb.

Further reading

Idelcik, I.E. (1969) *Memento des pertes de charge*. Eyrolles, Paris.

Jović, V. (1977) Non-steady flow in pipes and channels by finite element method. *Proceedings of XVII Congress of the IAHR*, 2: 197–204.

Jović, V. (1987) Modelling of non-steady flow in pipe networks. *Proceedings of the Int. Conference on Numerical Methods NUMETA 87*. Martinus Nijhoff Publishers, Swansea.

Jović, V. (1992) Modelling of hydraulic vibrations in network systems. *International Journal for Engineering Modelling*, 5: 11–17.

Jović, V. (1994) Contribution to the finite element method based on the method of characteristics in modelling hydraulic networks. *Zbornik radova 1. kongresa hrvatskog društva za mehaniku*, 1: 389–398.

Jović, V. (1995) Finite elements and method of characteristics applied to water hammer modeling. *International Journal for Engineering Modelling*, 8: 51–58.

8

Pumping Units

8.1 Introduction

Since the start of civilization, people have known how to harness the power of water by means of the water mill. The principles of turbine operation have been known since the Classical period as Hero's wheel (Alexandria); in truth a clever toy displayed in the temple, with no practical purpose. It is still not known when and where the first devices for water supply and boosting were invented, although it is known that devices like the bucket water wheel, Archimede's screw, and the Ctesibius firefighting piston pump were used in Antiquity. Development of these devices stagnated in the Middle Ages and did not continue until the Renaissance. One of the most famous pumping plants of the time was built in France in 1682. It had 14 water mill wheels of 12 m in diameter that operated 221 pumps, which delivered water from the Seine River to the royal palace of Versailles and the town of Marly-le-Roi and overcame 162 m of height difference. It was not until 1689 that Denis Papin (1647–c. 1712) constructed the first centrifugal pump consisting of straight vanes. In 1851, a British inventor by the name of John Appold established the curved vane centrifugal pump. Incredibly, Leonardo da Vinci (1452–1519) proposed a pump that operated on the centrifugal force principle, while Francesco di Giorgio Martini, an Italian engineer, was also associated with the prototype of that pump in 1475.

Figure 8.1a is a photograph of a wooden impeller of a water mill on the Cetina River (Croatia) that hangs as a decoration from the ceiling of a tavern in Blato na Cetini. It is testament to the high engineering sense of an uneducated water mill constructor. Figure 8.1b shows a sketch of an impeller from a modern turbine. The ingenuity of the water mill constructor is admirable.

8.2 Euler's equations of turbo engines

Pump and turbine. Hydrokinetic engines operate on the principle of interchange of the angular momentum of a liquid and the vanes that liquid flows through. Engines that operate on the principle of impeller rotation, such as pumps and turbines, are generally called turbo engines. One should note that those engines are often called centrifugal engines, which is true only for a radial impeller. For an axial impeller, the nomenclature is completely wrong.

Figure 8.2 shows the principle of turbo engine operation (i) as a pump and (ii) as a turbine. In the case of a pump, the impeller rotates at angular velocity ω and discharge Q flows towards the external perimeter of the impeller. In the case of a turbine, the impeller rotates in the opposite direction at angular velocity $-\omega$ while the discharge $-Q$ flows from the external towards the internal perimeter of the impeller. Torque at the impeller shaft is equal to the change of the angular momentum of a liquid.

Analysis and Modelling of Non-Steady Flow in Pipe and Channel Networks, First Edition. Vinko Jović.
© 2013 John Wiley & Sons, Ltd. Published 2013 by John Wiley & Sons, Ltd.

(a) Wooden impeller of a water mill.   (b) Impeller of a modern turbine.

Figure 8.1   Then and now.

Torque, power, and hydraulic efficiency. Torque at the impeller shaft is equal to the change of the angular momentum of a liquid; namely, the difference between the angular momentum at the impeller inlet and outlet

$$M = \dot{M}(w_2 r_2 - w_1 r_1), \tag{8.1}$$

where $\dot{M}$ is the mass flow, and w_1 and w_2 are the tangential components of the absolute flow velocity at the inlet and outlet profile of the impeller. It can be written as

$$M = \rho Q(w_2 r_2 - w_1 r_1). \tag{8.2}$$

Radial components of the absolute velocity define the turbo engine discharge

$$Q = 2\pi \, (rbq)_1 = 2\pi \, (rbq)_2 . \tag{8.3}$$

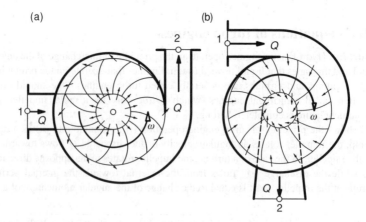

Figure 8.2   Turbo engine: (a) pump (b) turbine.

The power of the engine is defined by the torque and the angular velocity as

$$P = M\omega. \tag{8.4}$$

After the expression for torque is introduced, it can be written as

$$P = \rho Q \omega (w_2 r_2 - w_1 r_1). \tag{8.5}$$

Since the peripheral velocities are defined by the angular velocity

$$u = r\omega. \tag{8.6}$$

Then, the previous expression can be written in the form

$$P = \rho Q (w_2 u_2 - w_1 u_1). \tag{8.7}$$

After multiplication with and division by gH, the aforementioned expression will obtain the following form

$$P = \rho g Q H \frac{w_2 u_2 - w_1 u_1}{gH}, \tag{8.8}$$

where H is the difference between energy heads at the impeller inlet and outlet $H = H_2 - H_1$.

For an ideal turbo engine, the value of the fraction in the previous expression is equal to 1. Thus, the hydraulic efficiency of an engine is defined by the efficiency coefficient

$$\eta_h = \frac{(w_2 u_2 - w_1 u_1)}{gH}, \tag{8.9}$$

that is an ideal energy head at the turbo engine impeller has the form

$$H = \frac{1}{g}(w_2 u_2 - w_1 u_1). \tag{8.10}$$

Velocity diagrams. Absolute velocity, that is velocity seen by the observer from an absolute coordinate system, consists of the impeller transient velocity and relative velocity, that is velocity seen by an observer moving together with the impeller

$$\vec{V} = \vec{U} + \vec{V}^r. \tag{8.11}$$

Figure 8.3 shows the vectors of absolute, transient, and relative velocities at the impeller inlet and outlet, see Figure 8.3 and Figure 8.4. Based on the drawn geometric relations the following can be written

$$w = q \cot \alpha = u - q \cot \beta. \tag{8.12}$$

When applied to expression (8.10) of ideal energy head of a turbo engine, it is written as

$$H = \frac{1}{g}(q_2 u_2 \cot \alpha_2 - q_1 u_1 \cot \alpha_1). \tag{8.13}$$

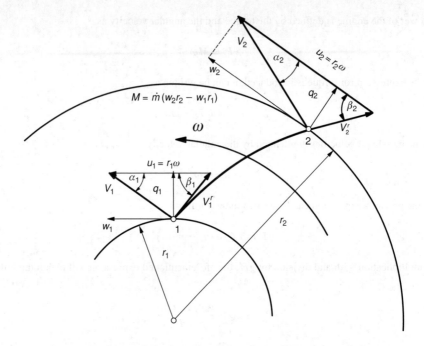

Figure 8.3 Pump impeller velocity diagram.

After arranging

$$H = \frac{1}{g}(u_2(u_2 - q_2 \cot \beta_2) - u_1(u_1 - q_1 \cot \beta_1)), \tag{8.14}$$

$$H = \frac{u_2^2 - u_1^2}{g} - \left(\frac{u_2 q_2}{g} \cot \beta_2 - \frac{u_1 q_1}{g} \cot \beta_1\right) \tag{8.15}$$

and introducing Eqs (8.3) and (8.6), an ideal energy head at the impeller is obtained

$$H = \omega^2 \frac{r_2^2 - r_1^2}{g} - \omega Q \left(\frac{\cot \beta_2}{2\pi b_2 g} - \frac{\cot \beta_1}{2\pi b_1 g}\right). \tag{8.16}$$

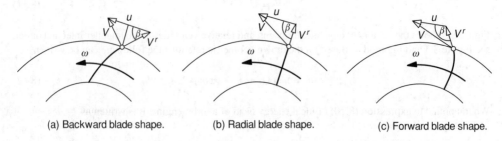

(a) Backward blade shape.   (b) Radial blade shape.   (c) Forward blade shape.

Figure 8.4 Blade shapes.

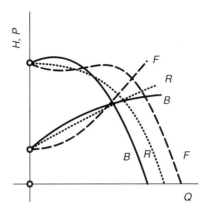

Figure 8.5 Characteristic curves of energy head H and power P B – backward shape, R – radial shape, F – forward shape.

Pump impeller blades are shaped so that, in general, the inlet angular momentum is equal to zero, that is the absolute velocity direction is radial; hence

$$H = \omega^2 \frac{r_2^2}{g} - \omega Q \frac{\cot \beta_2}{2\pi b_2 g}. \tag{8.17}$$

A real turbo engine consists of, apart from the impeller, other constructive components such as the casing, suction inlet, and pressure outlet components. From the inlet to the outlet, liquid must overcome resistances that are proportional to the squared velocity and can be expressed as $c^{\pm} |Q| Q$. If resistances are added to ideal terms, after grouping of constants, the difference of energy heads from the inlet to the outlet of a turbo engine can be written in the form:

$$H = a\omega^2 + b\omega Q + c^{\pm} \cdot Q |Q|. \tag{8.18}$$

The obtained expression is a good approximation of a radial turbo engine operation.

Similarly, turbines are shaped in a manner that the outlet angular momentum is equal to zero, that is the outflow velocity is as small as possible, which is obtained by gradual expansion of the outlet cross-section; namely, by installation of a diffuser.

Note that the energy head of a turbo engine depends on several parameters such as rotational speed, discharge, impeller diameter, construction of impeller blades, and so on, see Figure 8.5. Dependences between the energy head, power, and efficiency coefficient are tested on each turbo engine prototype and are called the turbo engine characteristics or performances.

8.3   Normal characteristics of the pump

Turbo engine characteristics are, generally, determined experimentally on a model or a prototype. Figure 8.6 shows a device that might be used for pump characteristic testing. Piezometers that measuring the H difference of piezometric heads are installed at the suction end (point 1) and the pressure end (point 2). This head is called a manometric head since it is the difference between the pressure readings at manometers at points 1 and 2. In the pressure pipe end there is a valve for discharge regulation. Discharge

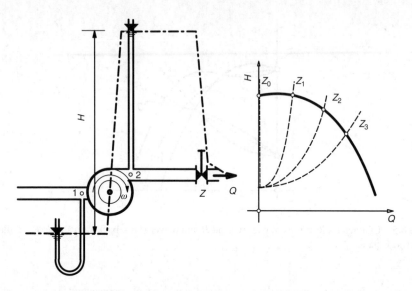

Figure 8.6   Testing of Q–H characteristics of the pump.

Q is measured at the end of the pipeline, for example by the volumetric method, using the gauging weir or some other standard flow metering device.

Apart from that, a dynamic torque balance for torque gauging at the shaft, as well as a device for measuring the number of revolutions of the pump, are installed at the pump shaft. If, for each valve opening Z, measured data Q and H are shown graphically as on the right side of the Figure 8.6, the normal Q–H pump characteristic is obtained.

The angular velocity is calculated from the number of revolutions

$$\omega\left[s^{-1}\right] = 2\pi \frac{n\left[rev/\min\right]}{60}. \tag{8.19}$$

The pump increases the hydraulic power of the flow

$$P_0 = \rho g Q H. \tag{8.20}$$

The pump efficiency coefficient is calculated from the hydraulic power P_0 and power P measured at the shaft

$$\eta = \frac{\rho g Q H}{P}. \tag{8.21}$$

It is a common practice to present normal pump characteristics as a diagram like that shown in Figure 8.7. Apart from the graphs H, P, η, the diagram also shows the normal number of revolutions for the given data.

The working point, where the pump operates at maximum efficiency with respect to the input power, can be determined from the normal pump characteristic. It is a *nominal or duty point*

$$Q_0, H_0, P_0, \eta_{\max}. \tag{8.22}$$

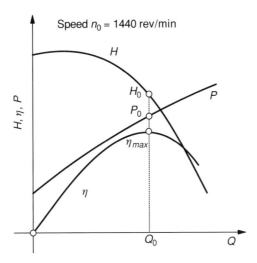

Figure 8.7 Normal characteristic of the pump.

Pumping block, pumps in parallel, and pumps in series. Figure 8.8a shows the Q–H and η curves for two pumps in a parallel connection. A parallel connection increases discharge.

Figure 8.8b shows the Q–H curve for two pumps in series. Pumps connected in a series increase the head. Several pumps connected in parallel or a series are called a pumping block.

System resistances. Figure 8.9 shows a longitudinal section of pipeline and a pumping block that pumps water from lower to higher elevations. Apart from the static head H_s, all resistances in the pipeline shall also be overcome.

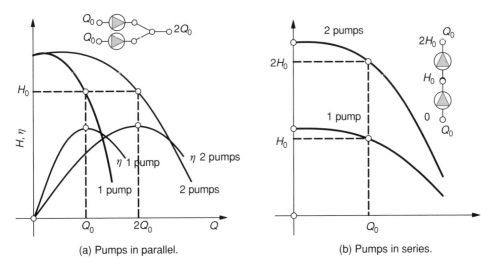

(a) Pumps in parallel. (b) Pumps in series.

Figure 8.8 Pumps in parallel and pumps in series.

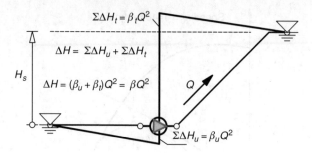

Figure 8.9   Pump system scheme.

Resistances in the inlet (suction) pipeline and the pressure pipeline are discharge dependant and can be expressed as:

$$\sum \Delta H_u = \beta_u Q^2$$
$$\sum \Delta H_t = \beta_t Q^2 \qquad \qquad (8.23)$$
$$\Delta H = (\beta_u + \beta_t) Q^2 = \beta Q^2$$

The resistance curve will have the following form:

$$H = H_s + \beta Q^2. \qquad \qquad (8.24)$$

Working point. Q_0, H_0, P_0, η is located at the intersection between the pipeline resistance curve and the Q–H curve of a pump. It is marked as point R in Figure 8.11. If the pump efficiency coefficient η has its maximum at the working point, then the working point is called the *optimum working point* or *duty point*.

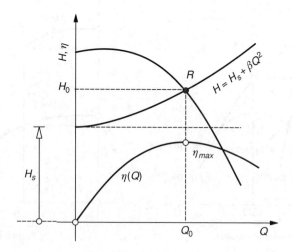

Figure 8.10   Pump working point with maximal efficiency – nominal or "duty point."

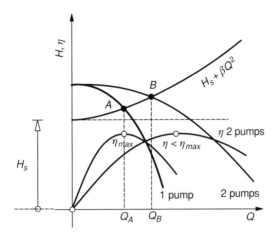

Figure 8.11   Working point for pumps in parallel; optimum operation with a single pump.

Figure 8.11 shows two working points A and B for a single pump and a parallel connection of two pumps of the same characteristics. If only one pump is operating and delivers discharge Q_A, then its performance is an optimum one since its efficiency coefficient is the maximum. Note that when the second pump is added, the discharge is not doubled. Similarly, if one pump has an optimum working point, two pumps in parallel cannot have the optimum working point.

Figure 8.12 shows two working points A and B for a single pump and parallel connection of two pumps of the same characteristics. The optimum operation is provided by two pumps in parallel. Note that operation with a single pump cannot be optimum.

Optimum sizing of pumping units is a complex problem that depends on several factors. One of the possible criteria for selection of pumping units is energy consumption. Thus, selection of optimum pumps in a hydraulic system is not entirely a hydraulic problem.

Suction head. Figure 8.13 shows energy heads expressed by absolute pressures in the pump suction pipeline.

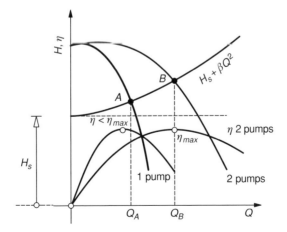

Figure 8.12   Working point for pumps in parallel; optimum operation with two pumps.

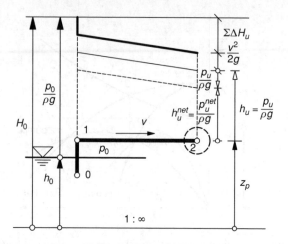

Figure 8.13 Pump suction pipeline.

From the start of the suction pipeline where, in point *0*, the energy is equal

$$H_0 = h_0 + \frac{p_0}{\rho g},$$

to point 2, located at the eye of the pump impeller, energy is lost to overcome local and friction resistances. Between these two points, Bernoulli's equation in head form can be written as

$$h_0 + \frac{p_0}{\rho g} = z_c + \frac{p_u}{\rho g} + \frac{v^2}{2g} + \sum \Delta H_u, \tag{8.25}$$

where p_u is the absolute suction pressure at the eye of the pump impeller, as shown in the figure. In terms of energy, the system can function as long as the suction head is $h_u > 0$. However, due to thermodynamic conditions, flow will be possible only until pressure p_u becomes close to the pressure of a saturated liquid,[1] that is as long as $p_u > p_v$. Above that, liquid boiling and cavitation will occur. Based on that, the *net positive suction head* h_u^{net} is defined as the difference between the gross suction head h_u and the saturated vapor pressure expressed as the height of the liquid's column

$$h_u^{net} = \frac{p_u - p_v}{\rho g} \tag{8.26}$$

that shall be greater than zero for pumping functioning. Note that in some parts within the pump velocities are significantly greater than those in the pipeline and respectively smaller pressures may occur. Thus, cavitation may be expected at critical pump locations, although pressure in the suction pipeline is still above the pressure of a saturated liquid. Similar phenomena occur during the pumping of a liquid that

---

[1]Note the difference between the saturated liquid and saturated vapor; see Chapter 5 Equations of non-steady flow in pipes.

contains diluted gases. Thus, the net positive suction head shall be greater than some suction head with enough reserve. This datum is called the *NPSH* curve, that is the required net suction head of the pump

$$h_u^{net} = \frac{p_u - p_v}{\rho g} \geq NPSH. \tag{8.27}$$

It can be found in pump catalogues together with the standard normal characteristics and is discharge dependent.

8.4   Dimensionless pump characteristics

Let us observe all turbo engines of similar construction defined by one linear geometric parameter; namely, impeller diameter D. These are the homologous turbo engines to which the laws of similarity apply. The main condition is that the Reynolds number be high enough so the viscosity influence of the flow pattern is negligible and the fluid is incompressible ($\rho = const$). Then, a general functional connection between discharge, manometric pressure, angular velocity, density, and impeller diameter can be assumed

$$f(Q, \Delta p, \omega, \rho, D) = 0. \tag{8.28}$$

Using the dimensionless analysis; namely, the Buckingham *Pi* method, two dimensionless variables are obtained

$$dimensionless\ discharge : cQ = \frac{Q}{\omega D^3}, \tag{8.29}$$

$$dimensionless\ pumping\ head : cH = \frac{\Delta p}{\rho \omega^2 D^2} = \frac{gH}{\omega^2 D^2}, \tag{8.30}$$

Apart from the fundamental dimensionless variables, derived ones can also be written

$$dimensionless\ power : cP = \frac{P}{\rho \omega^3 D^5}, \tag{8.31}$$

$$dimensionless\ torque : cM = \frac{M}{\rho \omega^2 D^5}. \tag{8.32}$$

The dimensionless pumping head and efficiency coefficient are shown in Figure 8.14 as functions of the dimensionless discharge. Dimensionless power is equal to the area of the square. Its maximum is at the point of maximum efficiency.

For the *constant impeller diameter*, laws of transformation of discharge, pumping head, torque, and power in reference to the rotational speed can be derived from the dimensionless characteristics

$$\frac{\omega_1}{\omega_2} = \frac{n_1}{n_2} = \frac{Q_1}{Q_2} = \left(\frac{H_1}{H_2}\right)^{\frac{1}{2}} = \left(\frac{M_1}{M_2}\right)^{\frac{1}{2}} = \left(\frac{P_1}{P_2}\right)^{\frac{1}{3}}. \tag{8.33}$$

Namely, the following can be written

$$\frac{Q_1}{Q_2} = \frac{n_1}{n_2}; \frac{H_1}{H_2} = \left(\frac{n_1}{n_2}\right)^2; \frac{M_1}{M_2} = \left(\frac{n_1}{n_2}\right)^2 i \frac{P_1}{P_2} = \left(\frac{n_1}{n_2}\right)^3. \tag{8.34}$$

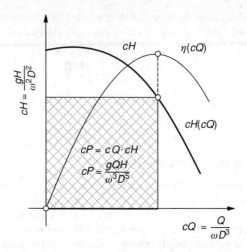

Figure 8.14   Dimensionless pump characteristic.

Homologous transformations of Q–H and Q–P pump characteristics for different rotational speeds are shown in

For the *constant rotational speed*, the laws of transformation of discharge, pumping head, torque, and power in reference to the impeller diameter can be derived from the dimensionless characteristics

$$\frac{Q_1}{Q_2} = \left(\frac{D_1}{D_2}\right)^3; \frac{H_1}{H_2} = \left(\frac{D_1}{D_2}\right)^2; \frac{M_1}{M_2} = \left(\frac{D_1}{D_2}\right)^5; \frac{P_1}{P_2} = \left(\frac{D_1}{D_2}\right)^5. \tag{8.35}$$

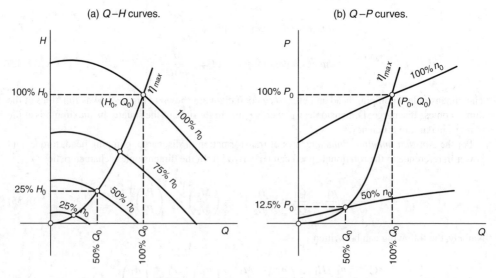

Figure 8.15   Transformation of pump characteristics.

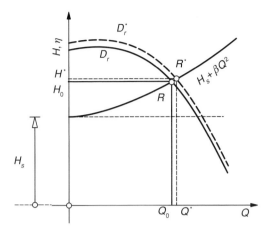

Figure 8.16 Pump impeller diameter adjustment.

Figure 8.16 shows how to obtain the required working point R by adjustment of the pump impeller diameter. The required reduction is obtained from the discharge ratio in working points

$$\frac{D_r}{D_r^*} = \sqrt[3]{\frac{Q_0}{Q^*}}.$$

The change of the pumping head will be

$$\frac{H_0}{H^*} = \left(\frac{D_r}{D_r^*}\right)^2.$$

The new pumping head will be close to the head required by the system resistance characteristic. In the case of a static head non-existence $H_s = 0$ an exact required value will be obtained because the transformation parabola and resistance parabola will be equal.

8.5 Pump specific speed

If a dimensionless product is observed

$$cH^3 \cdot cQ^{-2} = cons, \tag{8.36}$$

which is constant at the point of maximum efficiency for a turbo engine, an expression that does not depend on the impeller size D is obtained

$$\left(\frac{gH}{\omega^2 D^2}\right)^3 \cdot \left(\frac{\omega D^3}{Q}\right)^2 = \frac{g^3 H^3}{\omega^4 Q^2} = cons. \tag{8.37}$$

Table 8.1   Conversion of units for pump specific speed.

	US customary units	SI units
N	revolutions per minute	revolutions per minute
Q	gallons per minute	cubic meters per second
H	feet of head	meters of head

If expression (8.37) is written as a reciprocal value and the fourth root is computed, the following is obtained

$$\frac{\omega\sqrt{Q}}{g^{3/4}H^{3/4}} = cons. \tag{8.38}$$

This constant has a unique dimensionless value for any value of geometrically similar turbo engines.

It is common practice to include the factor $g^{3/4}$ in the constant. Thus, the obtained new constant becomes dimensionless and has the dimension of the angular velocity $[s^{-1}]$; namely, the number of revolutions [rev/min], and is called the *pump specific speed*

$$\omega_s = \frac{\omega\sqrt{Q}}{H^{3/4}} \quad \text{or} \quad N_s = \frac{N\sqrt{Q}}{H^{3/4}}. \tag{8.39}$$

Although previously defined specific speed can also be applied to turbine, the *turbine specific speed* is defined using the power P, which is, together with the head H, the main characteristic value of a turbine

$$\omega_s = \frac{\omega\sqrt{P}}{H^{5/4}} \quad \text{or} \quad N_s = \frac{N\sqrt{P}}{H^{5/4}}. \tag{8.40}$$

Table 8.1 shows the pump specific speed units expressed in *US customary units* and *SI units*.

Specific speed is a design value and it is used as an aid in the selection of turbo engine type, as shown in Figure 8.17.

The specific speed in SI units is obtained by multiplication of the specific speed expressed in US customary units by 51.64.

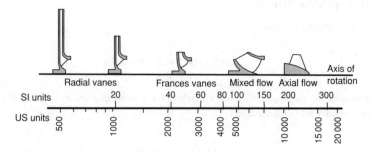

Figure 8.17   Pump type selection based on the specific speed.

8.6   Complete characteristics of turbo engine

8.6.1   *Normal and abnormal operation*

Normal pump and normal turbine operation are the steady states in which the four fundamental parameters – discharge Q, angular velocity ω, torque M, and pumping head H – are determined from normal pump or normal turbine characteristics and the characteristics of the hydraulic system within which they operate.

Should pump power outage occur, the engine start torque becomes zero, the pump slows down, discharge decreases and, finally, after some time, reverse flow and pump rotation in the opposite direction starts. A new steady state occurs, in which a new equilibrium is established, where discharge Q and angular velocity ω are negative, torque M is equal to zero, while manometric head H is positive.

A similar situation occurs at turbine shutdown. At turbine unloading, torque M becomes zero, the turbine speeds up and tends to the new equilibrium.

Therefore, between two steady states, the turbo engine or hydraulic system as an entity pass through numerous transient unsteady states. A definition of transient states requires knowledge of complete turbo engine characteristics. For example, let us assume that the pump impeller is blocked instantly. After transients of pressure and discharge, a reverse state will be established in the hydraulic system with the pump acting as a simple throttle. The same can be expected for turbine operation. In both cases for $\omega = 0$, the head characteristic will have the following form

$$H = c^{\pm} Q|Q|, \tag{8.41}$$

where $c^{\pm}$ is the asymmetric resistance coefficient at the impeller as a throttle, see pump characteristic (8.18).

8.6.2   *Presentation of turbo engine characteristics depending on the direction of rotation*

One way to present pressure and moment characteristics is to present them separately for positive and negative rotational speed (zero rotational speed is counted as positive); namely, for rotational speed $\omega \geq 0$ and $\omega < 0$. It is an extension of the normal pump operation characteristics to the domain of negative discharges and negative torque and manometric head values.

Figure 8.18 shows turbo engine characteristics for (a) non-negative rotation $\omega \geq 0$ and (b) negative rotation $\omega < 0$.

8.6.3   *Knapp circle diagram*

Knapp (1937) noted the need to distinguish between normal and abnormal operation of turbo engines. In their later works, Knapp and Swanson (1953), Donski (1961), and Stepanoff *et al.* determined eight possible turbo engine operations out of which four are normal and four are abnormal, see Figure 8.19.

More detailed descriptions of different operational modes of turbo engine operation were given by Martin (1983).

The diagram comprises normalized values of hydraulic torque M/M_0 and manometric head H/H_0 in a normalized coordinate system $(Q/Q_0, n/n_0)$, where index "$_0$" denotes the values at the pump nominal point: $n_0, Q_0, H_0, M_0, \eta_{\max}$. With respect to the coordinate system selection $(Q/Q_0, n/n_0)$, complete characteristics of a turbo engine are also called the *IV quadrant characteristics*. For clarity, only characteristic values of the torque and manometric head are shown for relative values $+1$,

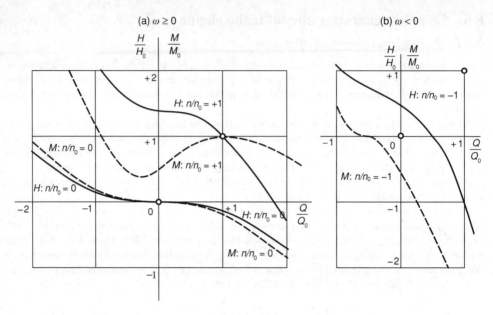

Figure 8.18 Normal pump characteristic for non-negative and negative rotational speeds.

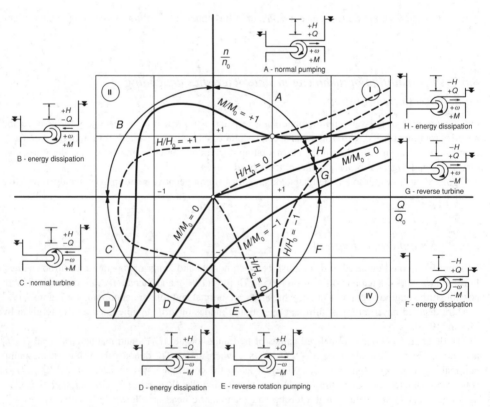

Figure 8.19 Knapp circle diagram.

0, and *−1*. In the absence of measured data, values can be interpolated using *affine (homologous)* transformations.

The diagram in Figure 8.19 is a circle diagram of a hypothetic radial pump. For an axial or mixed inflow, the branch H/H_0 is situated in the quadrant III. Thus the area E looks like that in Figure 8.20.

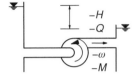

E - reverse rotating pumping
(mixed oraxial flow machines)

Figure 8.20

Between normal and abnormal operations there are also limited cases when different variables are equal to zero, for example at pump shutdown, the working point passes from normal zone A through abnormal zone H, then enters normal zone G from which it transfers into abnormal zone F to finish by a spiral trace at the boundary zone defined by the value $M=0$. A trace of the working point after pump shutdown (as modeled by `SimpipCore` software: see www.wiley.com/go/jovic) is shown in Figure 8.21.

The measurements, based on which all necessary characteristics of the engine will be determined in the laboratory on a model, are extremely difficult and expensive; thus, affine transformations are used. It is enough to know one of the curves H/H_0 and one of the curves M/M_0, while all the others can be determined based on the law of similarity.

For turbo engines with movable blades and reversible engines (pump-turbine) there is separate Knapp–Stepanoff diagram for each position of the stator blades. The number of diagrams is multiplied for engines with movable impeller blades, such as Kaplan turbines.

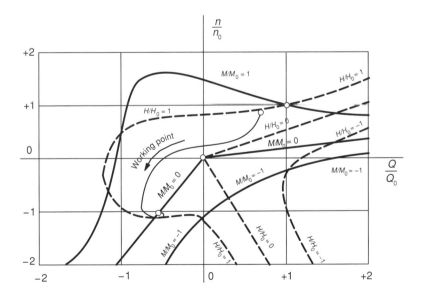

Figure 8.21   Working point trace after pump shutdown.

8.6.4   Suter curves

The four-quadrant turbo engine characteristics in the form of a Knapp diagram are not suitable for numerical modelling of transient states, regardless of the possibility of using the homologue transformations.

The problem can be solved by presentation of complete characteristics in the form of Suter curves, shown in Figure 8.22, where χ is the pump height variable and μ is the torque variable, both functions of an angle on the Knapp circle diagram.

All normal and abnormal turbo engine operation zones are marked on the Suter diagram.

The procedure of transformation of the pairs of values H/H_0 and M/M_0 from the Knapp circle diagram to the Suter curves χ, μ is the following. One point with normalized coordinates Q/Q_0, n/n_0 is selected on the Knapp circle diagram. Then, normalized torque M/M_0 and pump height H/H_0 values are read, see Figure 8.23. Polar coordinates ϑ and d are calculated from the following expressions:

$$\tan \vartheta = \frac{n/n_0}{Q/Q_0} \tag{8.42}$$

$$d^2 = \left(\frac{Q}{Q_0}\right)^2 + \left(\frac{n}{n_0}\right)^2 . \tag{8.43}$$

Values of the variables in the Suter diagram are calculated from the variables in the Knapp circle diagram as follows

$$\text{head variable} : \chi = \frac{\text{sgn}(H)}{d} \sqrt{\left|\frac{H}{H_0}\right|}, \tag{8.44}$$

$$\text{torque variable} : \mu = \frac{\text{sgn}(M)}{d} \sqrt{\left|\frac{M}{M_0}\right|} \tag{8.45}$$

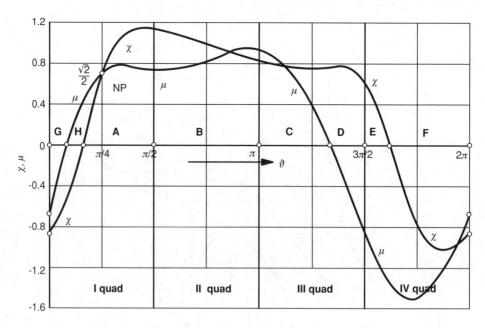

Figure 8.22   Suter curves.

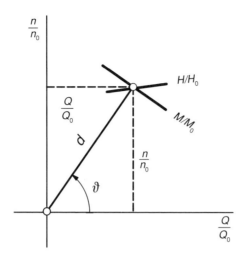

Figure 8.23   Transformation from the Knapp circle diagram to Suter diagram.

and vice versa, values of the normalized pump head H/H_0 and normalized torque M/M_0 are calculated from the readings on the Suter diagram for angle ϑ

$$normalized\ pump\ head: \quad \frac{H}{H_0} = \chi\,|\chi|\,d^2, \tag{8.46}$$

$$normalized\ torque: \quad \frac{M}{M_0} = \mu\,|\mu|\,d^2, \tag{8.47}$$

where angle ϑ and the distance of the point d are calculated from expressions (8.42) and (8.43).

Curves from Suter diagram $\chi(\vartheta)$ and $\mu(\vartheta)$ can be inversely mapped to the Knapp diagram in a manner such that for each absolute value $|H/H_0| = H^*$ coordinates are calculated in dependence on the angle ϑ:

$$\frac{Q}{Q_0} = \frac{\mathrm{sgn}\chi(\vartheta)}{\chi(\vartheta)}\sqrt{H^*}\cos(\vartheta), \tag{8.48}$$

$$\frac{n}{n_0} = \frac{\mathrm{sgn}(\chi)}{\chi}\sqrt{H^*}\sin(\vartheta). \tag{8.49}$$

Coordinates of the torque curve $|M/M_0| = M^*$ are formally defined in the same manner

$$\frac{Q}{Q_0} = \frac{\mathrm{sgn}\mu(\vartheta)}{\mu(\vartheta)}\sqrt{M^*}\cos(\vartheta), \tag{8.50}$$

$$\frac{n}{n_0} = \frac{\mathrm{sgn}\mu(\vartheta)}{\mu(\vartheta)}\sqrt{M^*}\sin(\vartheta). \tag{8.51}$$

Suter curves $\chi(\vartheta)$ and $\mu(\vartheta)$ each have two zero points related to the values $H/H_0{=}0$ and $M/M_0{=}0$ that determine inclinations of respective lines, see the Knapp circle diagram.

8.7   Drive engines

8.7.1   Asynchronous or induction motor

The asynchronous or induction motor is the most commonly used electrical motor nowadays. There are two types of induction motor depending on rotor design: the squirrel cage rotor and wound rotor (or slip ring) type motor. Wound rotor motors are more expensive compared to squirrel cage rotor motors and require maintenance of the slip rings and brushes. Wound rotor motors were the standard form for variable speed control before the advent of compact power electronic devices. The great advantage of the squirrel cage rotor induction motor is its simple design, rugged construction, low-price, and easy maintenance. This type of induction motor is nowadays used for pump drives. Directly connected to the standard electrical network, it runs essentially at constant speed from no-load to full load, rotating at a speed somewhat lower than the synchronous speed, that is the magnetic field rotational speed.

Nominal (rated) rotational speed ω_n is defined by the equilibrium of the motor torque and the load torque at the engine shaft, as shown in Figure 8.24 for pump operation.

A torque characteristic of the motor is drawn in a coordinate system with the dimensionless variable (s), called slip. It is the ratio of the difference between the synchronous speed and the rotor speed (slip speed) to the synchronous speed:

$$s = \frac{\omega^* - \omega}{\omega^*} = \frac{n^* - n}{n^*},$$
(8.52)

where ω is the angular velocity and ω^* is a synchronous angular velocity of electromagnetic field rotation (units s^{-1}), or expressed in revolutions per minute (mark n). An intersection between the torque curve of a motor and torque characteristic of a pump defines the nominal rotational speed of the engine (motor + pump). The synchronous speed, the speed of the electromagnetic field rotation, is determined by the frequency of an electrical network and the number of pole pairs as the following applies

$$f = p\frac{n}{60},$$
(8.53)

where n is the rotational speed expressed in revolutions per minute (unit rpm) while p is the number of pole pairs of an induction motor. Thus, a synchronous speed will be

$$n^* = 60\frac{f}{p}.$$
(8.54)

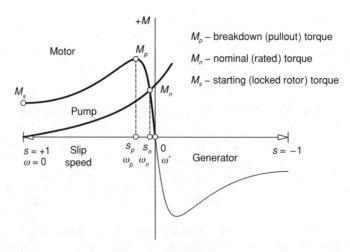

Figure 8.24   Torque characterisistics of an asynchronous motor and pump.

Table 8.2 shows synchronous speeds for frequencies 50 Hz (Europe) and 60 z (USA), depending on the number of pole pairs of an induction motor.

Table 8.2

Number pole pairs	50 Hz	60 Hz
1	3000 rev/min	3600 rev/min
2	1500 rev/min	1800 rev/min
3	1000 rev/min	1200 rev/min
4	750 rev/min	900 rev/min
5	600 rev/min	720 rev/min

The shape of the motor torque characteristic is defined by characteristic points; namely the starting motor torque M_s, which is the torque at zero speed; the breakdown torque M_p, which corresponds to the slip $s_p \approx 0.15$, and the rated torque M_n · which is equal to the pump rated torque which corresponds to the slip s_n or nominal speed ω_n. The torque characteristic of most of the motors can be approximated by the Kloss expression

$$M = 2M_p \frac{s \cdot s_p}{s^2 + s_p^2}. \tag{8.55}$$

The induction motor can be either single-phase or three-phase. If the pump and electric motor operate at variable rotational speed (frequency converter), a three-phase drive is recommended, which is also necessary for greater power. The three-phase induction motor has numerous advantages, its operation is "softer," change of the direction of rotation is simple, etc.

8.7.2 Adjustment of rotational speed by frequency variation

The rotational speed of an asynchronous motor is related to the frequency of its power source. Thus, with variation of the frequency of the power source, rotational speed can also be changed within a relatively wide range as synchronous speed is directly proportional to the frequency of the power source, see expression (8.53).

The working rotational speed is defined by the intersection between the torque curves of the motor and load and does not change linearly with a change in synchronous speed; thus, the problem is somewhat more complex.

If just the frequency of the source is changed, the magnetic flux of the asynchronous motor also changes and with it the motor characteristic. It means that during adjustment of the rotational speed by frequency variation, the voltage shall also be changed respectively, if the required mechanical characteristics of the motor are to be achieved.

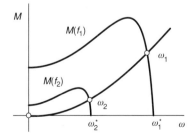

Figure 8.25

For purposes of numerical modelling of pumping unit operations, it can be considered that rotational speed will be proportional to frequency

$$\frac{\omega_1}{\omega_2} = \frac{f_1}{f_2}.$$

(8.56)

With a change of the rotational speed, all other turbo engine variables will change accordingly

$$\frac{\omega_1}{\omega_2} = \frac{n_1}{n_2} = \frac{Q_1}{Q_2} = \left(\frac{H_1}{H_2}\right)^{\frac{1}{2}} = \left(\frac{M_1}{M_2}\right)^{\frac{1}{2}} = \left(\frac{P_1}{P_2}\right)^{\frac{1}{3}}.$$

(8.57)

8.7.3 Pumping unit operation

During the transition from one to another steady operation, for example from rest to normal operation or vice versa, pumping units transfer through several electrical and hydraulic unsteady states. These hydraulic states are called transient states. Transient states are those of general unsteady states during normal operation of a hydraulic network.

In systems with pumping units there are *transient states of electromagnetic, electromechanical, and hydraulic phenomena*. In practical analyses the duration of the electromagnetic transient phenomena can be disregarded. When a switch is pushed on a control panel, pump operation starts, that is the electrical engine is connected to the power source and full voltage. The transient phenomena of establishing electromagnetic fields lasts for a very short time; namely, the motor starts to rotate until operational rotational speed is achieved under the full torque characteristic. Similarly, when a button is pushed to switch off the pump, the power supply is stopped instantly. There is no more electrical moment torque and the rotational velocity of the pump decreases. The pumping unit transition from the initial to the new steady rotational speed value is called the *transition state of the pumping unit rotation*.

An asynchronous electric motor torque is proportional to the square value of connected voltage. This fact enables controlled "soft" pump start and shutdown in the period of voltage increase or decrease from zero to the full working voltage and vice versa. Devices of this type are called *starters* and *brakers*.

Note that the pump power engine can be controlled either by change in frequency or by voltage change or both combined. In the numerical model of pump drive, two *operational (control) variables* are foreseen as time-dependent functions:

$$\text{voltage variable: } u(t) = \frac{U(t)}{U_0},$$

(8.58)

$$\text{frequency variable: } \varphi(t) = \frac{f(t)}{f_0}.$$

(8.59)

Implementation of these boundary conditions is by a subroutine `GetBDC` by calling the subroutine `SetSpeeds`.

The dynamic equation of the pumping unit rotation is:

$$I\frac{d\omega}{dt} = M_e(\omega, u, \varphi) - M_p(\omega, Q),$$

(8.60)

where $I\ [kgm^2]$ is the polar moment of rotating masses, ω is the angular velocity, M_e is the electrical motor torque, and M_p is the pump torque. Although the electric motor contributes more to the moment of inertia of the pump unit than the pump itself, manufacturers do not usually list it in pump catalogues.

Thus, it is usually an unknown for modelling purposes. An empirical formula for computation of the moment of inertia is inbuilt in the SimpipCore program solution:

$$I = (0.0008p^2 - 0.0005p + 0.0004) \, P_0^{1.3}$$

$$I \, [\text{kgm}^2]; \; P_0 \, [\text{kW}]$$

(8.61)

where p is number of poles and P_0 is the pumping unit power. The dynamic equation of rotation defines the variation of rotational speed of the engine where the electric motor depends on rotational speed as well as voltage and frequency variables, while pump torque depends on the rotational speed and hydraulic variables (pump head H and discharge Q).

Thus, for example, inertia of some unit can be defined in a manner to calculate the shutdown time from Eq. (8.60) assuming constant torque, that is constant power

$$I \frac{d\omega}{dt} = -M_0 = -\frac{P_0}{\omega_0},$$

from which the shutdown time is obtained:

$$T_z = \frac{I\omega_0^2}{P_0}.$$

(8.62)

Shutdown time can be used for evaluation of the duration of the transient states of the pumping unit. If a pumping unit is observed with the voltage variable in the form $u : 0 \rightarrow 1$, that is instantaneous voltage increase, while the frequency variable is equal to one, then the time necessary to achieve the full rotational speed ω_0 can be calculated from the integral:

$$T_{up} = I \int_0^{\omega_0} \frac{d\omega}{M_e - M_p} \approx I \sum_k \frac{\Delta\omega}{m_k^u}.$$

(8.63)

Similarly, shutdown time is calculated when the torque characteristic is equal to zero, as follows:

$$T_{down} = I \int_0^{\omega_0} \frac{d\omega}{M_p} \approx I \sum_k \frac{\Delta\omega}{m_k^d}.$$

(8.64)

Figure 8.26a shows the geometric definition of a subintegrand function in expressions (8.63) and (8.64).

Figure 8.26b shows curves of rotational speed at sudden pump start or shutdown. Note the asymptotic approaching the steady values. The reason is the zero value in the denominator of integrand functions (for the upper integration boundary).

If pump rotational velocity after shutdown and reflux preventer closing is analyzed, then a quadratic form of the torque characteristic is expected and the dynamic equation has the form:

$$I \frac{d\omega}{dt} = -\frac{P_0}{w_0^3} \omega^2 t$$

with the solution

$$\omega = \frac{1}{\dfrac{1}{\omega_0} + \dfrac{P_0}{\omega_0^3} t}.$$

(8.65)

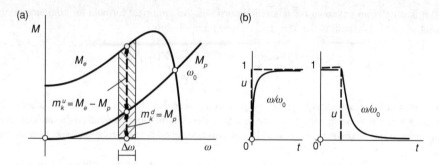

Figure 8.26   Duration of the transient state of pumping unit rotational speed.

This solution is also asymptotic, which does not correspond to the experience gained in practice, because pumping units shuts down in a finite time. The reason is found in the remaining magnetic field that additionally inhibits the electric motor rotor.

Pumping unit control by input variables with the following syntax is implemented in the `SimpipCore` program solution

```
Power pumpname Voltage(t)<Frequency(t[1]>
```

where frequency and voltage variables can be time-dependent variables, see www.wiley.com/go/jovic Section Input/Output syntax, Appendix B. Depending on the optional logic variable `SpeedTransients` it may or may not be possible to analyze the transient phenomena of pumping unit rotation.

8.8   Numerical model of pumping units

8.8.1   *Normal pump operation*

Normalized dimensionless characteristics

Torque and pressure characteristics of the pump can be presented in a normalized form with respect to the nominal point Q_0, H_0, M_0, ω_0 (point with the maximum efficiency coefficient), see Figure 8.27a. The torque characteristic is obtained by division of the power characteristic by the angular velocity. Thus, the normalized power curve is equal to the normalized torque curve.

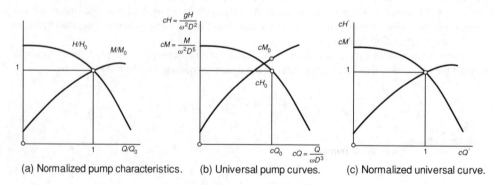

(a) Normalized pump characteristics.   (b) Universal pump curves.   (c) Normalized universal curve.

Figure 8.27   Normalized characteristics.

The universal dimensionless pump characteristic, see Figure 8.27b, can also be normalized with respect to the nominal working point, with the following that applies to the nominal point

$$dimensionless\ discharge\ :\quad cQ_0 = \frac{Q_0}{\omega_0 D^3}, \tag{8.66}$$

$$dimensionless\ pump\ head\ :\quad cH_0 = \frac{gH_0}{\omega_0^2 D^2}, \tag{8.67}$$

$$dimensionless\ torque\ :\quad cM_0 = \frac{M_0}{\omega_0^2 D^5}. \tag{8.68}$$

Figure 8.27c shows normalized universal curves with the discharge variable equal to

$$cQ' = \frac{Q}{Q_0}\frac{\omega_0}{\omega}. \tag{8.69}$$

Pressure head and torque variables have the following form

$$cH' = \frac{H}{H_0}\frac{\omega_0^2}{\omega^2}, \tag{8.70}$$

$$cM' = \frac{M}{M_0}\frac{\omega_0^2}{\omega^2}. \tag{8.71}$$

Note that the obtained normalized universal curves (c) correspond to normalized curves (a) since, in both cases, normalization is carried out by the same constant rotational speed $\omega = \omega_0$.

Input data are the manometric head and pump power

$$Q_i, H_i, P_i, \ i = 1, 2, 3, \cdots, n, \tag{8.72}$$

which are read in n points, see Figure 8.28. The first data are the values for $Q = 0$ and one of the data is the nominal working point data. The minimum number of points is 6. The input syntax has the form

```
Pump name p1 p2 Omega Qn Hn Pn Ip Dc
   Q    H    P
   ...  ...
```

after which it is added to the pump collection Pumps by the function

```
integer function AddPump(pmp)
```

while the data are added by the function

```
integer function AddPumpData(k,Q,H,P).
```

The following is prescribed in the title row: pumping unit name, nominal rotational speed in radians, nominal discharge, pump head and power, polar moment of inertia, and diameter of connecting pipe. The data are written in the next few rows.

The SimpipCore program solution contains the program module Pumps.f90 within which are all the necessary subroutines for pumping units. The functional subroutine

```
logical function FitPumpNormData(pump)
```

normalizes universal pump characteristics based on the input data.

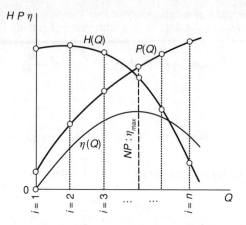

Figure 8.28   Normal pump characteristics.

Approximation by a polynomial

A polynomial with $3 \le m < n$ terms can be approximated (using the least square method) through n discrete data Q_i/Q_0, H_i/H_0, read from the Q–H pump characteristic

$$cH' = A_1 \cdot cQ'^0 + A_2 \cdot cQ'^1 + A_3 \cdot cQ'^2 + A_4 \cdot cQ'^3 + \cdots + A_m \cdot cQ'^{m-1} \tag{8.73}$$

with the additional condition that it passes through the nominal point (1,1). The least number of polynomial terms is 3; thus, the approximation is a square one, which approximates well the radial pump characteristic, see expression (8.18). Axial, that is pumps with the mixed flow, require an approximation of higher order. Thus, the order of a polynomial is determined according to the expression

$$m = \max\left[3, \text{int}(n-6)/2\right], \tag{8.74}$$

where n is the number of discrete points, the minimum $2m$. If Eqs (8.69) and (8.70) are introduced into polynomial (8.73) then it can be written

$$\frac{H}{H_0}\frac{\omega_0^2}{\omega^2} = A_1 \cdot \left(\frac{Q}{Q_0}\frac{\omega_0}{\omega}\right)^0 + A_2 \cdot \left(\frac{Q}{Q_0}\frac{\omega_0}{\omega}\right)^1 + A_3 \cdot \left(\frac{Q}{Q_0}\frac{\omega_0}{\omega}\right)^2 + A_4 \cdot \left(\frac{Q}{Q_0}\frac{\omega_0}{\omega}\right)^3$$
$$+ \cdots + A_m \cdot \left(\frac{Q}{Q_0}\frac{\omega_0}{\omega}\right)^{m-1} \tag{8.75}$$

and, after arranging, the approximation of the normalized pressure head is obtained in the following form

$$\frac{H}{H_0} = A_1 \cdot \left(\frac{\omega}{\omega_0}\right)^2 + A_2 \cdot \frac{\omega}{\omega_0}\frac{Q}{Q_0} + A_3 \cdot \left(\frac{Q}{Q_0}\right)^2 + A_4 \cdot \frac{\omega_0}{\omega}\left(\frac{Q}{Q_0}\right)^3$$
$$+ \cdots + A_m \cdot \left(\frac{\omega}{\omega_0}\right)^{3-m}\left(\frac{Q}{Q_0}\right)^{m-1} \tag{8.76}$$

or written as

$$\frac{H}{H_0} = \sum_{i=1}^{m} A_i \cdot \left(\frac{\omega}{\omega_0}\right)^{3-i} \left(\frac{Q}{Q_0}\right)^{i-1}.$$ (8.77)

Partial derivative of normalized manometric head by discharge is equal to

$$\frac{\partial}{\partial Q}\frac{H}{H_0} = \sum_{i=2}^{m} \frac{i-1}{Q_0} A_i \cdot \left(\frac{\omega}{\omega_0}\right)^{3-i} \left(\frac{Q}{Q_0}\right)^{i-2}.$$ (8.78)

Partial derivative of normalized manometric head by rotational speed is equal to

$$\frac{\partial}{\partial \omega}\frac{H}{H_0} = \sum_{i=1}^{m} \frac{3-i}{\omega_0} A_i \cdot \left(\frac{\omega}{\omega_0}\right)^{2-i} \left(\frac{Q}{Q_0}\right)^{i-1}.$$ (8.79)

A polynomial approximation of the torque characteristic

$$\frac{M}{M_0} = \sum_{i=1}^{m} A_i \cdot \left(\frac{\omega}{\omega_0}\right)^{3-i} \left(\frac{Q}{Q_0}\right)^{i-1}$$ (8.80)

and its partial derivatives are obtained by similar procedure

$$\frac{\partial}{\partial Q}\frac{M}{M_0} = \sum_{i=2}^{m} \frac{i-1}{Q_0} A_i \cdot \left(\frac{\omega}{\omega_0}\right)^{3-i} \left(\frac{Q}{Q_0}\right)^{i-2},$$ (8.81)

$$\frac{\partial}{\partial \omega}\frac{M}{M_0} = \sum_{i=1}^{m} \frac{3-i}{\omega_0} A_i \cdot \left(\frac{\omega}{\omega_0}\right)^{2-i} \left(\frac{Q}{Q_0}\right)^{i-1}.$$ (8.82)

Expressions (8.77), (8.78), (8.79), (8.80), (8.81), and (8.82) can be calculated for $\omega \geq 0$ if a parabolic approximation ($m = 3$) is selected. For $m > 3$ it can be applied only if $\omega > 0$.

Polynomial approximations of the pump head and torque are defined in subroutine `FitPumpNormData` by calling of a function

```
logical function FitConstrainedPoly (nData,x,y,mPoly,a,kData,ierr)
```

that can be found in the program module `FitPoly.t90`. The condition for a normalized polynomial approximation is that it passes exactly through the normalized nominal working point.

The following functional procedures were developed for computation of the normalized pump head and torque and their derivatives by rotational speed and discharge in normal pump operation.

(a) Normalized pump head:

```
    real*8 function PumpHHo (pump,O,Q,Oo,Qo),
  real*8 function PumpDeHHoDeO (pump,O,Q,Oo,Qo),
  real*8 function PumpDeHHoDeQ (pump,O,Q,Oo,Qo),
```

(b) Normalized torque:

```
    real*8 function PumpMMo (pump,O,Q,Oo,Qo),
  real*8 function PumpDeMMoDeO (pump,O,Q,Oo,Qo),
  real*8 function PumpDeMMoDeQ (pump,O,Q,Oo,Qo).
```

They can be found in the program module `Pumps.f90`. Real values of pump head and torque and their derivatives are obtained by multiplication of the values returned by these functions with the rated value H_0 or M_0.

8.8.2   Reconstruction of complete characteristics from normal characteristics

Knowledge of the complete pump characteristics is a necessity in high-quality modelling of a hydraulic system, even if the working point is within the normal pump operation area at every moment. Namely, due to necessary iterative solution of equations, iterative values of discharge Q and angular velocity ω can be temporarily found in completely different quadrants than the final solution quadrant.

Thus, for each Q and ω value, the manometric head and torque shall be defined

$$H = H(\omega, Q), \tag{8.83}$$

$$M = M(\omega, Q). \tag{8.84}$$

Unfortunately, pump manufacturers usually provide only normal characteristics, which is only a portion of data from the *zone A of the first quadrant*. All other data shall be reconstructed congruously.

Note: During interpretation of the results of transients, obtained based on the reconstructed complete characteristic, bear in mind their approximate quality of accuracy outside the normal operation area.

Naturally, the data on normal pump operation are insufficient for the reconstruction of complete characteristics.

A procedure of reconstruction of complete characteristics based on the proposal by J.A. Fox (1979) will be given here. The idea is to use Suter curves for the area outside the normal pump operation. Normal pump operation is the area A in which all the variables are positive $+\omega, +Q, +H, +M$, see Figure 8.19 and Figure 8.22. For this area, Suter variables χ, μ can be calculated from a polynomial approximation of normal pump characteristics; namely, the data for angles ϑ from the first zero-point of the curve χ to the angle $\vartheta = \pi/2$. The remaining area is unknown and shall be congruously reconstructed. Fox's idea is interpolation based on the pump specific speed between the known Suter curves for radial, axial, and mixed pump type. B. Donsky (1961) published Knapp circle diagrams for radial, axial, and mixed pump type, specific speeds (SI system) 35, 261, and 147 rev/min, from which the standard Suter curves shown in Figure 8.29, Figure 8.30, and Figure 8.31 were derived. Standard Suter curves are shown as a cubic natural spline through 17 characteristic points at equidistant distances $\Delta\vartheta = \pi/8$. The spline passes exactly through the nominal point, angle $\vartheta = \pi/4$. The data for spline computation for standard pumps are situated in the global fortran module `GlobalVars.f90`.

The interpolation parameter and necessary corrections are determined for the angle $\vartheta = \pi/2$. The applied interpolation is a parabola through three points based on a specific pump speed.

In the program module `Pumps.f90` there is a function

```
logical function SetSuterData(pump)
```

in which the described procedure is implemented. This subroutine is called from the subroutine `FitPumpNormData`.

The following functional procedures for computation of the pump head and torque in abnormal pump operation:

```
real*8 function SuterHead(pump,O,Q),
real*8 function SuterTorque (pump,O,Q)
```

can be found in the program module `Pumps.f90`.

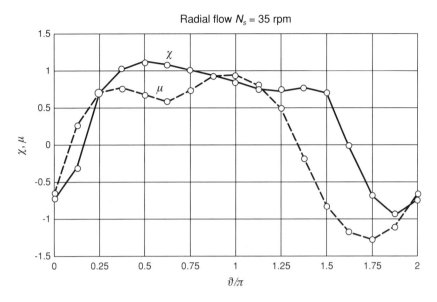

Figure 8.29 Radial flow.

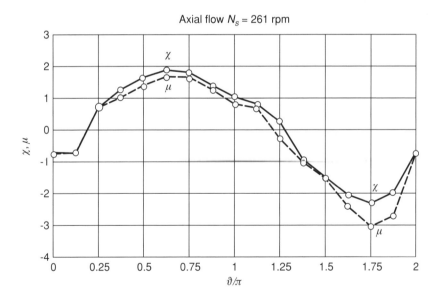

Figure 8.30 Axial flow.

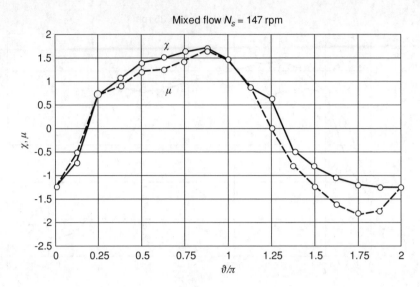

Figure 8.31   Mixed flow.

Computations of the manometric head and torque are implemented in the program module `Pumps.f90` as functional procedures:

```
real*8 function PumpHead(pump,O,Q,dHdO,dHdQ),
real*8 function PumpTorque(pump,O,Q,dTdO,dTdQ).
```

Functional procedure `PumpHead` calculates the manometric head and its derivatives by angular velocity and discharge for the prescribed angular velocity and discharge of the pump as follows:

```
if ω > 0 then ! positive rotation
  if Q ≥ 0 then ! positive flow
  ! normal pumping, use fitted polynom
  PumpHead = Ho*PumpHHo(pump,O,Q,Oo,Qo)
  dHdO = Ho*PumpDeHHoDeO(pump,O,Q,Oo,Qo)
  dHdQ = Ho*PumpDeHHoDeQ(pump,O,Q,Oo,Qo)
  else ! revers flow ϑ > π/2
  ! abnormal pumping, use Suter's curve
  PumpHead = SuterHead(pump,O,Q)
  dHdQ=(SuterHead(pump,O,Q+dQQ) -
       SuterHead(pump,O,Q-dQQ))/(2*dQQ)
  dHdO=(SuterHead(pump,O+dOO,Q) -
       SuterHead(pump,O-dOO,Q))/(2*dOO)
  endif
else ω=0 then ! zero rotation
  ! abnormal pumping, use Suter's curve
  PumpHead = SuterHead(pump,O,Q)
  dHdQ=(SuterHead(pump,O,Q+dQQ) -
       SuterHead(pump,O,Q-dQQ))/(2*dQQ)
  dHdO=(SuterHead(pump,O+dOO,Q) -
       SuterHead(pump,O-dOO,Q))/(2*dOO)
```

```
else ω < 0 then ! negative rotation
  ! abnormal pumping, use Suters curve
  PumpHead = SuterHead(pump,O,Q)
  dHdQ=(SuterHead(pump,O,Q+dQQ) -
        SuterHead(pump,O,Q-dQQ))/(2*dQQ)
  dHdO=(SuterHead(pump,O+dOO,Q) -
        SuterHead(pump,O-dOO,Q))/(2*dOO)
endif
```

Functional procedure `PumpTorque` calculates the manometric head and its derivatives by angular velocity and discharge for the prescribed angular velocity and discharge of the pump as follows:

```
if ω > 0 then ! positive rotation
  if Q ≥ 0 then ! positive flow
    ! normal pumping, use fitted polynom
    PumpTorque = Mo*PumpMMo(pump,O,Q,Oo,Qo)
    dTdO = Mo* PumpDeMMoDeO(pump,O,Q,Oo,Qo)
    dTdQ = Mo*PumpDeMMoDeQ(pump,O,Q,Oo,Qo)
  else ! revers flow ϑ > π/2
    ! abnormal pumping, use Suter's curve
    PumpTorque = SuterTorque(pump,O,Q)
    dTdQ=( SuterTorque(pump,O,Q+dQQ) -
          SuterTorque(pump,O,Q-dQQ))/(2*dQQ)
    dTdO =( SuterTorque(pump,O+dOO,Q) -
          SuterTorque(pump,O-dOO,Q))/(2*dOO)
  endif
else ω=0 then ! zero rotation
  ! abnormal pumping, use Suter's curve
  PumpTorque = SuterTorque(pump,O,Q)
  dTdQ=( SuterTorque(pump,O,Q+dQQ) -
        SuterTorque(pump,O,Q-dQQ))/(2*dQQ)
  dTdO =( SuterTorque(pump,O+dOO,Q) -
        SuterTorque(pump,O-dOO,Q))/(2*dOO)
else ω < 0 then ! negative rotation
  ! abnormal pumping, use Suter's curve
  PumpTorque = SuterTorque(pump,O,Q)
  dTdQ=( SuterTorque(pump,O,Q+dQQ) -
        SuterTorque(pump,O,Q-dQQ))/(2*dQQ)
  dTdO =( SuterTorque(pump,O+dOO,Q) -
        SuterTorque(pump,O-dOO,Q))/(2*dOO)
endif
```

Note that in abnormal operations, the values of the head and torque derivatives are calculated numerically, in the interval `2dQQ` and `2dOO`, while values `dQQ` and `dOO` are equal to 1% of the nominal values of the discharge and rotational speed.

8.8.3   Reconstruction of a hypothetic pumping unit

In practical modelling there is a need for reconstruction of the pumping unit based on knowledge of the nominal working point

$$Q_0, H_0, P_0, \omega_0.$$

If the nominal point is prescribed, then the specific speed can be calculated as

$$\omega_s = \frac{\omega\sqrt{Q}}{H^{3/4}} \tag{8.85}$$

and interpolation of the characteristics between the known Suter curves for radial, axial, and mixed pump type carried out (the data is shown in Figure 8.29, Figure 8.30, and Figure 8.31 and are found in the global module `SimpipVars.f90`). Thus, the following is prescribed in the title row for the previously defined pump: pumping unit name, nominal rotational speed in radians, nominal point data, polar moment of inertia, diameter of connecting pipe, and option USER:

```
Define Pump myPump Omega Qn Hn Pn Ipol Dc USER.
```

Then, registration of the new pump into collection `DefPumps` is called by the function:

```
integer function AddDefPump(obj).
```

Thus the formally defined pumping unit type can be used (repeatedly) by the command:

```
Pump pumpname1 p1 p2 myPump
Pump pumpname2 pX pY myPump
...
```

After which it is added to the pump collection `Pumps` by the function:

```
integer function AddPump(obj).
```

thus creating a real new pump using the subroutine

```
subroutine DefPumpToPump(idef,ipmp)
```

as well as each explicitly defined pump. Procedures `AddDefPump` and `DefPumpToPump` are situated in the program module `DefPumps.f90`.

8.8.4   Reconstruction of the electric motor torque curve

The nominal rotational speed of a pump, written on the name plate or pump characteristic, is a rotational speed recorded during measurements at the manufacturer's test table. However, since it is equal to the rotational speed at the equilibrium of the electric motor torque and the load torque, such a defined rotational speed is not constant and varies slightly around the mean value. Namely, a turbo engine torque depends not only on the rotational speed but also on the discharge, which is again dependent on the established pump working point in the system.

In numerical modelling of a steady or quasi unsteady pump operation in a hydraulic system, the error would not be great (under 1%) if it is assumed that the pump rotates at the declared nominal speed. However, in unsteady operation modelling, where transient states of the pump, from the initial one to the final rotational speed, shall be modeled, it is not possible to determine intermediate rotational speeds without knowledge of the torque characteristic of the electrical motor.

In the absence of the data required for numerical modelling, the torque characteristic of an electrical motor can always be reconstructed from the fact that the pump torque and the electrical motor torque are always in equilibrium at the pump shaft, even at the nominal rotational speed. Similarly, the mean value of the starting torque $M_s = \beta M_n$ is approx. $\beta = 70\%$ of the nominal torque.

If nominal angular velocity ω_n and nominal torque M_n are known for the prescribed nominal point of the steady pump operation, then, according to the Kloss expression, the breakdown torque $M_p = \alpha M_n$ and the breakdown relative rotational velocity can be computed from the condition that the torque curve passes through the points

$$s = 0; \quad M_s = \beta M_n, \tag{8.86}$$

$$s = s_n; \quad M_n = \frac{M_p}{\alpha}, \tag{8.87}$$

from which the following is calculated

$$s_p = \sqrt{\frac{s_n(\beta - s_n)}{1 - \beta s_n}}, \tag{8.88}$$

$$\alpha = \frac{1}{2}\beta\frac{s_p^2 + 1}{s_p}. \tag{8.89}$$

With the known breakdown torque $M_p = \alpha M_n$, the torque characteristic of the electrical motor according to the Kloss expression is

$$M = 2u^2\varphi^2 M_p \frac{s \cdot s_p}{s^2 + s_p^2}, \tag{8.90}$$

where u and φ are the voltage and frequency operational variables. Calculation of the electrical motor torque

$$M = M(\omega) \tag{8.91}$$

for the prescribed angular velocity and the torque derivative by angular velocity are implemented in a subroutine:

```
subroutine & Motor(freqfakt,voltfakt,omega_n,torque_n,omega,torque,torqder)
```

that can be found in the program module `Pumps.f90`.

8.9 Pumping element matrices

8.9.1 Steady flow modelling

In a subroutine `Steady` a call for a subroutine for computation of the steady flow matrix and vector for elemental pump type is added as follows:

```
select case(Elems(ielem).tip)
      case ...
           ...
      case (PUMP_OBJ)
           call SteadyPumpMtx(ielem)
      case ...
           ...
endselect
```

By calling the subroutine `SetSpeeds`, operational variables of the voltage u and frequency variables φ of the electrical motor are defined as well as the status variables that can be `POWER_ON` or `POWER_OFF`. For the values of operational variables of steady or unsteady operation, the equilibrium angular velocity is calculated from the dynamic equation of the engine for the equilibrium of the electric torque and the pump torque

$$M_e(\omega) - M_p(\omega, Q_0) = 0, \tag{8.92}$$

where Q_0 is the nominal pump discharge. The Newton–Raphson procedure is applied to calculate the equilibrium angular velocity

$$\frac{\partial M_e}{\partial \omega} \Delta\omega = -(M_e - M_p) \tag{8.93}$$

from which

$$\Delta\omega = -\frac{(M_e - M_p)}{\dfrac{\partial M_e}{\partial \omega}}. \tag{8.94}$$

Iteration has the following form

$$\omega^{(k+1)} = \omega^{(k)} + \Delta\omega, \tag{8.95}$$

where (k) marks the iteration. The initial iterative value is $\omega^{(0)} = \varphi \cdot u \cdot \omega_0$ where ω_0 is the nominal angular velocity of the pumping unit at the nominal working point. Computation is implemented as a functional subroutine

```
real*8 function CalcNominalSpeed(pump,voltFactor,freqFactor)
```

that is implemented in the program module Pumps.f90.

In a steady or quasi unsteady flow, namely for the optional logic variable SpeedTransients = .false., it is considered that the error is smaller than 1% if it is assumed the pumping units rotates at the equilibrium angular velocity. However, if steady state computation is followed by the computation of unsteady state (if transient states of pumping units are looked for, variable SpeedTransients has the value .true.) then an adapted equilibrium state of angular velocity shall be determined based on the equilibrium discharge. The dynamic equation of the pumping unit refers to equality of the electrical torque and the pump torque

$$M_e(\omega) - M_p(\omega, Q) = 0. \tag{8.96}$$

If the Newton–Raphson procedure is applied to the dynamic equation then

$$\left(\frac{\partial M_e}{\partial \omega} - \frac{\partial M_p}{\partial \omega}\right) \Delta\omega + \underbrace{\left(\frac{\partial M_e}{\partial Q} - \frac{\partial M_p}{\partial Q}\right)}_{=0} \Delta Q = -(M_e - M_p), \tag{8.97}$$

from which

$$\Delta\omega = -\frac{M_e - M_p}{\dfrac{\partial M_e}{\partial \omega} - \dfrac{\partial M_p}{\partial \omega}} + \frac{\dfrac{\partial M_p}{\partial Q}}{\dfrac{\partial M_e}{\partial \omega} - \dfrac{\partial M_p}{\partial \omega}} \Delta Q. \tag{8.98}$$

If the following marks are introduced

$$B^\omega = -\frac{M_e - M_p}{\dfrac{\partial M_e}{\partial \omega} - \dfrac{\partial M_p}{\partial \omega}} \quad \text{and} \quad C^\omega = \frac{\dfrac{\partial M_p}{\partial Q}}{\dfrac{\partial M_e}{\partial \omega} - \dfrac{\partial M_p}{\partial \omega}} \tag{8.99}$$

The expression for the angular velocity increment obtains the following form

$$\Delta \omega = B^\omega + C^\omega \Delta Q, \tag{8.100}$$

which is applied in the calculation of variables in the subroutine `IncVars`. Values are memorized in the structure `Pump_t`, see module `GlobalVars.f90`. For the pumping unit with the constant angular velocity $B^\omega = 0$ and $C^\omega = 0$.

Subroutine `SteadyPumpMtx`

Power on
The pump is active since there are non-zero values of operational variables. The dynamic equation of steady flow in the pump finite element is equal to

$$F_1 : \quad h_2 - h_1 - H_m(\omega, Q) = 0 \tag{8.101}$$

and is shown in Figure 8.32.
   When the Newton–Raphson procedure is applied to the dynamic equation then

$$- \Delta h_1 + \Delta h_2 - \frac{\partial H_m}{\partial Q} \Delta Q - \frac{\partial H_m}{\partial \omega} \Delta \omega = -(h_2 - h_1 - H_m). \tag{8.102}$$

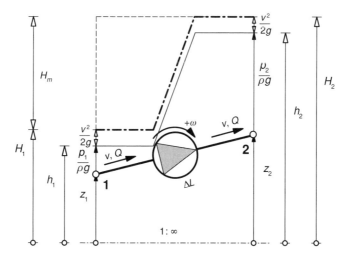

Figure 8.32   Pump finite element.

If the angular velocity increment from the dynamic equation of the engine (8.100) is introduced in the previous equation, after arranging, the following is obtained

$$- \Delta h_1 + \Delta h_2 + \left(-\frac{\partial H_m}{\partial Q} - \frac{\partial H_m}{\partial \omega} C^{\omega} \right) \Delta Q = - \left(h_2 - h_1 - H_m - \frac{\partial H_m}{\partial \omega} B^{\omega} \right), \qquad (8.103)$$

which is formally written using the matrix-vector operations

$$[\underline{H}] \cdot [\Delta h] + [\underline{Q}] \cdot [\Delta Q] = [\underline{F}]. \qquad (8.104)$$

Vector $[\underline{H}]$ has the form

$$[\underline{H}] = [-1 \quad 1]. \qquad (8.105)$$

Scalar $[\underline{Q}]$ is equal to

$$[\underline{Q}] = -\frac{\partial H_m}{\partial Q} - \frac{\partial H_m}{\partial \omega} C^{\omega} \qquad (8.106)$$

while scalar $[\underline{F}]$ is equal to

$$[\underline{F}] = - \left(h_2 - h_1 - H_m - \frac{\partial H_m}{\partial \omega} B^{\omega} \right). \qquad (8.107)$$

When the previous expression is multiplied by the inverse term $[\underline{Q}]^{-1}$ the following is obtained

$$[\underline{A}] \cdot [\Delta h] + [\Delta Q] = [\underline{B}], \qquad (8.108)$$

from which the value of the elemental discharge increment can be calculated as

$$[\Delta Q] = [\underline{B}] - [\underline{A}][\Delta h], \qquad (8.109)$$

where:

$$[\underline{A}] = [\underline{Q}s]^{-1} [\underline{H}], \qquad (8.110)$$

$$[\underline{B}] = [\underline{Q}]^{-1} [\underline{F}]. \qquad (8.111)$$

In the aforementioned expressions $[\underline{A}]$ is a two-term vector while $[\underline{B}]$ is a scalar. A process of elimination of elemental discharges from nodal equations of continuity defines the structure of the finite element matrix

$$A^e = \begin{bmatrix} +\underline{A} \\ -\underline{A} \end{bmatrix} \qquad (8.112)$$

and vector.

$$B^e = \begin{bmatrix} +Q \\ -Q \end{bmatrix} + \begin{bmatrix} +\underline{B} \\ -\underline{B} \end{bmatrix}. \qquad (8.113)$$

The procedure is carried out numerically.

Power off
The pumping unit without power (the pumping unit that is switched off) is physically isolated by a reflux preventer of some other closed valve.
   Modelling can be dual:

(a) The pumping unit is equipped with the respective valve, which is physically not included in the finite elements network configuration. Thus, in the case of power outage, the pumping unit shall split the system.
   In that case, elemental matrices shall ensure that discharge and angular velocity at the element are equal to $Q = 0$, $\omega = 0$, which implies that $[A] = 0$ and scalars $[B] = 0$, $B^\omega = 0$, $C^\omega = 0$, that is the finite element matrix $[A^e] = 0$ and vector $[B^e] = 0$ are equal to zero.
(b) The pumping unit is equipped with the respective valve, which is physically included in the finite elements network configuration. In that case, the system split is carried out by a valve finite element while the non-active pump does not break the system.

   In the case of an inactive pump, the piezometric heads of the first and second nodes of the pump finite element are equal while discharge and angular velocity are equal to zero. The following elemental equation can be applied to this case:

$$F_1 : \quad h_2 - h_1 = 0, \tag{8.114}$$

$$F_2 : \quad \omega = 0. \tag{8.115}$$

Thus, the zero elemental discharge increment is obtained when vector $[A] = 0$ and scalars are $[B] = 0$, while the zero angular velocity increment is obtained when $B^\omega = 0$ and $C^\omega = 0$, and equality of the second node piezometric head with the first one when:

$$A^e = \begin{bmatrix} 0 & 0 \\ -1 & 1 \end{bmatrix} \quad \text{and} \quad B^e = 0. \tag{8.116}$$

Let us remember that the steady state is an initial condition for unsteady modelling. If we are to be limited to the quasi unsteady state only, then the pumping units and protection against the reflux flow could be modeled using the procedure described under (a).

8.9.2   Unsteady flow modelling

In the subroutine `Unsteady` a call for a subroutine for computation of the steady flow matrix and vector for elemental pump type is added as follows:

```
select case(Elems(ielem).tip)
      case ...

            ...
      case (PUMP_OBJ)
            call UnsteadyPumpMtx(ielem)
      case ...

            ...
endselect
```

Subroutine `UnsteadyPumpMtx`

This subroutine calculates the pump finite element matrix and vector by calling the respective subroutine depending on the unsteady computation type:

```
if runmode = QuasiUnsteady then
  call QuPumpMtx(ielem)
else if runmode = RigidUnsteady then
  call RgdPumpMtx(ielem)
else if runmode = FullUnsteady then
   if SpeedTransients then
     call UnsPumpMtx(ielem)
   else
     call PumpMtx(ielem)
   endif
endif
```

Subroutine `QuPumpMtx`

In quasi unsteady modelling, transient stages of the pumping unit rotation are not calculated. Angular velocity depends on the equilibrium state for the prescribed operating variables. Angular velocity is not iterated in the integration time step; thus $B^\omega = 0$ and $C^\omega = 0$.

The dynamic equation of quasi unsteady flow in the pump finite element is equal to

$$h_2 - h_1 - H_m = 0. \tag{8.117}$$

Integration between the two time stages gives

$$(1 - \vartheta)\,\Delta t\,(h_2 - h_1 - H_m) + \vartheta\,\Delta t\,\left(h_2^+ - h_1^+ - H_m^+\right) = 0, \tag{8.118}$$

where $^+$ marks the values at the end of the time interval, while variables without that sign are those at the beginning of the time interval.

The Newton–Raphson iterative form is formally written using the matrix-vector operations

$$[\underline{H}] \cdot [\Delta h^+] + [\underline{Q}] \cdot [\Delta Q^+] = [\underline{F}], \tag{8.119}$$

where the scalar term $[\underline{F}]$ is equal to

$$[\underline{F}] = -\left[(1 - \vartheta)\,\Delta t\,(h_2 - h_1 - H_m) + \vartheta\,\Delta t\,\left(h_2^+ - h_1^+ - H_m^+\right)\right], \tag{8.120}$$

while the vector $[\underline{H}]$ has the form

$$[\underline{H}] = \vartheta\,\Delta t\,[-1 \quad 1]. \tag{8.121}$$

The scalar term $\left[\underline{Q}\right]$ is equal to

$$\left[\underline{Q}\right] = -\vartheta\,\Delta t\,\frac{\partial H_m^+}{\partial Q^+}. \tag{8.122}$$

When the previous expression is multiplied by the inverse term $\left[\underline{Q}\right]^{-1}$ the following is obtained

$$\left[\underline{A}\right]\cdot\left[\Delta h^+\right] + \left[\Delta Q^+\right] = \left[\underline{B}\right], \tag{8.123}$$

from which the value of the elemental discharge increment can be calculated as

$$\left[\Delta Q^+\right] = \left[\underline{B}\right] - \left[\underline{A}\right]\left[\Delta h^+\right], \tag{8.124}$$

where

$$\left[\underline{A}\right] = \left[\underline{Q}\right]^{-1}\left[\underline{H}\right], \tag{8.125}$$

$$\left[\underline{B}\right] = \left[\underline{Q}\right]^{-1}\left[\underline{F}\right]. \tag{8.126}$$

In the aforementioned expressions, $\left[\underline{A}\right]$ is a two-term vector while $\left[\underline{B}\right]$ is a scalar. A process of elimination of elemental discharges from nodal equations of continuity defines the structure of the finite element matrix

$$A^e = \vartheta\,\Delta t\begin{bmatrix} +\underline{A} \\ -\underline{A} \end{bmatrix} \tag{8.127}$$

and vector.

$$B^e = (1-\vartheta)\Delta t\begin{bmatrix} +Q \\ -Q \end{bmatrix} + \vartheta\,\Delta t\begin{bmatrix} +Q^+ \\ -Q^+ \end{bmatrix} + \vartheta\,\Delta t\begin{bmatrix} +\underline{B} \\ -\underline{B} \end{bmatrix}. \tag{8.128}$$

The procedure is carried out numerically.

Subroutine RgdPumpMtx

The special matrix of pumping units is not developed for the modelling of an unsteady incompressible fluid (rigid) fluid. Instead, the matrix of quasi unsteady flow QuPumpMtx is used. Namely, in unsteady flow of an incompressible fluid with the pumping unit as a boundary condition, some impossible combinations occur for which the solutions are unfeasible. For those cases that can be solved, it is sufficient to use the matrix QuPumpMtx. Otherwise, modelling of an incompressible (rigid) fluid shall be avoided since the problem can be solved as for a compressible (elastic) liquid without any major problem. Modelling of a compressible liquid shows no observed flaws in the modelling of a rigid fluid.

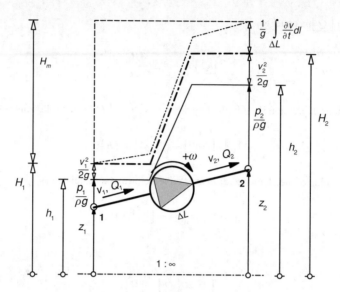

Figure 8.33 Pump finite element – unsteady flow.

Subroutine `UnsPumpMtx`

The pump finite element matrix and vector will be obtained by integration of the mass and specific mechanical energy conservation law between the initial and end state, see Figure 8.33. Apart from these laws, a dynamic equation of the pumping unit (8.60) is also used in the form

$$I\frac{d\omega}{dt} = M_e - M_p. \tag{8.129}$$

The mass conservation law on a finite element is

$$\frac{gA}{c^2} \int_{\Delta l} \frac{\partial h}{\partial t} dl + Q_2 - Q_1 = 0, \tag{8.130}$$

where the elastic accumulation of a pump is expressed by water hammer celerity c in an equivalent pipe with the diameter equal to the mean joint diameter.

The dynamic equation is obtained from the specific mechanic energy conservation law in the head form

$$\frac{1}{gA} \int_{\Delta l} \frac{\partial Q}{\partial t} dl + H_2 - H_1 - H_m = 0. \tag{8.131}$$

The first elemental equation is obtained by integration of Eq. (8.130) between the two time stages

$$F_1 : \quad \frac{gA}{c^2}\Delta l \left(\frac{h_1^+ + h_2^+}{2} - \frac{h_1 + h_2}{2}\right) + (1-\vartheta)\Delta t(Q_2 - Q_1) + \vartheta \Delta t(Q_2^+ - Q_1^+) = 0. \tag{8.132}$$

The second elemental equation is obtained by integration of Eq. (8.131) between the two time stages

$$F_2: \qquad \frac{\Delta l}{gA}\left(\frac{Q_1^+ + Q_2^+}{2} - \frac{Q_1 + Q_2}{2}\right) +$$
$$(1-\vartheta)\Delta t(H_2 - H_1 - \bar{H}_m) + \vartheta\,\Delta t(H_2^+ - H_1^+ - \bar{H}_m^+) = 0 \qquad (8.133)$$

The third elemental equation is obtained by integration of Eq. (8.129) between the two time stages

$$F_3: \qquad I\left(\omega^+ - \omega\right) - (1-\vartheta)\Delta t(M_e - M_p) + \vartheta\,\Delta t(M_e^+ - M_p^+) = 0. \qquad (8.134)$$

The sign $+$ marks the values at the end of the time interval, while variables without that sign are those at the beginning of the time interval.

The Newton–Raphson iterative form for the first two elemental equations is

$$\begin{bmatrix} \frac{\partial F_1}{\partial h_1^+} & \frac{\partial F_1}{\partial h_2^+} \\[2mm] \frac{\partial F_2}{\partial h_1^+} & \frac{\partial F_2}{\partial h_2^+} \end{bmatrix} \cdot \begin{bmatrix} \Delta h_1 \\[1mm] \Delta h_2 \end{bmatrix}^+ + \begin{bmatrix} \frac{\partial F_1}{\partial Q_1^+} & \frac{\partial F_1}{\partial Q_2^+} \\[2mm] \frac{\partial F_2}{\partial Q_1^+} & \frac{\partial F_2}{\partial Q_2^+} \end{bmatrix} \cdot \begin{bmatrix} \Delta Q_1 \\[1mm] \Delta Q_2 \end{bmatrix}^+ + \begin{bmatrix} 0 \\[1mm] \frac{\partial F_2}{\partial \omega^+} \end{bmatrix} \cdot \Delta\omega^+ = - \begin{bmatrix} F_1 \\[1mm] F_2 \end{bmatrix}$$
$$(8.135)$$

and for the third one

$$\left[\frac{\partial F_3}{\partial \omega^+}\right] \cdot \Delta\omega^+ + \left[\frac{\partial F_3}{\partial \bar{Q}^+}\right] \cdot \Delta \bar{Q}^+ = -F_3, \qquad (8.136)$$

where

$$\bar{Q}^+ = \frac{Q_1^+ + Q_2^+}{2} \quad \text{and} \quad \Delta \bar{Q}^+ = \frac{\Delta Q_1^+ + \Delta Q_2^+}{2}, \qquad (8.137)$$

from which

$$\Delta\omega^+ = -\left[\frac{\partial F_3}{\partial \omega^+}\right]^{-1} \cdot F_3 - \left[\frac{\partial F_3}{\partial \omega^+}\right]^{-1} \cdot \left[\frac{\partial F_3}{\partial \bar{Q}^+}\right] \cdot \Delta \bar{Q}^+. \qquad (8.138)$$

$$\Delta\omega^+ = B^\omega + C^\omega \Delta \bar{Q}^+ \qquad (8.139)$$

The angular velocity increment can be eliminated from expression (8.135) by the introduction of the previous expression (8.139). Thus, after arranging, the following is obtained

$$\begin{bmatrix} \frac{\partial F_1}{\partial h_1^+} & \frac{\partial F_1}{\partial h_2^+} \\[2mm] \frac{\partial F_2}{\partial h_1^+} & \frac{\partial F_2}{\partial h_2^+} \end{bmatrix} \cdot \begin{bmatrix} \Delta h_1 \\[1mm] \Delta h_2 \end{bmatrix}^+ + \begin{bmatrix} \frac{\partial F_1}{\partial Q_1^+} & \frac{\partial F_1}{\partial Q_2^+} \\[2mm] \frac{\partial F_2}{\partial Q_1^+} + \frac{\partial F_2}{\partial \omega^+}\frac{C^\omega}{2} & \frac{\partial F_2}{\partial Q_2^+} + \frac{\partial F_2}{\partial \omega^+}\frac{C^\omega}{2} \end{bmatrix} \cdot \begin{bmatrix} \Delta Q_1 \\[1mm] \Delta Q_2 \end{bmatrix}^+$$
$$= -\begin{bmatrix} F_1 \\[1mm] F_2 + \frac{\partial F_2}{\partial \omega^+} B^\omega \end{bmatrix}, \qquad (8.140)$$

which is formally written using the matrix operations

$$[\underline{H}] \cdot [\Delta h^+] + [\underline{Q}] \cdot [\Delta Q^+] = [\underline{F}].$$ (8.141)

Matrix $[\underline{H}]$ is equal to

$$[\underline{H}] = \begin{bmatrix} \dfrac{gA}{2c^2}\Delta l & \dfrac{gA}{2c^2}\Delta l \\ -\vartheta\,\Delta t & +\vartheta\,\Delta t \end{bmatrix}.$$ (8.142)

Matrix $[\underline{Q}]$ is equal to

$$[\underline{Q}] = \begin{bmatrix} \underline{Q}_{11} & \underline{Q}_{12} \\ \underline{Q}_{21} & \underline{Q}_{22} \end{bmatrix}$$ (8.143)

in which the terms are

$$\begin{aligned} \underline{Q}_{11} &= -\vartheta\,\Delta t \\ \underline{Q}_{12} &= +\vartheta\,\Delta t \end{aligned},$$ (8.144)

$$\begin{aligned} \underline{Q}_{21} &= \frac{\Delta l}{2gA} - \vartheta\,\Delta t\,\frac{Q_1^+}{gA^2} - \frac{\vartheta\,\Delta t}{2}\frac{\partial\bar{H}_m^+}{\partial\bar{Q}^+} - \frac{\vartheta\,\Delta t}{2}C^\omega\frac{\partial\bar{H}_m^+}{\partial\omega^+} \\ \underline{Q}_{22} &= \frac{\Delta l}{2gA} + \vartheta\,\Delta t\,\frac{Q_2^+}{gA^2} - \frac{\vartheta\,\Delta t}{2}\frac{\partial\bar{H}_m^+}{\partial\bar{Q}^+} - \frac{\vartheta\,\Delta t}{2}C^\omega\frac{\partial\bar{H}_m^+}{\partial\omega^+} \end{aligned}.$$ (8.145)

Vector $[\underline{F}]$ is equal to

$$[\underline{F}] = - \begin{bmatrix} F_1 \\ F_2 + \vartheta\,\Delta t\,\dfrac{\partial\bar{H}_m^+}{\partial\omega^+}B^\omega \end{bmatrix}.$$ (8.146)

When the matrix equation (8.141) is multiplied by the inverse term $[\underline{Q}]^{-1}$ the following is obtained

$$[\underline{A}] \cdot [\Delta h^+] + [\Delta Q^+] = [\underline{B}],$$ (8.147)

from which the value of the elemental discharge increment can be calculated as

$$[\Delta Q^+] = [\underline{B}] - [\underline{A}][\Delta h^+],$$ (8.148)

where

$$[\underline{A}] = [\underline{Q}]^{-1}[\underline{H}],$$ (8.149)

$$[\underline{B}] = [\underline{Q}]^{-1}[\underline{F}].$$ (8.150)

A process of elimination of elemental discharges from nodal equations of continuity defines the structure of the pump finite element matrix

$$A^e = \vartheta \, \Delta t \begin{bmatrix} +\underline{A}_{11} & +\underline{A}_{12} \\ -\underline{A}_{21} & -\underline{A}_{22} \end{bmatrix} \qquad (8.151)$$

and vector

$$B^e = (1 - \vartheta \, \Delta t) \begin{bmatrix} +Q_1 \\ -Q_2 \end{bmatrix} + \vartheta \, \Delta t \begin{bmatrix} +Q_1^+ \\ -Q_2^+ \end{bmatrix} + \vartheta \, \Delta t \begin{bmatrix} +\underline{B}_1 \\ -\underline{B}_2 \end{bmatrix}. \qquad (8.152)$$

Subroutine `PumpMtx`

For an optional variable `SpeedTransients = .false.` a pump finite element matrix and vector are used in which the angular velocity is calculated directly based on the operational variables that are prescribed in the input data for the pumping unit:

`Power pumpname Voltage(t)<Frequency(t[1]>,`

where the normalized voltage $u(t)$ and frequency variables $\varphi(t)$ can be time-dependent. Values of the variables and angular velocity for the equilibrium of the electric and hydraulic device are calculated at the time of boundary condition definition by the subroutine `SetSpeeds` calling. The matrix and vector are calculated by the same procedure as for the matrix in quasi unsteady modelling, with the difference that elemental discharges of unsteady flow are mutually equal.

Subroutine `PumpMtx` is implemented in the program module `Pumps.f90`.

8.10   Examples of transient operation stage modelling

Figure 8.34 shows a scheme of pump and valve connections for the purposes of protection against abnormal operation:

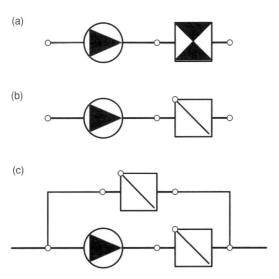

Figure 8.34   Pumping unit protection against the reflux.

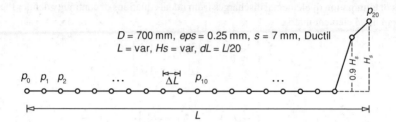

Figure 8.35   Test pressure pipeline data.

(a) by regulation valve;
(b) by automatic reflux preventer;
(c) by automatic reflux preventer and automatic bypass.

    In (a) the regulation valve inbuilt in the pressure branch of a pump regulates the pump start, operation, and shutdown, which prevents abnormal operation. A regulation valve is used for medium or large pumping unit power.

    In (b), protection against abnormal operation by an automatic reflux preventer is applied for small or medium pumping unit power. There are reflux preventers of different construction. In the `Simpip-Core` program solution a simple reflux preventer is modeled, which automatically opens or closes in time Δt depending on the nodal and elemental variables h, Q between the two computation steps.

    In (c) at the pumping unit's shutdown, pressure increases at the suction end and decreases at the pressure end. Then, the pressure difference opens the reflux preventer in the bypass though which the flow is redirected. In the return phase, pressure closes both automatic reflux preventers. This prevents reflux flow through the pump, while flow redirection into bypass greatly contributes to water hammer reduction.

    Transient phenomena are the values of unsteady variables ω, Q, H that occur between two steady states. Let us observe transient phenomena generated by pumping unit operations as the states between the operational variables $u(t), \varphi(t)$. The pumping unit quickly achieves the prescribed rotational speed, that is angular velocity determined by the torque equilibrium at the pump shaft. Thus, the transient period of rotation can be neglected for a normal pumping unit in operation, particularly if the pump is protected against reflux flow by a reflux preventer.

    Figure 8.35 shows a longitudinal section of the test pressure pipeline, prescribed by 20 points. Its dimensions are variable and defined by parameters L and Hs.

    Three test systems consisting of the pumping station and pressure pipeline were defined to test the pumping unit operation start and shutdown. The pressure pipeline shape is common for all three systems with differing pipeline lengths and static heads.

8.10.1   Test example (A)

(a) Figure 8.36a shows a short pressure pipeline with the directly connected pump at its start. Transient states will be tested at pump shutdown at time $t = 0$ and its restart after 40 seconds.
(b) Figure 8.36b shows a scheme of the pumping station connection. The pumping station is protected against the reflux by the reflux preventer. Transient states will be tested after sudden start and

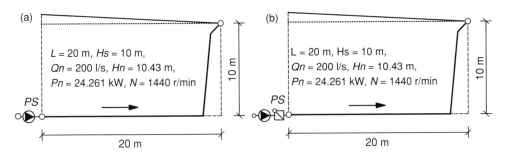

Figure 8.36 Test example (A).

shutdown at the time $t = 6$ seconds. Values with and without the modeled influence of transient values of angular velocity will be compared.

Modelling results under (a) are shown in the table in Figure 8.37. The table contains input data with the variable data prescribed parametrically. The pressure pipeline data can be found in `include file:` `ModelTestData`.

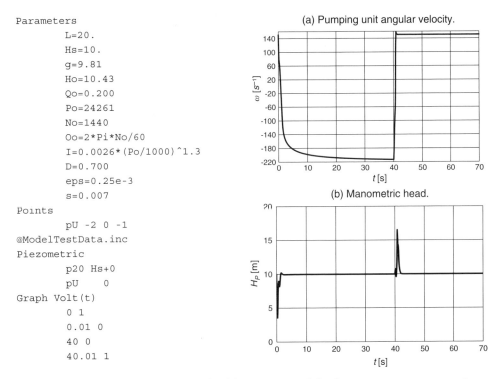

```
Parameters
        L=20.
        Hs=10.
        g=9.81
        Ho=10.43
        Qo=0.200
        Po=24261
        No=1440
        Oo=2*Pi*No/60
        I=0.0026*(Po/1000)^1.3
        D=0.700
        eps=0.25e-3
        s=0.007
Points
        pU -2 0 -1
@ModelTestData.inc
Piezometric
        p20 Hs+0
        pU    0
Graph Volt(t)
        0 1
        0.01 0
        40 0
        40.01 1
```

(a) Pumping unit angular velocity.

(b) Manometric head.

Figure 8.37 Test example (A). Alternative (a) pump start and shutdown, `File: Test example (A)-a.simpip` from www.wiley.com/go/jovic. (to be continued)

```
Pump
        pmp pU p0 Oo Qo Ho Po I D
        0.00*Qo 1.30*Ho 0.200*Po
        0.30*Qo 1.35*Ho 0.550*Po
        0.60*Qo 1.28*Ho 0.820*Po
        1.00*Qo 1.00*Ho 1.000*Po
        1.30*Qo 0.63*Ho 0.950*Po
        1.50*Qo 0.30*Ho 0.800*Po
Power pmp Volt(t)
@Longitudinal.inc
Options
        +SpeedTransients
        +NoPumpCheckValve
Steady 0
Unsteady 300 0.1 400 0.1
```

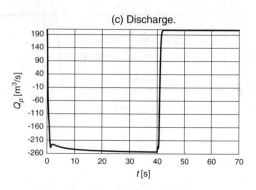

(c) Discharge.

Figure 8.37   (*Continued*)

Note two optional variables, SpeedTransients and NoPumpCheckValve, where the first enables complete modelling of angular velocity while the other enables annulment of the reflux preventer control. The angular velocity, manometric head, and pump discharge in the example of transition from the pump to turbine operation (no load, i.e. pumping unit runaway) and back to pump operation are shown.[2]

Modelling results under (b) are shown in the table in Figure 8.38. The table contains input data with the variable data prescribed parametrically. Pressure pipeline data can be found in include file: ModelTestData.

The optional variable SpeedTransients is prescribed when a full change of angular velocity is modeled. Results are shown as solid curves in the attached figures. The results obtained without that option are marked as a dashed curve in the same figures. The transient states occur both after pump start and pump shutdown. The pump achieves its speed quickly; thus velocity transient states can be neglected, that is angular velocity is defined by the equilibrium of torques at the pump shaft, after which the reflux preventer opens fast. At pump shutdown, when there is no more torque, the pressure drops; thus, the reflux preventer closes quickly and prevents reflux.

The discharge variation depends on the inertia of water in the pressure pipeline, that is, it depends on the pressure pipeline length. The transient phenomena of pumping unit rotation are almost always shorter than the water hammer cycle, particularly from the time of mass acceleration in the pressure pipeline. After pump shutdown there is no more load and the pumping unit slows down more than expected (see notes in Section 8.7.3, that is compare the change in the number of revolutions in the previous case).

8.10.2   Test example (B)

Figure 8.39 Test example (B) – shows a pressure pipeline of medium length with a pump and reflux preventer at its start. Transient states will be tested for pump operation start at time $t = 0$; and after

---

[2]This illustrates the possibilities of the program source SimpipCore even better than modelling of complex drives, since one should bear in mind all the reconstructions of complete pump and electric motor characteristics that have to be done!

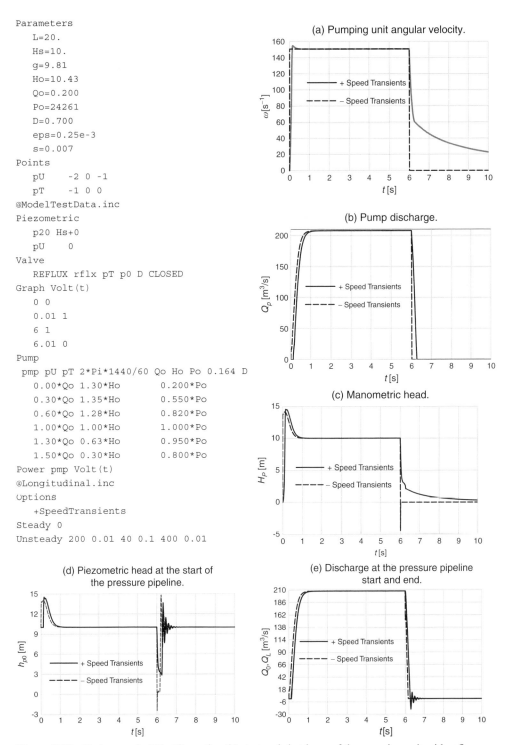

```
Parameters
  L=20.
  Hs=10.
  g=9.81
  Ho=10.43
  Qo=0.200
  Po=24261
  D=0.700
  eps=0.25e-3
  s=0.007
Points
  pU    -2 0 -1
  pT    -1 0 0
@ModelTestData.inc
Piezometric
  p20 Hs+0
  pU   0
Valve
  REFLUX rflx pT p0 D CLOSED
Graph Volt(t)
  0 0
  0.01 1
  6 1
  6.01 0
Pump
 pmp pU pT 2*Pi*1440/60 Qo Ho Po 0.164 D
  0.00*Qo 1.30*Ho         0.200*Po
  0.30*Qo 1.35*Ho         0.550*Po
  0.60*Qo 1.28*Ho         0.820*Po
  1.00*Qo 1.00*Ho         1.000*Po
  1.30*Qo 0.63*Ho         0.950*Po
  1.50*Qo 0.30*Ho         0.800*Po
Power pmp Volt(t)
@Longitudinal.inc
Options
  +SpeedTransients
Steady 0
Unsteady 200 0.01 40 0.1 400 0.01
```

(a) Pumping unit angular velocity.

+ Speed Transients

− Speed Transients

(b) Pump discharge.

+ Speed Transients

− Speed Transients

(c) Manometric head.

+ Speed Transients

− Speed Transients

(d) Piezometric head at the start of the pressure pipeline.

+ Speed Transients

− Speed Transients

(e) Discharge at the pressure pipeline start and end.

+ Speed Transients

− Speed Transients

Figure 8.38 Test example (A). Alternative (b) start and shutdown of the pumping unit with reflux preventer, File: `Test example (A)-b.simpip` from www.wiley.com/go/jovic.

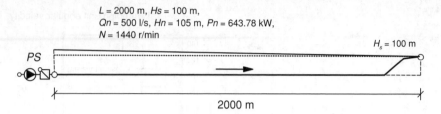

$L = 2000$ m, $Hs = 100$ m,
$Qn = 500$ l/s, $Hn = 105$ m, $Pn = 643.78$ kW,
$N = 1440$ r/min

Figure 8.39   Test example (B).

shutdown steadying after 90 seconds. Modelling is carried out without unsteady rotation of the pumping unit.

The modelling results are shown in the table in Figure 8.40. The table contains input data with the variable data prescribed parametrically. The pressure pipeline data can be found in `include file:` `ModelTestData`.

Note that transient states that occur at the pumping unit start are steadying fast. The transient state of the discharge and pressure disturbance after pump shutdown can last several minutes, depending on the pressure pipeline length, until the steady state is established.

```
Parameters
    L=2000
    Hs=100
    Zs=1
    g=9.81
    Ro=1000
    Qo=0.500
    D=0.700
    eps=0.25e-3
    s=0.007
    Dh=5
Points
    pU    -2 0 -1
    pT    -1 0 0
@ModelTestData.inc
Piezometric
    p20 Hs+0
    pU 0
Valve
    REFLUX rflx pT p0 D CLOSED
Graph Volt(t)
    0 0
    0.01 1
    90   1
    90.01   0
```

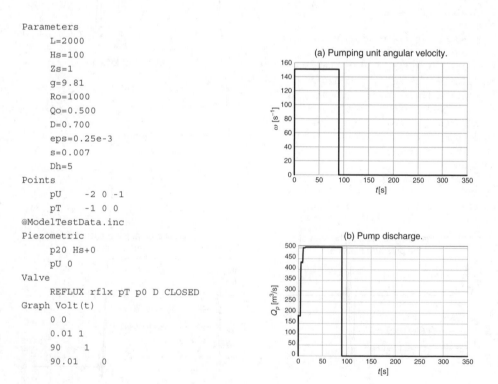

Figure 8.40   Test example (B) Pumping unit start and shutdown, `File:` `Test example` `(B).simpip` from www.wiley.com/go/jovic. (to be continued)

```
Parameters
    Ho=Hs+Dh
    Po=Ro*g*Qo*Ho/0.8
    I = 0.0026*(Po/1000)^1.3
    No=1440
Define Pump MojaCrpka 2*Pi*No/60 Qo
Ho Po I D
    0.00*Qo 1.30*Ho 0.200*Po
    0.30*Qo 1.35*Ho 0.550*Po
    0.60*Qo 1.28*Ho 0.820*Po
    1.00*Qo 1.00*Ho 1.000*Po
    1.30*Qo 0.63*Ho 0.950*Po
    1.50*Qo 0.30*Ho 0.800*Po
Pump
    pmp pU pT MojaCrpka
Power pmp Volt(t)
Steady 0
Unsteady 3000 0.1 300 0.01 270 0.1
```

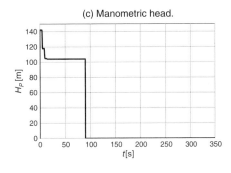

(c) Manometric head.

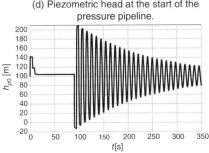

(d) Piezometric head at the start of the pressure pipeline.

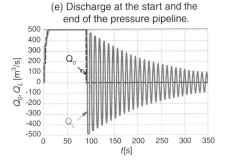

(e) Discharge at the start and the end of the pressure pipeline.

Figure 8.40 (*Continued*)

8.10.3 Test example (C)

Figure 8.41 shows a long, practically horizontal pressure pipeline with the pump and reflux preventer at its start. Transient states will be tested for pump operation start at time $t = 0$; and after steadying of the shutdown after 250 seconds.

Modelling results are shown in the table in Figure 8.42. The table contains input data with the variable data prescribed parametrically. The pressure pipeline data can be found in include file: ModelTestData.

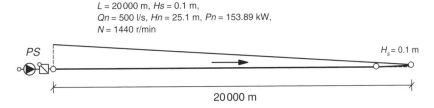

$L = 20\,000$ m, $Hs = 0.1$ m,
$Qn = 500$ l/s, $Hn = 25.1$ m, $Pn = 153.89$ kW,
$N = 1440$ r/min

PS

$H_s = 0.1$ m

20 000 m

Figure 8.41 Test example (C).

```
Parameters
        L=20000
        Hs=0.1
        g=9.81
        Ro=1000
        Qo=0.500
        D=0.700
        eps=0.25e-3
        s=0.007
Points
        pU -2 0 -1
        pT -1 0 0
@ModelTestData.inc
Piezometric
        p20 Hs+0
        pU 0
Valve
        REFLUX rflx pT p0 D CLOSED
Graph Volt(t)
        0 0
        0.01 1
        250        1
        250.01     0
Parameters
        Ho=Hs+25
        Po=Ro*g*Qo*Ho/0.8
        No=1440
        I = 0.0026*(Po/1000)^1.3
Pump pmp pU pT 2*Pi*No/60 Qo Ho Po I D
        0.00*Qo 1.30*Ho 0.200*Po
        0.30*Qo 1.35*Ho 0.550*Po
        0.60*Qo 1.28*Ho 0.820*Po
        1.00*Qo 1.00*Ho 1.000*Po
        1.30*Qo 0.63*Ho 0.950*Po
        1.50*Qo 0.30*Ho 0.800*Po
Power pmp Volt(t)
@Longitudinal.inc
Options
        +SpeedTransients
Steady 0
Unsteady 100 0.1 480 0.5 3000 0.1
```

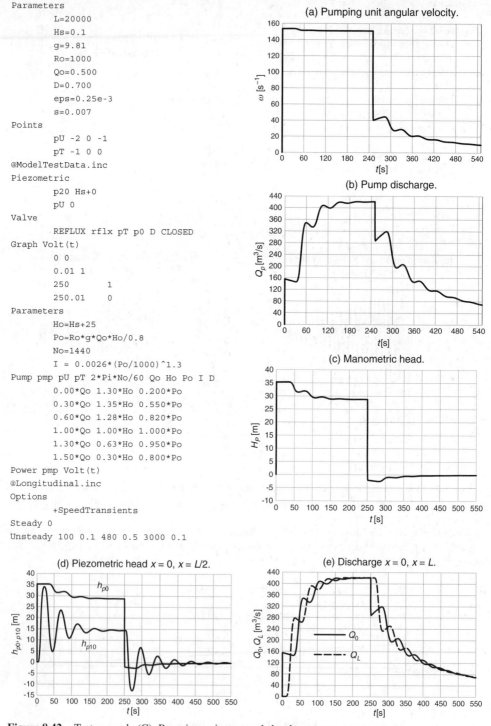

(a) Pumping unit angular velocity.

(b) Pump discharge.

(c) Manometric head.

(d) Piezometric head $x = 0$, $x = L/2$.

(e) Discharge $x = 0$, $x = L$.

Figure 8.42   Test example (C). Pumping unit start and shutdown, `File: Test example C).simpip` from `www.wiley.com/go/jovic`.

After the operation start and the achievement of the full rotational speed, the pump gradually accelerates water by overcoming inertia in a manner such that discharge increases after each cycle until equilibrium discharge is achieved. What is described is an analog to the water hammer that occurs at sudden forced inflow. Note that the transient state of discharge acceleration can take several minutes.

Transient states after pump shutdown are again characterized by the water hammer and inertia of a long pipeline as can be observed on the characteristic piezometric head and discharge graphs.

8.10.4   Test example (D)

Figure 8.43 shows a relatively long pressure pipeline with the input data given in the included file `sysdat-d).inc`, that can be supplied by gravity until discharge Q_g is achieved. Above that discharge boosting is necessary. The interpolated pumping station consists of two pumping units in booster connection. During gravity operation (pumping units are on standby) the entire discharge flows through the bypass (reflux preventer between the points pU and pT). When the pumping units are operating, the pressure difference closes the bypass and discharge is redirected to the pumps that are also equipped with reflux preventers. These valves serve to prevent reflux flow which occurs in the water hammer return phase. The pumping units are fitted with frequency converters for variation of pump rotation velocity. They allow a gradual discharge increment from 0 to Q_0 in time T_g.

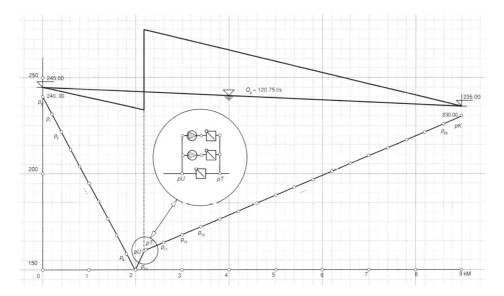

Figure 8.43   Test example (D). Booster station on the supply gravity pipeline.

Let us observe transient states that occur after the transition from operation by gravity into pumping as well as the pump shutdown. Figure 8.44 shows a working point position during gravity and pump operation at the moment when the pump takes the entire discharge. Until the discharge value Q_g is reached, the manometric head is negative, that is when the pump transfers to gravity flow the manometric head is equal to zero.

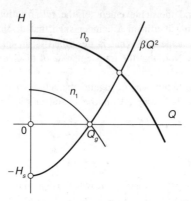

Figure 8.44   Transition from gravity to pump drive.

The transition to gravity flow is controlled by the frequency converter; namely, pumping units have a respective rotational speed n_1.

Figure 8.45 shows the results of the pump unit start and shutdown in the modelling of the transition to gravity flow. Gravity flow is the initial steady state; the pump starts at $t = 0$ and shuts down at $t = 100$ seconds.

The transient states of the pumping units are shown in Figure 8.45: (a) angular velocity, (b) discharge, (c) manometric head, and (d) discharges through reflux preventers. The transient state of the pump start

```
Parameters
        D = .500
        eps = 2/1000
        s = 0.025
@SysData-d).inc
Parameters
        No=1495
        Qo=0.300
        Ho=113.75
        Po=443.03*1000
        I=5
        Do=0.350
        Tg=0.1 ; 30
Define Pump Nova 2*Pi*No/60 Qo Ho Po I Do
        0.0000*Qo 1.1487*Ho 0.4894*Po
        0.3333*Qo 1.1343*Ho 0.5964*Po
        0.6667*Qo 1.0847*Ho 0.7847*Po
        0.8500*Qo 1.0425*Ho 0.9032*Po
        1.0000*Qo 1.0000*Ho 1.0000*Po
        1.3333*Qo 0.8801*Ho 1.1882*Po
Pumps
        pmp1 pc1 pc2 Nova
        pmp2 pc4 pc5 Nova
```

(a) Pumps angular velocity.

Figure 8.45   Test example (D). Booster station, `File: Test example (D) from` `www.wiley.com/go/jovic.` (to be continued)

```
Graph freq
        0.00 0
        Tg 1
        100.0 1
        100.1 0
Power
        pmp1 1 freq
        pmp2 1 freq
Valve
        REFLUX bypas pU pT D Open
        REFLUX rfx1 pc2 pc3 D Closed
        REFLUX rfx2 pc5 pc6 D Closed
Piezom p0 245
Piezom pK 235
@Unions-d).inc
Options
        +SpeedTransients
Steady 0
Unsteady 1200 0.1 100 1
```

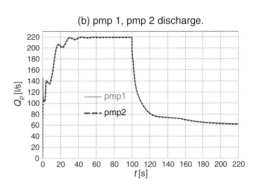

(b) pmp 1, pmp 2 discharge.

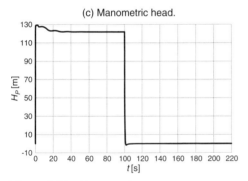

(c) Manometric head.

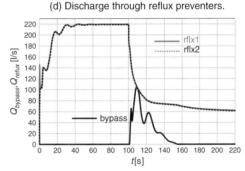

(d) Discharge through reflux preventers.

Figure 8.45   (*Continued*)

is characterized by gradual flow acceleration due to inertia. The working point gradually approaches the steady one. The transient state of pump shutdown and the return to gravity flow leaves the pumping unit in an abnormal state in the *I* quadrant, *H* sector with energy dissipation. Thus, the pump shall be additionally protected by switching off the *On–Off* valve at the pump branch.

If the optional variable SpeedTransients is excluded, switching on/off of the *ON–OFF* valve in the pump branches after the booster station, means shutdown is achieved.

The abnormal pump operation transfers to a normal operation in the I quadrant and ends with the zero angular velocity. The bypass takes the entire discharge and gravity flow is established again.

Figure 8.46 shows the envelopes of the highest and lowest piezometric heads together with the piezometric line for the gravity state in the transient states of booster station start and shutdown. Note that in the inlet pipe there is a dangerous under-pressure which occurs at the sudden pumping unit start.

The pumping unit start velocity is regulated by the parameter T_g. If it is increased to $T_g = 30$ s, there will be no more under-pressure in the inlet pipeline, see the modelling results in Figure 8.47.

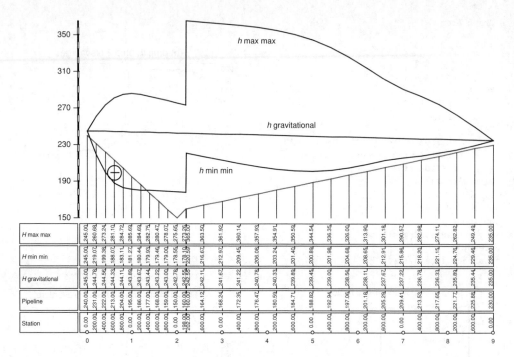

Figure 8.46 Fast start of pumping units, $Tg = 0.1\,\text{s}$ (option +SpeedTransients).

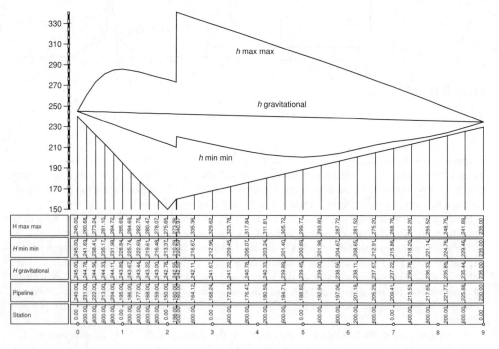

Figure 8.47 Slowed start of pumping units, $Tg = 30\,\text{s}$ (option +SpeedTransients).

8.11   Analysis of operation and types of protection against pressure excesses

8.11.1   Normal and accidental operation

An unsteady hydraulic state occurs during transition from one steady state to another in which pressure and velocity regularly exceed the values obtained in steady analyses. Transient states occur at flow energy changes:

- from the maximum to the minimum kinetic energy;
- from the minimum to the maximum energy;
- in mixed transitions during normal operation.

The bearing capacity of a pipeline, fittings, and other hydraulic network components is dimensioned to pressure states that occur during normal operation, which, apart from the steady ones, also includes transient states.

For a short time, the system can be in accident state. These accident operations occur in the case of power failure, damage, or similar and are called pressure excesses. Then, the pressures exceed (above or below) the normal ones that the system was sized to and the system is additionally protected. *The basic principle of protection against pressure excesses is to slow down the flow*, which is achieved by:

- system operation measures prescribed by the operation manual. This refers to the operation of valves, pumping units, and other devices;
- additional surge protection devices, namely vessels, surge tanks, air relief and flow relief valves, bypasses, and other devices.

There is no universal surge protection method or equipment! If inadequate protection is applied, money is spent in vain on unnecessary and expensive equipment, which generally increases the danger of damage or accident.

Procedures for protection against pressure excesses consist of the following:

- analyses of operation and definition of the respective transient states that cause pressure excesses;
- calculation of pressure excesses;
- testing of the possible protection measures and devices;
- selection of the most appropriate protection method.

8.11.2   Layout

An analysis of the operation and types of protection against pressure excesses will be shown on an example of a hypothetic water supply system. Figure 8.48 shows a longitudinal section of a 2-km long pipeline. The prescribed water quantity shall be raised from point A at elevation 100 m.a.s.l. to point B at 300 m.a.s.l. A pumping station is planned at chainage 900 m. There are two construction alternatives:

(a) conventional pipeline with suction basin;
(b) state-of-the-art solution, that is a booster station.

Different types of protection against pressure excesses will be analyzed for two alternatives, with no reference to design details.

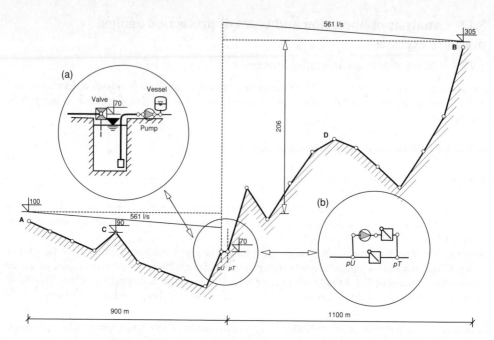

Figure 8.48   Pipeline alignment from point A to point B.

8.11.3   Supply pipeline, suction basin

Alternative (a) is a pumping station with a suction basin. Inflow into the suction basin is regulated by a regulation valve. Opening and closing of the regulation valve causes piezometric head variations in the supply pipeline. At any time, the piezometric head shall be above the pipeline; thus, point C of the observed supply pipeline is marked as the critical point. A regulation valve closes when the suction basin is full, and vice versa: when the suction basin is empty, a regulation valve opens.

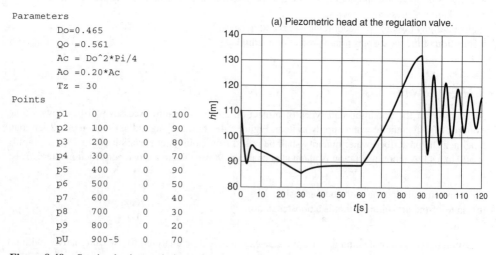

Figure 8.49   Suction basin regulation valve. File: Test example
(E) SuctionPipeline.simpip from www.wiley.com/go/jovic. (to be continued)

```
Pipes
  c1 p1 p2 Do 1.00E-04 0.001 steel
  c2 p2 p3 Do 1.00E-04 0.001 steel
  c3 p3 p4 Do 1.00E-04 0.001 steel
  c4 p4 p5 Do 1.00E-04 0.001 steel
  c5 p5 p6 Do 1.00E-04 0.001 steel
  c6 p6 p7 Do 1.00E-04 0.001 steel
  c7 p7 p8 Do 1.00E-04 0.001 steel
  c8 p8 p9 Do 1.00E-04 0.001 steel
Piezo
       p1 110
Graph A(t)
       0.0 0
       Tz Ao*sqrt(19.62)
       60. Ao*sqrt(19.62)
       60+Tz 0
Outlet GATE vlv pU A(t)
@Unions-e)a.inc
Steady 0.
Unsteady 600 0.1 600 0.1
```

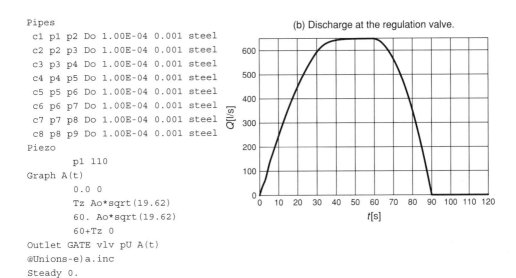

(b) Discharge at the regulation valve.

Figure 8.49 (*Continued*)

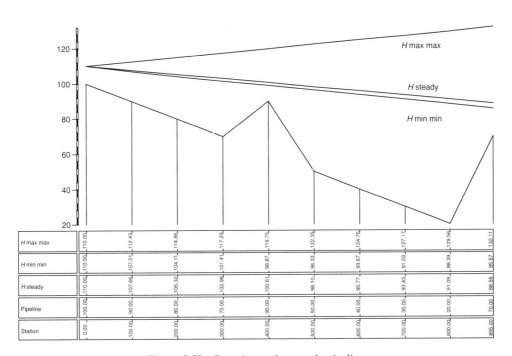

Figure 8.50 Pumping station supply pipeline.

Thus, the opening/closing law shall be defined so the pressure in point C is always greater than zero. A linear law of regulation valve outflow area will be adopted. When the regulation valve is completely open, discharge of $Q = 650\,l/s$ flows out through an area that is equal to 20% of the pipe cross-section.

The table in Figure 8.49 shows the input file for modelling the suction pipeline opening and closing in time T_z, that is set parametrically, where the regulation valve is modeled by the nodal function Outlet GATE, with the time-dependent area parameter.

The regulation valve axis corresponds to the axis of the inlet pipe at elevation 70 m.a.s.l. For the adopted closing time of 30 seconds the following is presented: (a) the piezometric head in front of the valve and (b) the time-dependent discharge.

Figure 8.50 shows a longitudinal section of the inlet pipeline with the envelopes of the maximum and minimum piezometric heads together with the piezometric head of a completely open valve.

8.11.4 Pressure pipeline and pumping station

In alternative (a) pumps are pumping water from the suction basin and are transporting it to the destination. The table shown in Figure 8.51 is the main input file of the pressure pipeline and pumping units. The main part of the input file is calling two files SysData-e)b.inc, which contains the pipeline data and Unions-e)b.inc which contains the longitudinal section data.

First, a subsequent pump start and shutdown is modeled for the pumping station and pressure pipeline without surge protection. In this alternative, the row marked Vessel ... is omitted. Modelling results are shown in Figure 8.51 as follows: (a) the piezometric head at the start, (b) the discharge at the start and end, and (c) the velocity at the start of the pipeline, as well as a longitudinal section of the pipeline, see Figure 8.52. Envelopes of maximum and minimum piezometric heads as well as a piezometric head in steady flow are shown in longitudinal section. Note that under-pressure occurs along the entire alignment as a consequence of surge after pumping unit shutdown. The under-pressure, that is the absolute pressure under the saturated vapor pressure, causes a break in the water body, which is as a result of the elastic model and shall be interpreted only as non-allowable state that shall be solved.

```
Parameters
        Do=0.465
        Qo=0.561
        Zs=77
        Hs=305-Zs
        Vo=2
        Ak=1.5
@SysData-e)b.inc
Point
        px   900 5 70
        p11 900-5        5.0        70
        p12 900+5        5.0        70
Pipes
c10 pU p11 Do 1.00E-04 0.001 steel
c12 p12 pT Do 1.00E-04 0.001 steel
Parameters
        Ho = 260;(328.5-75)
        Po=1000*9.81*Qo*Ho/0.80
        I = 0.0026*Po^1.3
```

(a) Piezometric head at the pipeline start.

Figure 8.51 Pressure pipeline and pumping station. File: Test example (E) PressurePipeline.simpip from www.wiley.com/go/jovic.

```
Pump
 pmp p11 px 2*Pi*1440/60 Qo Ho Po I Do
 0.000*Qo 1.176*Ho 0.448*Po
 0.118*Qo 1.201*Ho 0.527*Po
 0.235*Qo 1.216*Ho 0.607*Po
 0.354*Qo 1.216*Ho 0.682*Po
 0.470*Qo 1.205*Ho 0.749*Po
 0.522*Qo 1.198*Ho 0.779*Po
 0.587*Qo 1.183*Ho 0.813*Po
 0.704*Qo 1.147*Ho 0.873*Po
 0.822*Qo 1.099*Ho 0.928*Po
 0.939*Qo 1.037*Ho 0.978*Po
 1.000*Qo 1.000*Ho 1.000*Po
 1.057*Qo 0.963*Ho 1.020*Po
 1.174*Qo 0.875*Ho 1.052*Po
 1.292*Qo 0.777*Ho 1.080*Po
 1.358*Qo 0.714*Ho 1.092*Po
Valve
 REFLUX pmprfx px p12 Do CLOSED
Vessel vsl pT Ak Vo Zs Hs
Graph Volt
        0         0
        0.1       1
        60        1
        60.1      0
Power pmp         Volt
Piezo
        pU 68.5
        p24 305
@Unions-e)b.inc
Steady 0.
Unsteady 600 0.1 600 0.1
```

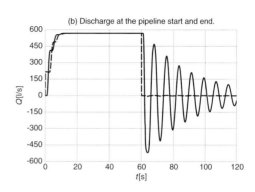

(b) Discharge at the pipeline start and end.

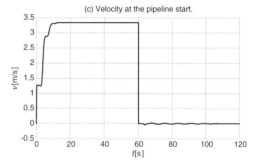

(c) Velocity at the pipeline start.

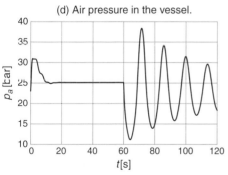

(d) Air pressure in the vessel.

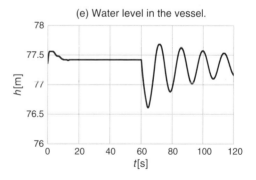

(e) Water level in the vessel.

Figure 8.51 (*Continued*)

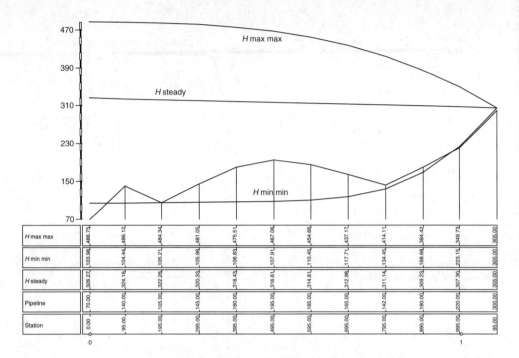

Figure 8.52   Pressure pipeline, no protection.

The vessel in the pumping station will be used as protection against under-pressure along the pipeline alignment.

The lines marked Vessel vsl pT Ak Vo Zs Hs add a vessel of selected dimensions. Thus, after repeated computation, the results are obtained, out of which those shown in Figure 8.51d: air pressure in the vessel, and e: water level in the vessel, are selected. Figure 8.53 shows effect of protection by vessel.

8.11.5   Booster station

In alternative (b) the pumping station is in booster connection. The suction end of the pumping unit is directly connected to the supply pipeline while the pumping end is connected to the pressure pipeline. Pressurized flow is uninterrupted.

A bypass is installed between the suction and pressure ends in order to extend the flow after the pump shutdown. The table in Figure 8.54 is the input file for the boosting alternative, which uses the included files: sysdata-e).inc for the supply and pressure pipeline data and unions-e).inc, unions-e)a.inc and unions-e)b.inc for longitudinal section data.

The results of the pump start and shutdown modelling are shown in Figure 8.54 as follows: (a) the piezometric head at the suction (point pU) and pressure end (point pT) of the pipeline, (b) discharge through the reflux valves of the pump and bypass, and (c) discharges at the start and end of the pressure pipeline. When the pressure at the suction end exceeds the pressure at the pressure end (in front and behind the bypass) the bypass opens. Simultaneously, the reflux preventer of the pumping unit closes (the default value of variable SpeedTransients is .false.). This extended flow through the bypass is sufficient to absorb the negative or, in return, the positive water hammer phases.

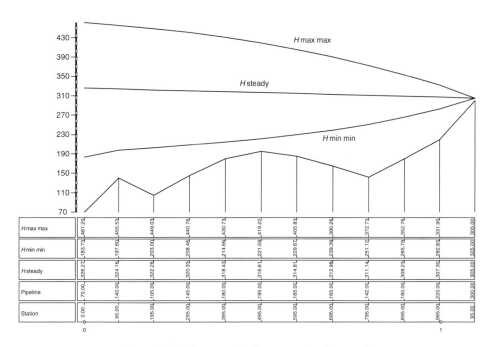

H max max	461.20	455.53	449.03	440.76	430.73	419.22	405.83	390.26	372.73	352.76	331.90	305.00
H min min	183.73	197.60	203.00	208.46	214.66	221.59	229.67	239.39	251.12	265.79	282.83	305.00
H steady	326.27	324.16	322.26	320.33	318.43	316.61	314.81	312.98	311.14	309.23	307.30	305.00
Pipeline	70.00	140.00	105.00	145.00	180.00	195.00	185.00	165.00	142.00	180.00	220.00	300.00
Station	0.00	95.00	195.00	295.00	395.00	495.00	595.00	695.00	795.00	895.00	995.00	95.00

Figure 8.53 Pressure pipeline, protection by vessel.

```
Parameters
        Do = 0.465
        Qo = 0.561
        Tg = 0.1;20
@SysData-e).inc
Parameters
        Ho = (328.5-92.61)
        Po=1000*9.81*Qo*Ho/0.80
        I = 0.0026*Po^1.3
Pump pmp p11 px 2*Pi*1440/60 Qo Ho Po
I Do
        0.000*Qo 1.176*Ho 0.448*Po
        0.118*Qo 1.201*Ho 0.527*Po
        0.235*Qo 1.216*Ho 0.607*Po
        0.354*Qo 1.216*Ho 0.682*Po
        0.470*Qo 1.205*Ho 0.749*Po
        0.522*Qo 1.198*Ho 0.779*Po
        0.587*Qo 1.183*Ho 0.813*Po
        0.704*Qo 1.147*Ho 0.873*Po
        0.822*Qo 1.099*Ho 0.928*Po
        0.939*Qo 1.037*Ho 0.978*Po
        1.000*Qo 1.000*Ho 1.000*Po
        1.057*Qo 0.963*Ho 1.020*Po
        1.174*Qo 0.875*Ho 1.052*Po
        1.292*Qo 0.777*Ho 1.080*Po
        1.358*Qo 0.714*Ho 1.092*Po
```

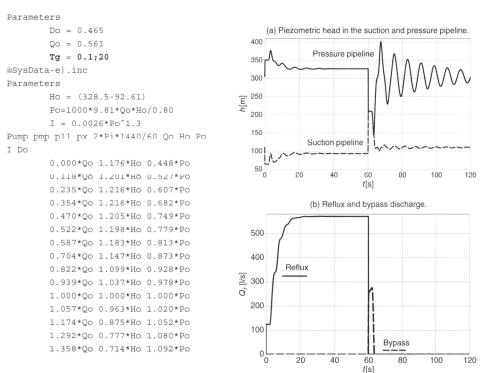

(a) Piezometric head in the suction and pressure pipeline.

(b) Reflux and bypass discharge.

Figure 8.54 Booster station. File: Test example (E)Booster.simpip from www.wiley.com/go/jovic. (to be continued)

```
Valve
        REFLUX pmprfx px p12 Do CLOSED
        REFLUX rfx pU pT Do CLOSED
Graph Volt
        0          0
        Tg 1
        60         1
        60.1       0
Power pmp           Volt
Piezo
        p1 110
        p24 305
@Unions-e).inc
@Unions-e)a.inc
@Unions-e)b.inc
;
Steady 0.
Unsteady 600 0.1 600 0.1
```

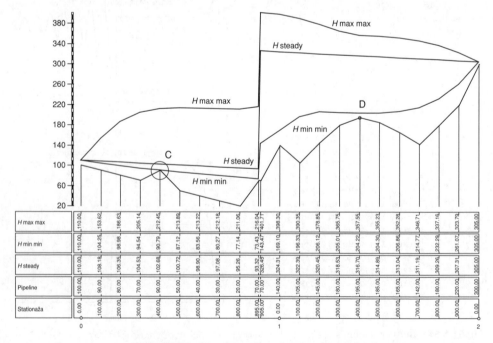

Figure 8.54   (*Continued*)

Very quickly, water suction from the supply pipeline causes a sudden pressure fall; thus, under-pressure occurs at high levels and reaches a dangerous value at point C. The pump start is a normal operation maneuver that can be controlled. Thus, for example, the pump can start when the valve at the pressure end is closed. When the pump reaches its full rotation, the valve starts to open gradually. Or, a slowed pumping unit start can be achieved by special devices, namely soft starters. Anyhow, a fast change is replaced by a slow one and the required start time shall be determined. In the input data, note the marked parametric value $Tg = 0.1$, namely the pump start time. Thus, the calculation will be repeated with the new value $Tg = 20$ seconds, which is sufficient for under-pressure elimination. Figure 8.55

Figure 8.55   Booster station, piezometric states.

shows a longitudinal section of the final pressure states of the solution with the booster station, that is alternative (b).

Figure 8.55 shows a longitudinal section of the observed pipeline with the envelopes of the maximum and minimum piezometric heads as well as the piezometric head in steady flow. Thus, the pressure pipeline of the prescribed alignment can be considered protected. A question may arise about a solution if, for example, a low pressure envelope is to intersect the alignment and there is under-pressure at point D.

One solution would be to install a small vessel in the pumping station of sufficient volume that would always provide positive pressure at the alignment. This is the classical solution. Another solution would be the installation of the vessel with air relief valves at point D of the alignment. Vessels installed at the pipeline alignment, either the regular ones or the ones with air relief valves, usually require a civil engineering structure. Thus, these solutions are avoided regardless of their simplicity and efficiency. And, finally, special air valves can be used instead of regular ones, as one solution to eliminate the under-pressure generated at the pump shutdown. These valves differ from the regular ones because after air suction the air is trapped in the return phase. These valves have, besides the big nozzle, a small one. In the return pressure phase, the large nozzle is closed, while the small one is left open for gradual air release. The trapped air becomes a water hammer absorber (similar to the vessel) and slows down the impact of water masses.

8.12   Something about protection of sewage pressure pipelines

Figure 8.56 shows a typical solution of a sewage pumping station that consists of:

- a wet well, of sufficient volume to balance discharge and receive water from the pressure pipeline;
- a supply pipeline;
- a protective weir that is activated in case of a long-term pumping station failure;

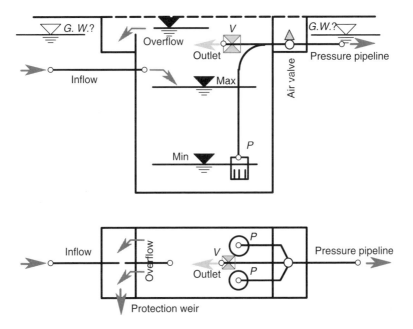

Figure 8.56   Typical sewage pumping station scheme.

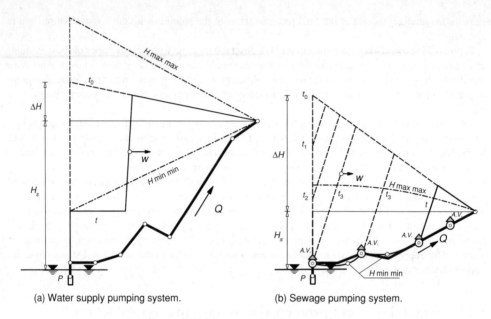

(a) Water supply pumping system.   (b) Sewage pumping system.

Figure 8.57   Pressure pipeline comparison.

- submersible pumping units;
- a pressure pipeline with the an air relief valve;
- in the case of groundwater presence, the structure shall be checked for uplift.

If water supply and sewerage pressure pipelines are compared, see Figure 8.57, the following can be observed:

- the static head of the sewerage pumping station is, in general, relatively small in comparison with the water supply one;
- the working pressure line of the sewage pressure pipeline has a lot larger gradient because of the required greater velocities (to prevent settling and washing);
- the water hammer that occurs in case of power outage causes:
  - greater pressure rise in the water supply pressure pipelines; thus, the pipeline shall be protected against pressure rise,
  - greater pressure fall in the sewage system pipeline; thus, the pipeline shall be protected against under-pressure.

The most common protection method against under-pressure is installation of special air valves, which, when under-pressure occurs, rapidly suck in large quantities of air that is held for a longer time in the pressure rise phase and absorbs the positive water hammer phase. Sewage pressure pipelines, which should be protected against under-pressure, should be made of pipe material resistant to under-pressure. Pipe material with joints sealed on the pre-pressure principle,[3] such as asbestos, cement, cast iron, ductile, and similar pipes are not recommended.

---

[3] If these pipe materials are insisted upon, the data on the under-pressure that the joints can be subjected to in repeated negative loading should be obtained from the manufacturer for the purpose of hydraulic calculations of unsteady (transition) states.

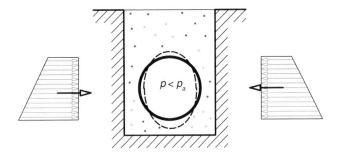

Figure 8.58 Pipeline with under-pressure.

It was proven that polyethylene pipes, which are jointed by welding, are a good choice for sewage pressure pipelines. The under-pressure that occurs in pressure pipelines can endanger the buckling stability of an underground thin-walled pipeline, just like a shell subjected to external pressure (atmospheric, water pressure, and earth pressure), see Figure 8.58.

8.13 Pumping units in a pressurized system with no tank

8.13.1 Introduction

The tank plays an important role in a pressure water supply system. Not only does it balance the daily consumption, but it also grants a stable pumping unit working point. If the tank has a fixed outflow elevation and there is no variable consumption connected to the pressure pipeline, the pump discharge will be constant and equal to the mean daily discharge. Then the required power and pumping unit power consumption will be smallest. If there are variable consumptions connected to the pressure pipeline, the pump duty point will vary, thus causing the pump discharge to oscillate around the main daily one for defined pumping requirements. Thus, a tank is always recommended when the spatial disposition of the pressure pipeline allows it.

When the spatial disposition of the pressure pipeline does not allow a tank, pumping units and other regulation devices have to be selected to secure variable consumption in regular operating conditions.

In this case, pumping units can be regulated by:

(a) pressure switches, applicable for low variable consumption;
(b) pressure switches with the vessel – namely hydrophor regulation – applicable for medium variable consumption;
(c) regulation of pumping unit rotational speed, applicable for highly variable consumption.

8.13.2 Pumping unit regulation by pressure switches

This is the simplest regulation that follows the outline *algorithm*:

• when the pressure at the pump pressure end falls to the lower regulation limit, the pumping unit is switched on;
• when the pressure at the pump pressure end exceeds the upper regulation limit, the pumping unit is switched off.

This regulation of pumping unit operation is applicable to short pressure pipelines. In case of a long pipeline, the transient hydraulic states occurring during the pumping unit switching on and off can generate significant over-pressures. Without detailed analysis of water hammer protection, consequences such as the breaking of a pipeline, reflux valve, or other fittings can be expected.

The ability of a pressure system to sustain variation of consumption can be presented by *elastic accumulation* of a pipeline. Namely, let us assume a pipeline of length L_c, cross-section area A_c, filled by discharge Q_0 and emptied by discharge Q_i, then the following continuity equation can be applied

$$\frac{dM}{dt} = \rho(Q_0 - Q_i), \tag{8.153}$$

where the water mass in the pipeline of volume V_c is equal to

$$M = \rho V_c = \rho A_c L_c.$$

After introduction into the previous equation, the following is obtained

$$L_c \frac{d\rho A_c}{dt} = \rho(Q_0 - Q_i). \tag{8.154}$$

If partial differentiation is applied to the left hand side numerator, then

$$d(\rho A_c) = A_c d\rho + \rho d A_c = \rho A_c \left(\frac{d\rho}{\rho} + \frac{d A_c}{A_c} \right).$$

Using the basic terms of liquid compressibility and the basic terms of the strength of materials, relative changes in the previous expression can be expressed by the elastic properties of water and pipeline, thus obtaining

$$d(\rho A_c) = \rho A_c \left(\frac{1}{E_v} + \frac{D}{s E_c} \right) dp.$$

Since the water hammer celerity is

$$c = \frac{1}{\sqrt{\rho \left(\dfrac{1}{E_v} + \dfrac{D}{s E_c} \right)}}$$

the left hand side in the initial equation (8.154) is written as

$$L_c \frac{d\rho A_c}{dt} = L_c \frac{A_c}{c^2} \frac{dp}{dt} = \frac{V_c}{c^2} \frac{dp}{dt},$$

thus, the equation of the elastic accumulation obtains the following form

$$\frac{V_c}{\rho c^2} \frac{dp}{dt} = Q_0 - Q_i, \tag{8.155}$$

where the pressure is expressed in *Pa*. If the pressure is expressed as the height of a water column $p = \rho g h$ then

$$\frac{g V_c}{c^2} \frac{dh}{dt} = Q_0 - Q_i. \tag{8.156}$$

Let the closed pipeline be filled to the pressure p_k. If the pipeline is not sealed, then a small discharge Q_i will outflow from the pipeline, depending on the pressure. The equation that explains exhaust from the pipeline is

$$\frac{V_c}{\rho c^2}\frac{dp}{dt} + Q_i = 0. \tag{8.157}$$

The discharge of exhaust will be thought of as discharge though the small hole of the cross-section A_i:

$$Q_i = \mu A_i\sqrt{\frac{2}{\rho}p},$$

where μ is the outflow coefficient, equal to the coefficient of contraction of the outflow jet through the small hole. After introduction into Eq. (8.157), the following is obtained:

$$\frac{V_c}{\rho c^2}\frac{dp}{dt} + \mu A_i\sqrt{\frac{2}{\rho}}\cdot p^{\frac{1}{2}} = 0.$$

The obtained equation will be integrated by the separation procedure as follows

$$\frac{V_c}{\rho c^2}p^{-\frac{1}{2}}dp + \mu A_i\sqrt{\frac{2}{\rho}}\cdot dt = 0.$$

When the upper and lower integration boundaries are applied, it is written

$$2\frac{V_c}{\rho c^2}(\sqrt{p} - \sqrt{p_k}) + \mu A_i\sqrt{\frac{2}{\rho}}\cdot t = 0,$$

from which the time period in which the pressure falls from the initial one p_k to the observed one p is calculated as

$$t = \frac{2\dfrac{V_c}{\rho c^2}}{\mu A_i\sqrt{\dfrac{2}{\rho}}}(\sqrt{p_k} - \sqrt{p}) = \sqrt{\frac{2}{\rho}}\frac{V_c}{\mu A_i c^2}(\sqrt{p_k} - \sqrt{p}).$$

The time in which the pressure will fall to atmospheric is

$$T = \frac{V_c}{\mu A_i c^2}\sqrt{\frac{2}{\rho}p_k} \tag{8.158}$$

or

$$T = \frac{Q_{i0}V_c}{\mu^2 A_i^2 c^2}, \tag{8.159}$$

where:

$$Q_{i0} = \mu A_i\sqrt{\frac{2}{\rho}p_k}. \tag{8.160}$$

Let us estimate the rate of pipeline exhaust in an example of the pressure test of a pipeline of 200 mm in diameter, 1000 m in length (volume 31.42 m^3), and water hammer celerity c $=$ 1000 m/s, with the pressure raised to 7 bars. Assume the hole size of (a) 0.01% of the pipe cross-section area and (b) 0.1% of the pipe cross-section area, which corresponds to holes of 2 and 6.3 mm in diameter. According to the calculations in (a) the pressure falls to atmospheric in 494 seconds while in (b) it is 49.4 seconds. This example illustrates the pressure pipeline exhaust, which makes this regulation procedure unacceptable for large variation in consumption, such as for water supply purposes. Even a small leak in the water supply system will cause the pump units to switch on and off frequently.

The observed equations can be used for assessment of the loss magnitude, that is to estimate the equivalent holes through which the water supply system is losing water.

Hydrostations, factory pre-adjusted devices consisting of one or several pumping units to build up the pressure, are operated on the principles of this regulation. To reduce pressure variations, small throttles in the form of spherical vessels with a membrane are applied, not as a protection against water hammers, but to ensure the stability of pressure switch operation.

8.13.3　Hydrophor regulation

The same outline *algorithm* can be applied to hydrophor regulation:

- when the pressure at the pump pressure end falls to the lower regulation limit, the pumping unit is switched on;
- when the pressure at the pump pressure end exceeds the upper regulation limit, the pumping unit is switched off.

A starting point for the *hydrophor equation* is the law of conservation of the water mass in the pipe and the vessel

$$\frac{d(M_c + M_k)}{dt} = \rho(Q_0 - Q_i), \tag{8.161}$$

where M_c is the mass of water in the pipe and M_k is the mass of water in the vessel. If the left hand side of the equation is divided into two terms, then it is written as

$$\frac{dM_c}{dt} + \frac{dM_k}{dt} = \rho(Q_0 - Q_i)$$

or in the form

$$L_c \frac{d\rho A_c}{dt} + \rho A_k \frac{dz}{dt} = \rho(Q_0 - Q_i),$$

where the change of water mass in the vessel is defined by the velocity of the water table displacement and the area of the horizontal cross-section of the vessel A_k. After division of the left and right hand sides of the equation by water density, and application of the previously given analyses of the first term in the equation of elastic accumulation of a pipe, it is written as

$$\frac{V_c}{\rho c^2} \frac{dp}{dt} + A_k \frac{dz}{dt} = Q_0 - Q_i. \tag{8.162}$$

The air mass in the vessel is defined by the initial filling, and remains constant during the oscillating process. Thus, the equation of state in the form of a polytrope can be applied to this closed thermodynamic system

$$p_{abs} V_z^n = const,$$

(8.163)

where p_{abs} is the absolute air pressure, V_z is the air volume, n is the exponent of a polytrope, which, determined experimentally on constructed vessels, is equal to $n = 1.25$. A constant in the equation of state is determined from the initial conditions

$$(p + p_0) \cdot V_z^n = (p_k + p_0) \cdot V_{z0}^n,$$

where p_k is the initial state of the gas and V_{z0} is the initial volume of the air in the vessel. Air volume at some prescribed pressure will be

$$V_z = V_{z0} \left(\frac{p_k + p_0}{p + p_0} \right)^{1/n}.$$

(8.164)

Also:

$$V_z = V_{z0} - A_k z,$$

where z is the water oscillation measured from the initial state while A_k is the cross-section of the vessel. The derivative of the above given expression in time gives

$$A_k \frac{dz}{dt} = -\frac{dV_z}{dt} = -\frac{dV_z}{dp} \frac{dp}{dt}.$$

The derivative of the air volume by pressure p will be

$$\frac{dV_z}{dp} = -V_{z0} \left(\frac{p_k + p_0}{p + p_0} \right)^{1/n} \frac{1}{n(p + p_0)};$$

namely

$$\frac{dV_z}{dp} = -\frac{V_{z0}}{n} (p_k + p_0)^{1/n} (p + p_0)^{-\frac{1+n}{n}}$$

thus

$$A_k \frac{dz}{dt} = \frac{V_{z0}}{n} (p_k + p_0)^{1/n} (p + p_0)^{-\frac{1+n}{n}} \frac{dp}{dt}.$$

After introduction into the equation of continuity (8.162) it is written

$$\frac{V_c}{\rho c^2} \frac{dp}{dt} + \rho \frac{V_{z0}}{n} (p_k + p_0)^{1/n} (p + p_0)^{-\frac{1+n}{n}} \frac{dp}{dt} = Q_0 - Q_i.$$

After arranging, a differential equation of pressure oscillations is obtained in the form

$$\left(\frac{V_c}{\rho c^2} + \frac{V_{z0}}{n} (p_k + p_0)^{1/n} (p + p_0)^{-\frac{1+n}{n}} \right) \frac{dp}{dt} = Q_0 - Q_i.$$

If marked

$$H(p) = \left(\frac{V_c}{\rho c^2} + \frac{V_{z0}}{n} (p_k + p_0)^{1/n} (p + p_0)^{-\frac{1+n}{n}} \right) \tag{8.165}$$

then the equation of oscillations has the following form

$$H(p) \frac{dp}{dt} = Q_0 - Q_i. \tag{8.166}$$

The obtained equation, called the *hydrophor equation*, enables computation of hydrophor vessel filling and emptying. The vessel, with its sufficient water and air volume, enables water compressibility and pipe expansion effects to be omitted.

8.13.4   Pumping unit regulation by variable rotational speed

The pumping unit (electric motor drive) is equipped with the rotational speed regulator; namely, the frequency regulator and pressure gauge in the pressure part of a pump. In general, the following outline algorithm applies to this regulation:

- when the pressure at the pump pressure end falls under the prescribed value (consumption is increasing), increase the number of revolutions of the pump;
- when the pressure at the pump pressure end exceeds the prescribed value (consumption has decreased), reduce the number of revolutions.

The velocity of the change in the number of revolutions has to be slow enough to absorb pressure waves in a pressurized system (water hammer).

A pump should be selected to obtain the maximum discharge with the full speed. Since, in the pressurized system, discharge ranges from zero to the maximum, which is defined by the diagram of daily consumption variation, so the number of revolutions also starts from zero to the full number of revolutions.

Since pump manufacturers prescribe the minimum allowable working discharge Q_{min} and the minimum rotational speed n_{min}, some limitations have to be applied to the previously described algorithm. Pumps have to be switched off when the rotational speed or discharge falls below the minimum value. Thus, the extended outline algorithm will be:

- when the pressure falls below the prescribed working pressure, increase the number of revolutions;
- when the pressure increases above the prescribed working pressure, decrease the number of revolutions:
  - if the number of revolutions or discharge are below the minimum value, switch off the pump.

During the night, water consumption is small. Also, if it is taken into account that there are always some water losses, pressurized systems are emptied faster after the pumps are switched off and the pressure falls rapidly. The air is sucked into the pressure system through the air valves and the water supply is interrupted.

A pressurized system should not be emptied during normal operation of the pumping station. Pumping units must be started again before the air is sucked in and the pressure should be raised to the normal level.

Since the pumping unit regulation algorithm is again testing the minimum rotational speed and minimum discharge, the units are switched off again. For small discharges, the pumping unit's switch on/switch off cycle is repeated until the system's consumption exceeds limitations.

The frequency of switch on/switch off cycles depends on the ability of elastic storage of water in the pressure system as described by the following equation (8.157)

$$\frac{V_c}{\rho c^2}\frac{dp}{dt} + Q_i = 0.$$

At the time of switching off, the pressure system – due to compressed water and expanded pipeline – still has some water storage volume at the pressure p_{radno}. System consumption can still go on, on behalf of the pressure reduction. The pressure reduction level depends on the spatial configuration of the pressure system. Thus, the minimum allowable pressure p_{min} for a regular water supply shall be prescribed. The time of constant minimum discharge withdrawal Q_{min} to reduce the pressure by pressure difference $\Delta p = p_{radno} - p_{min}$ shall be defined. The time can be found as the solution of the equation

$$T = \frac{V_c}{\rho Q_{min} c^2}\Delta p$$

or expressed as water column height

$$T = \frac{g V_c}{Q_{min} c^2}\Delta\left(\frac{p}{\rho g}\right).$$

What is critical is the minimum time of discharge withdrawal obtained for consumption that corresponds to the minimum allowable working discharge of the pump.

For example, imagine a pressurized system with a pipeline volume $V_c = 314.15$ m³ and elastic properties of the system (steel pipeline) to allow the water hammer celerity $c = 1000$ m/s. Then, the discharge of 1 l/s can be drained in a time of about $T = 62$ seconds to reduce the pressure by a 20 m water column. If the pipeline is made of softer material, for example polyethylene, $c = 300$ m/s possible draining time will be about $T = 685$ seconds.

The pressure shall be raised from p_{min} to p_{radno} by a rise in the number of revolutions from zero to the nominal one in the shortest possible time to allow the fastest possible pump transition through working limitations. It means that the pressure rise $\Delta p = p_{radno} - p_{min}$ occurs at a discharge that is equal to the maximum consumption discharge, that is nominal pump discharge Q_0. The time required for the pressure rise will be

$$T = \frac{g V_c}{Q_0 c^2}\Delta\left(\frac{p}{\rho g}\right).$$

Thus, the following is applied

$$\frac{T_{ga\check{s}enje}}{T_{paljenje}} = \frac{Q_0}{Q_{min}}.$$

The results obtained in the previous example are $T_{paljenje} = 4.7$ seconds for a steel pipeline and approximately 53 seconds for a polyethylene pipeline at ratio $Q_0/Q_{min} = 13$.

Note that the time between the pumping units switching off and on can be controlled by water accumulation in the system.

Frequent switching on and off of pumping units is not recommended; thus *a vessel of sufficient water and air volume should be installed within the pressurized system.* The volume of a vessel is determined based on the respective non-steady numerical modelling of pressure system operation.

Water consumption conditions within the pressure system are constantly altered, in compliance with water supply system development; thus, there is a possibility of a discharge greater than the maximum one that a pumping unit is sized to. Discharges occurring at a pipe break also count. Then, the pumping unit cannot provide the prescribed discharge and normal working pressure at a nominal rotational speed. The pressure starts to fall, the working point starts to move in the direction of the power increase (there is a danger of electrical motor failure), and the algorithm must be suspended and the causes tested. If the cause of increased discharge is a regular increase in consumption, pumping units should be replaced!

Algorithm. The final outline algorithm for regulation of pumping units with variable rotational speed is:

- pumping unit in operation (regular procedure):
  - when the pressure falls below the prescriber working pressure p_{radno}, increase the number of revolutions n:
    - if the number of revolutions is equal to the maximum and the working pressure is still not reached, suspend the algorithm and switch off the pumping unit,
  - when the pressure rises above the prescribed p_{radno}, reduce the number of revolutions n:
    - if the number of revolutions n is below the minimum one n_{min} or discharge is below the minimum Q_{min}, switch off the pump. Start the accident procedure
- otherwise (accident procedure):
  - if the pressure falls below the minimum p_{min} switch on the pumping unit with the full number of revolutions n_0 until the full working pressure p_{radno} is achieved. Then, start the regular procedure.

Reference

Donsky, B. (1962) Complete pump characteristics and the effects of specific speeds on transients. *Transactions of the ASME*, 685–689.

Fox, J.A. (1977) *Hydraulic Analysis of Unsteady Flow in Pipe Networks*. MacMillan Press Ltd., London.

Further reading

Holzenberger, K., Jung, K. (ed.) (1990) *Centrifugal Pump Lexicon*, 3th Edn. KSB Aktiengesellschaft, Frankenthal.

Jović, V. (1977) Non-steady flow in pipes and channels by finite element method. *Proceedings of XVII Congress of the IAHR*, 2, 197–204.

Jović, V. (1987) Modelling of non-steady flow in pipe networks. *Proceedings of the Int. Conference on Numerical Methods NUMETA 87*. Martinus Nijhoff Publishers, Swansea.

Jović, V. (1992) Modelling of hydraulic vibrations in network systems. *International Journal for Engineering Modelling*, 5: 11–17.

Jović, V. (1994) Contribution to the finite element method based on the method of characteristics in modelling hydraulic networks, *Zbornik radova 1. kongresa hrvatskog društva za mehaniku*, 1: 389–398.

Jović, V. (1995) Finite elements and method of characteristics applied to water hammer modelling. *International Journal for Engineering Modelling*, 8: 51–58.

Jović, V. (2006) *Fundamentals of Hydromechanics* (in Croatian: Osnove hidromehanike). Element, Zagreb.

Streeter, V.L. and Wylie, E.B. (1967) *Hydraulic Transients*. McGraw-Hill Book Co., New York, London, Sydney.

Walshaw, A.C. and Jobson, D.A. (1962) *Mechanics of Fluids*. Longmans, London.

Watters, G.Z. (1984) *Analysis and Control of Unsteady Flow in Pipe Networks*. Butterworths, Boston.

Sulzer, Bro. (1986) *Centrifugal Pump Handbook*, Winterthur, Switzerland Pump Division.

9

Open Channel Flow

9.1 Introduction

The problem of flow modelling in open channels is extremely complex due to the fact that there are two types of flow: *subcritical* and *supercritical*, as well as transitions from one flow regime to another. Thus, the author has limited himself to flow modelling in channel stretches with a predefined flow regime, in order to answer as simply as possible the majority of engineering tasks.

9.2 Steady flow in a mildly sloping channel

In general, open channel one dimensional flow does not differ much from pipe flow. Figure 9.1 shows energy relations in open channel flow. The flow is observed along the axis l that connects the centroids T of cross-sections perpendicular to the flow axis. Let us assume a flow with a developed boundary layer, namely a developed velocity profile to consider the flow to be one dimensional with the mean velocity $\bar{v} = Q/A$.

Except in exceptional circumstances, channels are mildly[1] sloping at small angles $\beta \approx 0$, thus $\cos \beta \cong 1$. Longitudinal variable l, set along the flow axis and connecting the cross-section centroids, can be replaced by the horizontal distance x (profile chainage or stations) while the piezometric head coincides with the water level h.

Cross-sections can be considered to be vertical; thus, the piezometric head is equal to the water level

$$h = z_l + \frac{p_T}{\rho g} = z_T + y_T - z_0 + y.$$
(9.1)

Specific energy in a mildly sloping channel is equal to

$$H = z_0 + y + \alpha \frac{\bar{v}^2}{2g} = h + \alpha \frac{\bar{v}^2}{2g}.$$
(9.2)

---

[1] The centerline of the flow slope is expressed by a tangent of the angle β in percent or permille, for example centerline slope $I_0 = 0.1\% = 1\‰$ means a slope of 1 m per kilometer. Angle $\beta = 5°$ gives $\cos \beta = 0.996$, $tg\beta = 0.0875$ and centerline slope of $87.5\‰$, which can still be considered mild. Steep slope channels with centerline slopes of flow are called *spillways*.

Analysis and Modelling of Non-Steady Flow in Pipe and Channel Networks, First Edition. Vinko Jović.
© 2013 John Wiley & Sons, Ltd. Published 2013 by John Wiley & Sons, Ltd.

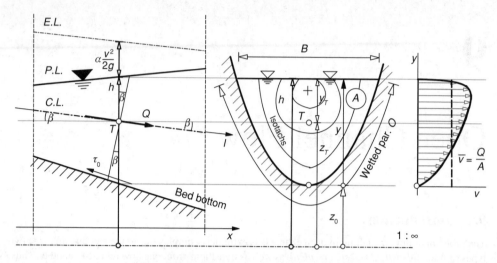

Figure 9.1   Flow in a mildly sloping channel.

Velocities are almost horizontal, and pressure distribution in the vertical direction is hydrostatic. The position of the maximum velocity depends on the shape of the cross-section, that is the shape of the isotachs (lines connecting points with equal velocities); thus, for a simple cross-section as shown in the figure, it is located somewhat below the water level, while in relatively broad channel it is positioned on the surface.

As a flow with resistances is observed, specific energy decreases along the channel length

$$\frac{dH}{dl} + \frac{\tau_0}{\rho g R} = \frac{dH}{dl} + J_e = 0, \tag{9.3}$$

where J_e is the gradient along the flow axis l. One shall differentiate between slope and gradient. The energy line gradient $J_e = -dH/dl$ is a measure of the energy line decrease along the flow axis, which, in steep channels is not equal to the slope: $I_e = -dH/dx$. Slope I_e is a tangent of an angle β enclosed by the flow axis and the horizontal plane. The relationship between the gradient and slope is derived from the differential ratio $dx = dl \cos \beta$, thus

$$\frac{d...}{dl} = \frac{d...}{dx}\frac{dx}{dl} = \cos \beta \frac{d...}{dx} \tag{9.4}$$

which gives $J_e = I_e \cos \beta$. If $\cos \beta \approx 1$, then $J_e = I_e$, and the energy equation for mildly sloping channels is written as

$$\frac{dH}{dx} + I_e = 0, \tag{9.5}$$

where the energy line slope is defined by mean friction τ_0 along the wetted channel section

$$I_e = \frac{\tau_0}{\rho g R} \tag{9.6}$$

or

$$\frac{dH}{dx} + \frac{\tau_0}{\rho g R} = 0. \tag{9.7}$$

Since friction can be expressed as a coefficient of friction in the developed boundary layer

$$\tau_0 = c_f \frac{1}{2} \rho \bar{v}^2, \tag{9.8}$$

where $\bar{v}$ is the mean velocity of the developed velocity profile and c_f is the mean friction coefficient along the wetted surface for a developed boundary layer, the energy line slope is written as

$$I_e = \frac{c_f}{R} \frac{\bar{v}^2}{2g}. \tag{9.9}$$

$$\bar{v} = Q/A$$

Mean velocity, obtained from the previous expression, can be written in the following form

$$\bar{v} = \frac{\sqrt{2g}}{\sqrt{c_f}} \sqrt{R I_e}. \tag{9.10}$$

9.3   Uniform flow in a mildly sloping channel

9.3.1   Uniform flow velocity in open channel

In gradually varied flow, channel parameters such as the cross-section shape and bottom slope are continuous.

The prismatic channel is a type of channel which has the same shape cross-section throughout its length. If a cross-section does not change along the flow axis, uniform velocity profile is expected. Because of this, not only does the mean velocity remain constant, but the Coriolis coefficient too. The water depth is constant, thus the water level slope is equal to the bottom slope and – due to the same velocity head in all cross-sections – also equal to the slope of the energy line

$$I = I_0 = I_e, \tag{9.11}$$

where $I = -dh/dx$ is the water level slope, $I_0 = -dz_0/dx$ is the bottom slope, and $I = -dH/dx$ is the energy line slope. Flow of the aforementioned properties in prismatic channels is called the *uniform flow*. In uniform flow, the mean profile velocity, expression (9.10) can be written in the form

$$\bar{v} = \frac{\sqrt{2g}}{\sqrt{c_f}} \sqrt{R I_0}. \tag{9.12}$$

In the following text the mean velocity will be written without the mathematical sign for average above the letter v. If

$$C = \frac{\sqrt{2g}}{\sqrt{c_f}}, \tag{9.13}$$

the *Chezý*[2] formula for the flow velocity in open channels will be obtained

$$v = C\sqrt{RI_0}. \tag{9.14}$$

The coefficient C is called the *Chezý coefficient*. In 1769 Chezý devised the formula when he was collecting experimental data from earth channels such as the canal of Courpalet and from the river Seine. He assumed that constant C can be assigned to each riverbed. Subsequently, more accurate measurements showed that C is not a constant of a cross-section. It was followed by a "flood" of Chezý coefficient formulas.[3]

Owing to boundary layer research, note that the Chezý coefficient is defined by the roughness coefficient c_f in the boundary layer; thus, similarly as for the flow in pipes, the following flow regimes can be distinguished:

- laminar flow in open channels;
- transient flow in open channels;
- turbulent flow in channels that can be:
  - turbulent rough flow,
  - turbulent smooth flow,
  - and turbulent transient flow.

Since development of the boundary layer depends not only on the Reynolds number and relative roughness, but also on the cross-section shape, it becomes clear that some simple law that could be generally applied to all channel shapes cannot be found.

Out of all formulas devised through history for velocity in an open channel, the most commonly applied formula is the *Manning*[4] formula (1889)

$$v = \frac{1}{n}R^{\frac{2}{3}}I_e^{\frac{1}{2}}, \tag{9.15}$$

where n is the channel roughness coefficient, R is the hydraulic radius, and I_e is the energy line slope. In the Manning formula, the hydraulic radius is given in meters; thus, the velocity will be calculated in m/s. The value of the Chezý coefficient, which corresponds to the Manning formula, is

$$C = \frac{1}{n}R^{\frac{1}{6}}. \tag{9.16}$$

The Manning formula is an approximation of the turbulent rough flow, which is accurate enough to have been used in calculations of flow in pipes for a long time; in particular in large tunnel pipelines. Simplicity and, above all, abundance of the data on natural and artificial channel roughness, as well as the accuracy that is sufficient for engineering calculations, makes the Manning formula a base for velocity calculations in channels of different shapes. Values of the Manning's roughness coefficient for different channel types are given in Chow (1959). Reciprocal values of the roughness coefficient $K = 1/n$ are also often used, which are a measure of the channel smoothness, also known as the Strickler coefficient (reciprocals are easier to remember!).

For *simple channel cross-sections*, instead of the roughness coefficient c_f in the Chezý coefficient, an equivalent pipe flow resistance coefficient $\lambda = 4c_f$ can be used, which is defined for the equivalent pipe

---

[2] Antoine Chezý (1718–1798), French engineer.
[3] Ganguillet and Kutter 1869, Strickler 1923 (Swiss engineers), Bazain 1897 (French engineer), Pavlovski 1925 (Russian engineer), and many others.
[4] Robert Manning, Irish engineer (1816–1897).

of diameter $D = 4R$ (equality of hydraulic radius) and relative roughness $k/4R$. Then, the formula for channel flow velocity is obtained in the form

$$v = \frac{\sqrt{8g}}{\sqrt{\lambda}} \sqrt{RI_e}, \tag{9.17}$$

where the resistance coefficient λ can be calculated from the Colebrook–White[5] equation

$$\frac{1}{\sqrt{\lambda}} = 1{,}14 - 2\log\left(\frac{k}{4R} + \frac{9{,}35}{R_e\sqrt{\lambda}}\right). \tag{9.18}$$

Although the Colebrook–White equation is not explicit for the resistance coefficient calculations, channel flow velocity can be calculated explicitly. Namely, if expressed from Eq. (9.17)

$$\frac{1}{\sqrt{\lambda}} = \frac{v}{\sqrt{8gRI_e}} \tag{9.19}$$

and substituted in the right hand side of expression (9.18), together with the Reynolds number $R_e = v4R/v$, where v is the coefficient of kinematic viscosity, the following is obtained after arranging

$$\frac{1}{\sqrt{\lambda}} = 1{,}14 - 2\log\left(\frac{k}{4R} + \frac{9{,}35v}{4R\sqrt{8gRI_e}}\right). \tag{9.20}$$

When Eq. (9.20) is introduced into velocity equation (9.17) the following is obtained

$$v = \left[1{,}14 - 2\log\left(\frac{k}{4R} + \frac{9{,}35v}{4R\sqrt{8gRI_e}}\right)\right] \cdot \sqrt{8gRI_e} \tag{9.21}$$

as well as the respective Chezý coefficient

$$C = \left[1{,}14 - 2\log\left(\frac{k}{4R} + \frac{9{,}35v}{4R\sqrt{8gRI_e}}\right)\right] \cdot \sqrt{8g}. \tag{9.22}$$

Note that the Chezý coefficient according to the Colebrook–White formula depends on the energy line slope (i.e. the Reynolds number). This formula also enables flow calculations in turbulent transient flow with an accuracy that corresponds to the accuracy of the applied simplification. The equality of slopes $I_e = I_0$ applies to the uniform flow; thus, the unit slope can be used in expression (9.22) for ordinary channel flow velocities and bottom slopes, with no large-scale errors.

The equivalent Manning's roughness coefficient n can be calculated from expressions (9.22) and (9.16) for the prescribed absolute hydraulic roughness k. However, it is not constant for constant k. Figure 9.2 shows a variation of the Manning's roughness coefficient as a function of circular cross-section fullness, where n_0 refers to the full filled pipe, for relative hydraulic roughness $k/D = 0.022$ and unit slope.

Note the almost constant n/n_0 value in the upper half of the pipe. In the bottom half of the pipe values are increasing and approaching infinity when a pipe is completely empty.

---

[5]Colebrook, C. F. and White, C. M. (1937). Experiments with fluid friction in roughened pipes. *Proceedings of the Royal Society of London. Series A, Mathematical and Physical Sciences* **161** (906): 367–381.

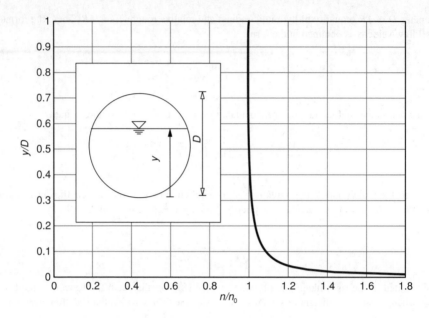

Figure 9.2   Variation of the Manning coefficient n/n_0.

9.3.2   Conveyance, discharge curve

Discharge curve of a simple cross-section

The calculation of uniform flow discharge for a prescribed channel cross-section, roughness, and bottom slope I_0 is explicit at prescribed depth y.

For the prescribed depth y, based on cross-section geometry, according to Figure 9.3a, the cross-section area A and wetted perimeter O are calculated, followed by calculation of the hydraulic radius. Based on the prescribed channel roughness n, the velocity is calculated according to the Manning formula as

$$v = \frac{1}{n} R^{\frac{2}{3}} \sqrt{I_0}. \qquad (9.23)$$

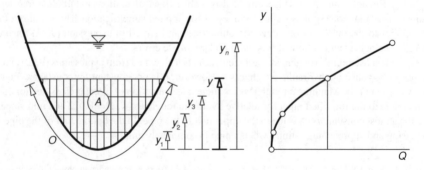

Figure 9.3   Discharge curve.

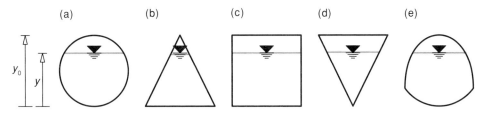

Figure 9.4   Closed channels.

Finally, discharge is calculated as $Q = Av$. However, an inverse task to calculate depth from the prescribed discharge is not explicit. Thus, the successive approach method (trial-and-error method, iterative solution) shall be applied or the discharge curve calculated, which gives an unambiguous correlation between the discharge and depth; see Figure 9.3b.

The depth, which corresponds to the prescribed discharge, is called the *normal depth*. The discharge curve of a channel can be expressed by the *conveyance* function K in the form

$$Q(y) = K(y)\sqrt{I_e}. \tag{9.24}$$

Channel conveyance $K\,[\mathrm{m^3/s}]$ is the discharge at unit slope of the energy line and a function of channel resistances and geometric properties. For cross-sections that do not narrow with depth it is a monotonically increasing function of depth, similar to the discharge curve.

Discharge curve of simple closed cross-sections

Figure 9.4 shows several simple closed channel cross-sections with a free surface or pressurized flow. For the given uniform flow slope I_0 and constant roughness (Manning's n, or absolute hydraulic roughness k) discharge curve $Q(y)$ has its maximum somewhere below the fully filled cross-section.

Figure 9.5 shows normalized discharge curves, namely conveyances of cross-sections (a), (b), (c), and (d) of Figure 9.4, calculated using the Manning formula. Almost the same results are obtained with the Colebrook–White formula.

The relative position of the maximum conveyance, framed in the figure for each cross-section, was derived by equating the next term's derivative

$$\frac{d}{dy} A \left(\frac{A}{O} \right)^{\frac{2}{3}} = \frac{d}{dy} \left(\frac{A^{\frac{5}{3}}}{O^{\frac{2}{3}}} \right) \tag{9.25}$$

with zero. Unlike sections (a) and (b) of Figure 9.4, discharge curves of sections (c) and (d) monotonically increase until a cross-section is completely filled. Discharge decrease occurs when the wetted perimeter increment becomes so large that the derivative (9.25) changes its sign, which is particularly notable for sections (c) and (d). For values $y > y_0$ the flow is pressurized; thus the discharge remains constant for the given slope $I_e = I_0$.

Discussion

A question is raised over the credibility of discharge curves for simple closed channels calculated using the formulas assuming uniformly distributed roughness along the wetted perimeter. Normally, higher discharge would always be expected for the same slope and greater depth. Figure 9.6 shows the imaginary

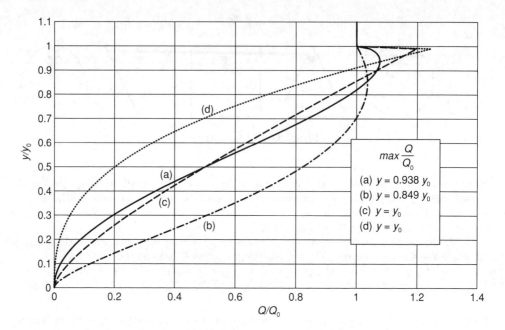

Figure 9.5 Discharge curves of closed channels.

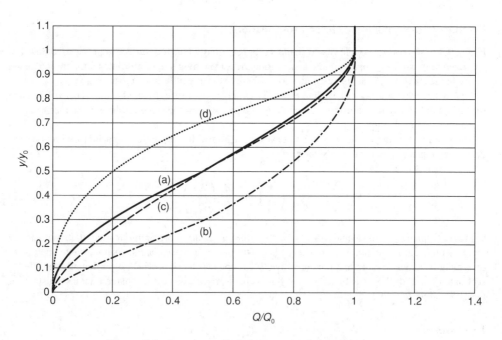

Figure 9.6 Imaginary discharge curves of closed channels.

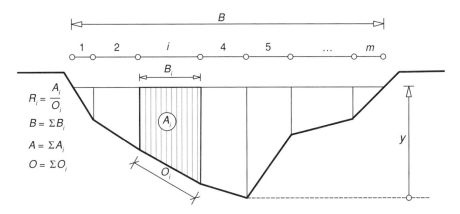

Figure 9.7   A compound cross-section.

monotonically increasing discharge curves of simple closed cross-sections (a), (b), (c), and (d) of Figure 9.4.

The shape of an imaginary discharge curve for cross-sections (a) and (b) is logical and can be explained by the unknown character of unexplored developed boundary layers in cross-sections that narrow with height. Namely, in his book (Chow, 1959); Ven Te Chow[6] mentioned circular sewers with non-uniform roughness distribution per depth, where discharge monotonically increased with depth. Most probably, roughness is constant and the developed boundary layer does not justify strict application of Manning's formula.

However, imaginary discharge curves for cross-sections (c) and (d) are obviously not possible, due to the fact that the cross-section either expands or remains constant until completely filled. When the cross-section is completely filled, the wetted perimeter changes suddenly. Because, up until that moment, these are simple sections for which the use of the Manning's formula is justified, an answer shall be sought in the character of the non-researched developed boundary layer.

Discharge curve of compound channels

The results of calculations on the compound channel discharge curve can be unacceptable if a cross section is observed as a simple cross-section, as explained in the textbooks, for example (Jovic, 2006). In compound channels, either artificial or natural, total discharge shall be calculated as a sum of discharge per segment of section; which, together with the prescribed constant energy line slope (or the bottom slope in uniform flow), means that total conveyance will be calculated as a sum of conveyances in all segments.

The procedure will be demonstrated on a general compound cross-section shown in Figure 9.7. The cross-section is divided into m simple segments per width. The roughness of each i-th segment is n_i.

Velocity, discharge, and conveyance in the i-th segment are calculated as

$$v_i = \frac{1}{n_i} R_i^{2/3} I_e^{1/2}, \quad Q_i = K_i I_e^{1/2}, \quad K_i = \frac{1}{n_i} A_i R_i^{2/3}. \tag{9.26}$$

---

[6]Ven Te Chow, Hangchow (1919–1981), Chinese and American engineer and professor.

The total discharge will be

$$Q = \sum_{i=1}^{m} Q_i = \sum_{i=1}^{m} K_i I_e^{1/2} = K I_e^{1/2},$$

(9.27)

where the channel conveyance is equal to the sum of conveyances in all segments

$$K = \sum_{i=1}^{m} K_i.$$

(9.28)

Similarly, the following applies to each i-th segment

$$Q_i = \frac{K_i}{K} Q.$$

(9.29)

9.3.3   Specific energy in a cross-section: Froude number

The specific energy of uniform flow in a cross-section (with reference to the channel bottom) is defined as

$$H_s = y + \alpha \frac{v^2}{2g}.$$

(9.30)

The specific energy curve has its minimum at depth y_c. This depth is also referred to as the *critical depth*. When the normal depth is equal to the critical one the flow is a *critical flow*.

When the actual depth is greater than the critical one, the flow is *subcritical*. When the actual depth is lower than the critical one, the flow is *supercritical*.

Specific energy of a simple cross-section

An analytical criterion for the minimum specific energy is

$$\frac{dH_s}{dy} = \frac{d}{dy}\left(y + \alpha \frac{Q^2}{2gA^2}\right) = 0$$

(9.31)

from which it is written

$$\frac{dH_s}{dy} = 1 - \alpha \frac{Q^2}{gA^3}\frac{dA}{dy} = 0.$$

(9.32)

A derivative of the cross-section area per depth is equal to the water level width, see Figure 9.8; thus, it is written

$$\alpha \frac{Q^2}{gA^3} B = 1.$$

(9.33)

The obtained expression is a critical flow criterion known as the Froude[7] number

$$F_r = \alpha \frac{Q^2}{gA^3} B$$

(9.34)

---

[7] William Froude (1810–1879), British engineer.

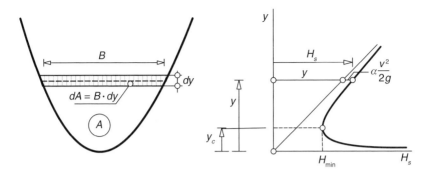

Figure 9.8 Specific energy at a channel cross-section.

and can be written in the following form

$$\frac{dH_s}{dy} = 1 - F_r.$$ (9.35)

The Coriolis coefficient value of 1 can be adopted for simple channel cross-sections. Based on the aforementioned, the Froude number defines the flow regime:

> $F_r < 1$ - *subcritical flow, relatively small velocities,*
>
> $F_r = 1$ - *critical flow,*
>
> $F_r > 1$ - *supercritical flow, relatively large velocities.*

Rectangular channel
The specific energy of a rectangular channel of width B is

$$H_s = y + \alpha \frac{Q^2}{2gB^2y^2}.$$ (9.36)

In critical flow, that is when $F_r = 1$, it can be written

$$\frac{v_c^2}{gy_c} = 1,$$

which gives the critical velocity

$$v_c = \sqrt{gy_c}$$ (9.37)

and the minimum specific energy

$$H_{min} = y_c + \frac{v_c^2}{2g} = y_c + \frac{y_c}{2} = \frac{3}{2}y_c = \frac{3}{2}\sqrt[3]{\frac{Q^2}{B^2g}}.$$ (9.38)

Critical flow is equal to

$$Q_c = B y_c v_c = B y_c \sqrt{g y_c} = B \sqrt{g} y_c^{\frac{3}{2}}. \tag{9.39}$$

Critical depth is equal to

$$y_c = \left(\frac{Q_c}{B\sqrt{g}}\right)^{2/3} = \frac{3}{2}\sqrt[3]{\frac{Q^2}{B^2 g}}. \tag{9.40}$$

For the given specific energy $H_s \geq H_{min}$, subcritical and supercritical flow depths can be calculated analytically as follows

$$y_m = \frac{H_s}{3}\left(1 + 2\sin\frac{\pi + \arcsin\varphi}{3}\right), \tag{9.41}$$

$$y_s = \frac{H_s}{3}\left(1 - 2\sin\frac{\arcsin\varphi}{3}\right), \tag{9.42}$$

where

$$\varphi = 1 - \frac{27}{4}\frac{Q^2}{B^2 g H_s^3}. \tag{9.43}$$

$$-1 \leq \varphi \leq +1$$

The solution does not exist if $H_s < H_{min}$, namely when $|\varphi| > 1$.

Specific energy and the Froude number of a compound channel

When calculating the specific energy of a compound channel consisting of several simple segments of cross-sections, the Coriolis coefficient α shall be calculated for the entire cross-section because of the variable distribution of kinetic energy, see Figure 9.9; thus

$$H_s = y + \alpha\frac{v^2}{2g}. \tag{9.44}$$

The Coriolis coefficient is calculated from the power of flow in a channel using the mean profile velocity Q/A.

The power of flow is defined by discharge and specific energy, where density and gravitational acceleration are constant

$$P = \rho g Q H_s. \tag{9.45}$$

Let us observe calculation of the product QH_s. Specific energy of the i-th segment is

$$H_{si} = y + \frac{v_i^2}{2g} \tag{9.46}$$

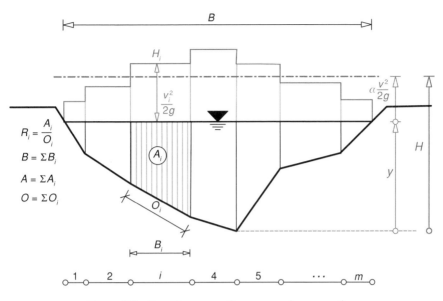

Figure 9.9   Specific energy of a compound cross-section.

thus it can be written

$$QH_s = \sum_{i=1}^{m} Q_i \left(y + \frac{Q_i^2}{2gA_i^2} \right) = yQ + \sum_{i=1}^{m} \frac{Q_i^3}{2gA_i^2}. \tag{9.47}$$

If Q_i is expressed by Eq. (9.29), it is written as

$$QH_s = yQ + \sum_{i-1}^{m} \frac{(K_i/K)^3}{2gA_i^2} Q^3. \tag{9.48}$$

When the right hand side term in the previous expression is multiplied by the squared cross-section area

$$QH_s = yQ + \sum_{i=1}^{m} \frac{(K_i/K)^3}{(A_i/A)^2} \frac{Q^2}{2gA^2} Q \tag{9.49}$$

the following is obtained

$$QH_s = \left(y + \alpha \frac{v^2}{2g} \right) Q, \tag{9.50}$$

where the Coriolis coefficient of a compound channel is equal to

$$\alpha = \sum_{i=1}^{m} \frac{(K_i/K)^3}{(A_i/A)^2} \tag{9.51}$$

and specific energy is

$$H_s = y + \alpha \frac{v^2}{2g}. \tag{9.52}$$

An analytical criterion for the flow regime in a compound channel is obtained from the minimum specific energy requirement

$$\frac{dH_s}{dy} = \frac{d}{dy}\left(y + \alpha \frac{Q^2}{2gA^2}\right) = 0. \tag{9.53}$$

When the Coriolis coefficient is introduced into the previous expression, it is written as

$$\frac{d}{dy}\left(y + \sum_{i=1}^{m} \frac{(K_i/K)^3}{(A_i/A)^2} \frac{Q^2}{2gA^2}\right) = 0. \tag{9.54}$$

After derivation by y (the term $A_i(y)$ first by variable A_i, then $dA_i/dy = B_i$) the following is obtained

$$\frac{Q^2}{g} \sum_{i=1}^{m} (K_i/K)^3 \frac{B_i}{A_i^3} = 1. \tag{9.55}$$

If the left hand side term is supplemented by A^3 and B, then

$$\frac{Q^2}{gA^3} B \sum_{i=1}^{m} \left(\frac{K_i}{K}\right)^3 \left(\frac{A}{A_i}\right)^3 \frac{B_i}{B} = 1. \tag{9.56}$$

When a correction factor is introduced

$$\gamma = \sum_{i=1}^{m} \left(\frac{K_i}{K}\right)^3 \left(\frac{A_i}{A}\right)^3 \frac{B_i}{B}, \tag{9.57}$$

the minimum specific energy requirement for a compound channel obtains the form

$$\gamma \frac{Q^2}{gA^3} B = 1. \tag{9.58}$$

Thus, the Froude number of a compound channel has the following form

$$F_r = \gamma \frac{Q^2}{gA^3} B. \tag{9.59}$$

Note: Application of the described procedure for flow regime definition in compound channels gives adequate solutions in most examples in practice; however, it does not ensure an acceptable solution in all cases, see (Jovic, 2006).

9.3.4 Uniform flow programming solution

Explicit tasks

A programming solution for uniform channel flow requires writing a series of simple procedures such as functional subroutines for calculation of geometric properties, namely cross-section area A, wetted perimeter O, hydraulic radius R, and channel width at water level B for the selected channel cross-section "presjek":

$$A=\text{A\_presjek}(y), \quad O=\text{O\_presjek}(y),$$
$$R=\text{R\_presjek}(y), \quad B=\text{B\_presjek}(y),$$

as well as a series of functional subroutines for the computation of hydraulic properties:computation of (normal) uniform flow discharge Q from a given depth, roughness, and bottom slope:

$$Q=\text{QN\_presjek}(y,n,I_0),$$

inverse function for normal depth computation

$$D_n=\text{DN\_presjek}(Q,n,I_0),$$

specific energy computation

$$H_s=\text{HS\_presjek}(Q,y),$$

critical discharge computation

$$Q_c=\text{QC\_presjek}(y),$$

critical depth computation

$$d_c=\text{DC\_presjek}(Q),$$

critical slope computation

$$I_c=\text{IC\_presjek}(Q),$$

computation of depth in subcritical flow

$$d_m=\text{DM\_presjek}(Q,H_s),$$

computation of depth in supercritical flow

$$d_s=\text{DS\_presjek}(Q,H_s).$$

Since all the aforementioned functional subroutines refer to the selected channel cross-section "presjek," it is advisable (not mandatory) to add the cross-section name to a subroutine name; for example a subroutine for computation of a "trapez" channel cross-section will be named *A_trapez* to distinct it from another for computation of a circular cross-section named *A_circle*.

The accompanying website – www.wiley.com/go/jovic, folder SimpipCore/SimpipCore project – contains sources for fortran modules Trapez, DBgraf, SDBgraf, and Circular. Module Trapez can be applied to the two most common channel cross-sections; namely, trapezoidal and rectangular cross-sections, DBgraf refers to compound, SDBgraf refers to compound closed, and Circular refers to circular cross-sections.

Implicit tasks

In the uniform flow analysis there are several inverse tasks, such as:

(a) calculation of a normal depth y_0 for a given discharge Q_o;
(b) calculation of critical depth y_c for a given discharge Q_o;
(c) calculation of a normal depth y_0 in a subcritical flow from a given specific energy H_s;
(d) calculation of a normal depth y_0 in a supercritical flow from a given specific energy H_s.

An iterative solution algorithm, which can be generally applied to all the aforementioned cases, will be shown for an example of normal depth calculation from the given discharge. The algorithm is based on the numerical computation of a non-linear equation null-point by the interval halving method.

Figure 9.10 shows a discharge curve $Q = Av$:

$$Q = \frac{1}{n} A(y) R^{2/3}(y) I_0^{1/2},$$
$$Q = Q(y)$$

(9.60)

which defines a normal depth y_0 for a given discharge Q_0; namely, in the inverse task

$$Q_0 = Q(y_0).$$

If a residual function $R(y)$ is formed as

$$R(y) = Q(y) - Q_0,$$

then a null-point of a function will be a solution of the inverse task (9.60).

Computation of a non-linear equation null-point by the interval halving method is feasible if two values y_1 and y_2 are selected in the vicinity of the null-point so the residual function will change its sign within the interval.

The next step will be to calculate the point y_p in the middle of the interval and the residual function value R_p. If the sign of R_p is equal to the sign of R_1 then the point y_1 is moved to the point y_p, otherwise the point from the other end of the interval is moved into it. By further halving of the interval, point y_p will approach the null-point y_0. The procedure shall be repeated until the prescribed accuracy is achieved, which can be written as $|R_p| < \varepsilon$ or $|y_2 - y_1| < \varepsilon$. Figure 9.11 shows a fortran source of the described algorithm for normal depth y_0 computation. Sources for other cases, listed under (b) to (d), can be easily written by simple alterations of the algorithm source.

9.4   Non-uniform gradually varied flow

9.4.1   Non-uniform flow characteristics

If some disturbance, for example a dam that raises water level above the normal depth for discharge Q, is inserted into the uniform flow, see Figure 9.12, the flow will become non-uniform because the depth y varies along the flow.

This is an example of a *backwater curve*. The developed disturbance gradually decreases upstream, thus the depth upstream of the disturbance asymptotically approaches the undisturbed state, that is normal depth.

A similar phenomenon occurs when water depth decreases because of construction of a channel bed with supercritical flow, for example; see Figure 9.13. This is an example of a *drawdown curve*.

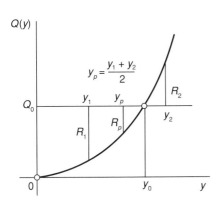

```
y1=0.
y2=Ymax
y = y1
res = Q(y) - Q
do while(abs(res) > epsQ)
        y = 0.5*(y1+y2)
        res = Q(y) - Q
        if(res > 0.) then
          y2 = y
        else if(res < 0.) then
                y1=y
        else
                exit
        endif
enddo
y0=y
```

Figure 9.10 Null-point computation by the interval halving method.

Figure 9.11 Fortran source of the interval halving.

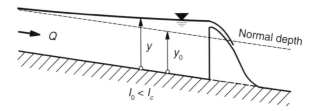

Figure 9.12 Backwater curve.

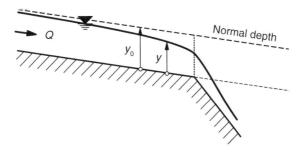

Figure 9.13 Drawdown curve.

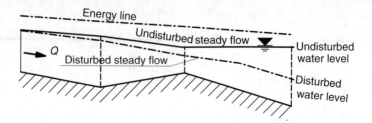

Figure 9.14   General non-uniform flow.

The developed disturbance decreases gradually and the depth upstream of the disturbance asymptotically approaches the undisturbed state, that is normal depth.

In the general case of a steady flow in a non-prismatic channel, the channel cross-section is varied along the flow with a variable bottom slope and the flow will be non-uniform, see Figure 9.14. This is an example of a general non-uniform flow.

If a disturbance in the form of a water level shift is added to the general non-uniform flow, it will cause a new non-uniform flow which asymptotically approaches the undisturbed previous non-uniform flow.

In general, disturbances caused by water level changes will asymptotically approach the undisturbed water level states.

9.4.2   Water level differential equation

The dynamic equation in the energy form can be applied to non-uniform gradually varied flow in a mildly sloping channel

$$\frac{dH}{dx} + I_e = 0, \tag{9.61}$$

where $H = z_0 + y + \alpha v^2 / 2g$ while the energy line slope is defined by, for example, the Manning formula

$$I_e = \frac{\tau_0}{\rho g R} = \frac{n^2 v^2}{R^{4/3}}. \tag{9.62}$$

Introducing $H = z_0 + H_s$, where $H_s = y + \alpha v^2 / 2g$ specific energy of a cross-section, Eq. (9.61) is written as

$$\frac{d}{dx}(z_0 + H_s) + I_e = 0 \tag{9.63}$$

or:

$$\frac{dH_s}{dx} + I_e - I_0 = 0. \tag{9.64}$$

After substituting $v = Q/A$ in the specific energy formula, the following is obtained

$$\frac{dH_s}{dx} = \frac{dy}{dx} + \frac{d}{dx}\left(\alpha \frac{Q^2}{2gA^2}\right). \tag{9.65}$$

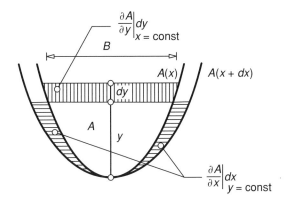

Figure 9.15 Differential variations in a non-prismatic channel.

Deriving the second term, first by variable A, then A by x, gives

$$\frac{dH_s}{dx} = \frac{dy}{dx} - \alpha \frac{Q^2}{gA^3} \frac{dA}{dx}. \tag{9.66}$$

Since the cross-section area $A(x,y)$ is gradually varied along the flow, there is a total differential as follows

$$dA = \left. \frac{\partial A}{\partial y} \right|_{x=const} dy + \left. \frac{\partial A}{\partial x} \right|_{y=const} dx \quad , \tag{9.67}$$

which gives (see Figure 9.15)

$$\frac{dA}{dx} = \frac{dy}{dx} \left. \frac{\partial A}{\partial y} \right|_{x=const} + \left. \frac{\partial A}{\partial x} \right|_{y=const} = B \frac{dy}{dx} + \left. \frac{\partial A}{\partial x} \right|_{y=const}. \tag{9.68}$$

After substitution in Eq. (9.66), it is written as

$$\frac{dH_s}{dx} = \frac{dy}{dx} - \alpha \frac{Q^2}{gA^3} \left(B \frac{dy}{dx} + \left. \frac{\partial A}{\partial x} \right|_{y=const} \right), \tag{9.69}$$

from which

$$\frac{dH_s}{dx} = \frac{dy}{dx} \left(1 - \alpha \frac{Q^2}{gA^3} B \right) - \alpha \frac{Q^2}{gA^3} \left. \frac{\partial A}{\partial x} \right|_{y=const}. \tag{9.70}$$

On the right hand side of expression (9.70) all terms refer to the slope; thus it can be written as

$$\frac{dH_s}{dx} = (1 - F_r) \frac{dy}{dx} - F_r I_b, \tag{9.71}$$

where the slope due to the non-prismatic channel, that is widening of the sides, is equal to

$$I_b = \frac{1}{B} \left. \frac{\partial A}{\partial x} \right|_{y=const}.$$

(9.72)

Substituting Eq. (9.71) into Eq. (9.64) it is written as

$$(1 - F_r)\frac{dy}{dx} - F_r I_b + I_e - I_0 = 0.$$

(9.73)

From the expression (9.73), the differential equation of depth variation along the flow in a non-prismatic channel is written as:

$$\frac{dy}{dx} = \frac{I_0 + F_r I_b - I_e}{1 - F_r}.$$

(9.74)

The obtained differential equation is called the *differential equation of water level in a non-prismatic channel.*

For prismatic channels, the second term in the numerator is equal to zero; thus, the water level differential equation has the following form:

$$\frac{dy}{dx} = \frac{I_0 - I_e}{1 - F_r}.$$

(9.75)

9.4.3 Water level shapes in prismatic channels

Water level shapes can be classified based on the water level Eq. (9.75); namely, its right hand side sign.

Subcritical flow

In subcritical flow, the Froude number is smaller than 1; thus the denominator of the fraction is always positive. The sign of the fraction depends on the sign of the nominator $I_0 - I_e$; thus, there are three possibilities, as shown in Figure 9.16.

(a) $F_r < 1$ Subcritical flow.

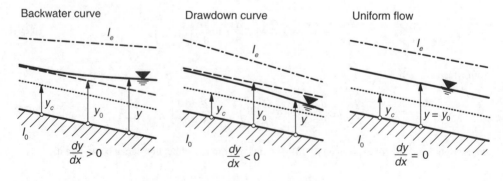

Figure 9.16 Water level shapes in subcritical flow.

(b) $F_r > 1$ Subcritical flow.

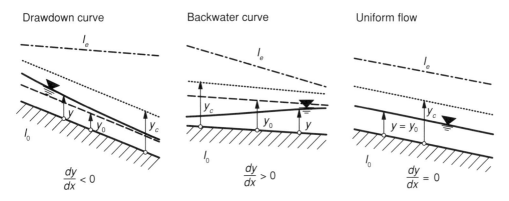

Figure 9.17   Water surface shapes in supercritical flow.

(a) The backwater curve is above the critical depth and above the normal depth that it approaches asymptotically in the upstream direction.
(b) The drawdown curve is above the critical depth and below the normal depth that it approaches asymptotically in the upstream direction.
(c) The uniform flow with normal depth is above the critical one.

Supercritical flow

In supercritical flow, the Froude number is greater than 1; thus, the denominator of the fraction is always negative. The sign of the fraction depends on the sign of the nominator $I_0 - I_e$; thus, there are three possibilities, as shown in Figure 9.17.

(a) The drawdown curve is below the critical depth and above the normal depth that it approaches asymptotically in the downstream direction.
(b) The backwater curve is below the critical and normal depth that it approaches asymptotically in the downstream direction.
(c) The uniform flow with normal depth is below the critical one.

Critical flow, critical slope

If the uniform flow is critical, the Froude number is equal to 1; thus, the denominator of the fraction is zero. In uniform flow $dy/dx = 0$, which is only possible if the right hand side of Eq. (9.75) is undetermined $0/0$; namely, if $I_0 - I_e = 0$. The bed slope of a channel with the uniform critical flow is called the critical slope. In a subcritical flow, the bed slope is smaller than the critical one, while in supercritical flow it is greater than the critical one.

9.4.4   Transitions between supercritical and subcritical flow, hydraulic jump

Figure 9.18 shows a channel with a transition from subcritical to supercritical flow and again from supercritical to subcritical flow. At the transition from subcritical to supercritical flow the water level is

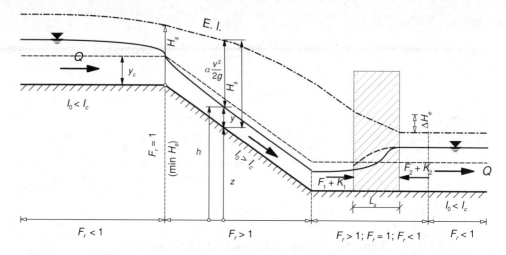

Figure 9.18 Transitional channel stretch.

continuous through the control section. In the control section, specific energy is the minimum and defines the upstream and downstream flow energy. It is a critical cross-section in which the Froude number is equal to 1. The water level shape upstream and downstream from the critical section is a drawdown curve of the respective flow regime. Specific energy in a cross-section is increasing upstream and downstream.

Supercritical flow extends to the downstream mildly sloping channel stretch where the backwater curve occurs. The Froude number decreases with depth increase and, when it reaches 1, specific energy is at its minimum. Subsequent flow over a horizontal bed would not be feasible because of energy dissipation. Only a subcritical flow could develop in the downstream channel stretch since the bed slope is smaller than the critical one $I_0 < I_c$. The water level is defined by downstream boundary conditions; namely, depth is above the critical one. Obviously, at the transition between the upstream supercritical flow and the subcritical flow, a hydraulic jump will develop in order to preserve enough energy for subsequent flow.

The position of a hydraulic jump is defined by the equilibrium of pressure forces and the momentum $F + K$ between two independent flows. The hydraulic jump starts at the front and ends behind a cross-section with the minimum specific energy.

Hydraulic jump conjugate depths

The length of a hydraulic jump L_s cannot be precisely determined due to the very strong turbulent flow. A normal hydraulic jump develops at the position where equilibrium between the pressure forces and the momentum between the upstream supercritical flow and the downstream subcritical flow is established.

The upstream cross-section is located at the start of the observed flow expansion, while the downstream cross-section is located at the position of flow separation into return upstream flow and downstream flow on the surface, see Figure 9.19. This position is not easily observable. Thus, in laboratory conditions a color is used to enhance the visibility of the phenomenon. The length of a hydraulic jump that is not well developed (small Froude number) is hard to determine.

Let us observe the equilibrium of pressure and momentum change in a normal hydraulic jump in a rectangular channel of width B. A hydraulic jump develops over the horizontal plane, thus there is no

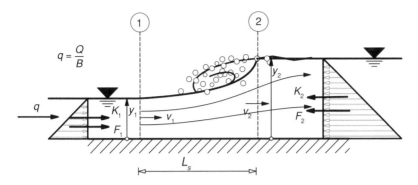

Figure 9.19   Normal hydraulic jump.

water weight contribution (volumetric force), and friction can be neglected due to short length. Thus, it is written as

$$F_1 + K_1 = F_2 + K_2,\qquad(9.76)$$

where

$$F_1 = \rho g \frac{y_1^2}{2} B \qquad\qquad \textit{pressure in upstream section,}$$

$$K_1 = \rho Q v_1 = \rho \frac{Q^2}{y_1 B} \qquad \textit{momentum in upstream section,}$$

$$F_2 = \rho g \frac{y_2^2}{2} B \qquad\qquad \textit{pressure in downstream section,}$$

$$K_2 = \rho Q v_2 = \rho \frac{q^2}{y_2 B} \qquad \textit{momentum in downstream section.}$$

After substitution into Eq. (9.76), equilibrium will be

$$g \frac{y_2^2}{2} B + \frac{Q^2}{y_2 B} = g \frac{y_1^2}{2} B + \frac{Q^2}{y_1 B}.\qquad(9.77)$$

After division by width B, it is written as

$$\frac{y_2^2}{2} + \frac{Q^2}{g y_2 B^2} = \frac{y_1^2}{2} + \frac{Q^2}{g y_1 B^2}.\qquad(9.78)$$

Grouping of terms gives

$$\frac{y_2^2 - y_1^2}{2} = \frac{Q^2}{g B^2} \frac{y_2 - y_1}{y_1 y_2}.\qquad(9.79)$$

When the rule of difference of squares is applied and after division by $y_2 - y_1$, the following is obtained

$$y_2 + y_1 = 2\frac{Q^2}{gB^2 y_1 y_2} = 2\frac{(v_1 B y_1)^2}{gB^2 y_1 y_2} = 2\underbrace{\frac{v_1^2}{gy_1}}_{F_{r1}}\frac{y_1^2}{y_2}. \tag{9.80}$$

Introducing the Froude number sign in the first cross-section, a quadratic equation is obtained

$$y_2^2 + y_1 y_2 - 2y_1^2 F_{r1} = 0. \tag{9.81}$$

The equation can be written using the depths ratio

$$\left(\frac{y_2}{y_1}\right)^2 + \frac{y_2}{y_1} - 2F_{r1} = 0. \tag{9.82}$$

Solution of the quadratic equation is

$$\frac{y_2}{y_1} = -\frac{1}{2} \pm \frac{\sqrt{1 + 8F_{r1}}}{2}. \tag{9.83}$$

Only the quadratic equation solution with a positive sign has a physical meaning; thus it is written as

$$y_2 = \frac{y_1}{2}\left(-1 + \sqrt{1 + 8F_{r1}}\right). \tag{9.84}$$

A hydraulic jump develops only at the transition from supercritical to subcritical flow, as can be seen from the analyses given hereunder.
     The Froude number in the second section is equal to

$$F_{r2} = \frac{v_2^2}{gy_2} = \frac{\left(\sqrt{1 + 8F_{r1}} + 1\right)^3}{64F_{r1}^2}. \tag{9.85}$$

If $F_{r1} > 1$, then expressions (9.83) and (9.85) give

$$\frac{y_2}{y_1} > 1 \text{ and } F_{r2} < 1. \tag{9.86}$$

If $F_{r1} = 1$, then expressions (9.83) and (9.85) give

$$\frac{y_2}{y_1} = 1 \text{ and } F_{r2} = 1. \tag{9.87}$$

If $F_{r1} < 1$, then expressions (9.83) and (9.85) give imaginary solutions.
     Depths y_1 and y_2 are called the *conjugate depths of a normal hydraulic jump*. The downstream conjugate depth of a hydraulic jump depends only on hydraulic values in the upstream conjugate cross-section.

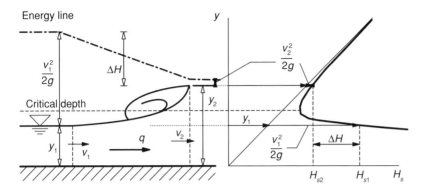

Figure 9.20   Energy dissipation in a normal hydraulic jump.

Hydraulic jump length

As already mentioned, the hydraulic jump length L_s cannot be precisely defined because of the very strong turbulent flow. The following expression, proposed by Smetana,[8] can be used in practice

$$L_s \approx 6(y_2 - y_1). \tag{9.88}$$

More accurate values can be obtained by tests presented in the form of a graph, see (Peterka, 2006[9]), or an approximation[10] of the graph by expression

$$\frac{L_s}{y_1} = 10(F_{r1}^2 - 1) - 0,0289(F_{r1}^2 - 1)^{2.3978}. \tag{9.89}$$

Hydraulic jump energy dissipation

Energy dissipation by a hydraulic jump, see the specific energy curve in Figure 9.20, is equal to the difference of specific energies in conjugate cross-sections

$$\Delta H = H_{s1} - H_{s2}; \tag{9.90}$$

namely, when the Coriolis coefficient is close to 1, then

$$\Delta H = y_1 + \frac{v_1^2}{2g} - y_2 - \frac{v_2^2}{2g}. \tag{9.91}$$

Energy dissipation in a normal hydraulic jump, in head form, is obtained based on the relationship between the conjugate cross-sections, after substitution into the previous expression and rearranging

$$\Delta H = \frac{(y_2 - y_1)^3}{4y_1 y_2}. \tag{9.92}$$

---

[8]J. Smetana, Czech hydraulic engineer.
[9]A. J. Peterka, Czech hydraulic engineer, emigrated to the USA.
[10]Approximation by V. Jović.

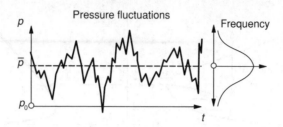

Figure 9.21　Pressure fluctuations in a hydraulic jump.

Due to the intensive turbulence in a hydraulic jump, where turbulent vortexes are relatively large, there is a great fluctuation of pressure and all other hydraulic properties. Pressure fluctuations, quantitatively shown in Figure 9.21, can reach values below atmospheric pressure (under-pressure).

Thus, particular care should be paid to energy dissipaters, and an adequate hard channel foreseen that can withstand fluctuating hydrodynamic loads. Construction of stilling basins belongs to the engineering discipline of water structures.

Hydraulic jump submergence and position

A normal hydraulic jump is defined by the equilibrium between the forces at conjugate depths. Upstream flow defines the minimum required values in the downstream conjugate depth to establish the equilibrium. However, downstream subcritical flow is defined only by downstream flow conditions. If the downstream flow gives the *tailwater* depth t_0 at the end of a hydraulic jump, which is equal to the second conjugate depth, equilibrium is established. Thus, this type of hydraulic jump is referred to as the *normal hydraulic jump*.

The hydraulic jump position defines the location of the equilibrium between the upstream and downstream flow forces, which can be expressed as *hydraulic jump submergence*.

On a horizontal plane, submergence can be expressed by the relation of *tailwater* water depth t and the second conjugate depth y_2; namely, by testing the relations in the second conjugate depth cross-section

$$t < y_2; \; t = y_2; \; t > y_2, \tag{9.93}$$

see Figure 9.22. If there is a discontinuity of the channel bottom, water levels above the reference level are tested.

A thrown away hydraulic jump develops when downstream flow at the position of the expected second conjugate depth gives the tailwater depth t_2, which is smaller than the second conjugate depth y_2. Then, the downstream subcritical flow forces cannot establish equilibrium; a hydraulic jump propagates in the downstream direction until upstream forces come into equilibrium with the downstream ones. A thrown away hydraulic jump is shown in Figure 9.23a.

A submerged hydraulic jump develops when the downstream flow at the position of the expected second conjugate depth gives the tailwater depth t_1, which is greater than the second conjugate depth y_2.

Then, the downstream subcritical flow forces are greater than forces required for equilibrium, and the hydraulic jump propagates upstream. A submerged hydraulic jump is shown in Figure 9.23b.

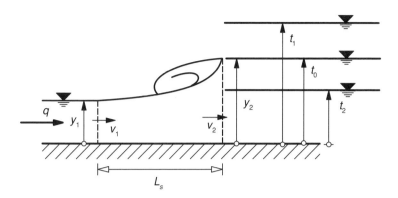

Figure 9.22   Hydraulic jump submergence.

If the downstream channel cross-section differs from the hydraulic jump cross-section, either by shape or size, testing of the hydraulic jump submergence shall be carried out for the entire discharge range, which can simply be done by drawing the respective discharge curves at the second conjugate depth cross-section, see Figure 9.24.

Figures 9.24a, b, and c show the same type of submergence within the entire discharge range. Figures 9.24d and e show submergence change when discharge exceeds a certain value.

Computational model deviations

The flow is critical when the Froude number is equal to 1; thus the denominator in Eq. (9.75) is equal to zero. For numerator values $I_0 - I_e \neq 0$ the gradient of the depth increment dy/dx becomes infinite; thus, in critical depth cross-sections, the water level tangent becomes a vertical line, as shown in Figure 9.25 right. If the nominator is equal to zero, that is in the uniform critical flow, the expression becomes undetermined.

Transition from the subcritical to the supercritical flow occurs at the cross-section where critical depth is expected. However, experience (measurements) show that the actual form of the transition from the

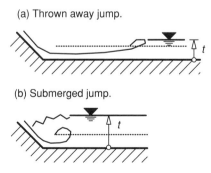

Figure 9.23   Hydraulic jump location.

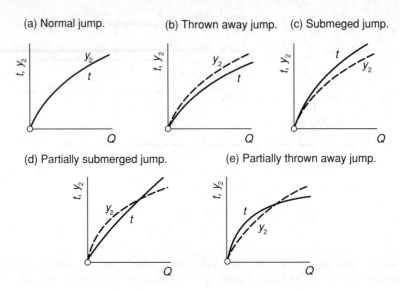

Figure 9.24   Hydraulic jump submergence testing.

subcritical to the supercritical flow is more likely to happen as shown in the figure (left); namely, critical depth develops a little more upstream from the bottom change.

A cause for water level shape discrepancy in a transition zone is a deviation of the computational model from the real state. Namely, one of the assumptions of the water level equation is parallelism of velocity vectors in the cross-section, that is hydrostatic distribution of the pressure, which is not fulfilled in the vicinity of the transition.

Spillway aeration

The beginning of a spillway is characterized by an accelerated flow which slows down the development of the boundary layer. When the boundary layer expands over the entire depth, spillway aeration occurs,

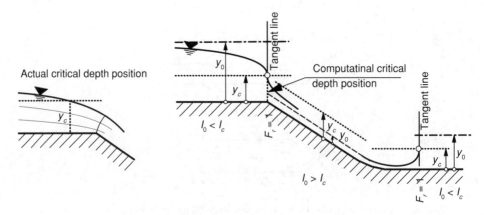

Figure 9.25   Water level shape in critical flow.

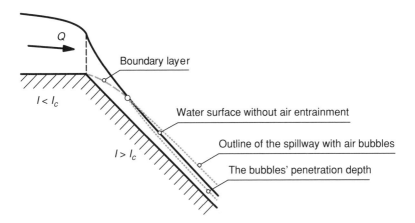

Figure 9.26   Spillway aeration.

see Figure 9.26. Aeration occurs in the form of air bubbles; thus in the subsequent flow there is a two-phase flow of air and water. The area occupied by a two-phase flow increases and the depth of the non-aerated spillway decreases. The water level obtained by non-uniform flow computation is in between the boundaries of spreading air bubbles.

The problem of spillway aeration and calculation of aerated spillway depth can be analyzed as flow of a two-phase liquid. However, apart from complex academic models there are no simple calculation methods acceptable for engineering purposes.

For a solution of spillway aeration in engineering practice, it is recommended to use the data obtained on already constructed engineering structures. The problem cannot be solved even by conventional model testing in a hydraulic laboratory because there is no possible analogy between the modeled basic flow and modeled aeration.

9.4.5   Water level shapes in a non prismatic channel

Water level shape in a non-prismatic channel is defined by differential equation

$$\frac{dy}{dx} = \frac{I_0 + F_r I_b - I_e}{1 - F_r}.$$

(9.94)

When compared to the previously described analysis for prismatic channels, quantitative analysis of potential water level shapes is a lot more complex because of the term $F_r I_b$. Equation (9.94) shows that a non-prismatic term, which contains the Froude number, either increases or decreases the longitudinal slope I_0, depending on the sign of the term I_b.

Open channel narrowing and widening

The influence of a non-prismatic channel on water level shape by application of the specific energy curve will be shown hereafter on examples of channel narrowing and widening. In order to describe the influence of a non-prismatic channel, a flow over a horizontal plane with no energy dissipation that the following equation applies to, will be analyzed

$$\frac{dy}{dx} = \frac{F_r I_b}{1 - F_r}.$$

(9.95)

The water level slope along the flow depends on the algebraic signs of the numerator and the denominator. In subcritical flow, the sign of the numerator is always positive, while in supercritical flow it is always negative. The sign of the numerator depends on the non-prismatic slope: it is positive for channel widening and negative for channel narrowing.

Figure 9.27 shows channel narrowing between the two close cross-sections. The right side of the figure shows specific energy curves in upstream and downstream cross-sections. Discharge and specific energy are constant along the observed stretch.

When the channel is narrowing, the nominator term $F_r I_b < 0$ is negative; thus, the depth is decreasing with length in the subcritical flow and increasing in the supercritical flow, as can be observed from intersecting points a and b of the subcritical flow, and a' and b' of the supercritical flow. Critical depth is increasing along the flow.

Figure 9.28 shows channel widening between the two close cross-sections. The right side of the figure shows specific energy curves in upstream and downstream cross-sections. Discharge and specific energy are constant along the observed stretch.

When the channel is widening, the nominator term $F_r I_b > 0$ is positive; thus, the depth is increasing with length in the subcritical flow and decreasing in the supercritical flow, as can be observed from the intersecting points a and b of the subcritical flow, and a' and b' of the supercritical flow. Critical depth is decreasing along the flow.

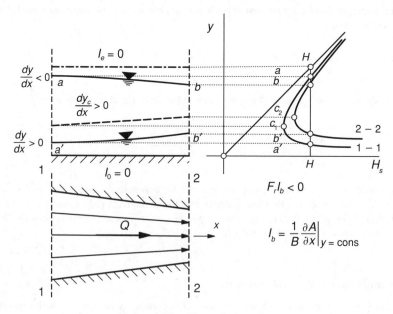

Figure 9.27 Channel narrowing.

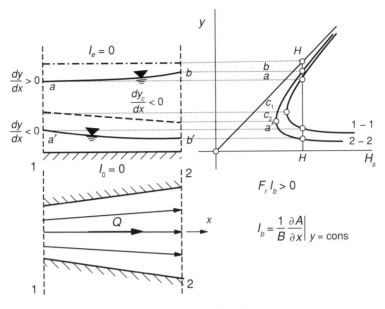

Figure 9.28   Channel widening.

Convergent spillway

In a convergent spillway, the Froude number decreases along the flow. Calculation is feasible until the Froude number reaches the value $F_r = 1$; namely, to the cross-section with the minimum specific energy. If there is no change of slopes I_0 and I_b in the downstream direction, a subsequent flow will be possible only at the minimum specific energy, in other words the flow will be critical, see Figure 9.29. Thus, neither supercritical nor subcritical flow will be possible without a specific energy increase.

Supercritical flow

Experience shows that the calculated water surface in supercritical flow differs from the real one, see experiments and theoretical analyses by Ippen[11] et al. and (Chow, 1959).

One cause for water level shape discrepancy in a transition zone is a deviation of the computational model from the real state. Namely, one of the assumptions of the water level equation is parallelism of velocity vectors in the cross-section, which is not fulfilled in supercritical flow in non-prismatic channels. This assumption is also not true for subcritical flow, but there is a good match between calculated and real water levels. This can be explained by the fact that the upstream and downstream motion of the two elementary waves (disturbances) influences the solution in the subcritical flow, while in the supercritical flow both elementary waves (disturbances) are always moving in a downstream direction.

The possible motion of supercritical flow waves in a non-prismatic channel is shown in Figure 9.30. Figure 9.30a shows a convergent transition where a wave bounces off the channel wall in the form of water level jump, both in transversal and longitudinal directions; and the flow becomes complex and two-dimensional. Figure 9.30b shows a divergent transition where elementary waves do not follow channel

---

[11] Arthur Thomas Ippen (1907–1974), professor, President of IAHR 1967.

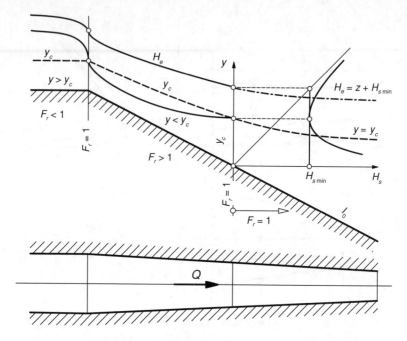

Figure 9.29   Convergent spillway.

widening. Thus, wave separation from the wall will be possible if the widening is sudden; the flow will not be one-dimensional anymore and the one-dimensional computation model cannot be applied.

Problems of supercritical flow in non-prismatic channels should be solved experimentally and by theoretical analyses that respect wave kinematics (disturbances, solution by the method of characteristics).

If aeration at large Froude numbers is also added, then model testing of supercritical flow also becomes questionable because it is susceptible to the model's scale.

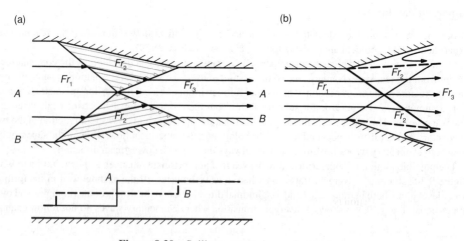

Figure 9.30   Spillway narrowing and widening.

9.4.6   Gradually varied flow programming solutions

Two differential equations may be used for gradually varied flow problem solving:

- the water level equation for non-prismatic channels, expression (9.74), which can also be applied to prismatic channels if $I_b = 0$, which is derived from,
- the energy equation, expression (9.61).

Gradually varied flow is solved for initial conditions, that is prescribed depth and discharge in the initial cross-section. One common initial condition is the minimum specific energy in a cross-section for a prescribed discharge, namely the initial depth is the critical depth. In this case, the nominator in Eq. (9.74) is equal to zero and calculation is infeasible (certainly without additional tricks such as profile displacement at a small distance or similar).

Thus, the use of an energy equation is recommended

$$\frac{dH}{dx} + I_e = 0 \tag{9.96}$$

with a simple integration between the two cross-sections

$$H_2 - H_1 + \int_{x_1}^{x_2} I_e dx = 0, \tag{9.97}$$

where index 1 denotes the cross-section with prescribed values and index 2 cross-section with the unknown values. For an arbitrary small distance between the cross-sections $\Delta x = x_2 - x_1$, the rule of an average value of integration applies

$$\int_{x_1}^{x_2} I_e dx = \frac{I_{1e} + I_{2e}}{2} \Delta x, \tag{9.98}$$

where the slope of the energy line is calculated using the discharge function $K(y)$

$$I_e = \frac{Q^2}{K^2}. \tag{9.99}$$

The positive orientation of the x axis is in the direction of flow. The calculation can be the downstream one with the positive integration step $\Delta x = x_2 - x_1$ or the downstream one with the negative integration step.

The problem can be solved in two ways. Both procedures are iterative and applicable to integration of the subcritical and supercritical flow. Both iterative procedures converge fast.

The transition from the subcritical to supercritical flow, that will illustrate the procedure, is shown in Figure 9.31. Cross-section 0 is a control section with the prescribed flow energy as well as all the other hydraulic variables because of the minimum specific energy $H_0 = z + H_{s\,min}$.

The first procedure, written in pseudo programming language, is shown in Figure 9.32. Computation starts from the cross-section 0 with the prescribed values (initial conditions). The slope of the energy line in the known cross-section is adopted as the first iteration value of the energy line slope in the subsequent cross-section (*upstream 1 in subcritical flow and downstream 1' in supercritical flow*). After computation of the energy head, all geometric and hydraulic parameters are calculated in a new cross-section for the selected flow regime using the uniform flow programming solutions, see Section 9.3.4, which are based

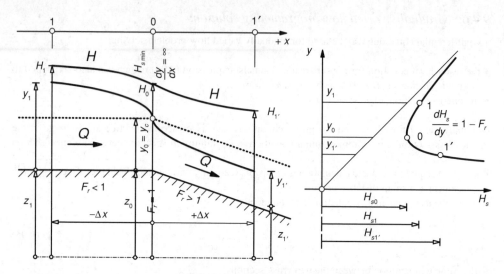

Figure 9.31   Integration steps.

on the specific energy curve. *The second procedure*, also written in the pseudo programming language, see Figure 9.33, is based on the iterative solution of the energy equation

$$H_2 - H_1 + \frac{I_{1e} + I_{2e}}{2}(x_2 - x_1) = 0 \tag{9.100}$$

using the Newton–Raphson method. Since the parameters are known in the first cross-section and unknown in the second, the previous expression can be written as a function of the unknown depth

$$F(y_2) = H_2 - H_1 + \frac{I_{1e} + I_{2e}}{2}(x_2 - x_1) = 0. \tag{9.101}$$

```
• select initial conditions and flow regime
• calculate geometric and hydraulic
  parameters of initial cross-section
do while next cross-section
  equalize slope Ie with the initial one
  do iter = 1 to Maxiter
  calculate H, Eq.(9.97)
  for H calculate:
  • depth for flow regime
  • geometric and hydraulic parameters
  • slope Ie at the cross-section
  if accuracy is achieved, exit loop iter
  end loop iter
end loop next
```

Figure 9.32   Pseudocode of the first procedure of energy equation integration.

```
    • select initial conditions and flow regime
    • calculate geometric and hydraulic
      parameters of initial cross-section
do while next cross-section
    set initial depth y
      do iter = 1 to Maxiter
      for y calculate:
          • value of the function F, expression ()
          • value of the derivative dF/dy, expression ()
          • value of the increment dy from expression ()
          • recalculate depth y = y+dy
      if dy less or equal to eps, exit loop
    end loop iter
  end loop next
```

Figure 9.33   Pseudocode of the second procedure of energy equation integration.

A solution is sought as a null-point of the function $F(y_2)$, creating an iteration procedure between two iteration steps k in the form

$$y_2^{k+1} = y_2^k + \Delta y_2, \tag{9.102}$$

where increment Δy_2 is calculated from the k-th iteration step

$$\frac{dF^k}{dy_2} \Delta y_2 = -F^k. \tag{9.103}$$

The derivative is equal to

$$\frac{dF^k}{dy_2} = \frac{dH_2}{dy_2} + \frac{(x_2 - x_1)}{2} \frac{dI_{2e}}{dy_2} = 1 - F_{r2} + \frac{(x_2 - x_1)}{2} \frac{\Delta I_{2e}}{\Delta y_2}, \tag{9.104}$$

where energy is expressed by the specific energy with the derivative in the y-direction equal to $1 - F_r$, and the derivative of the energy line calculated numerically over the interval δ, for example 1 cm

$$\frac{\Delta I_{2e}}{\Delta y_2} = \frac{I_e(y_2 + \delta) - I_e(y_2 - \delta)}{2\delta}. \tag{9.105}$$

For subcritical flow according to Figure 9.31 (upstream calculation) the negative step Δx applies, the initial prescribed cross-section is marked by index 0, and the unknown cross section is marked by index 1. The initial iteration depth y_1 shall be somewhat greater than or equal to the depth in the previous cross-section.

For supercritical flow according to Figure 9.31 (downstream calculation) the positive step applies, the initial prescribed cross-section is marked by index 0, and the unknown cross-section is marked by index 1'. The initial iteration depth $y_{1'}$ shall be smaller than the depth in the previous cross-section, for example $y_{1'} = y_0/2$.

The first procedure of unsteady flow calculation is implemented in a subroutine

```
logical function DoChannelInit(Channel),
```

while the second is implemented in a subroutine

```
logical function SteadyCalc(channel).
```

These subroutines are called depending on the logical value of the optional global variable *InitChannelMethod*:

```
if(InitChannelMethod) then
            Channels(channel).init=DoChannelInit(Channels(channel))
else
            Channels(channel).init=SteadyCalc(Channels(channel))
endif
```

in a subroutine which initializes the channel stretch

```
logical function & InitializeChannel(channel,kind,iPnt,Qinit,hInit,stsBDC)
```

Programming solutions can be found in the file `Channels.f90`, see www.wiley.com/go/jovic.

9.5   Sudden changes in cross-sections

Out of all the kinds of sudden changes in cross-sections along the flow, particular attention will be paid to lateral channel narrowing and widening that occurs when narrower channel stretches are added, and to weirs in mildly sloping or horizontal channels, see Figure 9.34.

A hydraulic situation for the incoming subcritical flow will be observed. Lateral channel narrowing and widening are causing flow acceleration, see Figure 9.35, accompanied by local energy loss ΔH_1. Separation of a boundary layer occurs after sudden channel narrowing thus causing a flow profile reduction to width B_c, which is smaller than the stretch width B_1. Energy dissipation occurs along the length of the separated boundary layer.

The water level is decreasing because the velocity head is increasing. Critical depths in the inlet (*mark* 0) and contracted channel stretches (*mark* 1) are shown in the figure as well as specific energy curves in cross-sections that serve as explanation of the phenomena. The water level shape in a narrowed stretch depends on the water level in subsequent channel widening (*mark* 2) and channel bed slope in the narrowed stretch.

- If the downstream water level (tailwater) submerges the critical depth of the narrowed stretch (elevations above the reference level are compared), a subcritical flow with water level curve marked **a** in Figure 9.35 will be established in the narrowed stretch. After sudden channel widening, the flow energy will decrease by ΔH_2, while energy dissipation will mainly occur in lateral eddies within the separated boundary layer.

(a) Lateral sharp narrowing and enlargement.   (b) Broad-crested weir.

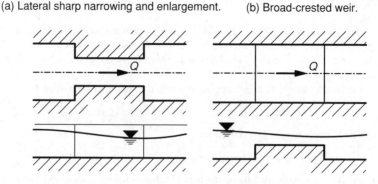

Figure 9.34   Types of sudden channel cross-section changes.

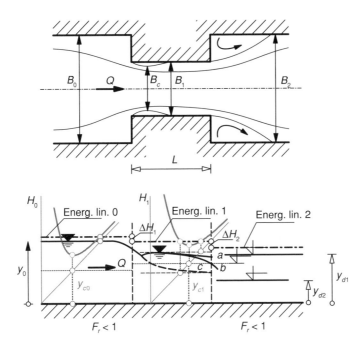

Figure 9.35   Lateral narrowing.

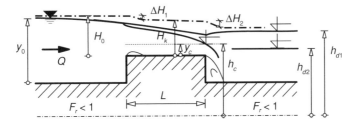

Figure 9.36   Flow over a broad-crested weir.

- If the tailwater does not submerge the critical depth of the narrowed stretch (elevations above the reference level are compared), a subcritical flow with water level curve marked **b** Figure 9.35 in will be established in the narrowed stretch: Namely, at the transition between the narrowed and newly widened stretch, a flow with minimum specific energy at the cross-section shall develop. At the transition between the inlet and narrowed stretch in a channel with a horizontal bottom there can be no critical depth because it would mean that from that point specific energy, and thus the total energy, increase in the downstream direction.

- Incoming subcritical flow can transform into supercritical flow in the narrowed channel stretch only if a slope of the narrowed stretch is greater than the critical one. Thus, total flow energy will decrease and specific energy will increase along the flow. Curve **c** in Figure 9.35 represents water level shape.

A similar analysis can be conducted for a broad-crested weir, Figure 9.36. Reduction due to a raised weir will cause boundary layer separation at the start of the weir and energy dissipation by ΔH_1. The water level will decrease due to increased velocity.

- If the tailwater level h_{d1} submerges at the critical depth at the weir (elevations above the reference level are compared); then an eddy will develop behind the weir in a separated boundary layer where energy dissipation occurs. Thus, the energy head will decrease by ΔH_2. Subcritical flow will be established at the weir. The downstream water level will influence the upstream flow.
- If the tailwater level h_{d2} does not submerge critical depth at the weir (elevations above the reference level are compared); then a critical depth will occur at the weir end. Tailwater will not influence the upstream flow. Thus, flow over a broad-crested weir will be defined by overflowing energy head H_k at the end of a weir; namely, there will be overflow over a cascade. Energy at the cascade will be defined by the minimum specific energy of a cross-section. The upstream state shall be defined by computation of water level at the weir.

Energy dissipation at sudden narrowing and widening occurs in a separated boundary layer, based on similar principles as described for sudden contractions and expansions in pipelines. Unlike the flow through sudden changes in a pipe cross-section, where the phenomena are studied well enough for engineering purposes, this is not a case in open channel hydraulics. A similarity with the pipe hydraulics can be used as the first assessment of energy losses:

- loss at sudden widening (divergent flow)

$$\Delta H_{div} = k_{div} \frac{(v_1 - v_2)^2}{2g},$$ (9.106)

- loss at sudden narrowing (convergent flow)

$$\Delta H_{conv} = k_{conv} \frac{v_2^2}{2g},$$ (9.107)

where, in the case of a sudden widening, v_1 denotes the flow velocity in front of the expansion, while in case of a sudden narrowing v_2 denotes the flow velocity behind the contraction. If there are no measured data, coefficients k_{pro} and k_{suz} shall be adopted as for pipes.

It is not hard to conclude that these are rough estimates that are allowed in the preliminary design phase only. *Otherwise, published data should be used or hydraulic model tests of real flow should be carried out.*

It is of utmost importance to know the flows in suddenly narrowed or widened stretches during the construction of hydraulic structures in rivers (Izbash and Khaldre, 1970), where, due to the construction of a hydraulic structure, there is river closure. Figure 9.37 shows a potential dam construction method in a lowland river stretch.

The dam construction plan consists of a cofferdam construction (*type of embankment*), for example the left one, within which the construction site will be established. When the left side of a dam is constructed, part of the left cofferdam will be destroyed to allow its use as a bypass channel during construction of the right cofferdam and a dam within it. Following the completion of construction, cofferdam debris will be removed.

Sudden flow narrowing due to bridge piers in the riverbed is also a similar type of problem.

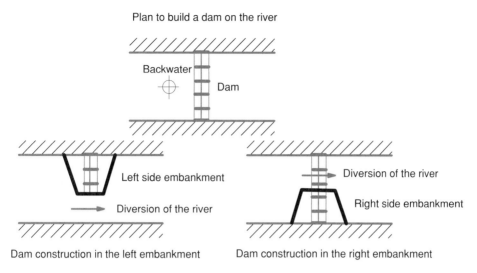

Figure 9.37   River partial closure and cofferdams in dam construction.

9.6   Steady flow modelling

9.6.1   Channel stretch discretization

A channel stretch is a hydraulic network branch that consists of several channel finite elements, see Figure 9.38; thus, it is a macro element. The channel stretch axis is defined by points p, which define spatial position (x, y, z_0), where z_0 is the channel bottom elevation. Cross-sections (*profiles*) are set through channel points approximately perpendicular to the channel axis.

A channel is defined as a series of pairs of points and profiles (p, prf) in the form of a collating sequence starting from the upstream and moving towards the downstream channel stretch end. Thus, the

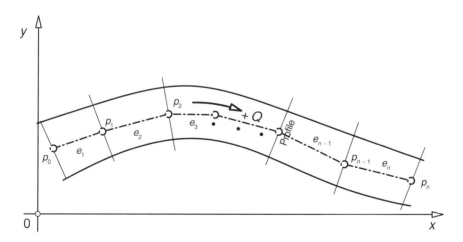

Figure 9.38   Channel stretch discretization.

positive flow direction is always from the upstream towards the downstream channel end, that is from the first towards the second finite element. In general, different types of cross-sections may be appointed to the points, but they can also be of the same type. Channel finite elements are, thus, implicitly defined, as an arranged channel stretch subdivision. The first channel element is located at the channel start and the last at the channel end.

Boundary conditions are assigned to boundary points of a channel stretch. Internal points are usually connections of channel elements and cannot have some other nodal object function.

9.6.2   Initialization of channel stretches

Channels can be a constituent of a complex hydraulic network. Before starting a calculation of steady or non-steady flow in a hydraulic network, the initial water level and discharge for iteration values should be initialized. The program solution `SimpipCore` initializes initial iterative values using the logical function `InitializeChannel`, which is called at the time of the input data processing in a logical function `Initialize(buf)`. A call syntax is

        Initialize Channel Name FlowKind pnt Qchannel <hValue> or <BDCstatus>

where:

- `Name` – channel name.
- `FlowKind` – `SUBCRITICAL` or `SUPERCRITICAL`.
- `pnt` – name of the point in which the water level boundary condition (piezometric heads) is prescribed.
- `Qchannel` – discharge.
- `hValue` – explicitly prescribed value at the point `pnt` or boundary condition status, which can be:
- `<SBL>`
  - `HS_MIN` or `DCRITICAL` – minimum specific energy (critical depth) as the boundary condition in the profile appointed to the point `pnt`,
  - `H_IMPLICIT` – implicit value h at the point `pnt`, initialized in previous initialization.

By calling the logical function, channel subdivision into finite elements is also carried out, thus adding them to the general finite element mesh.

In a hydraulic network, channels should be classified by flow regime. Flow regime in the channel is already prescribed by initialization. Channels in subcritical flow can be networked generally, just like other hydraulic network branches, as shown in Figure 9.39 and Figure 9.40. The initial discharge distribution in channels Q_{kanal}^{init} should be defined based on which steady flow will be calculated. This can be done accurately if there are no loops in the system.

Otherwise, initial discharges Q_{kanal}^{init} shall be estimated, based on which initialization of steady values h, Q will be made. These are the initial values for the final iteration of the network channels that will be made by the steady flow procedure `Steady`.

Initialization starts from the point with the prescribed piezometric head (point P_4 in the figure), and the initial discharge $Q_{K_4}^{init}$ in channel K_4

$$\text{Initialize Channel K4 SUBCRITICAL P4 QK4 h4} \qquad (9.108)$$

based on which all piezometric heads in all channel points will be calculated, point P_3 included. Since the piezometric head P_3 is an implicitly known value, initialization of the channel K_3 or channel K_2 may start:

        Initialize Channel K3 SUBCRITICAL P3 QK3 H_IMPLICIT,

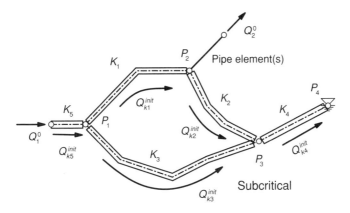

Figure 9.39   Hydraulic network with channels in subcritical flow.

then

```
Initialize Channel K2 SUBCRITICAL P3 QK2 H_IMPLICIT,
Initialize Channel K1 SUBCRITICAL P2 QK1 H_IMPLICIT,
Initialize Channel K5 SUBCRITICAL P1 QK5 H_IMPLICIT.
```

Logical procedure `InitializeChannel` sets the initial iteration values of the steady flow in a channel using the water level computation for the prescribed initial boundary conditions as defined in Section 9.4.6.

The boundary condition of the minimum specific energy (critical depth) in a profile can be prescribed at point P_4. Then, instead of initialization (9.108), it shall be written

```
Initialize Channel K4 SUBCRITICAL P4 QK4 HS_MIN.
```

The boundary condition of the minimum specific energy in a profile can be appointed only to the node with *one* channel connection in subcritical flow!

Some limitations apply to channel stretches with supercritical flow due to supercritical flow complexity and specific boundary conditions, see Section 9.10. This type of a channel can be networked with only

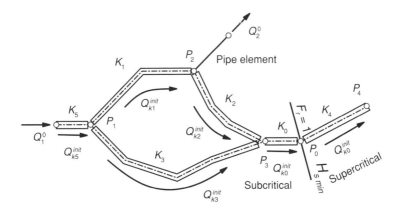

Figure 9.40   Hydraulic network with channels in subcritical and supercritical flow.

one upstream branch of the channel network with subcritical flow, as shown in Figure 9.40. At the transition from subcritical to supercritical flow there is a minimum specific energy at the cross-section which defines the energy state of the upstream and downstream flow.

Initialization of channel system, shown in Figure 9.40, is the following:

```
Initialize Channel K4 SUPERCRITICAL P0 QK0 HS_MIN,
Initialize Channel K0 SUBCRITICAL P0 QK0 HS_MIN,
Initialize Channel K3 SUBCRITICAL P3 QK3 H_IMPLICIT,
Initialize Channel K2 SUBCRITICAL P3 QK2 H_IMPLICIT,
Initialize Channel K1 SUBCRITICAL P2 QK1 H_IMPLICIT,
Initialize Channel K5 SUBCRITICAL P1 QK5 H_IMPLICIT.
```

In a subroutine `Steady` a call for a subroutine for computation of steady flow matrix and vector for the elemental finite element is added as follows:

```
select case(Elems(ielem).tip)
  case ...
    ...
  case (CHANNEL_OBJ)
    lookup=Elems(ielem).lookup
    Channel=Channels(lookup)
    if(Channel.kind.eq.SUBCRITICAL) &
      call SubCriticalSteadyChannelMtx(ielem)
    if(Channel.kind.eq.SUPERCRITICAL)
      call SuperCriticalSteadyChannelMtx(ielem)
  case ...
    ...
endselect
```

Depending on the flow regime initialized in the channel finite element, computation branches into subcritical or supercritical steady flow.

9.6.3   Subroutine `SubCriticalSteadyChannelMtx`

A channel finite element matrix and vector in steady flow are calculated in subroutine:

```
subroutine SubCriticalSteadyChannelMtx(ielem).
```

Starting from the energy equation integrated over a finite element

$$H_2 - H_1 + \int_{\Delta l} I_e dx = 0, \tag{9.109}$$

where

$$I_e = \frac{|Q| Q}{K^2}, \tag{9.110}$$

The elemental equation is

$$F_e : H_2 - H_1 + \frac{\Delta l}{2} (I_{e1} + I_{e2}) = 0. \tag{9.111}$$

The Newton–Raphson iterative form for a finite element is formally written using the matrix-vector operations

$$[\underline{H}] \cdot [\Delta h^+] + [\underline{Q}] \cdot [\Delta Q^+] = [\underline{F}], \tag{9.112}$$

where the vector is

$$[\underline{H}] = \left[-(1 - F_r)_1 + \frac{\Delta l}{2} \frac{\Delta I_{e1}}{\Delta h_1}, \quad (1 - F_r)_2 + \frac{\Delta l}{2} \frac{\Delta I_{e2}}{\Delta h_2} \right]. \tag{9.113}$$

Derivative of the energy line slope by head h is calculated numerically

$$\frac{\Delta I_e}{\Delta h} = \frac{I_e(Q, y + \delta) - I_e(Q, y - \delta)}{2\delta}, \tag{9.114}$$

where the interval δ is a small depth increment, for example one centimeter. Scalar values are on the right hand side

$$[\underline{F}] = -F_e, \tag{9.115}$$

while the discharge derivative is

$$[\underline{Q}] = -\alpha_1 \frac{Q}{gA_1^2} + \alpha_2 \frac{Q}{gA_2^2} + \Delta l \, |Q| \left(\frac{1}{K_1^2} + \frac{1}{K_2^2} \right). \tag{9.116}$$

The inverse value is equal to

$$[\underline{Q}]^{-1} = \frac{1}{\Delta l \, |Q| \left(\dfrac{1}{K_1^2} + \dfrac{1}{K_2^2} \right)} = \frac{K_1^2 K_2^2}{\Delta l \, |Q| \left(K_1^2 + K_2^2 \right)}. \tag{9.117}$$

When expression (9.112) is multiplied by the inverse term $[\underline{Q}]^{-1}$ the following is obtained

$$[\underline{A}] \cdot [\Delta h] + [\Delta Q] = [\underline{B}], \tag{9.118}$$

from which the value of the elemental discharge increment can be calculated as

$$[\Delta Q] = [\underline{B}] - [\underline{A}][\Delta h], \tag{9.119}$$

where

$$[\underline{A}] = [\underline{Q}]^{-1}[\underline{H}], \tag{9.120}$$

$$[\underline{B}] = [\underline{Q}]^{-1}[\underline{F}]. \tag{9.121}$$

In the aforementioned expressions $[A]$ is a two-term vector while $[B]$ is a scalar. A process of elimination of elemental discharges from nodal equations of continuity defines the structure of the finite element matrix

$$A^e = \begin{bmatrix} +\underline{A} \\ -\underline{A} \end{bmatrix}$$

(9.122)

and vector

$$B^e = \underbrace{\begin{bmatrix} +Q^e \\ -Q^e \end{bmatrix}}_{(1)} + \underbrace{\begin{bmatrix} +\underline{B} \\ -\underline{B} \end{bmatrix}}_{(2)}.$$

(9.123)

Term (1) is the existing right hand side term before elimination, while (2) is the increment after elimination of the elemental discharge from the nodal equation.

9.6.4   Subroutine `SuperCriticalSteadyChannelMtx`

In supercritical flow, upstream boundary conditions are always prescribed; thus, the unknown value Δh_2 can always be calculated from the elemental equation (9.111). The Newton–Raphson procedure is used again with the derivative by downstream level h_2

$$[\underline{H}] = (1 - F_r)_2 + \frac{\Delta l}{2} \frac{\Delta I_{e2}}{\Delta h_2}$$

(9.124)

from which an iterative increment is equal to

$$\Delta h_2 = \frac{-F_e}{(1 - F_r)_2 + \dfrac{\Delta l}{2} \dfrac{\Delta I_{e2}}{\Delta h_2}}.$$

(9.125)

Since the discharge is calculated from the upstream boundary conditions, and the elemental discharge increment $[\Delta Q] = 0$ is equal to zero, then it shall be

$$[\underline{B}] = 0 \text{ and } [\underline{A}] = 0$$

(9.126)

for the procedure `IncVar` to remain common to all finite elements. In supercritical steady flow, the finite element matrix is written as

$$A^e = \begin{bmatrix} 0 & 0 \\ 0 & [\underline{H}] \end{bmatrix},$$

(9.127)

while the vector is

$$B^e = \begin{bmatrix} 0 \\ -F_e \end{bmatrix}.$$

(9.128)

Note that for *internal points* of the channel in supercritical flow, nodal sums of discharges are not closed as in subcritical flow and the continuity of flow is secured within the subroutine by transfer of the discharge to the next connected channel element

$$Q^{e+1} = Q^e. \tag{9.129}$$

The channel finite element matrix and vector in steady flow are computed in a subroutine:

```
subroutine SuperCriticalSteadyChannelMtx(ielem)
```

that can be found in the Fortran module `Channels.f90`.

9.7 Wave kinematics in channels

9.7.1 Propagation of positive and negative waves

Figure 9.41 shows waves generated by the sudden partial lowering or lifting of a sluice gate which regulates a channel flow. For a sudden sluice gate lowering and discharge reduction, Figure 9.41a, downstream negative and upstream positive waves are generated. For a sudden sluice gate lifting and discharge increase, Figure 9.41b, downstream positive and upstream negative waves are generated.

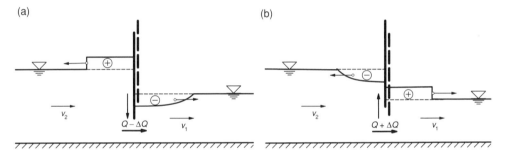

Figure 9.41 Waves generated by sudden partial discharge change.

The concept of positive and negative waves is defined by the character of water level change.[12] Generated disturbances extend throughout the entire cross section, pressure is hydrostatic in front of and behind the wave front, velocity vectors are parallel; thus, all the necessary assumptions for one-dimensional analysis are retained. A positive wave shape differs from a negative one, which will be explained in the following text.

9.7.2 Velocity of the wave of finite amplitude

According to Figure 9.42, an observer standing aside will see a propagation of the positive wave of finite amplitude in the channel, as a wave front motion at an absolute velocity w. If we imagine relative motion along the channel at coordinate system velocity w, then an observer standing away from it will see a steady flow at relative velocity $w - v$ in front and behind him. Mass flow towards the movable observer (wave front) is equal to the mass flow moving away from the observer.

---

[12] In wave kinematics, positive and negative waves refer to the motion of elementary waves in the direction of positive or negative channel axes.

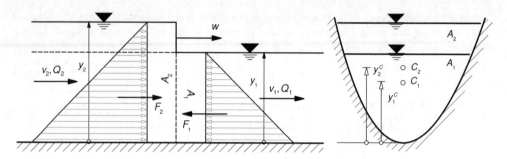

Figure 9.42   Relative motion of the wave of finite amplitude.

Discharges Q, average velocities v, areas A, and hydrostatic pressure in front and behind the wave front which define forces F are shown in Figure 9.42. Relative flow velocity in front of and behind the wave front, marked by indexes 1 and 2, will be

$$c_1 = w - v_1 \ i \ c_2 = w - v_2. \tag{9.130}$$

The continuity equation in relative motion is

$$A_2 (w - v_2) = A_1 (w - v_1) \tag{9.131}$$

from which the absolute velocity of a *finite wave* can be expressed as

$$w = \frac{A_2 v_2 - A_1 v_1}{A_2 - A_1} = \frac{Q_2 - Q_1}{A_2 - A_1} = \frac{\Delta Q}{\Delta A}. \tag{9.132}$$

The momentum equation of the relative motion of a finite wave front is

$$F_2 + \rho A_2 (w - v_2)^2 = F_1 + \rho A_1 (w - v_1)^2 \tag{9.133}$$

from which

$$F_2 - F_1 = \rho A_1 (w - v_1)^2 - \rho A_2 (w - v_2)^2. \tag{9.134}$$

If the following is written from Eq. (9.131)

$$w - v_2 = \frac{A_1}{A_2} (w - v_1). \tag{9.135}$$

and substituted into previous expression, then

$$F_2 - F_1 = \frac{A_1}{A_2} \rho (A_2 - A_1) \cdot (w - v_1)^2 \tag{9.136}$$

from which

$$w = v_1 \pm \sqrt{\frac{F_2 - F_1}{\rho \dfrac{A_1}{A_2}(A_2 - A_1)}}\,, \tag{9.137}$$

$$w = v_1 \pm c_1, \tag{9.138}$$

where relative wave celerity is

$$c_1 = \sqrt{\frac{F_2 - F_1}{\rho \dfrac{A_1}{A_2}(A_2 - A_1)}}. \tag{9.139}$$

Second relative wave celerity can be obtained by a similar procedure

$$c_2 = \sqrt{\frac{F_2 - F_1}{\rho \dfrac{A_2}{A_1}(A_2 - A_1)}}. \tag{9.140}$$

It can be used to express an absolute velocity

$$w = v_2 \pm c_2. \tag{9.141}$$

Equating expressions (9.138) and (9.141) gives

$$w = \frac{v_1 \pm c_1}{2} \pm \frac{v_2 \pm c_2}{2}. \tag{9.142}$$

9.7.3 *Elementary wave celerity*

The following applies to *differentially small disturbance* in limits $v_2 \to v_1 = v$, $A_2 \to A_1 = A$; namely, finite differences become derivatives $A_2 - A_1 \to dA$, $Q_2 - Q_1 \to dQ$, and $F_2 - F_1 \to dF$. Thus, expression (9.132) can be written in differential form

$$w = \frac{dQ}{dA}. \tag{9.143}$$

Similarly, expression (9.136) can also be written in differential form

$$dF = \rho dA(w - v)^2 \tag{9.144}$$

from which

$$w = v \pm \sqrt{\frac{dF}{\rho dA}} = v \pm c. \tag{9.145}$$

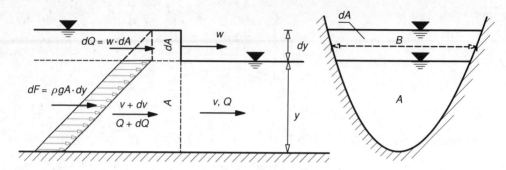

Figure 9.43   Elementary wave.

Hence, there are two waves, one propagating in the direction of the flow and the other moving in the opposite direction by relative celerity

$$c = \sqrt{\frac{dF}{\rho dA}}. \tag{9.146}$$

Since the hydrostatic force increment for an elementary wave is equal to

$$dF = \rho g A dy \tag{9.147}$$

and, based on cross-section geometry, see Figure 9.43,

$$dA = Bdy, \tag{9.148}$$

relative wave celerity, that is celerity of an elementary wave on still water, is equal to

$$c = \sqrt{g\frac{A}{B}} = \sqrt{g\bar{y}}, \tag{9.149}$$

where $\bar{y} = A/B$ is the equivalent depth of a prismatic channel of the same width B and cross-section area A.

The Froude number is a dimensionless number defined as the ratio between channel flow velocity and the celerity of the still water elementary wave

$$F_r = \frac{Q^2}{gA^3}B = \frac{v^2}{g\dfrac{A}{B}} = \frac{v^2}{c^2}, \tag{9.150}$$

namely

$$F_r = \frac{v^2}{g\bar{y}} \text{ or } F = \frac{v}{\sqrt{g\bar{y}}}. \tag{9.151}$$

Let us observe a velocity of generated waves $w_{1,2} = v \pm c$ with respect to the flow regime in the channel:

(a) subcritical flow $F_r < 1$, $c = \sqrt{g\bar{y}} \Rightarrow v < c$
   • wave velocity $w_1 = v + c > 0$,
   • wave velocity $w_2 = v - c < 0$,
(b) supercritical flow $F_r > 1$, $c = \sqrt{g\bar{y}} \Rightarrow v > c$
   • wave velocity $w_1 = v + c > 0$,
   • wave velocity $w_2 = v - c > 0$,
(c) critical flow $F_r = 1$, $c = \sqrt{g\bar{y}} \Rightarrow v = c$
   • wave velocity $w_1 = v + c > 0$,
   • wave velocity $w_2 = v - c = 0$.

Based on the aforementioned, note that the discharge or water level disturbances, which are propagating at elementary wave celerities, are traveling:

• in a subcritical flow, in both upstream and downstream directions, at different velocities;
• in a supercritical flow, in a downstream direction only, both waves are propagating in a downstream direction at different velocities;
• in a critical flow, in a downstream direction only, at a single velocity.

These facts play an important role in the analyses of weir submergence; namely, open channel flow with different flow regimes. Furthermore, they also play an important role in understanding boundary conditions when solving open channel flow differential equations.

9.7.4   Shape of positive and negative waves

Waves of finite amplitude can be observed as a superposition of generated elementary waves. A positive wave, Figure 9.44a, is characterized by a steep wave front. Velocities of elementary waves generated at greater depths are higher than velocities of waves previously generated at smaller depths. Thus, waves that were generated later catch up and flow over previously generated ones. Therefore, a positive wave keeps a steep wave front.

A negative wave, Figure 9.44b, is characterized by wave front expansion, because subsequently generated waves have smaller celerities than previously generated ones. Thus, a negative wave is increasingly "smudging" its shape.

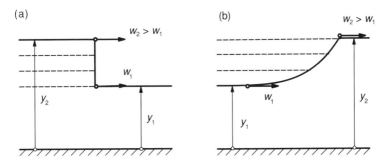

Figure 9.44   Positive and negative wave shape.

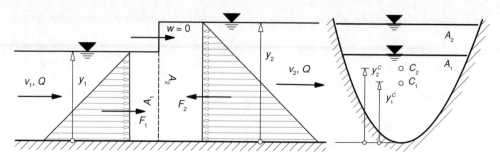

Figure 9.45 Standing wave.

9.7.5 Standing wave – hydraulic jump

A standing wave is a wave with zero absolute velocity. Figure 9.45 shows a standing wave of finite amplitude.

The continuity equation of relative motion of the finite wave front (9.131) gives

$$Q_2 = Q_1 + w\,(A_2 - A_1) \tag{9.152}$$

from which

$$Q_2 = Q_1 = Q. \tag{9.153}$$

The momentum equation of relative motion of the finite wave front (9.133) gives

$$F_2 - F_1 = \rho Q\,(v_1 - v_2) \tag{9.154}$$

from which it can be written

$$F_2 - F_1 + \rho Q^2 \left(\frac{1}{A_2} - \frac{1}{A_1} \right) = 0. \tag{9.155}$$

If hydrostatic pressure forces are expressed by pressure at a cross-section centroid

$$F = \rho g y^C A, \tag{9.156}$$

the following is obtained

$$g y_2^C A_2 - g y_1^C A_1 + \rho Q^2 \left(\frac{1}{A_2} - \frac{1}{A_1} \right) = 0 \tag{9.157}$$

or in arranged form

$$g y_2^C A_2 + \frac{Q^2}{A_2} = g y_1^C A_1 + \frac{Q^2}{A_1}. \tag{9.158}$$

The right hand side term contains the known values in the cross-section in front of the wave front while the left side term contains the unknown values at the cross-section behind the wave front for a prescribed discharge. Since the centroid position and the cross-sectional area are functions of depth, the unknown value y_2 can be calculated from equation (9.158) by one of the methods for non-linear equation solving.

Depths y_1 and y_2 are called standing wave conjugate depths. All the steps of the standing wave conjugate depth computation are equal to the conventional procedure for a hydraulic jump. Thus, a conventional hydraulic jump is in fact a standing wave, see analyses in Section 9.4.4. Furthermore, a standing wave is generated only in transition from supercritical to subcritical flow.

9.7.6   Wave propagation through transitional stretches

The solid line in Figure 9.46 is the water level curve for discharge Q in a channel with transitions from subcritical to superficial and back to subcritical flow. A hydraulic jump is normal; thus, the second conjugate depth is equal to the tailwater level. A hydraulic jump position is marked by hatched rectangle.

Let us observe positive and negative waves *propagating downstream* in a channel. Positive and negative waves, generated due to a sudden discharge increase or decrease in the upstream stretch in subcritical flow, are increasing and decreasing water levels, respectively. At the transition from subcritical to supercritical flow, a specific energy is always the minimum, which is a boundary condition for the upstream and downstream flow.

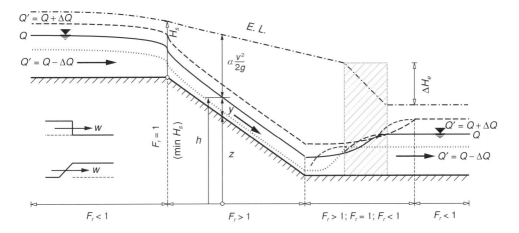

Figure 9.46   Positive and negative wave downstream.

In the downstream subcritical stretch, the water depth will increase or decrease due to propagation of a positive or negative wave.

At the hydraulic jump stretch, a new equilibrium will be established between the incoming supercritical flow forces and the downstream subcritical flow forces. A positive wave will propagate downstream or upstream depending whether the wave is positive or negative. The steady water level is marked by dashed line.

Propagation of a positive wave upstream, due to downstream significant discharge decrease, is shown in Figure 9.47. Let us assume that a significant discharge decrease is established by the lowering of a downstream sluice gate; thus, for steady outflow discharge $Q' = Q - \Delta Q$ the tailwater level shall be above the level established in critical cross-section (level h_c). In the tailwater subcritical stretch, the positive wave will increase water depth when propagating at an absolute velocity w. When it reaches a hydraulic jump position, the equilibrium of forces in front of and behind a hydraulic jump will change, the hydraulic jump will be pushed upstream, as a finite positive wave thus reducing the supercritical

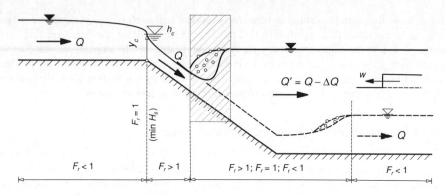

Figure 9.47   Positive wave upstream.

flow zone and expanding the subcritical flow zone. Since the upstream flow is undisturbed, there will be no equilibrium of discharges in front of and behind the wave, which will cause an increase in the tailwater level.

Propagation of the hydraulic jump, namely the finite positive wave, continues upstream until the upstream transition cross-section is reached. When the tailwater level submerges the level at the critical cross-section (level h_c), due to discharge decrease and level increase, the specific energy will become greater than the minimum one, and positive wave will continue to propagate upstream in subcritical flow.

9.8   Equations of non-steady flow in open channels

9.8.1   Continuity equation

Integral form of mass conservation law

A mass conservation law in the form of mass flow $\dot{M} = \rho Q$ will be set on a stretch between the two stations x_1, x_2, as shown in Figure 9.48:

$$\dot{M}_a + \dot{M}_2 - \dot{M}_1 = 0,$$
(9.159)

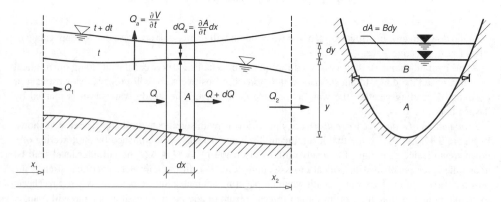

Figure 9.48   Mass conservation law.

where $\dot{M}_1$ is the inflow and $\dot{M}_2$ is the outflow discharge through the control volume while $\dot{M}_a$ is the mass discharge of accumulation. Incompressible water of constant density is observed; thus, a volume discharge can be used

$$Q_a + Q_2 - Q_1 = 0. \tag{9.160}$$

The volume accumulated due to the water level shift in time defines the accumulation discharge

$$Q_a = \frac{\partial V}{\partial t} = \frac{\partial}{\partial t} \int_{x_1}^{x_2} A dx. \tag{9.161}$$

When an accumulation discharge is introduced into the previous expression, an integral mass conservation law for open channel flow is obtained

$$\frac{\partial V}{\partial t} + Q_2 - Q_1 = 0 \tag{9.162}$$

or

$$\frac{\partial}{\partial t} \int_{x_1}^{x_2} A dx + Q_2 - Q_1 = 0. \tag{9.163}$$

The integral law applies to arbitrary selected stations x_1, x_2.

Differential form of mass conservation law

Using the usual operations the integral law (9.163) can be written in the form

$$\int_{x_1}^{x_2} \frac{\partial A}{\partial t} dx + \int_{x_1}^{x_2} \frac{\partial Q}{\partial x} dx = \int_{x_1}^{x_2} \left(\frac{\partial A}{\partial t} + \frac{\partial Q}{\partial x} \right) dx = 0. \tag{9.164}$$

Since the previous form of the law is valid for arbitrary selected stations x_1, x_2, then the integrand function is equal to zero:

$$\frac{\partial A}{\partial t} + \frac{\partial Q}{\partial x} = 0. \tag{9.165}$$

The obtained equation is the differential form of the continuity equation of non-steady flow in channels. It describes the law of conservation of volume discharges on an elementary volume of the length dx in Figure 9.48.

Special form of the continuity equation

When the chain rule for partial derivatives is applied on the continuity equation, then

$$\frac{dA}{dh} \frac{\partial h}{\partial t} + \frac{\partial Q}{\partial x} = 0. \tag{9.166}$$

Since $dA = Bdh$, it is written

$$B\frac{\partial h}{\partial t} + \frac{\partial Q}{\partial x} = 0. \tag{9.167}$$

If the first term is written as

$$\frac{gA}{gA}B\frac{\partial h}{\partial t} = \frac{gA}{c^2}\frac{\partial h}{\partial t}, \tag{9.168}$$

where c is the elementary wave celerity, the continuity equation is obtained in the form

$$\frac{gA}{c^2}\frac{\partial h}{\partial t} + \frac{\partial Q}{\partial x} = 0. \tag{9.169}$$

which is formally equal to the continuity equation for pipes. The continuity equation (9.169) can be integrated between two stations

$$\int_{x_1}^{x_2} \frac{gA}{c^2}\frac{\partial h}{\partial t} dx + Q_2 - Q_1 = 0, \tag{9.170}$$

where the integral is the other form of the accumulation discharge

$$Q_a = \frac{\partial V}{\partial t} = \int_{x_1}^{x_2} \frac{gA}{c^2}\frac{\partial h}{\partial t} dx. \tag{9.171}$$

9.8.2 Dynamic equation

Figure 9.49 vividly presents relationships between dynamic equation terms in head form over a finite, as well as differentially small, channel stretch.

Energy head form

The dynamic equation of non-steady flow in channels, formally equal to the dynamic equation for pipes, written in the head form

$$\frac{1}{g}\frac{\partial v}{\partial t} + \frac{\partial H}{\partial x} + I_e = 0, \tag{9.172}$$

where the energy head H can be expressed, as required, in one of the following forms

$$H = z + y + \alpha\frac{v^2}{2g} = z + H_s = h + \alpha\frac{v^2}{2g}.$$

The slope of the energy line is equal to

$$I_e = \frac{\tau_0}{\rho g R} = \frac{c_f}{2g R}|v|\, v = \frac{|Q|\, Q}{K^2}, \tag{9.173}$$

where $K = K(y)$ is the channel conveyance function.

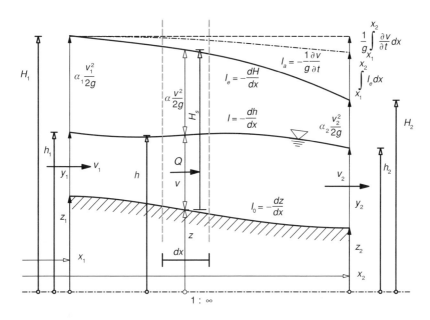

Figure 9.49   Head relations for the dynamic equation of non-steady flow.

Differential form

The differential form of a dynamic equation is formally equal to the equation for pipes where the pressure head is equal to water depth in the channel

$$\frac{\partial v}{\partial t} + v \frac{\partial v}{\partial x} + g \frac{\partial y}{\partial x} + g (I_e - I_0) = 0. \tag{9.174}$$

Integral energy form

The dynamic equation (9.172) can be integrated between two stations

$$H_2 - H_1 + \int\limits_{x_1}^{x_2} I_e dx + \frac{1}{g} \int\limits_{x_1}^{x_2} \frac{\partial v}{\partial t} dx = 0. \tag{9.175}$$

It is an integral form of the energy equation in head form.

9.8.3   Law of momentum conservation

Derivation

Starting from Newton's second law, which refers to the dynamic equilibrium between the momentum change rate and the acting force

$$\frac{d}{dt} m \vec{v} = \sum \vec{F} \tag{9.176}$$

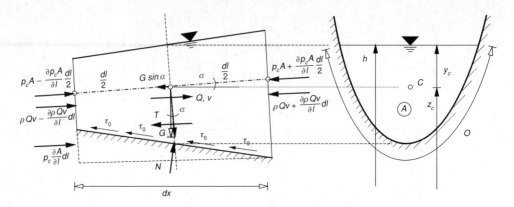

Figure 9.50   Channel element forces.

the law of momentum conservation will be established on a differential control volume, shown in Figure 9.50. The flow is one-dimensional along the axis which connects the cross-section centroids. Cross-sections are perpendicular to the axis while imaginary streamlines are parallel to it. Pressure distribution in a cross-section is hydrostatic. The flow axis is a spatial continuous curve $l(x, y, z)$, which closes an angle α with the horizontal. Thus, the vector term (9.176) is divided into two independent directions; namely, the equilibrium is established parallel and perpendicular to the flow axis.

Normal and shear surface forces and weights are acting on the control volume mass $\rho A dl$. In the direction perpendicular to the flow axis, the problem is reduced to the hydrostatic equilibrium of curved flow.

Let us observe the equilibrium of the momentum rate and total forces in the direction of flow. Total control volume momentum change, which consists of the momentum change within the volume and the difference generated due to momentum flow through cross-sections, can be written in the form

$$\frac{d}{dt}mv = \frac{\partial \rho A v}{\partial t}dl + \left(\rho Qv + \frac{\partial \rho Qv}{\partial l}\frac{dl}{2}\right) - \left(\rho Qv - \frac{\partial \rho Qv}{\partial l}\frac{dl}{2}\right),\tag{9.177}$$

where v is the mean velocity in a cross-section. After arranging, the rate of momentum change of the control volume mass is obtained

$$\frac{d}{dt}mv = \rho\left(\frac{\partial Q}{\partial t} + \frac{\partial Qv}{\partial l}\right)dl.\tag{9.178}$$

This expression is obtained for a uniform flow of momentum, that is the Boussinesq coefficient $\beta = 1$, see discussion.

Weight is projected in the flow direction

$$-G\sin\alpha = -\rho g A\frac{\partial z_c}{\partial l}dl.\tag{9.179}$$

Normal pressure forces consist of the cross-sectional forces

$$dN_a = \left(p_c A - \frac{\partial p_c A}{\partial l}dl\right) - \left(p_c A + \frac{\partial p_c A}{\partial l}dl\right) = -\frac{\partial p_c A}{\partial l}.\tag{9.180}$$

and the normal compressive force acting on the wetted surface of a control volume, projected in the flow direction as follows

$$dN_b = p_c \frac{\partial A}{\partial l} dl, \tag{9.181}$$

where p_c is the pressure at the cross-section centroid. Note that the term is not zero even for the prismatic channel. The total normal force is obtained by the addition

$$dN = dN_a + dN_b = -\frac{\partial p_c A}{\partial l} dl + p_c \frac{\partial A}{\partial l} dl. \tag{9.182}$$

The derivation of complex terms

$$dN = -A \frac{\partial p_c}{\partial l} dl - p_c \frac{\partial A}{\partial l} dl + p_c \frac{\partial A}{\partial l} dl \tag{9.183}$$

and reduction, gives the total pressure force

$$dN = -A \frac{\partial p_c}{\partial l} dl. \tag{9.184}$$

The projection of the friction force on the wetted surface of the control volume, in the flow direction, is approximately equal to

$$-T = -\tau_0 O dl. \tag{9.185}$$

(because of the assumed parallelism of streamlines with the flow axis). Total force is equal to

$$\sum F = dN - G \sin \alpha - T, \tag{9.186}$$

namely, after substituting Eqs (9.184), (9.179), and (9.185) into Eq. (9.186) and arranging, it is written as

$$\sum F = -\rho g A \left(\frac{\partial}{\partial l} \frac{p_c}{\rho g} + \frac{\partial z_c}{\partial l} + \frac{\tau_0}{\rho g R} \right) dl. \tag{9.187}$$

Since $h = z_c + \dfrac{p_c}{\rho g} = z_0 + y$ and the energy line gradient $\tau_0 = \rho g R$, it is written

$$\sum F = -\rho g A \left(\frac{\partial h}{\partial l} + J_e \right) dl. \tag{9.188}$$

Finally, the law on momentum conservation in a control volume is

$$\rho \left(\frac{\partial Q}{\partial t} + \frac{\partial (Qv)}{\partial l} \right) dl + \rho g A \left(\frac{\partial h}{\partial l} + J_e \right) dl = 0, \tag{9.189}$$

from which a differential form is derived

$$\frac{\partial Q}{\partial t} + \frac{\partial (Qv)}{\partial l} + g A \left(\frac{\partial h}{\partial l} + J_e \right) = 0. \tag{9.190}$$

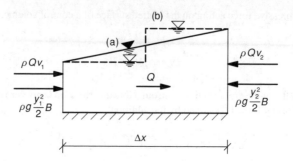

Figure 9.51   Finite control volume.

Since in mildly sloping channels gradients can be replaced by slopes, then

$$\frac{\partial Q}{\partial t} + \frac{\partial (Qv)}{\partial x} + gA\left(\frac{\partial h}{\partial x} + I_e\right) = 0. \tag{9.191}$$

Discussion

Note that the dynamic equation (9.191) is a one-dimensional model of complex spatial non-steady flow in open channels, established based on many assumptions for the purposes of simplicity of analysis, above all, and monitoring of energy and other flow parameters using the mean profile velocity $v = Q/A$. It is an acceptable description of a real flow in engineering terms.[13] The derivation of complex terms will show that the momentum conservation law (9.191) is equivalent to the energy equation, assuming the Boussinesq number $\beta = 1$ and the Coriolis number $\alpha = 1$:

$$\frac{\partial Av}{\partial t} + \frac{\partial (Qv)}{\partial x} + gA\left(\frac{\partial h}{\partial x} + I_e\right) = 0. \tag{9.192}$$

Following the grouping, the continuity equation can be identified (9.165)

$$A\frac{\partial v}{\partial t} + \underbrace{v\frac{\partial A}{\partial t} + v\frac{\partial Q}{\partial x}}_{\equiv 0} + Av\frac{\partial v}{\partial x} + gA\left(\frac{\partial h}{\partial x} + I_e\right) = 0. \tag{9.193}$$

Since the sum of the second and the third term is equal to zero, after division by gA, the energy equation in head form is obtained

$$\frac{1}{g}\frac{\partial v}{\partial t} + \frac{\partial}{\partial x}\left(\frac{v^2}{2g}\right) + \frac{\partial h}{\partial x} + I_e = 0. \tag{9.194}$$

In engineering practice, the momentum conservation law is often used in its original form on *finite control volumes*. In that case, a critical review is needed because the obtained results do not guarantee energy conservation.

Let us observe a finite control volume and steady flow without friction, according to Figure 9.51; namely, (a) with a continuous water level and (b) with a discontinuous water level. In both cases there

---

[13] See the discussion on dynamic equations for pipes given in Chapter 5.

is equilibrium of pressure forces and the momentum between the upstream and downstream cross-section

$$\Delta N + \Delta K = 0, \tag{9.195}$$

where the differences are equal to

$$\Delta N = \rho g B \frac{y_1^2 - y_2^2}{2}, \tag{9.196}$$

$$\Delta K = \rho Q (v_1 - v_2). \tag{9.197}$$

When in limits $\Delta x \to dx$, a differential law on momentum conservation applies

$$\rho \underbrace{\frac{\partial (Qv)}{\partial x} dx}_{\substack{dK = \lim \Delta K \\ \Delta x \to 0}} + \underbrace{\rho g A \frac{\partial h}{\partial x} dx}_{\substack{dN = \lim \Delta K \\ \Delta x \to 0}} = 0.$$

(a) Assuming the water level according to scheme (a) on Figure 9.51 discharge Q can be expressed by mean area and mean velocity; thus the momentum change in limits when $\Delta x \to dx$ will be

$$\Delta K = \rho Q (v_1 - v_2) = \rho B \underbrace{\frac{y_1 + y_2}{2} \frac{v_1 + v_2}{2}}_{\lim \Delta x \to dx = Q} (v_1 - v_2)$$

which, together with compressive forces gives

$$\Delta N + \Delta K = \rho g B \frac{y_1 + y_2}{2} (y_1 - y_2) + \rho B \frac{y_1 + y_2}{2} \frac{(v_1 + v_2)}{2} (v_1 - v_2) = 0.$$

After reduction it is written as

$$g (y_1 - y_2) + \frac{(v_1 + v_2)}{2} (v_1 - v_2) = 0,$$

from which the energy form is obtained according to which energy is conserved between two cross-sections

$$y_1 + \frac{v_1^2}{2g} = y_2 + \frac{v_2^2}{2g}.$$

(b) If a discontinuous water surface is assumed according to scheme (b) discharge Q cannot be expressed by mean area and mean velocity, and the momentum change is written in the original form

$$\Delta K = \rho Q (v_2 - v_1) = \rho \left(y_2 v_2^2 - y_1 v_1^2 \right),$$

$$\Delta N = \rho g \frac{y_1^2 - y_2^2}{2} = \rho g \frac{(y_1 + y_2)}{2} (y_1 - y_2).$$

This fact is used when deriving conjugate depths of a hydraulic jump on a horizontal surface, with energy dissipation between them, see Section 9.7.5.

9.9   Equation of characteristics

The general procedure of transformation of differential hyperbolic equations is described in Chapter 5 Equations of non-steady flow in pipes.

9.9.1   Transformation of non-steady flow equations

Equations of non-steady flow in channels will be expressed by primitive variables such as the depth y and velocity v. The continuity equation (9.165) will be dismembered by partial derivatives, thus

$$\frac{dA}{dy}\frac{\partial y}{\partial t} + A\frac{\partial v}{\partial x} + v\frac{\partial A}{\partial x} = 0. \tag{9.198}$$

Since the cross-section area $A(x,y)$ is slightly variable along the flow, there is a total derivative

$$dA = \left.\frac{\partial A}{\partial y}\right|_{x=const} dy + \left.\frac{\partial A}{\partial x}\right|_{y=const} dx, \tag{9.199}$$

from which a spatial increment in the direction of the x axis is obtained

$$\frac{dA}{dx} = \left.\frac{dy}{dx}\frac{\partial A}{\partial y}\right|_{x=const} + \left.\frac{\partial A}{\partial x}\right|_{y=const} = B\frac{dy}{dx} + \left.\frac{\partial A}{\partial x}\right|_{y=const}. \tag{9.200}$$

A partial derivative can be, accordingly, expressed by spatial changes as

$$\frac{\partial A}{\partial x} = B\frac{\partial y}{\partial x} + \left.\frac{\partial A}{\partial x}\right|_{y=const}. \tag{9.201}$$

Introducing Eq. (9.201) into the continuity equation (9.198), together with $B = dA/dy$, the following is obtained

$$B\frac{\partial y}{\partial t} + A\frac{\partial v}{\partial x} + v\left(B\frac{\partial y}{\partial x} + \left.\frac{\partial A}{\partial x}\right|_{y=const}\right) = 0. \tag{9.202}$$

Division by width B gives

$$\frac{\partial y}{\partial t} + \frac{A}{B}\frac{\partial v}{\partial x} + v\left(\frac{\partial y}{\partial x} + \left.\frac{1}{B}\frac{\partial A}{\partial x}\right|_{y=const}\right) = 0. \tag{9.203}$$

If the term is marked as slope I_b due to lateral widening of a non-prismatic channel

$$I_b = \left.\frac{1}{B}\frac{\partial A}{\partial x}\right|_{y=const} \tag{9.204}$$

then, after arranging, a dismembered continuity equation is obtained

$$\frac{\partial y}{\partial t} + v\frac{\partial y}{\partial x} + \frac{A}{B}\frac{\partial v}{\partial x} + vI_b = 0. \tag{9.205}$$

The respective modified dynamic equation, expressed by variables y, v will be

$$\frac{\partial v}{\partial t} + v\frac{\partial v}{\partial x} + g\frac{\partial y}{\partial x} + g\left(I_e - I_0\right) = 0. \tag{9.206}$$

9.9.2 Procedure of transformation into characteristics

A procedure of transformation of Eqs (9.205) and (9.206) into characteristics is carried out according to the general procedure described in Chapter 5. Values of the coefficients in the general form of the equations system are obtained first:

	$U = y$		$V = v$		
	a	b	c	d	e
(1)	1	v	0	A/B	vI_b
(2)	0	g	1	v	$g\left(I_e - I_o\right)$

Then, the determinants are calculated:

$$A = \begin{vmatrix} 1 & 0 \\ 0 & 1 \end{vmatrix} = 1, \quad 2B = \begin{vmatrix} 1 & A/B \\ 0 & v \end{vmatrix} + \begin{vmatrix} v & 0 \\ g & 1 \end{vmatrix} = 2v, \quad C = \begin{vmatrix} v & A/B \\ g & v \end{vmatrix} = v^2 - g\frac{A}{B},$$

$$D = \begin{vmatrix} 1 & v \\ 0 & g \end{vmatrix} = g, \quad E = \begin{vmatrix} v & 0 \\ g & 1 \end{vmatrix} \begin{bmatrix} b & c \end{bmatrix} = v, \quad F = \begin{vmatrix} 1 & vI_b \\ 0 & g\left(I_e - I_0\right) \end{vmatrix} = g\left(I_e - I_0\right) \text{ and}$$

$$G = \begin{vmatrix} v & vI_b \\ g & g\left(I_e - I_0\right) \end{vmatrix} = gv\left(I_e - I_0\right) - vgI_b.$$

Calculation of the slope of characteristics is defined by the absolute wave velocity

$$w^{\pm} = v \pm \sqrt{g\frac{A}{B}} = v \pm c. \tag{9.207}$$

When $dU = dy$ and $dY = dv$ are substituted into general wave function equations

$$\begin{aligned} \Gamma^+: \quad & DdU + \left(Aw^+ - E\right)dV + \left(Fw^+ - G\right)dt = 0 \\ \Gamma^-: \quad & DdU + \left(Aw^- - E\right)dV + \left(Fw^- - G\right)dt = 0 \end{aligned} \tag{9.208}$$

the following is obtained

$$gdy + \left[(v \pm c) - v\right]dv + \left[g\left(I_e - I_0\right)(v \pm c) - vg\left(I_e - I_0\right) + vgI_b\right]dt = 0.$$

After arranging, differential equations of wave functions of the characteristics in open channel flow are obtained

$$dv \pm \frac{g}{c}dy + g\left(I_e - I_0 \pm \frac{v}{c}I_b\right)dt = 0. \tag{9.209}$$

9.10   Initial and boundary conditions

Initial conditions for solving a non-steady flow problem, described by differential equations, are prescribed states of selected hydraulic variables at time t_0, for example $y(t_0)$, $v(t_0)$ or $h(t_0)$, $Q(t_0)$ that all other hydraulic parameters can be calculated from. From the initial state, that is at time $t > t_0$, a solution is defined by the dynamic equation and the continuity equation based on boundary conditions that can be natural (prescribed boundary discharge) and essential (prescribed depth or level). Boundary conditions shall be selected to provide an unambiguous solution of the non-steady flow problem.

Since in subcritical flow disturbance (waves) are propagating downstream and upstream at wave velocities $w^{\pm}$, see Figure 9.52a, upstream and downstream disturbances influence the solution at point P. In a channel stretch with subcritical flow, boundary conditions are prescribed at the upstream and downstream end, as time-dependant functions that wave disturbances can be unambiguously defined from. For example an upstream boundary condition is discharge $Q(t)$ while the downstream one is level $h(t)$.

In supercritical flow disturbances (waves) are always moving downstream at wave velocities $w^{\pm}$, see Figure 9.52b, only upstream disturbances influence the solution at point P. In a channel stretch with supercritical flow, natural and essential boundary conditions are prescribed at the upstream channel end only.

At the transition from subcritical to supercritical flow, see Figure 9.52c, upstream disturbances are transferred downstream, passing through the critical section. Since the specific energy is always the minimum at the critical section, only one boundary condition is prescribed at the upstream end; namely, discharge $Q(t)$ or level $h(t)$.

At the transition from supercritical to subcritical flow, see Figure 9.52d, both boundary conditions are prescribed at the beginning of the spillway. A functional relationship $h(Q)$ shall be known at the downstream end of the channel stretch in subcritical flow. Namely, at the start of subcritical flow, discharge is calculated from the upstream supercritical flow stretch that the hydraulic jump passes through, while

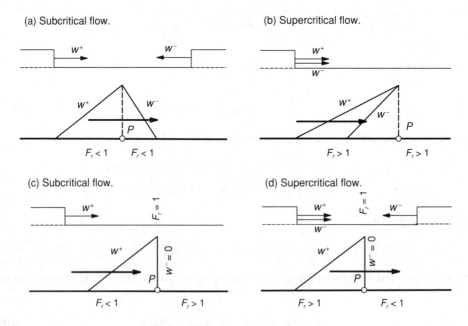

Figure 9.52   Boundary conditions.

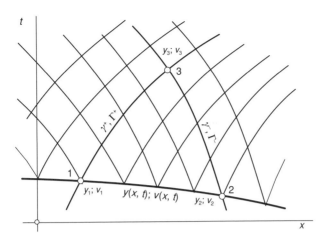

Figure 9.53   Integration along characteristics.

downstream disturbances cannot be transferred upstream from subcritical to supercritical flow. A finite positive wave that submerges the upstream supercritical flow is analyzed in Section 9.7.6.

9.11   Non-steady flow modelling

9.11.1   Integration along characteristics

A curve of the prescribed initial state of the open channel flow in the plane x, t is marked as a thick line in Figure 9.53. Selected variables are depth y and velocity v, but other pairs of state variables can be selected. For each x, t value of the solution $y(x, t)$, $v(x, t)$ are on the curve. From each point with the known state, two elementary waves propagate at velocities $w^{\pm}$ with their trajectories described by positive γ^{+} and negative γ^{-} characteristic, as shown by the mesh of characteristics. If two points 1 and 2 are selected on the initial state curve, then the intersection of the positive and negative characteristic will be point 3. In a curvilinear triangle, defined by these three points, the state is under the influence of the known state between points 1 and 2.

Positive and negative wave trajectories $\gamma^{\pm}$ are described by two equations

$$\gamma^{\pm}: \quad \frac{dx}{dt} = v \pm c \tag{9.210}$$

along which a hydraulic state is defined by two equations

$$\Gamma^{\pm}: \quad \frac{dv}{dt} \pm \frac{g}{c}\frac{dy}{dt} + g\left(I_e - I_0 \pm \frac{v}{c}I_b\right) = 0 \tag{9.211}$$

The equations of characteristics (9.210) can be integrated between points a and b lying on one of the trajectories

$$\int_a^b dx = \int_a^b (v \pm c)dt \tag{9.212}$$

as well as wave functions and Eq. (9.211)

$$\int_a^b \left(dv \pm \frac{g}{c}dy\right) + g\int_a^b \left(I_e - I_0 \pm \frac{v}{c}I_b\right)dt = 0. \tag{9.213}$$

For two points that are sufficiently close, such as points *1* and *3* as points *a,b* or *2* and *3* as points *a,b*, integration of Eq. (9.212) will give algebraic equations

$$F_1: \ x_3 - x_1 - \frac{1}{2}\left[(v_1 + c_1) + (v_3 + c_3)\right](t_3 - t_1) = 0, \tag{9.214}$$

$$F_2: \ x_3 - x_2 - \frac{1}{2}\left[(v_2 - c_2) + (v_3 - c_3)\right](t_3 - t_2) = 0 \tag{9.215}$$

while integration of Eq. (9.213) will give algebraic equations

$$F_3: \ (v_3 - v_1) + \frac{2g}{c_1 + c_3}(y_3 - y_1) + \frac{g}{2}\left(S_1^+ + S_3^+\right)(t_3 - t_1) = 0, \tag{9.216}$$

$$F_4: \ (v_3 - v_2) - \frac{2g}{c_2 + c_3}(y_3 - y_2) + \frac{g}{2}\left(S_2^- + S_3^-\right)(t_3 - t_2) = 0. \tag{9.217}$$

The second integral in Eq. (9.213) is calculated using the mean integral value, as well as abridged designations for integrand functions

$$S^+ = I_e - I_0 + \frac{v}{c}I_b, \tag{9.218}$$

$$S^- = I_e - I_0 - \frac{v}{c}I_b. \tag{9.219}$$

Equations (9.214), (9.215), (9.216), and (9.217) form a system of four non-linear equations with the unknowns x_3, t_3, y_3, v_3:

$$F_1(x_3, t_3, y_3, v_3) = 0, \tag{9.220}$$

$$F_2(x_3, t_3, y_3, v_3) = 0, \tag{9.221}$$

$$F_3(x_3, t_3, y_3, v_3) = 0, \tag{9.222}$$

$$F_4(x_3, t_3, y_3, v_3) = 0 \tag{9.223}$$

that can be calculated by one of the methods for non-linear equation solving.

The domain of non-steady channel flow problem solving is limited to interval $[0, L]$; thus, the characteristics are influenced by the boundary conditions, see Figure 9.54. In the left boundary points $(0, t)$ there is only a negative characteristic; thus, in order to determine the state in point *3* the positive characteristic is replaced by a boundary condition; namely the functional relationship between velocity and depth. Similarly, the negative characteristic that is missing on the right boundary (L, t) shall be replaced by another, independent, functional relationship between velocity and depth.

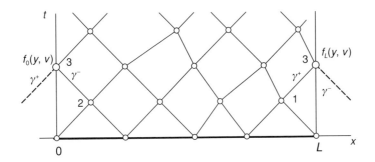

Figure 9.54 Boundary conditions.

Thus, for example, if velocity $v_0(t)$ or discharge $Q_0(t)$ is prescribed on the left boundary, the system of equations is written in the form

$$F_1 : \; x_3 = 0, \tag{9.224}$$

$$F_2 : \; x_2 - \frac{1}{2}\left[(v_2 - c_2) + (v_0 - c_3)\right](t_3 - t_2) = 0, \tag{9.225}$$

$$F_3 : \; v_3 - v_0 = 0 \text{ or } Av_3 - Q_0 = 0, \tag{9.226}$$

$$F_4 : \; (v_3 - v_2) - \frac{2g}{c_2 + c_3}(y_3 - y_2) + \frac{g}{2}\left(S_2^- + S_3^-\right)(t_3 - t_2) = 0. \tag{9.227}$$

If depth y_0 is prescribed on the right boundary the equations will be

$$F_1 : \; L - x_1 - \frac{1}{2}\left[(v_1 + c_1) + (v_3 + c_3)\right](t_3 - t_1) = 0, \tag{9.228}$$

$$F_2 : \; x_3 - L = 0. \tag{9.229}$$

$$F_3 : \; (v_3 - v_1) + \frac{2g}{c_1 + c_3}(y_3 - y_1) + \frac{g}{2}\left(S_1^+ + S_3^+\right)(t_3 - t_1) = 0. \tag{9.230}$$

$$F_4 : \; y_3 - y_0 = 0. \tag{9.231}$$

9.11.2 Matrix and vector of the channel finite element

Integration of the conservation law on a finite element

The integral equation of continuity (9.163) written for a finite element of length Δl will be

$$\frac{\partial}{\partial t}\int_{\Delta l} A\, dx + Q_2 - Q_1 = 0, \tag{9.232}$$

where the first term is the accumulation discharge, written as the volume change in time

$$\frac{\partial V}{\partial t} + Q_2 - Q_1 = 0. \tag{9.233}$$

Integration of Eq. (9.233) in time interval Δt between the two time stages gives

$$V^+ - V + \int_{\Delta t} (Q_2 - Q_1)\, dt = 0. \tag{9.234}$$

Using the time integration rule, the first elementary equation is written in the form

$$F_1: \quad V^+ - V + (1 - \vartheta)\Delta t\, (Q_2 - Q_1) + \vartheta \Delta t \left(Q_2^+ - Q_1^+\right) = 0, \tag{9.235}$$

where the volume is equal to

$$V = \int_{\Delta l} A\, dx \doteq \frac{A_1 + A_2}{2} \Delta l. \tag{9.236}$$

Starting from the integral form of the energy conservation law (9.175) on a finite element of length $\Delta l = [x_1, x_2]$, it is written as

$$H_2 - H_1 + \int_{\Delta l} I_e dx + \frac{1}{g} \int_{\Delta l} \frac{\partial}{\partial t} \frac{Q}{A} dx = 0. \tag{9.237}$$

Integration of Eq. (9.237) in time interval Δt between two time stages gives

$$\frac{\Delta l}{2g} \int_{\Delta t} \left(\frac{\partial}{\partial t} \frac{Q_1}{A_1} + \frac{\partial}{\partial t} \frac{Q_2}{A_2} \right) dt + \int_{\Delta t} (H_2 - H_1)\, dt + \frac{\Delta l}{2} \int_{\Delta t} (I_{e1} + I_{e2}) dt = 0. \tag{9.238}$$

When the mean value theorem of the integral is applied, the second elementary equation is obtained in the form

$$F_2: \quad \frac{\Delta l}{2g} \left[\left(\frac{Q_1^+}{A_1^+} + \frac{Q_2^+}{A_2^+} \right) - \left(\frac{Q_1}{A_1} + \frac{Q_2}{A_2} \right) \right] +$$
$$+ (1 - \vartheta)\Delta t\, (H_2 - H_1) + \vartheta \Delta t \left(H_2^+ - H_1^+ \right) + \tag{9.239}$$
$$+ (1 - \vartheta)\Delta t\, (I_{e1} + I_{e2}) \frac{\Delta l}{2} + \vartheta \Delta t \left(I_{e1}^+ + I_{e2}^+ \right) \frac{\Delta l}{2} = 0.$$

In a subroutine Unsteady a call for a subroutine for computation of the non-steady flow matrix and vector for elemental channel finite element is added.

```
select case(Elems(ielem).tip)
   case ...

      ...

   case (CHANNEL_OBJ)
      lookup=Elems(ielem).lookup
      Channel=Channels(lookup)
      if(Channel.kind.eq.SUBCRITICAL) &
        call SubCriticalChannelMtx(ielem)
      if(ExplicitSuperCritical) then
        if(Channel.kind.eq.SUPERCRITICAL) &
          call explSuperCriticalChannelMtx(ielem)
        elseif(Channel.kind.eq.SUPERCRITICAL) &
```

```
        call SuperCriticalChannelMtx(ielem)
    endif
  case ...
      ...
endselect
```

Computation is branching according to the flow regime in a finite element. For subcritical flow, a subroutine `SubCriticalChannelMtx` is called to calculate the non-steady flow matrix. In supercritical flow, computation branches, depending on the optional variable, into explicit integration by calling the subroutine `explSuperCriticalChannelMtx` or implicit integration when subroutine `SuperCriticalChannelMtx` is called.

Subcritical flow: Subroutine `SubCriticalChannelMtx`

The Newton–Raphson iterative form for a finite element is formally written using the matrix-vector operations

$$[\underline{H}] \cdot [\Delta h^+] + [\underline{Q}] \cdot [\Delta Q^+] = [\underline{F}], \tag{9.240}$$

where the matrix

$$[\underline{H}] = \begin{bmatrix} \dfrac{\Delta l}{2} B_1^+ & \dfrac{\Delta l}{2} B_2^+ \\[2ex] \left[-\dfrac{\Delta l}{2g} B_1^+ \dfrac{Q_1^+}{A_1^{2+}} - \vartheta \Delta t \left(1 - F_{r1}^+\right) \right] & \left[-\dfrac{\Delta l}{2g} B_2^+ \dfrac{Q_2^+}{A_2^{2+}} + \vartheta \Delta t \left(1 - F_{r2}^+\right) \right] \end{bmatrix} \tag{9.241}$$

and matrix

$$[\underline{Q}] = \begin{bmatrix} -\vartheta \Delta t & +\vartheta \Delta t \\[2ex] \left(\dfrac{\Delta l}{2gA_1^+} - \vartheta \Delta t \alpha_1^+ \dfrac{Q_1^+}{gA_1^{2+}} + \vartheta \Delta t \dfrac{\Delta l}{K_1^{2+}} |Q_1^+| \right) & \left(\dfrac{\Delta l}{2gA_2^+} + \vartheta \Delta t \alpha_2^+ \dfrac{Q_2^+}{gA_2^{2+}} + \vartheta \Delta t \dfrac{\Delta l}{K_2^{2+}} |Q_2^+| \right) \end{bmatrix} \cdot$$

$$\tag{9.242}$$

The right hand side vector is

$$[\underline{F}] = - \begin{bmatrix} F_1 \\ F_2 \end{bmatrix}. \tag{9.243}$$

Elemental discharge increments are calculated from the Newton–Raphson form of the elemental equations

$$[\underline{H}] \cdot [\Delta h^+] + [\underline{Q}] \cdot [\Delta Q^+] = [\underline{F}]$$

in a manner such that the equation is multiplied by the inverse matrix $[\underline{Q}]^{-1}$

$$[\Delta Q^+] = [\underline{F}] \cdot [\underline{Q}]^{-1} - [\underline{H}] \cdot [\underline{Q}]^{-1} \cdot [\Delta h^+]. \tag{9.244}$$

Introducing the symbols

$$[\underline{A}] = [\underline{H}][\underline{Q}]^{-1},$$ (9.245)

$$[\underline{B}] = [\underline{F}][\underline{Q}]^{-1},$$ (9.246)

an expression for elemental discharge increments is obtained

$$[\Delta Q^+] = [\underline{B}] - [\underline{A}] \cdot [\Delta h^+].$$ (9.247)

The finite element matrix A^e and vector B^e for non-steady modelling have the form

$$A^e = \vartheta \, \Delta t \begin{bmatrix} +\underline{A}_{11} & +\underline{A}_{12} \\ -\underline{A}_{21} & -\underline{A}_{22} \end{bmatrix},$$ (9.248)

$$B^e = (1 - \vartheta)\Delta t \begin{bmatrix} +Q_1 \\ -Q_2 \end{bmatrix} + \vartheta \, \Delta t \begin{bmatrix} +Q_1^+ \\ -Q_2^+ \end{bmatrix} + \vartheta \, \Delta t \begin{bmatrix} +\underline{B}_1 \\ -\underline{B}_2 \end{bmatrix}.$$ (9.249)

The unknown elementary discharge increments are calculated from Eq. (9.244) following the computation of the unknown piezometric head nodal increments, subroutine `IncVar`.

The entire procedure is implemented numerically in a subroutine

> `subroutine SubCriticalChannelMtx(ielem)`

in a program module `Channels.f90`.

Supercritical flow: Subroutine `SuperCriticalChannelMtx`

Since in supercritical flow both boundary conditions are the upstream ones, the unknowns h_2^+ and Q_2^+ occur on each finite element in the downstream node, which can be calculated from elemental equations (9.235) and (9.239). The Newton–Raphson procedure of calculation of the unknowns is iterative, with the increments of the unknowns Δh_2^+, ΔQ_2^+ calculated from the system

$$\begin{bmatrix} J_{11} & J_{12} \\ J_{21} & J_{22} \end{bmatrix} \cdot \begin{bmatrix} \Delta h_2^+ \\ \Delta Q_2^+ \end{bmatrix} = - \begin{bmatrix} F_1 \\ F_2 \end{bmatrix}$$ (9.250)

where the Jacobian matrix $[J]$ members are equal to

$$J_{11}: \quad \frac{\partial F_1}{\partial h_2^+} = \frac{\Delta l}{2} B_2^+$$ (9.251)

$$J_{12}: \quad \frac{\partial F_1}{\partial Q_2^+} = +\vartheta \, \Delta t$$ (9.252)

$$J_{21}: \quad \frac{\partial F_2}{\partial h_2^+} = \left[-\frac{\Delta l}{2g} B_2^+ \frac{Q_2^+}{A_2^{2+}} + \vartheta \, \Delta t \left(1 - F_{r2}^+ \right) \right]$$ (9.253)

$$J_{22}: \quad \frac{\partial F_2}{\partial Q_2^+} = \frac{\Delta l}{2gA_2^+} + \vartheta \, \Delta t \alpha_2^+ \frac{Q_2^+}{gA_2^{2+}} + \vartheta \, \Delta t \frac{\Delta l}{K_2^{2+}} |Q_2^+|$$ (9.254)

If $\left[\underline{J}\right]$ is the inverse matrix of $[J]$, a solution is obtained in the form

$$\begin{bmatrix} \Delta h_2^+ \\ \Delta Q_2^+ \end{bmatrix} = - \begin{bmatrix} \underline{J}_{11} & \underline{J}_{12} \\ \underline{J}_{21} & \underline{J}_{22} \end{bmatrix} \cdot \begin{bmatrix} F_1 \\ F_2 \end{bmatrix} \tag{9.255}$$

The finite element matrix A^e and vector B^e for non-steady modelling of superficial flow have the form

$$A^e = \begin{bmatrix} 0 & 0 \\ 0 & 1 \end{bmatrix} \tag{9.256}$$

$$B^e = \begin{bmatrix} 0 \\ \left(-\underline{J}_{11}F_1 - \underline{J}_{12}F_2\right) \end{bmatrix} \tag{9.257}$$

which enables a frontal solver to calculate the increment Δh_2^+ while increment ΔQ_2^+ is calculated from Eq. (9.255) and the auxiliary elemental matrix $\left[\underline{A}\right]$ and vector $\left[\underline{B}\right]$ in a form suitable for subroutine `IncVar`

$$\left[\underline{A}\right] = 0 \tag{9.258}$$

$$\left[\underline{B}\right] = \begin{bmatrix} 0 \\ -\underline{J}_{21}F_1 - \underline{J}_{22}F_2 \end{bmatrix} \tag{9.259}$$

Note that for the *internal points* of the channel in supercritical flow, nodal sums of discharges are not closed as in the subcritical flow and the continuity of flow is secured within a subroutine by transfer of the discharge to the next connected channel element

$$Q_1^{e+1} = Q_2^e \tag{9.260}$$

The entire procedure is implemented numerically in the subroutine

subroutine SuperCriticalChannelMtx(ielem),

which is located in the program module `Channels.f90`.

9.11.3   Test examples

Subcritical flow comparison test

The validity of modelling non-steady flow in a channel with subcritical flow will be shown by a comparison test with the solution obtained by the method of characteristics. A simple prismatic channel of rectangular cross-section and horizontal bed was selected due to the relatively simple solution by the method of characteristics. The test channel is shown in Figure 9.55.

The initial state in the channel is hydrostatic; the initial depth is $y = y_0$. The left edge boundary condition $Q(t)$ is shown in the graph while the right edge level is defined by depth y_0. The channel is discretized into 50 m long finite elements. Discretization nodes are also nodes of positive and negative characteristics at time $t = 0$.

Figure 9.56 shows water level change in the time of two hours for $x = 0$, that is at the start of the channel. The continuous line refers to the solution obtained by `SimpipCore` software (www.wiley.com/go/jovic) and the dashed line to the method of characteristics. With respect to the

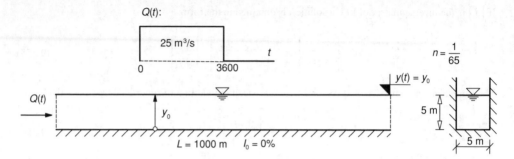

Figure 9.55 Comparison test channel.

channel's simplicity and possibly simple implementation of the method of characteristics, it is considered that the solution obtained by this method is accurate in the framework of numerical integration.

Note the great similarity between non-steady flows in an open channel with the sudden filling of a pipe with great friction. One should also notice a smoothing of the waves obtained by the finite element method, which is a consequence of the value of the time integration parameter ϑ. A value of the parameter ϑ closer to 0.5 significantly improves the picture, although not in all real examples of non-steady flow modelling. Thus, a value of $\vartheta = 0.75$ was adopted as an optimum for integration of the conservation law of non-steady flow in channels.

Figure 9.57 shows a comparison of solutions for the discharge at the end of the channel as well as a boundary condition at the start of the channel.

Non-steady flow modeled by integration of the conservation law on finite elements gives entirely reliable results in engineering terms. In real hydraulic networks it secures conservation of important parameters in the modeled system, such as mass and energy, which is not a case with possible implementation of the method of characteristics.

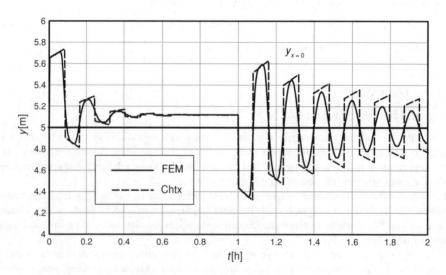

Figure 9.56 Subcritical flow, Fem vs. Chtx, water level at the start of the channel.

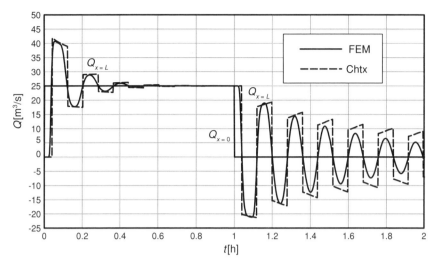

Figure 9.57   Subcritical flow, Fem vs. Chtx, discharge at the start and end of a channel.

Supercritical flow test

A test of non-steady supercritical flow modelling in a spillway, as shown in Figure 9.58, was carried out. A rectangular cross-section channel has an absolute hydraulic roughness of 22 mm, approximately corresponding to the Manning coefficient of 1/50. The initial state is a steady water level for discharge $Q = 10\,m^3/s$. Discharge $Q(t)$ enters the channel at its start at the minimum specific energy.

The channel is discretized into 10 finite elements of 10 m length, calculation with a $\Delta t = 2\,s$ time step gives 25 time stages.

Water depth development in time at the spillway start, middle, and end is shown in Figure 9.59.

Figure 9.60 shows discharge development in time at the spillway start, middle, and end.

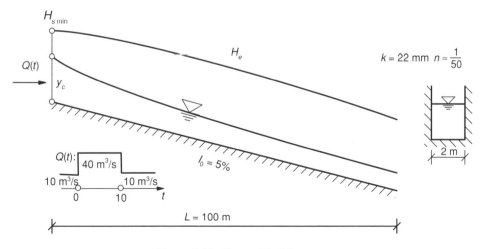

Figure 9.58   Supercritical flow test.

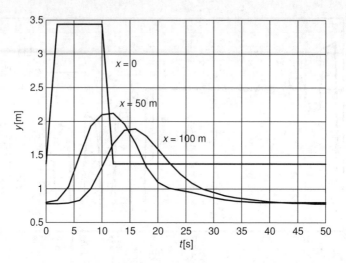

Figure 9.59   Depths.

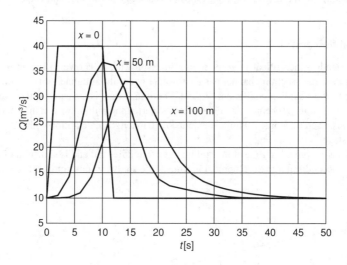

Figure 9.60   Discharges.

Examples of modelling of these and other channel flows can be found at www.wiley.com/go/jovic: Program sources & tests\ Test simpip files\9 Chapter Channels tests (see Programs Chtx characteristics).

References

Chow, V.T. (1959) *Open-Channel Hydraulics*. McGraw-Hill Kogakusha Ltd, Tokyo.

Izbash, S.V. and Khaldre Kh, Yu. (1970) *Hydraulics of River Channel Closure*. Butterworths, London.

Jović, V. (2006) *Fundamentals of Hydromechanics* (in Croatian: Osnove hidromehanike), Element, Zagreb.

Peterka, A.J. (2006) *Hydraulic Design of Stilling Basins and Energy Dissipators*. U.S.B.R Eng. Monograph No. 25.

Further reading

Abbot, M.B. (1979) *Computational Hydraulics – Elements of the Theory of Surface Flow.* Pitman, Boston.

Agroskin, I.I., Dmitrijev, G.T., and Pikalov, F.I. (1969) *Hidraulika* (Hydraulics). Tehnička knjiga, Zagreb.

Beraković, B., Franković, B. and Jović, V. (1983) Simulation of experiment performed on a long stretch of the Drava River. *XX IAHR Congress*, Moscow.

Bogomolov, A.I. and Mihajlov, K.A. (1972) *Gidravlika*. Stroiizdat, Moskva.

Budak, B.M., Samarskii, A.A., and Tikhonov, A.N. (1980) *Collection of Problems on Mathematical Physics* [in Russian], Nauka, Moscow.

Cunge, J.A., Holly, F.M., and Verwey, A. (1980) *Practical Aspects of Computational River Hydraulics.* Pitman Advanced Publishing Program, Boston.

Davis, C.V. and Sorenson, K.E. (1969) *Handbook of Applied Hydraulics*, 3th edn, McGraw-Hill Co., New York.

Dracos, Th. (1970) Die Berechnung istatationärer Abfüsse in offenen Gerinnen beliebiger Geometrie. *Schweizerische Bauzeitung*, 88. Jahrgang Heft 19.

Elevatorski, E.A. (1959) *Hydraulic Energy Dissipators*. McGraw-Hill Book Co., New York.

Filipovic, M., Geres, D., Vranjes, M., Jovic, V. (2000) Flood control planning for the Sava River basin in Croatia. 4th International Conference Hydromatics 23–27 July 2000, Iowa USA.

Fox, J.A. (1977) *Hydraulic Analysis of Unsteady Flow in Pipe Networks.* The MacMillan Press Ltd., London.

Jović, V. (1977) Non-steady flow in pipes and channels by finite element method. *Proceedings of XVII Congress of the IAHR*, 2: 197–204.

Jović, V. (1987) Modelling of non–steady flow in pipe networks. *Proceedings of the Int. Conference on Numerical Methods NUMETA 87*. Martinus Nijhoff Pub., Swansea.

Jović, V. (1992) Modelling of hydraulic vibrations in network systems. *International Journal for Engineering Modelling*, 5: 11–17.

Jović, V. (1994) Contribution to the finite element method based on the method of characteristics in modelling hydraulic networks. *Proceedings of the 1st Congress of the Croatian Society for Mechanics*, 1: 389–398.

Jović, V. (1995) Finite elements and method of characteristics applied to water hammer modelling. *International Journal for Engineering Modelling*, 8: 51–58.

Rouse, H. (1969) *Engineering Hydraulics* (Tehnička hidraulika, translation in Serbian). Građevinska knjiga, Beograd.

Streeter, V.L. and Wylie, E.B. (1967) *Hydraulic Transients*. McGraw–Hill Book Co., New York, London, Sydney.

Watters, G.Z. (1984) *Analysis and Control of Unsteady Flow in Pipe Networks*. Butterworths, Boston.

Vranješ, M., Jović, V., Braun, M., and Filipović, M. (1994) Analysis of flood control solution by applying a mathematical model. *XVIIth Conference of the Danube Countries*, 237–243 Budapest.

Vranješ, M., Bilač, P., Jović, V., and Vidoš, D. (1999) Rješenje izmjene vode u jezeru Birina kod Ploča, 2nd Croatian Conference on Waters, Dubrovnik, 19–22 May.

10

Numerical Modelling in Karst

10.1 Underground karst flows

10.1.1 Introduction

Within the geologically younger rocks such as limestone, dolomite, chalk, gypsum, and others, special types of aquifers have developed, generally referred to as karst aquifers. Karst aquifers have surface and groundwater basins that usually are not congruous.

The totality of the flows takes place in the karst catchment, whose boundaries are difficult to determine (there is mutual overlapping of karst catchments, as well as overlapping of underground catchments and catchments of surface flows). The main feature of karst catchments is negligible surface flows, as compared to underground flows, since precipitation almost entirely penetrates under the ground.

The underground water flows in karst areas are very complex and are the result of the mutual activities of tectonics and long-term activity of water erosion in the area. Tectonic development predetermines the global direction of underground flow. Due to the discrepancy between the gradients of piezometric potential and the direction of the main faults, the erosive action of water creates lateral directions of flow much faster than tectonics.

Also, the changes in erosion base, that is sea level, significantly influence the development of flows in the lower parts of karst catchments.

10.1.2 Investigation works in karst catchment

The study of flows in karst begins with investigation works. The scope of the works depends on the objective to be achieved. The basic investigation works are:

- Hydrogeological investigation works, defining the ground and surface watershed, including mapping of all swallow-holes, sinkholes, and caves, that is, determining the paths of rainfall runoff into the underground system, extended with speleological investigation works. Various geophysical investigation works, aimed at defining karstification at the depth for locating the main channel systems, including heat detection and satellite imagery for determination of micro fissures.
- Drilling of deep wells in order to monitor the piezometric levels of the channel and diffuse (between channel) space.
- Tracing groundwater flows, using the known sinkholes, caves, and wells, in typical hydrological conditions (wet and dry period), including monitoring of geochemical (hydrochemical) tracers.

Analysis and Modelling of Non-Steady Flow in Pipe and Channel Networks, First Edition. Vinko Jović.
© 2013 John Wiley & Sons, Ltd. Published 2013 by John Wiley & Sons, Ltd.

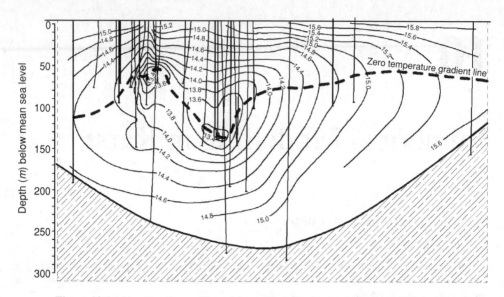

Figure 10.1   Locating the position of the main channel system by thermal detection.

- Setting up a representative network of meteorological stations.
- Setting up a monitoring system of meteorological stations, piezometric wells, and spring water levels, with a resolution of hourly readings for sufficient precision, needed for proper analysis of the phenomena that occur in individual rain episodes.

An example of locating the position of the main channel system through geophysical investigation works (thermal field)[1] is shown in Figure 10.1. It refers to the channel system of the Ombla River spring. There are two main channels, which were subsequently proved by drilling. Around the channel system there is a part of karstified space where diffuse flow takes place, that is flow with the characteristics of classical seepage.

Further investigation works have proved that a larger canal runs about 95% of the water of the Ombla River spring.

10.1.3   The main development forms of karst phenomena in the Dinaric area

Besides the fact that the channel and diffuse space are mutually interwoven, in karst we can find some other specific objects, such as karren, sinkholes, pits, caves, caverns (can be filled with air), *poljes*, flood spaces with estavels, sinks, and other formations, which make the karst area exceptionally complex. These are the main ways of recharging; the smaller part is drained through a fine porous interspace. Sinkholes appear as special sinking places. They are predisposed erosion places, usually with a hidden sink. Pools are also karst sinkholes without an adjacent sink. Sinkholes and similar formations appear in the upper parts of the catchment. Flood spaces with estavels are a transitive phase between sinkholes and fields.

---

[1]Used in research works for the HPP Ombla Project, Croatia.

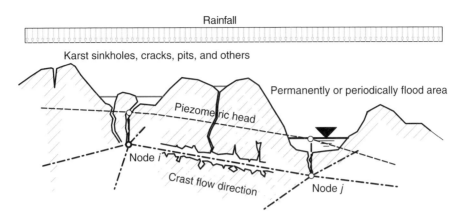

Figure 10.2   Karst sinkholes, karren, and flood spaces of the channel system.

A brief description of some of these, mostly from the hydraulic point of view, is presented hereinafter. Interested readers can gain broader information from Bonacci (1987), Marijanović (2008), and Roglić (2004, 2005) where numerous karst phenomena in the Dinarides and the wider world are described.

Sinkholes, karrens, pools, and flood space

Figure 10.2 shows the position of the channel system in relation to the ways of recharge through the system of karst sinkholes, cracks, pits, and others. In the upper parts of the catchment the piezometric head is regularly below the ground surface, however, at the turn of the middle, uplift above the ground is possible. If the piezometric head is constantly above the ground, a permanent karst lake occurs.

Periodical rising of the piezometric head above the surface creates a temporary lake. Charging and discharging of the temporary lake is mostly through the estavel (sink-spring). Short-term water drainage is at the outlet of the underground system and the water that pours out returns to the underground system through an estavel or a sink at a lower elevation.

Springs, sinks

Springs are discharge places of the underground catchment and can be permanent or temporary. Sinks are sinking places of surface flows, and can also be permanent or temporary.

Springs and sinks mostly have the form of an ascending channel. Figure 10.3 shows the hydraulic scheme of a karst spring. The spring discharge is determined by conveyance K of connection to the main channel system and the piezometric difference between the end points

$$Q_0 = K\sqrt{\frac{h_i - h_0}{L}}, \tag{10.1}$$

where L is the length of the ascending channel part, h_0 is the water level at the spring, and h_i is the piezometric head in node i.

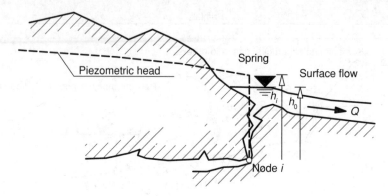

Figure 10.3   Karst spring.

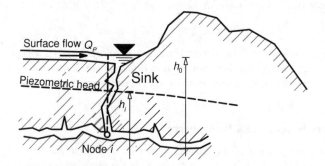

Figure 10.4   Sink of the karst surface flow.

Figure 10.4 shows the hydraulic scheme of a karst sink. The sink flow is defined by conveyance K of connection to the main channel system and by the piezometric difference between the end points

$$Q_p = K \sqrt{\frac{h_0 - h_i}{L}},$$
(10.2)

where L is the length of the descending channel part and h_0 is the water level of the sink. The equation refers to a submerged sink. In the case of an unsubmerged sink an adequate expression for spilling along the sink perimeter should be applied.

Polje

A *polje*[2] is a complex karst formation that always has one or more springs, one or more sinks, and permanent or temporary surface flow – *matica*. It is sinking river. A *polje* is also the flood space of underground channel flow.

---

[2]*Polje* is a Croatian word for a complex karst structure, which is not translated here, see Figure 10.5. The same applies for the surface flow *matica*.

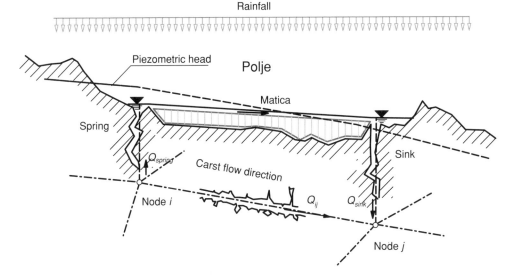

Figure 10.5 *Polje*, complex phenomena.

In the lower parts of the karst catchment *maticas* are permanent flows, rivers that have a specific name or several names. In the higher parts of the catchment they are temporary and often unnamed.

Figure 10.5 shows the hydraulic scheme of a *polje*. Under the *polje* is the main channel system through which the majority of the groundwater flows. In the upstream part of the *polje* the piezometric head is above the ground surface, causing drainage of the channel system through springs. If the piezometric head is constantly above the surface, the springs are permanent, otherwise they are temporary. Permanent or temporary surface flow disappears in the downstream sinks, where the piezometric head is below the ground surface.

During abundant rainfall the piezometric head of the main channel system rises significantly, and leads to flooding of the *polje* that can last throughout the entire winter period, because sinks, for the same piezometric reasons, have a reduced capacity. The flooding of the *polje* is also possible through sinks.

Vrulje – submerged springs

After the end of the last ice age the sea level rose, according to the dynamics shown in Figure 10.6 (Forenbaher, 2002). According to this, sunken springs are created by the rising of the sea level in newer geological history.

Flooding of the whole *polje* creates one or more submerged springs in a row, because, apart from former springs, former sinks can also become submerged springs, depending on the degree of blockage of the connecting karst channels.

The hydraulic scheme of a submerged spring is shown in Figure 10.7. The discharge calculation is necessary in order to calculate the equivalent piezometric sea elevation, because of the difference in density of fresh and salt water which depends on the depth of the throat of the submerged spring.

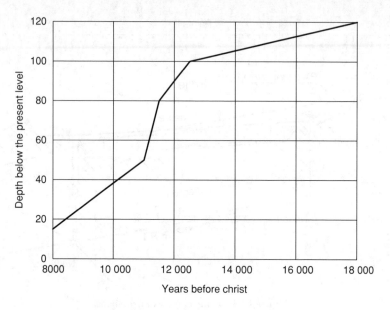

Figure 10.6   Rising of the Adriatic Sea level.

The flow of the submerged spring is determined by conveyance K of the connection to the main channel system and by the piezometric difference between the end points

$$Q_v = K \sqrt{\frac{h_i - h_0}{L}},$$

where L is the length of the ascending channel part, h_0 is the equivalent sea level, and h_i is the piezometric head in node i.

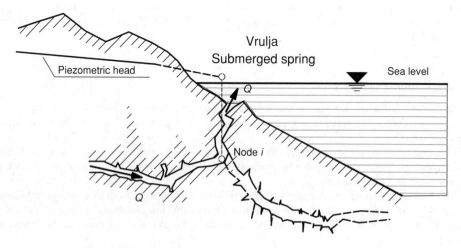

Figure 10.7   Submerged spring, sunken spring in sunken *polje*.

10.1.4   The size of the catchment

Sinking. Rain is expressed by discharge per unit area as the velocity obtained from the specific volume V_r fallen in a certain time

$$q_k = \frac{dV_r}{dt} \ [\mathrm{ms^{-1}}] \quad \text{or} \quad [\mathrm{m^3/m^2/s^{-1}}]. \tag{10.3}$$

The rainfall discharge in the catchment area includes two sinking flows (we are observing karst catchment where surface retention and surface runoff equals zero, rainfall is the only way of recharging, and springs are places of catchment discharge)

$$q_k = q_1 + q_2, \tag{10.4}$$

where q_1 is the specific discharge of direct sinking through the system of karren, sinkholes with adjacent sinks, pits, caves, large or small faults that lead directly to the underground karst channels; and q_2 is the specific discharge of sinking through fertile soil that occurs in limited areas on rocky ground, such as filled karren, sinkholes, depressions, and *poljes*.

Figure 10.8 shows the cross-section in the vertical plane through karst and schematization of the flow in vertical balance. Fertile soil and rocks characterized by porosity and linear laws of seepage (diffusion area) are shown as an equivalent layer of grain (granular) formation of equivalent thickness M^*. Of the total rainfall, waterflow q_2 flows through this layer, and here the laws of flow in porous media should be applied. The water table, that is the surface where there is atmospheric pressure, is far below the surface layer. Flow in the upper fertile karst layer is regularly unsaturated, except for a short time during a rain episode. Figure 10.9 shows the moisture profile in the capillary area, that is the dependence of saturation on the capillary height.

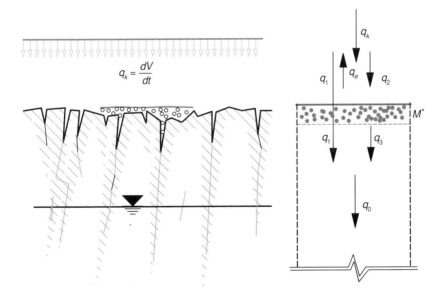

Figure 10.8   Components of vertical underground flow.

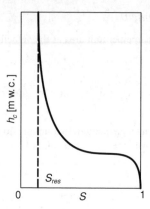

Figure 10.9   Steady saturation profile in capillary area.

When the rain falls, the moisture content within the layers changes, so when the saturation state is reached, drainage discharge q_3 occurs, this is summed up with the direct discharge into the overall sinking discharge

$$q_0 = q_1 + q_3. \tag{10.5}$$

With the cessation of rain the layer is drained, while humidity decreases to the limit of residual saturation S_r. Residual saturation is the lower limit of moisture, that is saturation, which can be achieved in the flow under the influence of gravitational forces. Further reduction of saturation or a decrease in moisture content is governed by the influence of thermodynamic forces. This is the phenomenon of evapotranspiration that occurs as a component of flow q_e in vertical balance.

Therefore, the principle of mass conservation is valid for the unsaturated medium

$$n_0 M^* \frac{dS}{dt} = q_2 - q_3 - q_e, \tag{10.6}$$

where n_0 is the geomechanical porosity of the layer (for fertile soil it is 0.3 to 0.4). From Eqs (10.4) and (10.5) the following is obtained

$$q_k = q_0 + q_2 - q_3. \tag{10.7}$$

By expressing $q_2 - q_3$ from Eq. (10.5) and inserting into the previous expression, the discharge of rain is obtained

$$q_k = q_0 + n_0 M^* \frac{dS}{dt} + q_e. \tag{10.8}$$

The size of the catchment. If Eq. (10.8) is integrated in the time of the hydrological cycle T (one year) and on the surface of the catchment

$$\int_A \left(\int_0^T q_k dt \right) dA = \int_{A_0} \left(\int_0^T q_0 dt \right) dA + \int_A \left(\int_0^T n_0 M^* \frac{dS}{dt} dt \right) dA + \int_A \left(\int_0^T q_e dt \right) dA$$

the annual volume equation is obtained

$$
\begin{aligned}
Ap_k &= V_0 + n_0 A M^* (S_T - S_0) + A p_e \\
&= V_0 + A \left[n_0 M^* (S_T - S_0) + p_e \right]
\end{aligned}
\tag{10.9}
$$

where p_k is the total annual sedimentation of overall rain per unit of catchment area A, V_0 is total annual volume drained from the catchment at springs (measured), p_e is the total annual evaporation per unit of catchment area A, and the expression $\Delta S = (S_T - S_0)$ is the annual saturation difference. The size of the catchment can be obtained from the annual volume balance

$$
A = \frac{V_0}{p_k} + A \frac{n_0 M^* (S_T - S_0) + p_e}{p_k}.
\tag{10.10}
$$

The first element in Eq. (10.10) is the catchment area A_0 where rain sinks directly to karst channels, through the sinking system, and is expressed as follows

$$
A = A_0 + A \frac{n_0 M^* (S_T - S_0) + p_e}{p_k}.
\tag{10.11}
$$

By inserting α for the fraction on the right hand side

$$
\alpha = \frac{n_0 M^* (S_T - S_0) + p_e}{p_k}
\tag{10.12}
$$

the following is obtained

$$
A = A_0 + \alpha A
\tag{10.13}
$$

therefore, the catchment area is

$$
A = \frac{A_0}{1 - \alpha}.
\tag{10.14}
$$

Coefficient α can be called a climate parameter that can be evaluated as follows. As after the end of the rain episode the moisture returns to the value of residual saturation, this difference can be considered negligible in the calculation of the catchment area of the karst spring. The following applies

$$
\alpha = \frac{p_e}{p_k}.
\tag{10.15}
$$

The catchment area A_0, where karst channels are directly recharged and springs are drained, is determined based on this measurement (discharge of the springs and rain sediment), and can be considered a reliable parameter. It should be noted that this data is not constant, but is subject to climate changes. The total catchment area A is determined by investigation works; its accuracy is congruent to the quality of hydrogeological investigation works and is considered somewhat less accurate information.

Evapotranspiration calculation

Evapotranspiration is the sum of vaporization (evaporation) and transpiration (releasing water vapor through plant leaves during absorption of carbon dioxide in photosynthesis). It is a complex process that includes water loss through atmospheric evaporation and evaporative loss of water through the

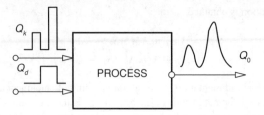

Figure 10.10   Transformation of input into output flow.

life processes of plants. Potential evapotranspiration is the amount of water that could evaporate if there is water in the observed area. Evapotranspiration is measured by devices called lysimeters.[3] A precise determination of the amount of evapotranspiration is very complex, but it can be reasonably well estimated using the simple method of estimating the dependent quantities. It is the potential amount of water that can be discharged from the soil in conditions of measurement of rainfall and temperature. For the observed problem of determining the catchment surface of karst springs, two equations will be shown hereinafter: the *Penman* and *Thornthwaite* equations.[4] The Penman formula is one of the most complete theoretical formulas for evapotranspiration. Evapotranspiration is associated with radiation that occurs at the surface. It requires mean daily values of mean temperature, wind speed, relative humidity, and solar radiation and is used in cases where such parameters are measured. Otherwise, evapotranspiration should be estimated by less theoretical formulas.

10.2   Conveyance of the karst channel system

10.2.1   Transformation of rainfall into spring hydrographs

A part of the catchment or the entire catchment is observed as a control volume of the flow. General conservation principles, for example, the principle of mass conservation, can be applied to the control volume. If the volume of water in the aquifer is equal to V, the principle of conservation of mass can be written

$$\frac{dV}{dt} = Q_k + Q_d - Q_0, \tag{10.16}$$

which shows that the volume change rate equals the difference between the input $Q_k + Q_d$ and the output Q_0 discharges, where Q_k is the discharge due to rainfall and Q_d is inflow from other catchments. The input flow is transformed into the output according to the laws of hydrodynamics, and can be viewed as a process, shown in Figure 10.10.

   The quality of the transformation depends on the selection of process transformation functions. For a chosen class of functions, unknown parameters are optimized to minimize the difference between the calculation results in relation to those actually measured.

   It is best to choose process transformations from the class of functions that are solutions of differential equations that describe the nature of the process. The analytical form of the solution is possible for a very small number of simple processes in a simple catchment, which when applied to more complex catchments give significantly different solutions to those expected. Fortunately, through the development

---

[3] From the Greek *lysis* means freeing, decomposing.
[4] Ven Te Chow: *Handbook of Applied Hydrology*, McGraw – Hill Book Comp., New-York, Toronto, 1964. See the chapter on Evapotranspiration, Table 11.6.

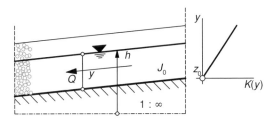

Figure 10.11   Uniform filtration in an underground porous channel.

of numerical methods, it is possible to seek process transformations as numerical solutions of differential equations that describe the nature of the process.

The following presentation shows the simplest model of numerical transformation of input flows, based on the numerical solution of the corresponding hydrodynamic equations. The model is suitable for basic investigation works, although the level of modelling can be raised to higher levels that require further expansion of investigation works in the catchment.

Disregarding the details of the genesis of karst formation, karst areas with underground flows, from the hydrodynamic point of view, can be divided into areas with *channel* and *diffuse* flows.

The diffuse area is a karst region (minor cracks and pores) distributed around the channel area. In this area the diffuse laws of flows prevail, similar to those in granular media, that is the linear laws of hydrodynamic resistance (classical seepage).

The term "channel area" refers to the area where water runs through one or more underground channels with turbulent rough hydrodynamic resistances. The nature of the flow through underground karst channels is similar to the flow through a perforated irregular tube or surface channel in karst and porous rock. In the hydrodynamic sense, the channels are mainly pressurized; however some effects of flow in open riverbeds are also possible.

10.2.2   Linear filtration law

Darcy's linear filtration law, which connects filtration velocity and the slope of the piezometric head, is applied to water flow through porous, granular areas

$$v = kJ, \tag{10.17}$$

where k is the filtration coefficient. Darcy's law is a linear law which is the consequence of laminar water flow in a porous media. The slope[5] of the piezometric head is expressed as the proportionality

$$J = -\frac{dh}{dl} \propto v, \tag{10.18}$$

which represents the linear resistance law. In an underground porous channel, Figure 10.11, the flow will be determined by the linear flow law

$$Q = kByJ = K(y)J, \tag{10.19}$$

where B is width and $K(y) = kBy$ is the equation for relative channel conveyance. The conveyance curve is a discharge curve for the channel unit slope. This is an expression for *uniform filtration* where the piezometric function slope is equal to the channel slope $J = J_0$.

---

[5]The slope is equal to the negative gradient.

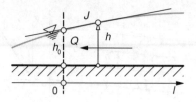

Figure 10.12   Non-uniform filtration in an underground porous channel.

The calculation of discharge depending on the piezometric head h using relative conveyance is as follows

$$Q(h) = K(h - z_0)J = K(y)J, \tag{10.20}$$

where $K(y)$ is equal in all channel cross-sections.

Non-uniform filtration in a horizontal underground channel, Figure 10.12, is discharge where the piezometric line slope is not equal to the channel slope $J \neq J_0$ and is described in the equation

$$Q = kBhJ = kBh\frac{dh}{dl}. \tag{10.21}$$

The profile of the piezometric line in non-uniform discharge is obtained by integration starting with the known profile

$$\int_{l_0}^{l} \frac{Q}{kB} dl = \int_{h_0}^{h} h\,dh, \tag{10.22}$$

which gives the equation for the profile piezometric height along the channel axis

$$\frac{Q}{kB}(l - l_0) = \frac{1}{2}\left(h^2 - h_0^2\right), \tag{10.23}$$

that is the equation for discharge in the channel

$$Q = \frac{kB}{2}\frac{h^2 - h_0^2}{l - l_0} = kB\frac{h + h_0}{2}\frac{h - h_0}{l - l_0} = \bar{K}\frac{h - h_0}{l - l_0}. \tag{10.24}$$

Function (10.24) shows that discharge can be expressed by the average conveyance $\bar{K}$ between the end cross-sections 1 and 2, length of column Δl, and piezometric level difference Δh:

$$Q = \bar{K}\frac{\Delta h}{\Delta l}, \tag{10.25}$$

where

$$\bar{K} = \frac{K_1 + K_2}{2} = kB\frac{h_1 + h_2}{2}. \tag{10.26}$$

Radial filtration. Figure 10.13 shows flow in the radial channel. In this case the equation for discharge is as follows

$$Q = kBhJ = k\alpha rh\frac{dh}{dr}. \tag{10.27}$$

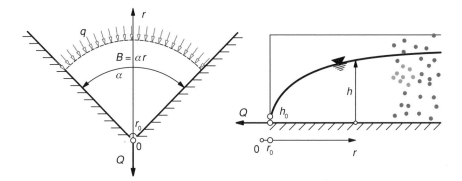

Figure 10.13   Radial seepage.

The integration of Eq. (10.27) from the initial to any cross-sectional area gives the expression for discharge

$$Q = k \frac{\alpha}{2} \frac{h^2 - h_0^2}{\ln \dfrac{r}{r_0}}. \tag{10.28}$$

10.2.3   Turbulent filtration law

Darcy's filtration law in the linear form (10.17) is valid up to the critical Reynolds number

$$R_e = \frac{v d_s}{\nu} \leq 1, \tag{10.29}$$

where v is the filtration velocity, d_s the grain diameter, and ν the coefficient of kinematic viscosity, above which turbulence appears. At values $R_e > 1$ transition and turbulent flow develop where resistance is non-linear, and the piezometric height slope is expressed as

$$J \propto v^n, \tag{10.30}$$

where $1 < n \leq 2$. The turbulent flow is $n = 2$. The appropriate filtration law in turbulent flows is as follows

$$v = k_t \sqrt{J}, \tag{10.31}$$

where k_t is the proportionality coefficient and is called the turbulent filtration coefficient. Non-linear filtration can also be expressed as

$$v = aJ + b\sqrt{J}, \tag{10.32}$$

where a and b are corresponding constants. The constant values $a = 0, b = k_t$ give turbulent filtration and the values $b = 0, a = k$ give clear laminar filtration. The combination of the constants $a > 0, b > 0$ describes transitional non-linear filtration.

Table 10.1 shows turbulent filtration coefficients for a rockfill, where d is the equivalent grain diameter and depends on the porosity n of the rock fill (karst).

Table 10.1

	d [m]	0.05	0.10	0.15	0.20	0.25	0.30	0.35	0.40	0.45	0.50
k_t [m/s]	$n = 0.40$	0.15	0.23	0.30	0.35	0.39	0.43	0.46	0.50	0.53	0.56
	$n = 0.46$	0.17	0.26	0.33	0.39	0.44	0.48	0.52	0.56	0.60	0.63
	$n = 0.50$	0.19	0.29	0.37	0.43	0.49	0.53	0.58	0.62	0.66	0.70

The flow in *uniform turbulent filtration* within a karst underground channel is calculated from the cross-sectional area A and filtration velocity v

$$Q = Av = k_t A\sqrt{J} = K(y)\sqrt{J}, \tag{10.33}$$

where the slope is equal to the channel slope $J = J_0$ and $K(y)$ is the relative conveyance function.

Non-uniform turbulent filtration in a horizontal underground channel, in Figure 10.14, is described by the equation

$$Q = K\sqrt{J}. \tag{10.34}$$

This further gives

$$J = -\frac{dh}{dl} = \frac{Q^2}{K^2} = \frac{Q^2}{(k_t Bh)^2}. \tag{10.35}$$

Upon the separation of variables, the derived expression is integrated between two cross-sections as follows

$$\int_{l_0}^{l} \left(\frac{Q}{k_t B}\right)^2 dl = \int_{h_0}^{h} h^2 dh \tag{10.36}$$

thus giving the expression for discharge

$$Q = k_t B\sqrt{\frac{h^3 - h_0^3}{3(l - l_0)}}. \tag{10.37}$$

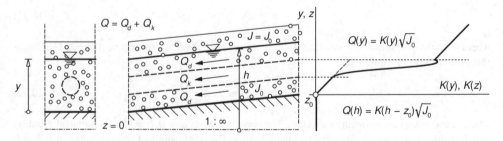

Figure 10.14   Conveyance in complex seepage.

From the formula for discharge we can derive the expression for the profile of the piezometric height along the channel at the distance l from the beginning

$$h = \sqrt{h_0^3 + 3 \left(\frac{Q}{k_t B} \right)^2 [l - l_0]}. \tag{10.38}$$

Radial turbulent filtration. Let us observe the case of turbulent filtration as shown in Figure 10.13. Turbulent filtration is

$$Q = k_t B h \sqrt{J} = k_t \alpha r h \sqrt{\frac{dh}{dr}}. \tag{10.39}$$

The separation of variables is followed by the integration between two cross-sections

$$\left(\frac{Q}{\alpha k_t} \right)^2 \int_{r_0}^{r} \frac{dr}{r^2} = \int_{h_0}^{h} h^2 dh, \tag{10.40}$$

which leads to the equation for discharge calculation

$$Q = \alpha k_t \sqrt{\frac{r_0 r}{3 (r - r_0)}} \left(h^3 - h_0^3 \right). \tag{10.41}$$

10.2.4   Complex flow, channel flow, and filtration

Figure 10.14 shows a flow where the major part of the discharge passes through a tubular part of the channel, and a minor part within the space around it. The total flow is summed

$$Q = Q_d + Q_k, \tag{10.42}$$

where Q_d is the diffuse flow and Q_k is flow through the tubular channel. The longitudinal slope of piezometric height J is the same for diffuse and tubular channel flow.

Both discharges can be expressed by the function of relative conveyance

$$Q = K_d^l(y)J + K_d^t(y)\sqrt{J} + K_k^t(y)\sqrt{J}. \tag{10.43}$$

where the first element refers to laminar filtration in a diffuse area, the second element refers to turbulent filtration in a diffuse area, and the third element refers to turbulent flow in a tubular channel. In order to calculate turbulent discharge in a tubular channel one can apply the Darcy–Weisbach or Manning equations for channel velocity calculation. The relative function of conveyance is as follows (for Manning's velocity formula)

$$Q_k = \frac{A_k(y)}{n} R(y)^{\frac{2}{3}} \sqrt{J} = K_k(y)\sqrt{J}. \tag{10.44}$$

As the laminar diffuse flow is negligible in relation to the turbulent flow, the first element can be neglected thus obtaining

$$Q = K(y)\sqrt{J}, \tag{10.45}$$

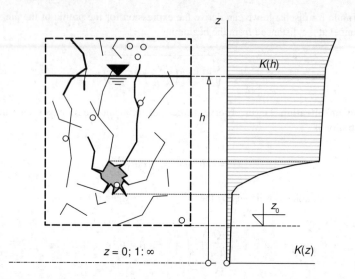

Figure 10.15   Cumulative conveyance of a karst channel system.

where $K(y)$ is the relative cumulative function of conveyance in a complex channel, see the graph on the right. The relative cumulative function of conveyance for the sloping channel system J_0, can be written as $K(z) = K(h - z_0)\sqrt{J}$.

If we imagine that the cross-sectional area of the channel system contains one or more natural underground channels instead of a tubular one, as shown in Figure 10.15, then the discharge is calculated from the general cumulative conveyance and the longitudinal slope of piezometric height

$$Q(h) = \mathrm{sgn}J \cdot K(h)\sqrt{|J|}. \tag{10.46}$$

The positive discharge is in the direction of a positive slope and in the direction of the negative gradient. The inverse form of Eq. (10.46) reads

$$J = -\frac{dh}{dl} = \frac{|Q|\,Q}{K^2(h)}. \tag{10.47}$$

Uniform channel system

The observed channel is one in which its relative conveyance function does not change along the flow. In this case, the expression (10.47) can give the equation for steady flow in a channel by integration between two cross-section areas

$$\int_1^2 dh + \int_{l_1}^{l_2} \frac{|Q|\,Q}{K^2(h)}dl = 0 \tag{10.48}$$

thus obtaining the equation for steady discharge in a karst channel

$$h_2 - h_1 + \frac{|Q|\,Q}{\bar{K}^2}L = 0, \tag{10.49}$$

where the second integral is expressed by the mean value of the integrals, or by conveyance

$$\bar{K} = K \left(\frac{h_1 + h_2}{2} \right),$$

(10.50)

that is the value that corresponds to the average piezometric level.

Non-uniform channel system

If we observe a channel in which the conveyance changes along the flow axis, the integration of expression (10.47) is as follows

$$\int_1^2 dh + \int_{l_1}^{l_2} \frac{|Q| Q}{K^2(l, h)} dl = 0.$$

(10.51)

Once again, the mean value of integrals is used

$$h_2 - h_1 + \frac{|Q| Q}{\bar{K}^2} L = 0,$$

(10.52)

where the mean value of conveyance is

$$\bar{K} = \frac{K(l_1, h_1) + K(l_2, h_2)}{2}$$

(10.53)

equal to the arithmetic average between conveyance at the end of the section.

10.3   Modelling of karst channel flows

10.3.1   Karst channel finite elements

The program SimpipCore (see: www.wiley.com/go/jovic) implements channel finite elements which are given through a simple syntax:

```
!          ...   ...   ...
! syntax:
!     Kanal name pnt1 pnt2 conv1 <conv2 <relative>
!
```

where name is the karst channel name, pnt1, pnt2 are the names of the start and end points, conv1, conv2 are conveyance at the start and end of the finite element, and relative is the optional parameter. Conveyance conv1, can be given as *numeric* values or given as a previously defined Graph object.

```
Graph conv
        0        20.5
        32.4     53.4
        100      450
Kanal odIzvora p1 p2 255.6 conv
```

Depending on the number of data on conveyance, the syntax implies that the data refers to *uniform* or *non-uniform* karst channel spaces. For instance, if we write:

```
        Kanal name pnt1 pnt2 conv
```

it is implied that the karst channel has "constant" conveyance on the element (the same conveyance function value $K(z)$ at the beginning and end of the channel), so that the program involves the use of Eq. (10.50) for the calculation of conveyance $\bar{K}(h)$. Or, if we write:

```
Kanal name pnt1 pnt2 conv1 conv2
```

it is implied that the karst channel has non-uniform conveyance on the element (different value $K_1(z)$ at the beginning and $K_2(z)$ at the end of the finite element), so that the program involves the use of Eq. (10.53) for the calculation of conveyance $\bar{K}(h)$.

Conductivities, whether given numerically or in graph charts, are presented as absolute cumulative functions $K(z)$. The option relative enables the inputting of relative values $K(y)$, so the program SimpipCore, in calculating conveyance $K(h) = K(h - z_0)$, uses the z_0 coordinate of the corresponding node. This option does not refer to *numeric* values given conveyance.

10.3.2   Subroutine SteadyKanalMtx

The equation for steady flow in a karst channel (dynamic equation) at the finite element of L length is as follows

$$h_2 - h_1 + L\frac{Q^2}{\bar{K}^2} = 0, \tag{10.54}$$

where $\bar{K} = K(h_1, h_2)$ is the mean value of conveyance between the beginning and the end of the channel. Equation (10.54) will be written in the form appropriate for numerical modelling of the finite element:

$$F^e: \quad h_2 - h_1 + \bar{\beta}\,|Q|\,Q = 0, \tag{10.55}$$

where

$$\bar{\beta} = \frac{L}{\bar{K}^2} \tag{10.56}$$

is the parameter of resistance of the karst channel. The Newton–Raphson iterative form for the elemental equation F^e is as follows

$$\left[\frac{\partial F^e}{\partial h_1} \quad \frac{\partial F^e}{\partial h_2}\right] \cdot \left[\begin{array}{c} \Delta h_1 \\ \Delta h_2 \end{array}\right] + \frac{\partial F^e}{\partial Q}\Delta Q = -F^e. \tag{10.57}$$

The dynamic equation which describes flow in a karst channel is a complex equation in which conveyance is given in graphs $K_1(h), K_2(h)$. In this case exact partial derivations are substituted by approximate ones

$$\left[\frac{\Delta F^e}{\Delta h_1} \quad \frac{\Delta F^e}{\Delta h_2}\right] \cdot \left[\begin{array}{c} \Delta h_1 \\ \Delta h_2 \end{array}\right] + \frac{\Delta F^e}{\Delta Q}\Delta Q = -F^e, \tag{10.58}$$

that is they are calculated numerically. Numerical calculating is simple because a Fortran function has been written for the elemental equation (10.55)

$$F^e: \quad \text{real*8 function FeKanal(ielem,h1,h2,Q),}$$

which calculates the mean finite differences

$$\frac{\Delta F^e}{\Delta h_1} = \frac{F^e(h_1 + \Delta h_1, h_2, Q) - F^e(h_1 - \Delta h_1, h_2, Q)}{2\Delta h_1}, \tag{10.59}$$

$$\frac{\Delta F^e}{\Delta h_2} = \frac{F^e(h_1, h_2 + \Delta h_2, Q) - F^e(h_1, h_2 - \Delta h_2, Q)}{2\Delta h_2}, \tag{10.60}$$

$$\frac{\Delta F^e}{\Delta Q} = \frac{F^e(h_1, h_2, Q + \Delta Q) - F^e(h_1, h_2, Q - \Delta Q)}{2\Delta Q}. \tag{10.61}$$

The program contains partial values of increments

$$\Delta h_1 = \Delta h_2 = \Delta Q = 0.1, \tag{10.62}$$

which are suitable for practical needs. The matrix and the vector of the finite element of the karst channel for steady flow are calculated numerically. Applying the Newton–Raphson iterative form (10.58), which is formally used with matrix-vector operations, we can write

$$\left[\underline{H}\right] \cdot [\Delta h] + \left[\underline{Q}\right] \cdot [\Delta Q] = \left[\underline{F}\right]. \tag{10.63}$$

By multiplying the previous expression with the inverse element $\left[\underline{Q}\right]^{-1}$ on the left, we obtain

$$\left[\underline{A}\right] \cdot [\Delta h] + [\Delta Q] = \left[\underline{B}\right] \tag{10.64}$$

hence to calculate the value of increase of the elemental discharge

$$[\Delta Q] = \left[\underline{B}\right] - \left[\underline{A}\right][\Delta h], \tag{10.65}$$

where

$$\left[\underline{A}\right] = \left[\underline{Q}\right]^{-1}\left[\underline{H}\right], \tag{10.66}$$

$$\left[\underline{B}\right] = \left[\underline{Q}\right]^{-1}\left[\underline{F}\right]. \tag{10.67}$$

In the previous expressions $[\underline{A}]$ is a dual vector, and $[\underline{B}]$ is a scalar quantity. The process of elimination of the increase of elemental discharges from continuity nodal equations determines the structure of the matrix

$$A^e = \begin{bmatrix} +\underline{A} \\ -\underline{A} \end{bmatrix} \tag{10.68}$$

and the vector of the finite element.

$$B^e = \underbrace{\begin{bmatrix} +Q^e \\ -Q^e \end{bmatrix}}_{(1)} + \underbrace{\begin{bmatrix} +\underline{B} \\ -\underline{B} \end{bmatrix}}_{(2)}. \tag{10.69}$$

Element (1) is an existing element on the right prior to elimination, and (2) is the result upon elimination of the elemental discharge from the nodal equation.

The matrix and the vector of the finite element of the karst channel for steady flow are calculated in the subroutine:

<code>subroutine SteadyKanalMtx (ielem),</code>

which is implemented in the Fortran module `carst.f90`. The subroutine `Steady` uses the subroutine `SteadyKanalMtx` for the calculation of the matrix and vector of steady flow in elemental types of karst channels:

```
select case(Elems(ielem).tip)
      case ...

            ...

      case (KANAL_OBJ)
            call SteadyKanalMtx(ielem)
      case ...

            ...

endselect
```

10.3.3 Subroutine UnsteadyKanalMtx

The water that flows in underground karst channels can be regarded as incompressible so that density is a constant value $\rho = const$. The wave nature of flow in underground karst channels is largely a result of rain episodes and the accumulation capacity of the karst structure and less a result of the compressible nature of the channel bed. Intensive changes of flow in karst happen before and during rain episodes which are measured in hours or longer periods. Unsteady water flow in karst channels is moderately changeable in time. This is called a *quasi unsteady* flow.

Modelling of changeable flow in time demands the integration of the steady flow equation (10.55) between two different time states $[t_1, t_2]$

$$R^e : \int_{t_1}^{t_2} F^e(t)dt = \int_{t_1}^{t_2} \left(h_2 - h_1 + \bar{\beta} |Q| Q \right) dt = 0. \tag{10.70}$$

The integral (10.70) is written applying the principles of numerical integration in the form of

$$R^e : (1 - \vartheta) \Delta t \underbrace{\left(h_2 - h_1 + \bar{\beta} |Q| Q \right)}_{F^e} + \vartheta \Delta t \underbrace{\left(h_2^+ - h_1^+ + \bar{\beta}^+ |Q^+| Q^+ \right)}_{F^{e+}} = 0, \tag{10.71}$$

which consists of the appropriate result of the equation F^e at the beginning and at the end of the time stages (without and with the symbol "+"). The Newton–Raphson iterative form is applied to the integrated elemental equation (10.71)

$$\left[\frac{\partial R^e}{\partial h_1^+} \quad \frac{\partial R^e}{\partial h_2^+} \right] \cdot \left[\begin{array}{c} \Delta h_1 \\ \Delta h_2 \end{array} \right] + \frac{\partial R^e}{\partial Q^+} \Delta Q = -R^e. \tag{10.72}$$

Introducing Eq. (10.71) into Eq. (10.72) we obtain

$$\vartheta \Delta t \left[\frac{\partial F^{e+}}{\partial h_1^+} \quad \frac{\partial F^{e+}}{\partial h_2^+} \right] \cdot \left[\begin{array}{c} \Delta h_1 \\ \Delta h_2 \end{array} \right] + \vartheta \Delta t \frac{\partial F^{e+}}{\partial Q^+} \Delta Q = - \left[(1 - \vartheta) \Delta t F^e + \vartheta \Delta t F^{e+} \right]. \tag{10.73}$$

After substituting exact partial derivations for approximate ones

$$\vartheta \, \Delta t \left[\frac{\Delta F^{e+}}{\Delta h_1^+} \quad \frac{\Delta F^{e+}}{\Delta h_2^+} \right] \cdot \left[\begin{array}{c} \Delta h_1 \\ \Delta h_2 \end{array} \right] + \vartheta \, \Delta t \frac{\Delta F^{e+}}{\Delta Q^+} \Delta Q = - \left[(1 - \vartheta) \, \Delta t \, F^e + \vartheta \, \Delta t \, F^{e+} \right]. \qquad (10.74)$$

It is observed that it is possible to use the Fortran equation `FeKanal(ielem,h1,h2,Q)` in the calculation of the matrix and vector of quasi unsteady flow. Namely, this function models the elemental equation $F^e(t)$ of the karst channel for a certain time, the data on it given as $h_1(t)$, $h_2(t)$, $Q(t)$. So, the approximate partial derivations as mean changes are as follows:

$$\frac{\Delta F^{e+}}{\Delta h_1^+} = \frac{F^e(h_1^+ + \Delta h_1, h_2^+, Q^+) - F^e(h_1^+ - \Delta h_1, h_2^+, Q^+)}{2\Delta h_1}, \qquad (10.75)$$

$$\frac{\Delta F^{e+}}{\Delta h_2^+} = \frac{F^e(h_1^+, h_2^+ + \Delta h_2, Q^+) - F^e(h_1^+, h_2^+ - \Delta h_2, Q^+)}{2\Delta h_2}, \qquad (10.76)$$

$$\frac{\Delta F^{e+}}{\Delta Q^+} = \frac{F^e(h_1^+, h_2^+, Q^+ + \Delta Q) - F^e(h_1^+, h_2^+, Q^+ - \Delta Q)}{2\Delta Q}. \qquad (10.77)$$

In order to calculate approximate partial derivations we use the value of increments

$$\Delta h_1 = \Delta h_2 = \Delta Q = 0.1. \qquad (10.78)$$

The Newton–Raphson iterative form (10.74) for elemental equations is formally written using matrix-vector operations

$$[\underline{H}] \cdot [\Delta h] + [\underline{Q}] \cdot [\Delta Q] = [\underline{F}], \qquad (10.79)$$

where the vector value is

$$[\underline{H}] = \vartheta \, \Delta t \left[\frac{\Delta F^{e+}}{\Delta h_1^+} \quad \frac{\Delta F^{e+}}{\Delta h_2^+} \right] \qquad (10.80)$$

and the scalar member is

$$\lfloor \underline{Q} \rfloor = \vartheta \, \Delta t \frac{\Delta F^{e+}}{\Delta Q^+}, \qquad (10.81)$$

whereas the value on the right is

$$[\underline{F}] = - \left[(1 - \vartheta) \, \Delta t \, F^e + \vartheta \, \Delta t \, F^{e+} \right]. \qquad (10.82)$$

Multiplying the expression (10.79) by the inverse element $\left[\underline{Q} \right]^{-1}$ on the left, we obtain

$$[\underline{A}] \cdot [\Delta h] + [\Delta Q] = [\underline{B}], \qquad (10.83)$$

which further leads to the calculation of the value of elemental discharge change

$$[\Delta Q] = [\underline{B}] - [\underline{A}][\Delta h], \qquad (10.84)$$

where

$$[\underline{A}] = [\underline{Q}]^{-1}[\underline{H}],$$

(10.85)

$$[\underline{B}] = [\underline{Q}]^{-1}[\underline{F}].$$

(10.86)

In the above expressions $[\underline{A}]$ is a dual vector value, and $[\underline{B}]$ is a scalar value. The required structure of the finite element matrix for quasi unsteady flow is as follows

$$A^e = \vartheta \Delta t \begin{bmatrix} +\underline{A} \\ -\underline{A} \end{bmatrix}.$$

(10.87)

The finite element vector is as follows

$$B^e = (1 - \vartheta \Delta t) \underbrace{\begin{bmatrix} +Q \\ -Q \end{bmatrix} + \vartheta \Delta t \begin{bmatrix} +Q^+ \\ -Q^+ \end{bmatrix}}_{(1)} + \underbrace{\vartheta \Delta t \begin{bmatrix} +\underline{B} \\ -\underline{B} \end{bmatrix}}_{(2)}.$$

(10.88)

Element (1) in (10.88) is an existing value on the right prior to elimination, and value (2) is the result upon elimination of the elemental discharges in nodal equations.

The matrix and vector of the finite element of a karst channel for unsteady flow is calculated in the subroutine

subroutine UnsteadyKanalMtx(ielem),

which is implemented in the Fortran module carst.f90. The subroutine Unsteady uses a subroutine for the calculation of the matrix and vector of unsteady flow for the elemental types of karst channels:

```
select case(Elems(ielem).tip)
      case  ...
            ...
      case  (KANAL_OBJ)
            call UnsteadyKanalMtx(ielem)
      case  ...
            ...
endselect
```

10.3.4   Tests

Task 1

Let us observe an underground porous sloping channel with a $J_0 = 1\%$ slope, $B = 25$ m wide, within which the water flow is steady at the depth $y_0 = 32$ m. It is necessary to calculate the water flow in the channel for uniform flow for filtration variants if we are given the following data:

(a) Darcy's filtration coefficient is $k = 10^{-2}$ m/s, which corresponds to filtration through large gravel.

(b) The turbulent filtration coefficient is $k = 0.25$ m/s, which corresponds to 10 cm average sized karst.

ad (a): Eq. (10.19) gives:

$$Q = kBy_0J_0$$
$$Q = 10^{-2} \cdot 25 \cdot 32 \cdot 0.01 . \tag{10.89}$$
$$Q = 0.080 \, (\text{m}^3/\text{s})$$

ad (b): Eq. (10.33) gives:

$$Q = K\sqrt{J_0} = k_t By_0\sqrt{J_0}$$
$$Q = 0.25 \cdot 25 \cdot 32 \cdot \sqrt{0.01} . \tag{10.90}$$
$$Q = 20 \, (\text{m}^3/\text{s})$$

It can be observed that in the case of turbulent filtration there is a significantly larger discharge. The discharge ratio is $20{:}0.08 = 250$, that is turbulent filtration gives a 250 times higher discharge than diffuse filtration, which could have been expected.

Task 2

We observe a horizontal underground porous channel ($J_0 = 0$), from which water springs at a flow rate of $Q = 20$ m^3/s. The relative conveyance function is given:

```
Graph      Conv
  0       6.25*0.1
100       6.25*100
```

which can be written in the form of an expression $K = 0.625 + 6.24375h$.
It is necessary to calculate the profile of the piezometric height in a channel 10 km long, if the piezometric head at the spring is $h(0) = 2$ m, thus applying:

(a) the SimpipCore program for uniform discretisation for finite elements 1000 m long,

(b) the SimpipCore program with discretization for finite elements between nodes 0, 10, 20, 40, 80, 160, 320, 640, 1280, 2560, 5120, 8000, 10 000,

(c) analytical calculation.

Figure 10.16 shows the input file with the layout of nodes ad (a) in the left column and the results of modelling ad (a) in the right-hand column. We can observe that the modelling results are acceptable even for rough channel discretization in an area of large gradients, see the dashed curve SimpipCore $\Delta l = 1000$ m curve. The solid curve is the analytical result of the equation

$$Q = K(h)\sqrt{\frac{dh}{dl}}, \tag{10.91}$$

which is obtained by separation of variables

$$Q^2 dl = K^2(h) dh.$$

```
Input file: KANAL D1 - profil.simpip
(from www.wiley.com/go/jovic:Program
solutions & tests/Test simpip files)


Parameters
        Q = 20
        Hizv=2
Points
        p0          0          0          0
        p1          1000       0          0
        p2          2000       0          0
        p3          3000       0          0
        p4          4000       0          0
        p5          5000       0          0
        p6          6000       0          0
        p7          7000       0          0
        p8          8000       0          0
        p9          9000       0          0
        p10         10000      0          0
Graph Conv
        0           6.25*0.1
        100         6.25*100
Kanal
        k1          p0         p1         Conv
        k2          p1         p2         Conv
        k3          p2         p3         Conv
        k4          p3         p4         Conv
        k5          p4         p5         Conv
        k6          p5         p6         Conv
        k7          p6         p7         Conv
        k8          p7         p8         Conv
        k9          p8         p9         Conv
        k10         p9         p10        Conv
Charge p10 Q
Piezo p0 Hizv
Option
 run quasi
Steady 0
Print
        solStage 0
```

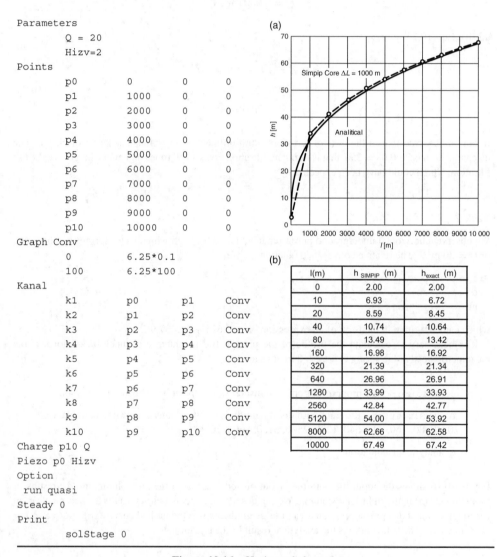

(a)

(b)

l(m)	h $_{SIMPIP}$ (m)	h $_{exact}$ (m)
0	2.00	2.00
10	6.93	6.72
20	8.59	8.45
40	10.74	10.64
80	13.49	13.42
160	16.98	16.92
320	21.39	21.34
640	26.96	26.91
1280	33.99	33.93
2560	42.84	42.77
5120	54.00	53.92
8000	62.66	62.58
10000	67.49	67.42

Figure 10.16   Horizontal channel.

On introducing the expression for $K(h)$ and limits of integration, we write

$$Q^2 \int_0^L dl = \int_{h_0}^h (0.625 + 6.24375 \cdot h)^2 \, dh. \tag{10.92}$$

After calculating the specified integrals we obtain the analytical equation for the profile of the piezometric level in the following form

$$lQ^2 = 0.390625 \cdot (h - h_0) + 3.90234375 \cdot (h - h_0)^2 + 12.99480469 \cdot (h - h_0)^3. \tag{10.93}$$

The form $l(h)$ in Eq. (10.93) can give us the form $h(l)$ through a numerical calculation. The table in Figure 10.16 shows the results of discretization modelling from (b) and the corresponding analytical results.

Task 3

An underground porous sloping channel $J_0 = 1\%$, from which water springs at a discharge of $Q = 20\ \text{m}^3/\text{s}$. The relative conveyance function is given:

```
Graph      Conv
  0        6.25*0.1
100        6.25*100
```

It is necessary to calculate:

(a) the normal depth of uniform flow,

(b) the backwater curve for piezometric 120 m height of the spring by modelling using the program SimpipCore and the integration of the fourth-order Runge–Kutta method.

ad (a): The uniform flow will be calculated from Eq. (10.33) thus obtaining:

$$Q = K(y)\sqrt{J_0}$$

$$K(y) = \frac{Q}{\sqrt{J_0}} = \frac{20}{\sqrt{0.01}} = 200$$

$$K(y) = 0.625 + 6.24375 y_0 = 200 \tag{10.94}$$

$$y_0 = \frac{200 - 0.625}{6.24375} = 31.932\,(\text{m})$$

$$K = \frac{Q}{\sqrt{J_0}} = \frac{20}{\sqrt{0.01}} = 200.$$

ad (b): The backwater curve.

Figure 10.17 shows the input file on the left, whereas the results can be seen on the right. On the right under (a) we can see the backwater curve for piezometric height of the spring at 120 m, and under (b) we can see the calculated values of profiles along the channel.

```
        Input file:
    Kanal uspor - profil.simpip
(from www.wiley.com/go/jovic: Program
solutions & tests/Test simpip files)


Parameters
  Q = 20   Hizv = 120
  Jo = 0.01
Points
  p0   0 0 0*Jo
  p1   1000  0 1000*Jo
  p2   2000  0 2000*Jo
  p3   3000  0 3000*Jo
  p4   4000  0 4000*Jo
  p5   5000  0 5000*Jo
  p6   6000  0 6000*Jo
  p7   7000  0 7000*Jo
  p8   8000  0 8000*Jo
  p9   9000  0 9000*Jo
  p10 10000  0 10000*Jo
Graph Conv
  0.1 6.25*0.1
  100 6.25*100
Kanal
  k1  p0  p1  Conv  Conv relative
  k2  p1  p2  Conv  Conv relative
  k3  p2  p3  Conv  Conv relative
  k4  p3  p4  Conv  Conv relative
  k5  p4  p5  Conv  Conv relative
  k6  p5  p6  Conv  Conv relative
  k7  p6  p7  Conv  Conv relative
  k8  p7  p8  Conv  Conv relative
  k9  p8  p9  Conv  Conv relative
  k10 p9  p10 Conv  Conv relative
Charge p10 Q
Piezo p0 Hizv
Option
  run quasi
Steady 0
Print
  solStage 0
```

(a)

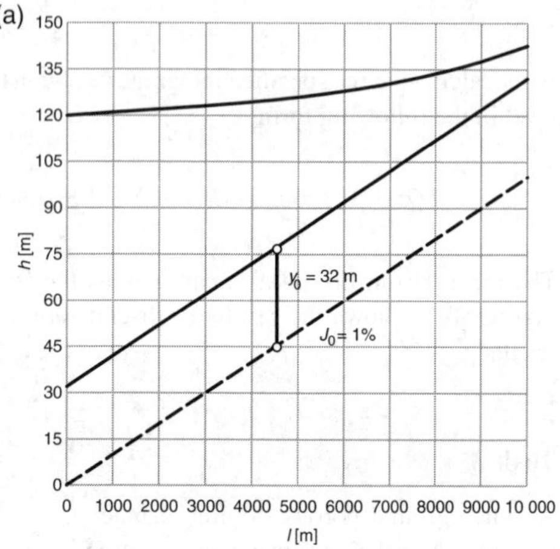

(b)

I [m]	Simpip	Runge–Kutta
0	120.00	120.00
1000	121.02	120.77
2000	122.05	121.68
3000	123.15	122.76
4000	124.45	124.08
5000	126.04	125.68
6000	128.01	127.68
7000	130.49	130.19
8000	133.64	133.38
9000	137.63	137.43
10000	142.66	142.51

Figure 10.17   Backwater curve in a sloping channel.

10.4   Method of catchment discretization

10.4.1   Discretization of karst catchment channel system without diffuse flow

Elementary catchment

The main investigation works should determine the distribution of the channel and diffuse system. At this level of investigative work the flows in diffuse space are almost always negligible compared to the flows in the channel system, so that in this (basic) analysis, only the channel systems in which the entire groundwater flow takes place are observed.

Figure 10.18 shows the elementary catchment or sub-catchment. A minimum scope of investigation works is performed at the catchment, consisting of the following:

- Measurement of input discharge Q_k by rainfall and the corresponding area A_s.
- Measurement of output discharge Q_0.
- Measurement of the piezometric level at two points h_s and h_0, one is the center of the channel system and the other is the point of the output hydrograph.

Therefore, from the known values of the input and output hydrograph, it is necessary to calculate two unknown transformation functions, such as porosity $S(z)$ and conveyance $K(z)$. It is therefore necessary to use two level curves, so the task can be completed in a mathematical sense.

The model of elementary catchment, that is sub-catchment, is thus defined. It should be noted that the sub-catchment is a finite and not differentially small value and depends on the degree of investigation works.

Depending on the degree of exploration of the karst aquifer, the aquifer can be discretized into a number of elementary catchments or sub-catchments, as shown in Figure 10.19. The finite elements of the channel system are connected in the protruding points of the sub-catchment.

Calculation of porosity and conveyance, model calibration

The initial numerical model of the karst aquifer is the model in which the entire catchment is observed as an elementary sub-catchment for which there are minimal investigation works and appropriate measurements. Typical water waves, generated during rainy episodes, stand out among the set of measured quantities. Porosity of the basin is determined based on the measurements of the level curve in a

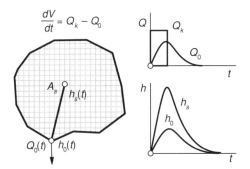

Figure 10.18   Elementary catchment or sub-catchment.

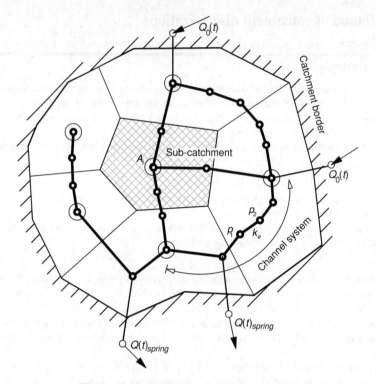

Figure 10.19   Discretization of karst aquifer.

representative piezometer and the output hydrograph in the recession part of the curve. Specifically, if in the catchment continuity equation

$$\frac{dV}{dt} = Q_k - Q_0,\qquad(10.95)$$

(Q_k is input discharge, and Q_0 is output discharge), time interval t_1, t_2 is observed, when Q_k equals zero, Eq. (10.95) can be integrated in the form

$$\int_1^2 \frac{dV}{dt}dt = -\int_1^2 Q_0 dt.\qquad(10.96)$$

The upper part of Figure 10.20 shows the output hydrograph and its discharge volume, and the bottom part shows the piezometric level and integration points 1 and 2. The integral on the left hand side of expression (10.96) will be written based on the piezometric head and porosity, and the integral on the right hand side will be expressed as the difference of the discharged volume, therefore, the following is obtained

$$\int_1^2 S(h)\frac{dh}{dt}Adt = -(V_2 - V_1).\qquad(10.97)$$

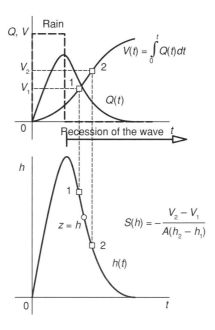

Figure 10.20 Determining porosity.

By using the average porosity between points 1 and 2, the integral can be written as

$$\bar{S}(h) \int_{1}^{2} \frac{dh}{dt} A \, dt = \bar{S}(h) (h_2 - h_1) A = -(V_2 - V_1) \tag{10.98}$$

from where

$$\bar{S}(h) = -\frac{V_2 - V_1}{A (h_2 - h_1)}. \tag{10.99}$$

The presented calculation procedure avoids the numerical calculation of derivatives, which is burdened with larger errors than the process of integration of measured values. The conveyance of the elementary catchment is obtained from measurements of discharge and piezometric difference, as shown in Figure 10.21. The average conveyance of the channel system between the two piezometers, located at distance L, is equal to

$$K(h_1, h_2) = \frac{Q_0}{\sqrt{\dfrac{h_1 - h_2}{L}}}. \tag{10.100}$$

The obtained data can be interpreted as uniform or non-uniform channel conveyance, in the sense of Section 10.2.4.

The calculated functions of porosity $S(z)$ and conductibility $K(z)$, obtained at the elementary catchment, can be used for calibration of a more complex model of a karst aquifer. The calibration process starts with the choice of the initial values of porosity and conveyance of new sub-catchments and channels

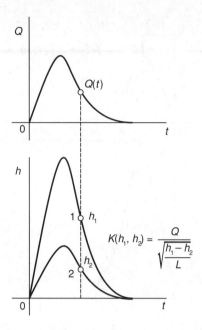

Figure 10.21 Determining conveyance.

on the basis of measured values, after which a model is created, in which the input hydrograph is transformed into the output hydrograph and piezometric level curves. By comparing the model output "signals" with those measured, we can decide how to modify the assumed functions in order to minimize the differences. The calibration process is tedious but necessary for understanding and quantifying the hydraulic behavior of karst aquifers.

10.4.2 Equation of the underground accumulation of a karst sub-catchment

The catchment is discretized into *sub-catchments* where precipitation (rain) charges the underground channel system. The area of direct penetration of rain A_0, the graph of the total residue of rainfall $p_k(t)$(rain volume per unit area), and porosity $S(z)$, dependent on the elevation, are associated with each sub-catchment. Charging of the underground karst volume is calculated by the equation of continuity

$$\frac{dV}{dt} = Q_k + Q_a, \tag{10.101}$$

where V is water in the karst system of cracks and pores, $Q = dV/dt$ is accumulation discharge, Q_k is rain discharge, and Q_a is net discharge coming from the channel system, see Figure 10.22.

Discharge Q_a in the node of the channel system is equal to an unbalanced discharge from all associated elements

$$Q_a = \sum_p {}^r Q_e. \tag{10.102}$$

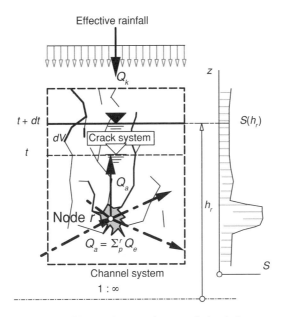

Figure 10.22 Underground accumulation in karst.

The nodal equation of the node that includes the catchment is

$$\sum_{p} {}^{r} Q_e = \frac{dV}{dt} - Q_k = A_0 S(h_r)\frac{dh_r}{dt} - A_0 \frac{dV_r}{dt}. \tag{10.103}$$

Update of the fundamental system

The volume increment of nodal accumulation is

$$\Delta V_a = \int_{\Delta t} Q_a dt = \int_{\Delta t} \left(A_0 S(h_r)\frac{dh_r}{dt} - A_0 \frac{dV_r}{dt} \right) dt. \tag{10.104}$$

After integration in a time step, the following is obtained

$$\Delta V_a = A_0 \bar{S}(h_r^+ - h_r) - A_0 \left(V_r^+ - V_r \right), \tag{10.105}$$

where $\bar{S}$ is porosity in interval $\left[h_r, h_r^+ \right]$ at the beginning and end of the time step. The volume increment is added to the global system vector (positive volume increment ΔV_k means reduction of water from the system)

$$F_r^{new} = F_r^{old} + \Delta V_a. \tag{10.106}$$

The elements that complement the fundamental system vector are functions of the main variables, therefore upgrading of the base system matrix with corresponding deviations is also required

$$A_{r,r}^{new} = A_{r,r}^{old} - \frac{\partial \Delta V_a}{\partial h_r^+} = A_{r,r}^{old} - A_0 \bar{S}. \qquad (10.107)$$

The update of the fundamental system for nodal function of karst sub-catchment is implemented in the subroutine

subroutine CarstNode(inode,iactiv),

which is invoked in the prefrontal subroutine subroutine FrontU(ielem,lun) for nodal type

```
...
case (CARST_PNT)
    call CarstNode(inode,iactiv)
...
```

The syntax for setting the sub-catchment is also simple:

```
!
! ...
! sintax:
!    Carst Name Point Area grfKisa <grfEvap <hInit>
!        Z1  Porosity1
!        Z2  Porosity2
!          ...
```

where Name is the name of the sub-catchment, Point is the connection point of the catchment and the channel system, Area is the catchment area, and grfKisa is the rain graph. The graph of evapotranspiration and the initial values of the piezometric head can be optionally set. For now, grfEvap equal to zero should be used, because the solution with evapotranspiration has not been implemented, but only serves as a venue for future upgrade of the program SimpipCore (see: www.wiley.com/go/jovic).

As the catchment node equation (10.103) is a differential equation with initial conditions, it is necessary to know the initial condition. Therefore, the initial piezometric head hInit can be optionally set in the catchment central point which will remain unchanged in the calculation of the steady flow. Otherwise, the required value will have to be calculated at the time of invoking by the sub-program Steady. After this line the data pairs for $S(z)$ follow: z, Porosity.

10.5   Rainfall transformation

10.5.1   Uniform input hydrograph

Task 4

A karst spring at elevation $z_0 = 10$ m.a.s.l. drains the aquifer through a 10 km long channel, its slope being $J_0 = 0.1\%$, as shown in Figure 10.23. In the catchment area of 4 km^2 it rains continuously for 24 hours, with uniform intensity of 10 mm per hour, according to the default hydrograph which is shown in Figure 10.24. Maximum inflow Q_k is 11.11 (m^3/s). It is necessary to examine the transformation of rain in the output hydrograph and the corresponding level curve for:

(a) a uniformly karstified aquifer in which the conveyance is set by the graph based on the formula $K(y) = 6.25y$, porosity is uniform S $= 0.1\%$ (graph a, Figure 10.23);

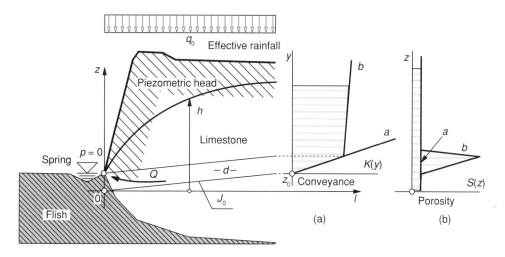

Figure 10.23 Test karst aquifer.

(b) a non-uniform aquifer similar to the natural karst aquifer with a pronounced channel system at the bottom: layer d, shown in Figure 10.23. Conveyance of the channel layer d is equal to $K(y) = 6.25y$ and is equal to that in (a). Above this layer conveyance is slightly altered so that at the depth $y = 250$ m it is $K(y) = 0.0625y$. The porosity of the channel system is ten times higher, (graph b, Figure 10.23).

The transformation will be tested if the pressure at the spring is:

• equal to normal atmospheric pressure (free discharge at the spring), the piezometric head is $h = z_0 = 10$ a. s. l.,
• equal to pressure of 110 m of water column (submerged discharge at spring), that is for a piezometric head of $h = 120$ a. s. l.

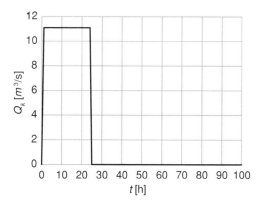

Figure 10.24 Input rainfall hydrograph.

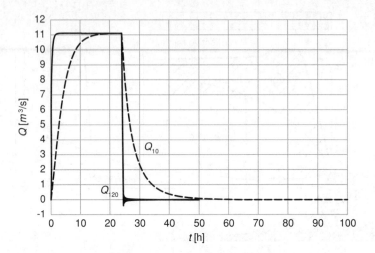

Figure 10.25   Output spring hydrograph
Q_{10} – hydrograph of unsubmerged spring
Q_{120} – hydrograph of submerged spring.

Input file: Kanal AB kisa.simpip (from www.wiley.com/go/jovic: Program solutions & tests\Test simpip files).

ad (a): the first aquifer (uniformly karstified aquifer)

Figure 10.26 show the output hydrographs and the corresponding level curves obtained by numerical modelling for the described first aquifer.

From the results it is concluded that the output hydrograph approaches the input rainfall hydrograph if the pressure at the spring is substantially increased. At the same time, with the increase in pressure

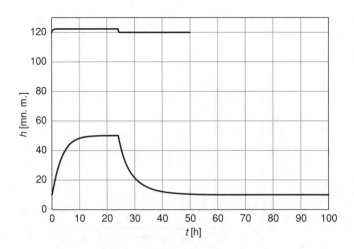

Figure 10.26   Level curve in the centre of catchment
h_{10} – unsubmerged spring (below)
h_{120} - submerged spring (above).

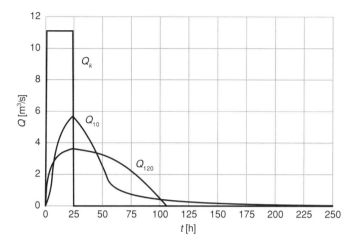

Figure 10.27 Input and output hydrographs:
Q_k – input rainfall hydrograph
Q_{10} – hydrograph of unsubmerged spring
Q_{120} – hydrograph of submerged spring.

of the submerged spring, the level curve decreases significantly as a result of increased conveyance for higher levels of uniformly karstified aquifer.

ad (b): the second aquifer (non-uniformly karstified aquifer)

Figure 10.27 and 10.28 show the input and output hydrograph and the corresponding level curves, obtained by numerical modelling for the second aquifer.

From the results it is concluded that the output hydrograph is transformed in relation to the rainfall hydrograph, much as in the first aquifer, for the unsubmerged spring, but with a pronounced delay in

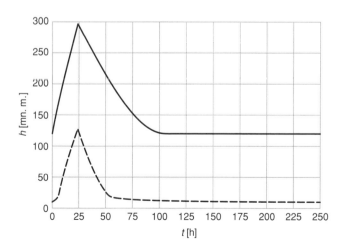

Figure 10.28 Level curve in the center of the catchment:
h_{10} – unsubmerged spring (dashed line)
h_{120} – submerged spring (solid line).

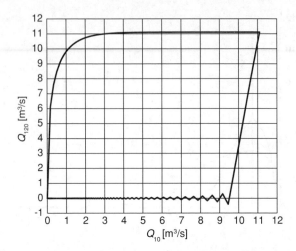

Figure 10.29   ad (a): correlation of normal and recessive hydrograph.

achieving the maximum. If the pressure at the spring is substantially increased, greater transformation occurs, the maximum is even smaller, and the hydrograph is stretched.

At the same time, with the increase in pressure of the submerged spring, the level curve increases significantly in relation to the first case, as a result of smaller conveyance for higher levels of non-uniformly karstified aquifer.

Figures 10.29 and 10.30 show a hydrograph correlation at the spring between the free and submerged discharge. In the first case, where the flow equals zero, numerical oscillations are observed, caused by instability in small relations between Δt and the length of the finite element, which is considered a solvable problem. It can be seen that a simple correlation between the hydrographs is not possible, but that the functional connection is complex and should be determined by solving hydrodynamical equations.

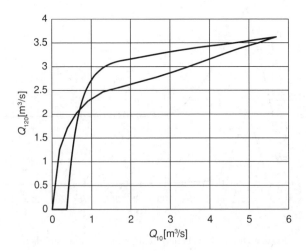

Figure 10.30   ad (b): correlation of normal and submerged hydrograph.

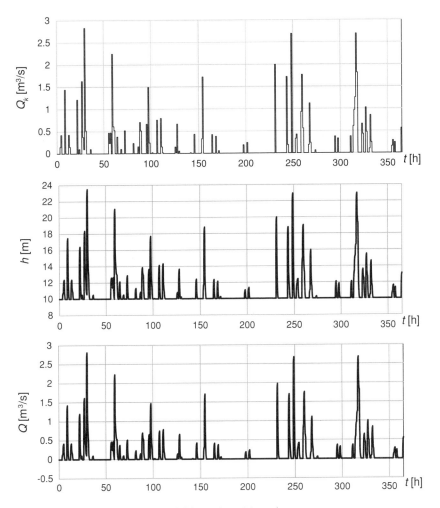

Figure 10.31   Rain at (a) catchment.

10.5.2   Rainfall at the catchment

Input file: `Kanal AB - kisa Dicmo.simpip` (from www.wiley.com/go/jovic: Program solutions & tests\Test simpip files).

Figure 10.31 and 10.32 show the modelling results of the two aquifers described in Task 4. Instead of a uniform impulse rain hydrograph, a real hydrograph was modeled, obtained for the meteorological station Dicmo (Croatia) for the year 2009.

By comparing the level curve in the center of catchment (a) (uniformly karstified aquifer) and catchment (b) (non-uniformly karstified aquifer) it can be observed that water levels in case (b) are generally lower, except in the isolated case of day 320, when the water level in aquifer (b) was higher than in aquifer (a), due to previously high water levels and slower discharge at the spring.

In case (b), recessive parts of the discharge curves are longer, but the greatest discharges are always smaller than in the case of aquifer (a).

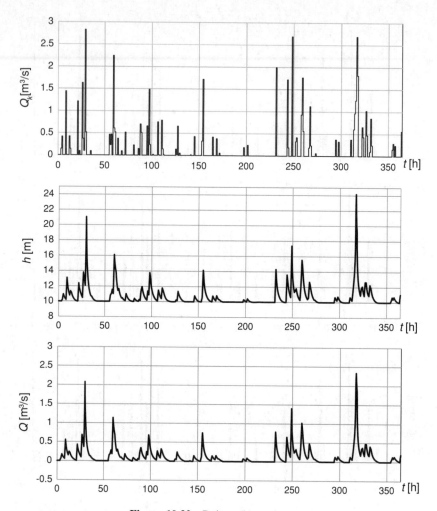

Figure 10.32   Rain at (b) catchment.

10.6   Discretization of karst catchment with diffuse and channel flow

The diffuse flow is modeled by three-dimensional finite elements (see Figure 10.33). For the diffuse space the filtration equation can be used

$$S\frac{\partial h}{\partial t} + \frac{\partial}{\partial x_i} K_{ij}\frac{\partial h}{\partial x_j} = f$$

$$i, j = 1, 2, 3,$$

(10.108)

where S is the elastic storage coefficient, K is the tensor of the diffusion, that is equivalent Darcy's filtration coefficients, and f is external inflow.

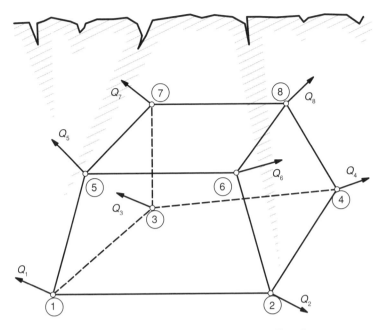

Figure 10.33   Sketch of finite element of diffuse flow.

The appropriate weak formulation of Eq. (10.108) is

$$\int_V \left(S\frac{\partial h}{\partial t} + \frac{\partial}{\partial x_i} K_{ij} \frac{\partial h}{\partial x_j} - f \right) w\, dV = 0, \tag{10.109}$$

where w is the test function. For the chosen variant of discretization on finite elements, the approximate solution of the piezometric head will be written as a linear combination of base vectors, constructed from the shape functions

$$h = h_r(t)\varphi_r(x_i)$$
$$r = 1, 2, 3 \cdots m, \tag{10.110}$$

where nodal values $h_r(t)$ appear as time functions. If an approximation base is used as test base (Galerkin formulation)

$$w = \varphi_s(x_i)$$
$$s = 1, 2, 3 \cdots m, \tag{10.111}$$

the discrete form of weak formulation as an equation system is obtained

$$\frac{dh_r}{dt} \int_\Omega S\varphi_r\varphi_s\, d\Omega + h_r \int_\Omega k_{ij}\frac{\partial \varphi_r}{\partial x_i}\frac{\partial \varphi_s}{\partial x_j}\, d\Omega = \int_\Omega f\varphi_s\, d\Omega + \int_\Gamma q\varphi_s\, d\Gamma$$
$$\tag{10.112}$$
$$r, s = 1, 2, 3 \cdots m$$
$$i, j = 1, 2, 3,$$

where the marked integral is written as

$$C_{rs} = \int_{\Omega} S\varphi_r\varphi_s d\Omega \text{ – global capacity matrix,} \tag{10.113}$$

$$D_{rs} = \int_{\Omega} k_{ij}\frac{\partial\varphi_r}{\partial x_i}\frac{\partial\varphi_s}{\partial x_j}d\Omega \text{ – global conductivity matrix,} \tag{10.114}$$

$$F_s = \int_{\Omega} f\varphi_s d\Omega \text{ – global vector of external inflow,} \tag{10.115}$$

$$Q_s = \int_{\Gamma} q\varphi_s d\Gamma \text{ – global vector of nodal discharges.} \tag{10.116}$$

By using these denominations, the system is

$$C_{rs}\frac{dh_r}{dt} + D_{rs}h_r = F_s + Q_s. \tag{10.117}$$

The obtained equations are ordinary differential equations with initial conditions. The global system is compliant to the process of assembling from the corresponding finite element matrices and vectors.

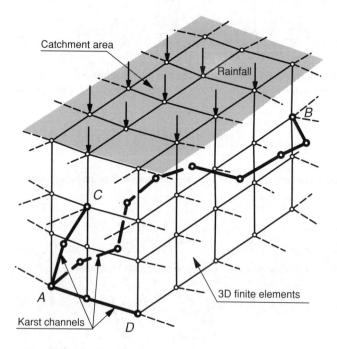

Figure 10.34　3D discretization of diffuse and channel space.

The corresponding form of the matrix system for a typical finite element is

$$C_{rs}^e \frac{dh_r}{dt} + D_{rs}^e h_r = F_s^e + Q_s^e, \tag{10.118}$$

where the element capacity matrix, element conductivity matrix, element external inflow vector, and element vector of nodal discharges appear, formally equal in expressions (10.113), (10.114), (10.115), and (10.116). The base functions φ_r should be replaced with shape functions ϕ_r.

Figure 10.34 shows part of the diffuse space of discretization of the karst aquifer on finite elements. The catchment area comprises the upper sides of finite elements. The finite elements of the channel systems are superimposed on the displayed finite element mesh. These are linear three-dimensional elements that connect close or distant points in the aquifer space. For example, Figure 10.34 shows the karst channel using the linear finite elements that connect the diffuse space between nodes A and B. Therefore, they are two points where the channel and diffuse space exchange water and all other nodes are ordinary connections of the channel system.

The nodes of the channel system can also be nodes of the diffuse space discretization. In that case the compatibility of elements should be taken into account. For example, the finite elements of the channel from A to D are also the edges of the finite elements of the diffuse space. The finite elements of the channel from A to C are connected to the diffuse space at nodes A and C, while there is no direct communication between the middle node and diffuse space. This discretization is called "the sponge model," where diffuse space elements mostly accumulate water while the adjacent water flow is insignificant in comparison to the transportation capabilities of the channel system.

References

Bonnaci, O. (1987) *Karst Hydrology with Special Reference to the Dinaric Karst.* Springer Series in Physical Environment, Springer-Verlag, New York.

Forenbaher, S. (2002) Prehistoric populations of the Island Hvar – An overiew of archaeological evidence. *Collegium Antropologicum*, 26: 361–378.

Marijanović, P. (2008) *Geostatistics Karst Dinarides* (in Croatian). Građevinski fakultet Sveučilišta u Mostaru, Mostar.

Roglić, J. (2004) Krš i njegovo značenje, Sabrana djela, Geografsko društvo Split, Hrvatsko geogr. društvo Zadar, Geografski odsjek PMF – Zagreb, Zagreb.

Roglić, J. (2005) Geomorfološke teme, Sabrana djela, Geografsko društvo Split, Hrvatsko geogr. društvo Zadar, Prirodno matematički fakultet Zagreb, Sveučilište u Zadru, Zagreb.

Further reading

Alfirević, S. (1969) Jadranske vrulje u vodnom režimu Dinarskog primorskog krša i njihova problematika, *Krš Jugoslavije*, 6.

Bojanić, D. (1994) Identification of hydrological-hydraulic elements of the runoff in karst areas. Master's thesis (in Croatian). Zagreb.

Denić-Jukić, V. (2002) Hydrological point of view of runoff in the karst areas. Doctoral dissertation (in Croatian).

Jović, V. (2003a) Numerical model of flows in the underground storage reservoir of the Ombla HE plant – Natural state and the state planned by the project, Hydro 2003 – The Way Forward For Hydropower. Proc. Vol I, Dubrovnik, 2003.

Jović, V. (2003b) Numerical models of karst flows, Hydro 2003 – The Way Forward For Hydropower. Proc. Vol I, Dubrovnik, 2003.

Jukić, D. and Denić-Jukić, V. (2009) Groundwater balance estimation in karst by using a conceptual rainfall-runoff model. *Journal of Hydrology* 373: 302–315.

Jukić, D. (2005) The role of transfer functions for creating balance and modelling runoff in karst areas. Thesis (in Croatian).

Paviša, T. (2003) HPP Ombla in Croatia – proposed use of energy from groundwater in karst aquifer, Hydro 2003 – The Way Forward For Hydropower. Proc. Vol I. Dubrovnik.

Ravnik, D. and Rajver, D. (1998) The use of inverse geotherms for determining underground water flow at the Ombla Spring near Dubrovnik – Croatia. *Journal of Applied Geophysics*, July.

11

Convective-dispersive Flows

11.1 Introduction

In addition to hydraulic flow as the primary process in the network, a secondary process may occur, such as heat transfer or transfer of the solution of a substance, that can be expressed as a concentration. In terms of solution, these two processes can be independent or coupled. An example of a coupled process is hydraulic flow as a primary process in which heat transfer occurs as a secondary process, where many physical values of fluid depend on the temperature. In the case of small temperature differences it can be considered that the basic flow is independent of heat transfer. The processes will be considered independent hereinafter.

Figure 11.1 shows an example of a hydraulic network where the branches are pipes, channels, and similar connecting elements. The solution to the hydraulic problem is described by the piezometric height and discharge

$$h(l, t), \ Q(l, t), \qquad (11.1)$$

where l is spatial and t is time-variable. This is the primary process. The hydraulic field is the holder of a secondary flow. Heat transfer is considered a secondary process and the obtained results can be applied without major difficulties to other types of transfer processes.

11.2 A reminder of continuum mechanics

Conservation law of the extensive field. Conservation law of the extensive field, such as mass, momentum, heat, and energy, can easily be understood if we observe a material body or a bounded part of the continuum in motion. Such a bounded continuum represents a mechanical and thermodynamic system for which constant mass in motion applies, or the change of momentum equals the impulse of forces affecting it. Physical quantities, such as mass, momentum, and energy, are characteristics of every one of its parts and they have additive properties. For example, the mass of the entire continuum equals the sum of the masses of its parts, momentum of the entire continuum equals the sum of momentum of all parts, and so on. Conservation law can be applied to extensive fields, which says that a change in the extensive field E is equal to the production F of a field in the unit of time t:

$$dE = Fdt. \qquad (11.2)$$

Analysis and Modelling of Non-Steady Flow in Pipe and Channel Networks, First Edition. Vinko Jović.
© 2013 John Wiley & Sons, Ltd. Published 2013 by John Wiley & Sons, Ltd.

(a) Hydraulic network configuration.

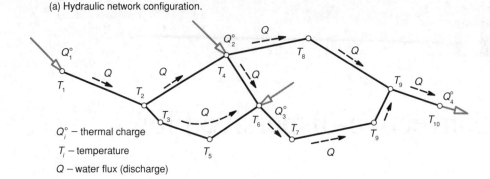

Q_i° – thermal charge

T_i – temperature

Q – water flux (discharge)

(b) Channel section.   (c) Pipe section.

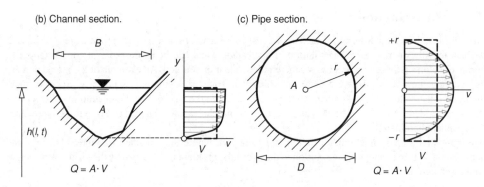

Figure 11.1 Hydraulic network of primary and secondary processes.

For example, for the mass ($E = M$) which remains constant for a continuum in motion, the production of field F equals zero; for momentum ($E = K_i$) the production of the field is equal to the impulse of all forces, etc. The extensive field conservation law is expressed as rate of change content E in time

$$\frac{d}{dt} E(V, t) = F(V, t)$$ (11.3)

which is valid in every moment t regardless of the size of the body of volume V.

The extensive field density e is defined,[1] based on which the content of the extensive field in volume V is calculated by

$$E(V, t) = \int_V e\, dV.$$ (11.4)

Field production density f is defined in a similar way, based on which the production in volume V is calculated

$$F(V, t) = \int_V f\, dV.$$ (11.5)

---

[1] The extensive field density in a continuum is $e = dE/dV$.

Therefore, the conservation law is expressed

$$\frac{d}{dt} \int_V e \, dV = \int_V f \, dV.$$ (11.6)

Convective flow. If the motion is observed in the control volume V, the Reynolds' transport theorem applies, based on which the general law on extensive field conservation is expressed

$$\frac{\partial}{\partial t} \int_V e \, dV + \int_A e v_i n_i \, dA = \int_V f \, dV,$$
$$i = 1, 2, 3$$ (11.7)

where v_i is the velocity vector, n_i is the outer unit normal vector, and A is the surface enclosing the control volume. This integral formulation of general conservation law will be written symbolically in the form of

$$\frac{\partial}{\partial t} E(V, t) + Q(A, t) = F(V, t),$$ (11.8)

where transfer flux is

$$Q(A, t) = \int_A e v_i n_i \, dA$$
$$i = 1, 2, 3$$ (11.9)

and has a positive sign for discharge (a positive scalar product of velocity with the outer unit normal). The general law on extensive field conservation will be expressed by the words: *the change of the extensive field quantity, which occurs in time within the control volume, plus the change that occurs due to flow through the control surface, is equal to the change that occurs due to the field production.*

Table 11.1 shows the density of some extensive fields and the density of their production.

Table 11.1

12 Extensive field E	13 Field density e	Production density f
Mass	ρ	0
Solution	ρc	r
Momentum	ρv_i	f_i
Heat	$\rho C T$	w
Energy	e	w

Differential form of the conservation law. The differential form of the conservation law can be obtained from the integral form (11.7)

$$\frac{\partial e}{\partial t} + \frac{\partial (e v_i)}{\partial x_i} = f; \quad i = 1, 2, 3.$$ (11.10)

Diffuse flow. Many physical phenomena have a diffuse nature, spreading from the place of greater concentration towards smaller concentration, even when the liquid is not in motion. This is due to intermolecular processes and the phenomenon is called molecular diffusion. Diffuse flow is generally anisotropic and is proportional to the field density gradient. Diffuse flow of an extensive value is obtained by integrating the specific diffuse flow on the surface A that encloses the control volume

$$Q(A, t) = -\int_A D_{ij}\frac{\partial e}{\partial x_j}n_i dA; \quad i, j = 1, 2, 3, \tag{11.11}$$

where D_{ij} is a diffusion tensor of dimensions $\left[L^2 T^{-1}\right]$. It is a constant proportionality in the most general linear form. By inserting diffuse flow into the conservation law (11.10) for an extensive value of diffuse spreading, it can be written as follows

$$\frac{\partial e}{\partial t} + \frac{\partial(e v_i)}{\partial x_i} = \frac{\partial}{\partial x_i}\left(D_{ij}\frac{\partial e}{\partial x_j}\right) + f; \quad i, j = 1, 2, 3 \tag{11.12}$$

which represents the differential form of the general transfer convective-diffuse field conservation.

Diffusion of momentum. If momentum density $e = \rho v_i$ is inserted in Eq. (11.12) and the isotropic diffuse process of momentum $D_{ij} = \delta_{ij} D$ is considered, where δ_{ij} is Kronecker unit tensor, we have

$$\frac{\partial \rho v_i}{\partial t} + \frac{\partial(\rho v_i v_j)}{\partial x_j} = \frac{\partial}{\partial x_j}\left(D\frac{\partial \rho v_i}{\partial x_j}\right) + f_i; \quad i, j = 1, 2, 3.$$

The following is obtained by analyzing the obtained equation

$$\rho\frac{\partial v_i}{\partial t} + \underbrace{v_i\left(\frac{\partial \rho}{\partial t} + \frac{\partial \rho v_j}{\partial x_j}\right)}_{\equiv 0} + \rho v_j\frac{\partial v_i}{\partial x_j} = \underbrace{\rho D}_{\mu}\frac{\partial^2 v_i}{\partial x_j \partial x_j} + \underbrace{f}_{\Sigma \text{ spec. forces}}.$$

In the resulting expression the continuity equation is evident, which is equal to zero, so the Navier–Stokes equation is recognized from the remaining members, where production f is the resultant of specific volume b_i and surface forces $-\frac{1}{\rho}\frac{\partial p}{\partial x_i}$, while the product $\mu = \rho D$ is equal to the dynamic viscosity coefficient

$$\frac{\partial v_i}{\partial t} + v_j\frac{\partial v_i}{\partial x_j} = D\frac{\partial^2 v_i}{\partial x_j \partial x_j} + b_i - \frac{1}{\rho}\frac{\partial p}{\partial x_i}. \tag{11.13}$$
$$i, j = 1, 2, 3$$

Therefore, the coefficient D of diffusion of momentum is equal to the coefficient of kinematic viscosity so it can be said that the viscosity is equal to the diffusion of momentum on the molecular level.

Something similar applies to the diffusive heat transfer, where i is the distributed heat source

$$\frac{\partial T}{\partial t} + v_j\frac{\partial T}{\partial x_j} = D\frac{\partial^2 T}{\partial x_j \partial x_j} + \frac{i}{\rho C}. \tag{11.14}$$
$$j = 1, 2, 3$$

Turbulent diffusion. Unlike the molecular diffusion that occurs in calm fluid or in laminar flow, in turbulent flow the spread increases several times due to turbulence. In turbulent flow all hydrodynamic variables are stochastic, including temperature. They are expressed by averaged and fluctuating values

$$v_i = \bar{v}_i + v_i'$$
$$T = \bar{T} + T'. \tag{11.15}$$
$$\cdots$$

After insertion into the Navier–Stokes equation, the Reynolds averaged equations are obtained

$$\frac{\partial \bar{v}_i}{\partial t} + \bar{v}_j \frac{\partial \bar{v}_i}{\partial x_j} = D \frac{\partial^2 \bar{v}_i}{\partial x_j^2} - \frac{\partial \overline{v_i' v_j'}}{\partial x_j} + b_i - \frac{1}{\bar{\rho}} \frac{\partial \bar{p}}{\partial x_i} \tag{11.16}$$
$$i, j = 1, 2, 3$$

where fluctuating velocity components occur. Fluctuating members of the Reynolds equations can be described as virtual stresses created by the transfer of momentum of pulsed turbulent vortices, or according to Prandtl's theory of turbulent vortices, the following applies

$$\tau_{ij}' = -\rho \, \overline{v_i' v_j'} = \rho \varepsilon_{jk} \frac{\partial \bar{v}_i}{\partial x_k}, \tag{11.17}$$

where ε_{jk} is the tensor of the turbulent diffusion coefficient. If temperature is expressed similarly, by averaged and fluctuating value, the transfer equation for averaged temperature is obtained

$$\frac{\partial \bar{T}}{\partial t} + \bar{v}_j \frac{\partial \bar{T}}{\partial x_j} = -\frac{\partial \overline{(v_j' T')}}{\partial x_j} + D \frac{\partial^2 \bar{T}}{\partial x_j \partial x_j} + \frac{i}{\rho C}. \tag{11.18}$$

The first member on the right hand side of the obtained equation occurs due to diffusion of turbulent vortices, which additionally spread the temperature, therefore

$$\overline{\rho (v_j' T')} = -\rho \varepsilon_j \frac{\partial \bar{T}}{\partial x_j}, \tag{11.19}$$

or expressed as

$$\frac{\partial \bar{T}}{\partial t} + \bar{v}_j \frac{\partial \bar{T}}{\partial x_j} = \varepsilon_j \frac{\partial \bar{T}}{\partial x_j} + D \frac{\partial^2 \bar{T}}{\partial x_j \partial x_j} + \frac{i}{\rho C}, \tag{11.20}$$

where ε_j is in the general anisotropic coefficient of turbulent diffusion. In isotropic turbulence turbulent diffusion is equal in all directions. Turbulent diffusion coefficients are many times larger than the molecular diffusion coefficient, so that in many practical cases, molecular diffusion can be neglected in relation to turbulent mixing. The coefficients of turbulent and molecular diffusion are determined experimentally, by measuring the velocity, temperature, and concentration in turbulent mixing processes.

11.3   Hydrodynamic dispersion

Turbulent diffusion in uniform flow in a channel or pipe varies in two cases, presented in Figure 11.2a and b. Average velocity is equal in both cases and is $V = Q/A$. In time t_0 a colored solution is added

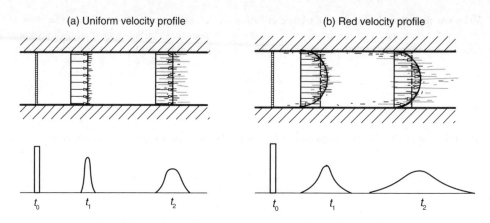

Figure 11.2   Diffusion and dispersion.

along the whole channel section and its spread along the flow is observed, that is the longitudinal section of concentration c.

In the case of an imaginary uniform velocity profile, the solution is transferred by convection and diffuse spreading. Unlike the uniform velocity profile, in the case of a real turbulent velocity profile, the solution is retained by the walls of the channel. Retaining the solution by the walls causes lateral diffusion and the mixing is stronger. This phenomenon is called *hydrodynamic dispersion*. For reasons of similar concentration profiles, the process of turbulent dispersion can be described by the equation

$$\frac{\partial c}{\partial t} + V \frac{\partial c}{\partial x} = \frac{\partial}{\partial x}\left(\varepsilon_t \frac{\partial c}{\partial x}\right),$$
(11.21)

where V is mean profile velocity, c is mean concentration in the profile, and ε_t is turbulent dispersion coefficient. Coefficient ε_t of longitudinal turbulent dispersion exceeds the mean turbulent diffusion coefficient by approximately 200 times.

G. I. Taylor (Taylor, 1954) determined the longitudinal dispersion coefficient in uniform turbulent flow in pipes,

$$\varepsilon_t = 10.1 \cdot r_0 \sqrt{\frac{\tau_0}{\rho}},$$
(11.22)

where r_0 is the pipe radius and τ_0 is friction on the pipe sheath. The Fischer formula for the longitudinal dispersion coefficient can be used for flows in open channels

$$\varepsilon_t = 20 R v_* = 20 R \sqrt{\frac{\tau_0}{\rho}},$$
(11.23)

where R is the hydraulic radius and v_* is the friction velocity. The expressions (11.23) and (11.22) differ by a constant, and are best determined by measuring the dispersion of a tracer in the nature. A standard mark D for coefficients of molecular and turbulent diffusion, or dispersion, will be used hereinafter.

11.4   Equations of convective-dispersive heat transfer

It is assumed that temperature changes are small enough not to affect the basic hydrodynamic conditions (11.1). Furthermore, full mixing and equalizing of the water temperature in cross-section is assumed, so that all thermal conditions are fully described by the function

$$T(l, t), \tag{11.24}$$

that is by averaged temperature T in the profile at station l in time t. As temperature T determines the thermal state of the mass, the thermal energy conservation law will be set, from which the characteristic equation of convective dispersion of one branch of hydraulic network will be developed, such as a pipe element or channel. Thermal energy of water flow between two profiles is

$$E_T(l_1, l_2, t) = \int_{l_1}^{l_2} \rho CTA \, dl \quad [J], \tag{11.25}$$

where $\rho\,[kg/m^3]$ is the water density, C is the specific heat or specific heat capacity of water $[JK^{-1}kg^{-1}]$, T is the average (absolute) water temperature in the profile at station l in time t in $[K]$, and A is the area of the cross-section.

Heat flux in a profile consists of a convection and dispersion part which is obtained by integrating the specific flux

$$Q^\circ(A, t) = \int_A (\rho CTv - \rho CDgradT) dA = \rho CQT - \rho CDAgradT \tag{11.26}$$

of dimensions $[W] = [J\,s^{-1}]$, where D is the averaged longitudinal coefficient of turbulent dispersion in the observed profile.

In addition to heat flux in cross-sections, there is thermal exchange between the water body and the environment. This thermal exchange occurs in the sheath of hydraulic branch which generally consists of a wetted sheath and water table, depending on the observed pipe or channel branch of the hydraulic network. The heat flux, due to changes in the water table and wetted perimeter, is

$$F(l_1, l_2, t) = \int_{l_1}^{l_2} \phi dl = \int_{l_1}^{l_2} \left(\Phi^a \cdot B + \Phi^b \cdot O \right) dl, \quad [W] = \left[J s^{-1} \right], \tag{11.27}$$

where $\phi\,[W/m]$ is the specific heat flux per unit of branch length, $\Phi^a = \Phi^a(l, t)$ is the specific heat flux on the water table $[W/m^2]$, averaged along the water table width B [m] in the profile, $\Phi^b = \Phi^b(l, t)$ is the specific heat flux distributed on the wetted sheath $[W/m^2]$, where O [m] is the wetted perimeter. The exchange of heat flux with the environment is positive in the case of leakage (release) of heat into the atmosphere or into the ground.

According to the energy conservation law, the total heat change is zero

$$\left\{ \int_{l_1}^{l_2} \rho CTA dl \right\} \Big|_{t_1}^{t_2} + \int_{t_1}^{t_2} \left\{ \left[\rho CTQ - \rho cDA \frac{\partial T}{\partial l} \right] \Big|_{l_1}^{l_2} \right\} dt + \int_{t_1}^{t_2} \left\{ \int_{l_1}^{l_2} \phi dl \right\} dt = 0 \tag{11.28}$$

which, due to the absolute continuity of the sub-integral function, can be written as follows

$$\int_{t_1}^{t_2}\int_{l_1}^{l_2}\left[\frac{\partial}{\partial t}(\rho CTA) + \frac{\partial}{\partial l}\left(\rho CTQ - \rho CDA\frac{\partial T}{\partial l}\right) + \phi\right]dldt \;\; [J].$$

(11.29)

This expression can be written as the heat flux equation

$$\frac{\partial}{\partial t}\int_{l_1}^{l_2}\rho CTAdl + \left[\rho CTQ - \rho CDA\frac{\partial T}{\partial l}\right]\Bigg|_{l_1}^{l_2} + \int_{l_1}^{l_2}\phi dl = 0 \;\; [W].$$

(11.30)

Transferring the integral into differential form, the following is obtained

$$\frac{\partial}{\partial t}(\rho CTA) + \frac{\partial}{\partial l}(\rho CTQ) = \frac{\partial}{\partial l}(\rho CDA\frac{\partial T}{\partial l}) + \phi \;\; \left[\frac{W}{m}\right]$$

(11.31)

which is the equation of convective-dispersive heat transfer in a typical branch of the hydraulic network.

As we assume that density ρ and specific heat c are constant for the observed temperature range, Eq. (11.31) can be written

$$\frac{\partial}{\partial t}(TA) + \frac{\partial}{\partial l}(TQ) - \frac{\partial}{\partial l}(DA\frac{\partial T}{\partial l}) + \frac{\phi}{\rho C} = 0.$$

(11.32)

Furthermore, by partial derivation of complex members we have

$$A\frac{\partial T}{\partial t} + T\frac{\partial A}{\partial t} + T\frac{\partial Q}{\partial l} + Q\frac{\partial T}{\partial l} - A\frac{\partial}{\partial l}\left(D\frac{\partial T}{\partial l}\right) - D\frac{\partial T}{\partial l}\frac{\partial A}{\partial l} + \frac{\phi}{\rho C} = 0$$

or

$$A\frac{\partial T}{\partial t} + T\underbrace{\left[\frac{\partial A}{\partial t} + \frac{\partial Q}{\partial l}\right]}_{\equiv 0} + \frac{\partial T}{\partial l}\left[Q - D\frac{\partial A}{\partial l}\right] - A\frac{\partial}{\partial l}\left(D\frac{\partial T}{\partial l}\right) + \frac{\phi}{\rho C} = 0$$

The term in the first brackets represents the equation of continuity, and disappears. After dividing the expression with the cross-section area A, the following is obtained

$$\frac{\partial T}{\partial t} + \frac{\partial T}{\partial l}\left[\frac{Q}{A} - \frac{D}{A}\frac{\partial A}{\partial l}\right] = \frac{\partial}{\partial l}\left(D\frac{\partial T}{\partial l}\right) - \frac{\phi}{\rho CA}$$

(11.33)

which is a final form of the equation for convectively dispersed flow in the hydraulic network branch. It should be noted that in the literature the form of equation is often used

$$\frac{\partial T}{\partial t} + V\frac{\partial T}{\partial l} = D\frac{\partial^2 T}{\partial l^2} + r,$$

(11.34)

where $V = \frac{Q}{A}$ is the average water velocity and $r = \frac{\phi}{\rho C A}$ is the heat exchange parameter and can be applied only for uniform flow where $\frac{\partial A}{\partial l} = 0$ and for constant coefficient D of longitudinal dispersion, or when

$$V \gg \frac{D}{A} \frac{\partial A}{\partial l} = 0 \quad \text{and} \quad \frac{\partial D}{\partial l} = 0. \tag{11.35}$$

If an equivalent velocity $V_e = V_e(l, t)$ is inserted into Eq. (11.34), where

$$V_e = \frac{Q}{A} - \frac{D}{A} \frac{\partial A}{\partial l} \tag{11.36}$$

assuming that $D = const$ it is also possible to appraise the convective-dispersion flow in a non-prismatic branch of the hydraulic network by a simple equation (11.34).

11.5   Exact solutions of convective-dispersive equation

11.5.1   Convective equation

We will observe the equation

$$\frac{\partial u}{\partial t} + V \frac{\partial u}{\partial l} = 0 \tag{11.37}$$

which describes the convection of temperature, concentration, or other similar quantity in a field of velocity V along the axis l. This equation belongs to the hyperbolic partial differential equations. It can be obtained from Eq. (11.34) if the coefficient of dispersion is equal to zero. Comparing the equation and material derivative of the variable u

$$\frac{du}{dt} = \frac{\partial u}{\partial t} + V \frac{\partial u}{\partial l} = \frac{\partial u}{\partial t} + \frac{dl}{dt} \frac{\partial u}{\partial l} = 0 \tag{11.38}$$

results that on characteristic curves defined by the simple differential equation

$$\frac{dl}{dt} = V \tag{11.39}$$

the total change of u equals zero

$$\frac{du}{dt} = 0. \tag{11.40}$$

By integrating the equation of characteristics (11.39) from the known point (l_0, t_0) to a point (l, t) in plane l, t, it can be written as follows

$$\int_{l_0}^{l} dl = \int_{t_0}^{t} V \, dt. \tag{11.41}$$

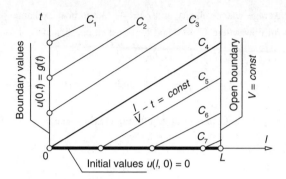

Figure 11.3 Characteristics or path of wave front.

For the constant velocity V the straight line is obtained

$$(l - l_0) = V(t - t_0),$$

or

$$\left(\frac{l}{V} - t\right) = \left(\frac{l_0}{V} - t_0\right) = const. \tag{11.42}$$

This is the path of wave front u starting from point (l_0, t_0). Figure 11.3 shows some characteristics drawn from initial and boundary conditions. The general solution of the convective equation is the function

$$f\left(\frac{l}{V} - t\right), \tag{11.43}$$

which satisfies the equation, which can be checked by insertion into Eq. (11.37)

$$\frac{\partial u}{\partial t} : \quad -f'; \ V\frac{\partial u}{\partial x} : V\frac{f'}{V}; \ \text{results}: \ \frac{\partial u}{\partial t} + V\frac{\partial u}{\partial x} = -f' + f' = 0.$$

From the known values of initial conditions $u(l, 0) = 0$ and boundary conditions $u(0, t) = g(t)$, the value of the solution $u(l, t)$ can easily be calculated, because Eq. (11.40) applies, that is

$$du = 0 \quad \text{or} \quad u = const. \tag{11.44}$$

11.5.2 Convective-dispersive equation

If it is assumed that, according to Figure 11.4, the mean fluid flow velocity through a pipe or channel is $V = Q/A$, Eq. (11.34)

$$\frac{\partial u}{\partial t} + V\frac{\partial u}{\partial l} = D\frac{\partial^2 u}{\partial l^2} \tag{11.45}$$

can be solved analytically for the initial and boundary conditions:

$$u(l, 0) = 0, \ 0 \le l < \infty,$$

$$u(0, t) = u_0, \ u(\infty, t) = 0$$

for production member $r = 0$.

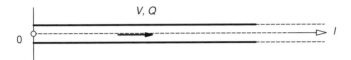

Figure 11.4

The analytical solution is

$$u(l, t) = \frac{u_0}{2} \left[e^{\frac{v \cdot l}{D}} erfc \left(\frac{l + v \cdot t}{2\sqrt{D \cdot t}} \right) + erfc \left(\frac{l - v \cdot t}{2\sqrt{D \cdot t}} \right) \right], \tag{11.46}$$

where $erfc(x) = 1 - erf(x)$ function complementary to $erf(x)$ function and erf is the error function

$$erf(x) = \frac{2}{\sqrt{\pi}} \int_0^x e^{-t^2} dt = \Phi\left(x\sqrt{2}\right) \tag{11.47}$$

which is calculated from the Gaussian integral of probability Φ. These functions are shown in Figure 11.5 for the argument value $-3 < x < +3$.

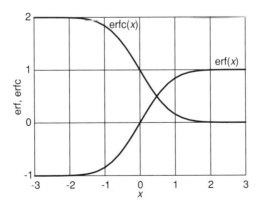

Figure 11.5   erf(x) and erfc(x) functions.

11.5.3   Transformation of the convective-dispersive equation

Equation (11.45) which describes the problem of convective-dispersed flow

$$\frac{\partial u}{\partial t} + v \frac{\partial u}{\partial x} = \mu \frac{\partial^2 u}{\partial x^2} \tag{11.48}$$

with initial $u(x, 0) = 0$ and boundary conditions: $u(0, t) = U(t)$, $u(l, t) = D(t)$ is by transformation

$$u(x, t) = \psi(x, t) e^{\frac{vx}{2\mu} - \frac{v^2 t}{4\mu}} \tag{11.49}$$

transformed into equation

$$\frac{\partial \psi}{\partial t} = \mu \frac{\partial^2 \psi}{\partial x^2} \tag{11.50}$$

and the corresponding initial $\psi(x, 0) = 0$ and boundary conditions

$$\psi(0, t) = U(t) e^{\frac{v^2 t}{4\mu}}, \tag{11.51}$$

$$\psi(l, t) = D(t) e^{\frac{v^2 t}{4\mu} - \frac{vl}{2\mu}}. \tag{11.52}$$

11.6   Numerical modelling in a hydraulic network

Here we propose a method which will, in relation to the known methods of finite differences or finite elements, improve accuracy and make the model very flexible without a serious increase in the number of computational operations. Numerical modelling of convective-dispersed flow in a hydraulic network is based on the fact that Eq. (11.31) can be decomposed into two

$$\frac{\partial}{\partial t}(\rho CAT) + \frac{\partial}{\partial l} Q^\circ + \phi = 0, \tag{11.53}$$

$$Q^\circ = \rho CQT - \rho CDA \frac{\partial T}{\partial l}, \tag{11.54}$$

where the first expresses the continuity of flow, and the second expresses convective-dispersive law for heat flux. This is actually a natural description of the process being observed.

Thus, instead of describing it with one function $T(l, t)$, the method uses two functions

$$\left\{ \begin{array}{c} T(l, t) \\ Q^\circ(l, t) \end{array} \right\}. \tag{11.55}$$

11.6.1   The selection of solution basis, shape functions

According to Figure 11.6a, a temperature field on a finite element will be expressed by Hermite polynomials

$$T = T_1 H_1^{(0)} + T_2 H_2^{(0)} + T_1' H_1^{(1)} + T_2' H_2^{(1)}, \tag{11.56}$$

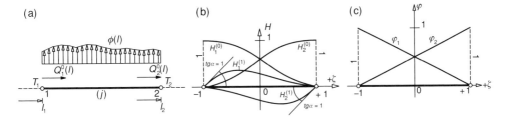

Figure 11.6   Typical finite element and shape functions.

where T_1, T_2, and T_1', T_2' are temperatures and temperature gradients on the boundaries of the finite element, $H_1^{(0)}$, $H_2^{(0)}$ are Hermite polynomials of zero order of the first kind, and $H_1^{(1)}$, $H_2^{(1)}$ are Hermite polynomials of the first order, first kind, associated to nodes 1 and 2.

The shape function of temperature field T on the finite element in the shape of Hermite polynomials of the first kind are shown in a standard interval, Figure 11.6b.

The expressions for Hermite polynomials of the first kind in standard interval $[-1, 1]$ are

$$H_1^{(0)} = \frac{1}{4}(2 + \xi)(1 - \xi)^2; \; H_2^{(0)} = \frac{1}{4}(2 - \xi)(1 + \xi)^2, \tag{11.57}$$

$$H_1^{(1)} = \frac{1}{4}(1 + \xi)(1 - \xi)^2; \; H_2^{(1)} = \frac{1}{4}(1 + \xi)^2(\xi - 1). \tag{11.58}$$

The value of the Hermite polynomials in an interval $[l_1, l_2]$ which determines the finite element can easily be calculated from the value of the standard interval.

Heat fluxes on the final element will be expressed in the form

$$Q^{\circ(j)} = Q_1^{\circ(j)}\phi_1 + Q_2^{\circ(j)}\phi_2, \tag{11.59}$$

where $Q_1^{\circ(j)}$, $Q_2^{\circ(j)}$ are heat fluxes on element (j) in nodes 1 and 2; and ϕ_1, ϕ_2 are shape functions on the element, according to Figure 11.6c. It should be noted that

$$T_1 = T(l_1, t); \; T_2 = T(l_2, t), \tag{11.60}$$

$$T_1' = \left. \frac{\partial T}{\partial l} \right|_{l=l_1}; \; T_2' = \left. \frac{\partial T}{\partial l} \right|_{l=l_2} \tag{11.61}$$

are unknown functions associated to nodes, and

$$Q_1^{\circ(j)} = Q^{\circ(j)}(l_1, t); \; Q_2^{\circ(j)} = Q^{\circ(j)}(l_2, t) \tag{11.62}$$

are unknown functions associated to model configuration elements. It will be understood hereinafter that Q_1° and Q_2° are associated to the element, so that (j) will be omitted. T_1' and T_2' can be eliminated from the expression for temperature approximation (11.56) and can be expressed by temperature and flux of the element

$$Q_1^\circ = (\rho CQT)_1 - \left(\rho cCDA \frac{\partial T}{\partial l} \right)_1, \tag{11.63}$$

$$Q_2^\circ = (\rho CQT)_2 - \left(\rho cCDA \frac{\partial T}{\partial l} \right)_2, \tag{11.64}$$

from which

$$T_1' = \frac{(\rho CQT)_1 - Q_1^\circ}{(\rho CDA)_1}; \; T_2' = \frac{(\rho CQT)_2 - Q_2^\circ}{(\rho CDA)_2}. \tag{11.65}$$

By inserting Eq. (11.65) into Eq. (11.56) and tabulating, an approximation of the temperature field on the element is obtained

$$T = T_1\left(H_1^{(0)} + \frac{Q_1}{D_1 A_1} H_1^{(1)}\right) + T_2\left(H_2^{(0)} + \frac{Q_2}{D_2 A_2} H_2^{(1)}\right) - \frac{H_1^{(1)}}{\rho C D_1 A_1} Q_1^\circ - \frac{H_2^{(1)}}{\rho c C D_2 A_2} Q_2^\circ. \tag{11.66}$$

11.6.2   Elemental equations: equation integration on the finite element

Equation (11.54) for the convective-dispersive flow can be written in the integral form, by initially integrating over the element

$$\frac{\partial}{\partial t} \int_{l_1}^{l_2} (\rho CAT)dl + Q_2^\circ - Q_1^\circ + \int_{l_1}^{l_2} \phi dl = 0, \tag{11.67}$$

$$\int_{l_1}^{l_2} Q^\circ dl = \int_{l_1}^{l_2} (\rho CQT)dl - \int_{l_1}^{l_2} \left(\rho CDA\frac{\partial T}{\partial l}\right) dl \tag{11.68}$$

and then the first equation can be integrated by time in the time interval $[t_1, t_2]$

$$\left\{\int_{l_1}^{l_2} (\rho CAT)dl\right\}\Bigg|_{t1}^{t2} + \int_{t_1}^{t_2} (Q_2^\circ - Q_1^\circ) \, dt + \int_{t_1}^{t_2}\int_{l_1}^{l_2} \phi dl dt = 0. \tag{11.69}$$

For a small time step $\Delta t = [t_1, t_2]$ the well-known implicit–explicit integration scheme is used, where the parameter time integration is $0 \le \vartheta \le 1$

$$\int_{l_1}^{l_2} \rho C(A^+ T^+ - AT)dl +$$

$$+(1 - \vartheta)\Delta t\left\{(Q_2^\circ - Q_1^\circ) + \int_{l_1}^{l_2} \phi dl\right\} + \vartheta \Delta t\left\{(Q_2^{\circ+} - Q_1^{\circ+}) + \int_{l_1}^{l_2} \phi^+ dl\right\} = 0. \tag{11.70}$$

where all values referring to the end of the interval are marked "+", and the values without this mark refer to the beginning of the time interval.

An approximate integration of Eq. (11.70), according to the theorem of the mean value of integrals, gives

$$\rho C \frac{A_1^+ + A_2^+}{2} \int_{l_1}^{l_2} T^+ dl - \rho C \frac{A_1 + A_2}{2} \int_{l_1}^{l_2} T dl +$$

$$+ (1 - \vartheta) \Delta t \left\{ (Q_2^\circ - Q_1^\circ) + \frac{\phi_1 + \phi_2}{2} \Delta l \right\} +,$$ (11.71)

$$+ \vartheta \Delta t \left\{ (Q_2^{\circ +} - Q_1^{\circ +}) + \frac{\phi_1^+ + \phi_2^+}{2} \Delta l \right\} = 0$$

where $\Delta l = (l_2 - l_1)$ is the length of the element.

Integrals from formulae (11.71), upon insertion of Eq. (11.66) show the Hermit polynomial integrals, which are

$$\int_{l_1}^{l_2} H_1^{(0)} dl = \frac{\Delta l}{2}; \int_{l_1}^{l_2} H_2^{(0)} dl = \frac{\Delta l}{2}; \int_{l_1}^{l_2} H_1^{(1)} dl = \frac{\Delta l^2}{12}; \int_{l_1}^{l_2} H_2^{(1)} dl = -\frac{\Delta l^2}{12}$$ (11.72)

and integrals of Hermit polynomial derivations

$$\int_{l_1}^{l_2} H_1^{(0)\prime} dl = -1; \int_{l_1}^{l_2} H_2^{(0)\prime} dl = 1; \int_{l_1}^{l_2} H_1^{(1)\prime} dl = 0; \int_{l_1}^{l_2} H_2^{(1)\prime} dl = 0.$$ (11.73)

Hence

$$\int_{l_1}^{l_2} T^+ dl =$$

$$\left(\frac{\Delta l}{2} + \frac{\Delta l^2}{12} \frac{Q_1^+}{D_1^+ A_1^+} \right) T_1^+ + \left(\frac{\Delta l}{2} - \frac{\Delta l^2}{12} \frac{Q_2^+}{D_2^+ A_2^+} \right) T_2^+$$ (11.74)

$$- \frac{\Delta l^2}{12 \rho C D_1^+ A_1^+} Q_1^{\circ +} + \frac{\Delta l^2}{12 \rho C D_2^+ A_2^+} Q_2^{\circ +}$$

and analogously

$$\int_{l_1}^{l_2} T dl =$$

$$\left(\frac{\Delta l}{2} + \frac{\Delta l^2}{12} \frac{Q_1}{D_1 A_1} \right) T_1 + \left(\frac{\Delta l}{2} - \frac{\Delta l^2}{12} \frac{Q_2}{D_2 A_2} \right) T_2 - .$$ (11.75)

$$\frac{\Delta l^2}{12 \rho C D_1 A_1} Q_1^\circ + \frac{\Delta l^2}{12 \rho C D_2 A_2} Q_2^\circ$$

By integrating Eqs (11.74) and (11.75) into Eq. (11.71) the result can be written in a shorter form

$$a_{11}^+ T_1^+ + a_{12}^+ T_2^+ + b_{11}^+ Q_1^{\circ+} + + b_{12}^+ Q_2^{\circ+} + a_{11} T_1 + a_{12} T_2 + b_{11} Q_1^{\circ} + b_{12} Q_2^{\circ} + c_1^+ + c_1 = 0, \quad (11.76)$$

where

$$a_{11}^+ = \rho C \frac{A_1^+ + A_2^+}{2} \left(\frac{\Delta l}{2} + \frac{\Delta l^2}{12} \frac{Q_1^+}{D_1^+ A_1^+} \right), \quad (11.77)$$

$$a_{12}^+ = \rho C \frac{A_1^+ + A_2^+}{2} \left(\frac{\Delta l}{2} - \frac{\Delta l^2}{12} \frac{Q_2^+}{D_2^+ A_2^+} \right), \quad (11.78)$$

$$b_{11}^+ = -\frac{A_1^+ + A_2^+}{2} \cdot \frac{\Delta l^2}{12 D_1^+ A_1^+} - \vartheta \Delta t, \quad (11.79)$$

$$b_{12}^+ = \frac{A_1^+ + A_2^+}{2} \cdot \frac{\Delta l^2}{12 D_2^+ A_2^+} + \vartheta \Delta t, \quad (11.80)$$

$$a_{11} = -\rho C \frac{A_1 + A_2}{2} \left(\frac{\Delta l}{2} + \frac{\Delta l^2}{12} \frac{Q_1}{D_1 A_1} \right), \quad (11.81)$$

$$a_{12} = -\rho C \frac{A_1 + A_2}{2} \left(\frac{\Delta l}{2} - \frac{\Delta l^2}{12} \frac{Q_2}{D_2 A_2} \right), \quad (11.82)$$

$$b_{11} = \frac{A_1 + A_2}{2} \cdot \frac{\Delta l^2}{12 D_1 A_1} - (1 - \vartheta) \Delta t, \quad (11.83)$$

$$b_{12} = -\frac{A_1 + A_2}{2} \cdot \frac{\Delta l^2}{12 D_2 A_2} + (1 - \vartheta) \Delta t, \quad (11.84)$$

$$c_1^+ = \vartheta \Delta t \frac{\phi_1^+ + \phi_2^+}{2} \Delta l, \quad (11.85)$$

$$c_1 = (1 - \vartheta) \Delta t \frac{\phi_1 + \phi_2}{2} \Delta l. \quad (11.86)$$

Equation (11.68) stands for any time state; however we are especially interested in its integration at the end of the time interval

$$\int_{l_1}^{l_2} \rho C Q^+ T^+ dl - \int_{l_1}^{l_2} \rho C D^+ A^+ \frac{\partial T^+}{\partial l} dl - \int_{l_1}^{l_2} Q^{\circ+} dl = 0. \quad (11.87)$$

An approximate integration of Eq. (11.87) gives

$$\rho C \frac{Q_1^+ + Q_2^+}{2} \int_{l_1}^{l_2} T^+ dl - \rho C \frac{D_1^+ A_1^+ + D_2^+ A_2^+}{2} (T_2^+ - T_1^+) - \frac{Q_1^{\circ+} + Q_2^{\circ+}}{2} \Delta l = 0. \quad (11.88)$$

By integrating Eq. (11.74) into Eq. (11.88) the result can be written in a shorter form

$$a_{21}^+ T_1^+ + a_{22}^+ T_2^+ + b_{21}^+ Q_1^{\circ+} + b_{22}^+ Q_2^{\circ+} + \\ + a_{21} T_1 + a_{22} T_2 + b_{21} Q_1^{\circ} + b_{22} Q_2^{\circ} + c_2^+ + c_2 = 0 \qquad (11.89)$$

where

$$a_{21}^+ = \rho C \frac{Q_1^+ + Q_2^+}{2} \left(\frac{\Delta l}{2} + \frac{\Delta l^2}{12} \frac{Q_1^+}{D_1^+ A_1^+} \right) + \rho C \frac{D_1^+ A_1^+ + D_2^+ A_2^+}{2}, \tag{11.90}$$

$$a_{22}^+ = \rho C \frac{Q_1^+ + Q_2^+}{2} \left(\frac{\Delta l}{2} - \frac{\Delta l^2}{12} \frac{Q_2^+}{D_2^+ A_2^+} \right) - \rho C \frac{D_1^+ A_1^+ + D_2^+ A_2^+}{2}, \tag{11.91}$$

$$b_{21}^+ = -\frac{Q_1^+ + Q_2^+}{2} \cdot \frac{\Delta l^2}{12 D_1^+ A_1^+} - \frac{\Delta l}{2}, \tag{11.92}$$

$$b_{22}^+ = \frac{Q_1^+ + Q_2^+}{2} \cdot \frac{\Delta l^2}{12 D_2^+ A_2^+} - \frac{\Delta l}{2}, \tag{11.93}$$

$$a_{21} = 0; \quad a_{22} = 0; \quad b_{21} = 0; \quad b_{22} = 0; \quad c_2^+ = 0; \quad c_2 = 0. \tag{11.94}$$

The resulting equations for the finite element can be expressed in a shorter form

$$F^e : \quad a_{ij}^+ T_j^+ + b_{ij}^+ Q_j^{\circ+} + a_{ij} T_j + b_{ij} Q_j^\circ + c_i^+ + c_i = 0$$

$$i = 1, 2 \quad j = 1, 2 \tag{11.95}$$

where for each index i summation is performed over index j. The obtained equations are non-linear due to the dependence of item c_i^+ on the result.

11.6.3 Nodal equations

The thermal conditions of the hydraulic network are determined by temperatures in N nodes $T_r(t)$, $r = 1, 2, 3, \ldots, N$, which enable the calculation of temperatures or thermal flows at each point of the network. Similarly to the calculation of unknown piezometric heads where we apply nodal equations of flow continuity in nodes, for the calculation of N unknown nodal temperatures it is necessary to form N independent equations of nodal thermal continuity in nodes

$$F_r = \sum_p {}^r Q_{k(p)}^{\circ e} = 0 \tag{11.96}$$

which can be solved by the Newton–Raphson iterative procedure. The iterative procedure for Eq. (11.96) shows an increase in element thermal flows $\Delta Q_k^{\circ e}$ which are to be expressed by corresponding increases of nodal temperatures. We thus apply the developed assembling algorithm of matrix and vector of finite element.

11.6.4 Boundary conditions

When dealing with the problem of transfer flow, that is the problem of thermal convective-dispersive transfer, it is necessary to know the *essential boundary condition* of the given temperature in at least one node. In such a case, the previously formed equation for the r node in the fundamental system is replaced by the equation $T_r = T_{bdc}$.

Natural boundary conditions are a supplement to the fundamental system of thermal discharges $Q_r^{\circ 0} = \rho C Q_r^0 T_{bdc}$. In steady flows, the natural boundary conditions supplement the base system with an additional element of thermal flux $Q_r^{\circ 0}$. Two cases shall be analysed:

(a) explicit form $Q_r^{\circ 0} = \rho C Q_r^0 T_{bdc}$ where water flow is Q_r^0 and temperature T_{bdc},
(b) implicit form $Q_r^{\circ 0} = \rho C Q_r^0 T_r$ where outflow Q_r^0 is known and temperature T_r should be determined by the result of the system. This is an open boundary of the model.

The nodal equation is as follows

$$\sum_p{}^r Q_e^{\circ} + Q_r^{\circ 0} = 0. \tag{11.97}$$

A supplement to the global vector (right hand side of the system of equations) in the steady state is simple, due to the additive feature of the thermal flux, and is as follows

$$F_r^{new} = F_r^{old} - Q_r^{\circ 0}. \tag{11.98}$$

In the implicit case when flux $Q_r^{\circ 0}$ depends on nodal temperature T_r, it is also necessary to supplement the global system matrix with the corresponding partial derivation

$$G_{r,r}^{new} = G_{r,r}^{old} + \rho C Q_r^0. \tag{11.99}$$

In non-steady flow, the initial fundamental system stands for the balance of heat in time, so that the system must be supplemented by the corresponding nodal volume

$$\Delta V_r^{\circ 0} = \int_{\Delta t} Q_r^{\circ 0} dt = (1 - \vartheta)\Delta t\, Q_r^{\circ 0} + \vartheta\, \Delta t\, Q_r^{\circ 0+}, \tag{11.100}$$

which is further added to the global system vector

$$F_r^{new} = F_r^{old} - \Delta V_r^{\circ 0}. \tag{11.101}$$

A corresponding supplement of the global matrix is in the form of

$$G_{r,r}^{new} = G_{r,r}^{old} + \frac{\partial \Delta V_r^{\circ 0}}{\partial T_r^{+}}. \tag{11.102}$$

11.6.5   Matrix and vector of finite element

Starting from elemental equations (11.95), the Newton–Raphson iterative form for the finite element e is applied and is formally noted by matrix-vector operations

$$[\underline{H}] \cdot [\Delta T] + [\underline{Q}] \cdot [\Delta Q^{\circ}] = [\underline{F}], \tag{11.103}$$

where the matrices

$$[\underline{H}] = a_{ij}^{+}; \; [\underline{Q}] = b_{ij}^{+} \tag{11.104}$$

and vectors are

$$[\underline{F}] = F^e. \tag{11.105}$$

The Newton–Raphson form for elemental equations (11.103) calculates the increments of elemental thermal flows, by multiplying the previous equation by the inverse matrix $[\underline{Q}]^{-1}$

$$[\Delta Q^+] = [\underline{Q}]^{-1}[\underline{F}] - [\underline{Q}]^{-1}[\underline{H}] \cdot [\Delta T^+]. \tag{11.106}$$

By introducing labels

$$[\underline{A}] = [\underline{Q}]^{-1}[\underline{H}], \tag{11.107}$$

$$[\underline{B}] = [\underline{Q}]^{-1}[\underline{F}]. \tag{11.108}$$

we get the formula for the increase of elemental thermal flows

$$[\Delta Q^+] = [\underline{B}] - [\underline{A}] \cdot [\Delta T^+]. \tag{11.109}$$

Matrix A^e and vector B^e of the finite element for non-steady modelling of thermal flow are as follows

$$A^e = \vartheta \, \Delta t \begin{bmatrix} +\underline{A}_{11} & +\underline{A}_{12} \\ -\underline{A}_{21} & -\underline{A}_{22} \end{bmatrix}, \tag{11.110}$$

$$B^e = (1-\vartheta)\Delta t \begin{bmatrix} +Q_1^\circ \\ -Q_2^\circ \end{bmatrix} + \vartheta \, \Delta t \begin{bmatrix} +Q_1^{\circ +} \\ -Q_2^{\circ +} \end{bmatrix} + \vartheta \, \Delta t \begin{bmatrix} +\underline{B}_1 \\ -\underline{B}_2 \end{bmatrix}. \tag{11.111}$$

The unknown increments of elemental thermal fluxes are calculated from equation (11.109) after having calculated the unknown nodal temperature increments.

The SimpipCore program, published on the website www.wiley.com/go/jovic, does not include the implementation of Fortran modules for modelling transfer disperse flows. The interested reader should apply his own knowledge and existing Fortran solutions in the book and add his own solutions for modelling convective-disperse flows of any physical phenomenon.

11.6.6 Numeric solution test

Task

A channel (or pipe), Figure 11.7, of length $L = 1000\,\text{m}$, has a uniform flow at an average speed $V = 1.734\,\text{m/s}$. At the beginning of the channel, temperature rises by $T_0 = 100\,°\text{C}$ above the initial one at a moment $t = 0$ and remains constant. The dispersion coefficient is $10\,\text{m}^2/\text{s}$. Compare the analytical solution with the numerical one for discretization values $\Delta l = 100$, 10, and 1 m.

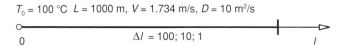

$T_0 = 100\,°\text{C}$ $L = 1000$ m, $V = 1.734$ m/s, $D = 10\,\text{m}^2/\text{s}$

0

$\Delta l = 100;\ 10;\ 1$

l

Figure 11.7

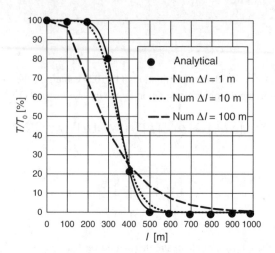

Figure 11.8   Temperature wave front $t = 200$ s.

    The domain of definition of the numerical model is limited and determined by the length of the channel, which differs from the domain where the analytical solution was obtained. Thus, we apply the implicit natural boundary condition to the downstream (open) boundary of the numerical model, that is the unknown convective outflow of heat $V \cdot T$, where T is the unknown temperature on the downstream boundary.

    Figure 11.8 shows the heat wave at moment $t = 200$ s for discretization $\Delta l = 100$, 10, and 1 m as well as the analytical solution according to the formula (11.46).

    It can be observed that the choice of discretization size influences the accuracy of the numerical solution. The error of numerical models is reflected in an increased smearing the wave front, called *numerical diffusion*. A rougher discretization causes larger numerical diffusion.

    Besides the appearance of numerical diffusion in certain cases, there is also a fluctuation behind the wave front in the shape of creases, that is waves. The size of numerical diffusion and creasing depends on the Péclet number

$$P_e = \frac{v \cdot B}{D},\qquad(11.112)$$

where B is the channel width B, or another length indicator [m], v the water velocity [m/s], D the dispersion coefficient [m²/s], and it depends on the chosen method of numerical integration. At higher Péclet numbers, numeric diffusion can be significantly higher than reality. Numeric diffusion is decreased by fine wave front discretization. For example, discretization $\Delta l = 100$ m causes serious front wave smear, whereas discretization $\Delta l = 1$ m gives results practically equal to the analytical ones.

    Figure 11.9 shows a heat wave at $t = 100, 200, 300, 400, 500, 600$, and 700 seconds for the analytical (discontinuous) and numeric (full line) solutions for discretization $\Delta l = 10$ m. The analytical solution that stands out is the one at $t = 300$ seconds together with the one with numeric discretization $\Delta l = 1$ m.

    Besides spatial discretization, the numeric solution is affected by time discretization. This method has no limitation regarding stability in the relations of certain time or spatial steps. However, Figure 11.10 shows some preferred relations between time and spatial steps. It is preferred that the wave front in a time step moves for the distance of one spatial step $V\Delta t \approx \Delta l$. The size of the spatial discretization should be such that the shape of the wave front approximates sufficiently by polygonal line.

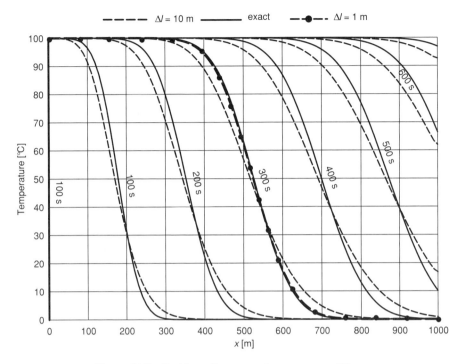

Figure 11.9   Position of heat wave in a sequence of times.

Figure 11.11 shows the temperature of the wave front in the position $l = Vt$, that is the temperature of the front. It can be observed that the temperature in the middle of the front is always above 50%, and this is a value to which it converges in infinity.

11.6.7   Heat exchange of water table

Heat exchange on the water table is the sum of basic heat fluxes, (Brocard and Harleman, 1976):

$$\Phi_n = \Phi_{sn} + \Phi_{an} - \Phi_{br} - \Phi_e - \Phi_c - \Phi_a \left[W/m^2 \right], \tag{11.113}$$

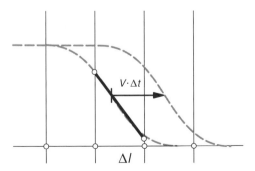

Figure 11.10   Spatial and time discretization.

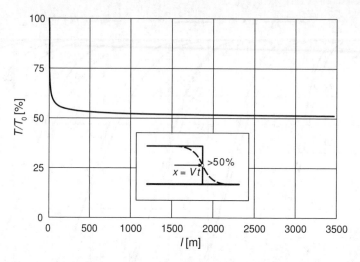

Figure 11.11   Temperature in front path.

where:

Φ_n = is net thermal flow on the water table,

Φ_{sn} = is net short-wave solar radiation, measured by the radiometer,

Φ_{an} = $5, 18 \cdot 10^{-13} \left(1 + 0, 17c^2\right) T_a^6$ is net atmospheric thermal counter radiation, where c is the cloudiness coefficient ranging from 0 for a clear sky to 1 for a completely dark sky, T_a [K] is the absolute temperature of the atmosphere,

Φ_{br} = is long-wave radiation from the water table expressed by the Stefan–Boltzmann law[2] in the function of absolute temperature of the water table T_s, that is $\Phi_{br} = 5, 44 \cdot 10^{-8} T_s^4$,

Φ_e = is thermal flow caused by evaporation is $\Phi_e = \rho c w L (e_s - e_a)$, where c is a constant $1,564 \cdot 10^{-9}$ that is determined experimentally, w is wind speed (m/s), usually measured 2 m above the water table, ρ is water density (kg/m^3), L latent evaporation heat depends on the water temperature, and is determined by the formula $L = 2, 494 \cdot 10^6 - 2281(T - 273, 16)$, where T [K] is absolute water temperature, e_s is pressure of saturated water vapor at water table temperature (mb), e_a is pressure of atmospheric water vapor (mb),

Φ_c = is thermal flow caused by conduction $\Phi_c = 0.61 \cdot 10^{-3} p \cdot \Phi_e$, p [mb] is atmospheric pressure,

Φ_a = is thermal flow caused by advection $\Phi_a = c_v \cdot \Phi_e \frac{T_s - T_B}{L}$, where c_v is the specific water heat, that is specific thermal capacity ($c_v = 4185$ kJ/kg/K), Φ_e is thermal flux caused by evaporation, T_s is absolute water temperature, T_B is base temperature which should be taken near the natural water temperature, L is latent evaporation heat.

11.6.8   Equilibrium temperature and linearization

The net thermal flux Φ_n in thermal exchange through the water surface depends on the meteorological conditions and water table temperature. The expression Φ_n is simplified for the purpose of obtaining

---

[2] Jozef Stefan, Slovene physicist and poet, (1835–1893); Ludwig Boltzmann, Austrian physicist (1844–1906).

closed solutions of convective-dispersive equations of thermal flux in a watercourse. The usual approach is linearization Φ_n in relation to balance temperature T_E. Equilibrium temperature T_E is defined as the temperature of a water table with given meteorological conditions without thermal exchanges through the water table. It can thus be written as follows:

$$\Phi_n = K(T_s - T_E). \tag{11.114}$$

The thermal coefficient K should be determined by obtaining the approximation of the thermal flux in a predicted temperature range, thus K is dependent on the temperature of the water table.

11.6.9   Temperature disturbance caused by artificial sources

For any constant hydraulic or meteorological conditions, the thermal process in a watercourse will strive towards a balanced temperature so that the thermal flux exchange on the water table will trend towards zero. The current water temperature at point zero, which occurs at natural meteorological and hydraulic conditions, is defined as the *watercourse natural temperature*.

The natural temperature of a watercourse changes constantly, with changes in meteorological and hydraulic parameters and change of equilibrium temperature.

As the natural temperature strives to reach the balance temperature in the asymptotic sense, fluctuations of natural temperature are always smaller than fluctuations of the equilibrium temperature.

By adding an artificial thermal source to the system, a redistribution of the temperature field is obtained. However, as soon as the artificial source is removed, the temperature field strives towards the natural temperature of the watercourse.

The method of determining the size of the temperature of disturbance caused by the addition of artificial heat sources in the system was first revealed by H. E. Jobson (Jobson, 1973).

The Jobson method is based on the fact that the natural watercourse temperature is also the solution of the convective-dispersed equation for a given set of natural meteorological and hydraulic initial and boundary conditions. The concept of natural initial and boundary conditions can be used over a range of artificial sources which exist in a certain space during the overall observed period of time.

In this way, the concept of natural temperature can be applied to any referential watercourse temperature which arises as the solution of a convective-dispersive equation (11.31).

The Jacobson method will be shown here, but slightly modified. The advantage of the method is that it enables the determination of changes in watercourse temperature in relation to the natural temperature. The solution is not burdened with a large number of unreliable parameters that appear in components of the thermal flux at thermal exchange on the water surface. The temperature rise dissipation will be strongly influenced by wind speed.

As the natural temperature of a watercourse is determined by measurements during the activity of all existing artificial or natural thermal sources, the influence of a new artificial source on the formation of watercourse temperature will be as follows

$$T = T_0 + T', \tag{11.115}$$

where T is the watercourse actual temperature in new conditions, T_0 is the watercourse natural temperature, and T' is the temperature disorder due to the presence of a new artificial source. If Eq. (11.115) is applied in the equation of convective-dispersive flow (11.31)

$$\frac{\partial}{\partial t}\left[\rho C A(T_0 + T')\right] + \frac{\partial}{\partial l}\left[\rho C Q(T_0 + T')\right] = \frac{\partial}{\partial l}\left[\rho C D A \frac{\partial(T_0 + T')}{\partial l}\right] + \Phi^a B$$

and the rule of derivation of sum function is applied, we thus obtain

$$
\frac{\partial}{\partial t}(\rho CAT_0) + \frac{\partial}{\partial l}(\rho c CQT_0) - \frac{\partial}{\partial l}(\rho CDA\frac{\partial T_0}{\partial l}) + \Phi^a B +
$$

$$
+ \frac{\partial}{\partial t}(\rho CAT') + \frac{\partial}{\partial l}(\rho c CQT') - \frac{\partial}{\partial l}(\rho CDA\frac{\partial T'}{\partial l}) = 0. \tag{11.116}
$$

In Eq. (11.116) thermal flux Φ^a depends on watercourse actual temperature

$$
\Phi^a = \Phi^a(T) = \Phi^a(T_0 + T'),
$$

so it can be expanded in series around the natural temperature

$$
\Phi^a(T) = \Phi^a(T_0) + \frac{1}{1!}\frac{\partial \Phi^a}{\partial T}\Big|_{T=T_0} dT + \frac{1}{2!}\frac{\partial^2 \Phi^a}{\partial T^2}\Big|_{T=T_0} dT^2 + \cdots. \tag{11.117}
$$

In fact, minor disorders will be observed so that all members further than the linear ones can be ignored, hence $dT \cong T'$, that is:

$$
\Phi^a = \Phi_0^a + \frac{\partial \Phi_0^a}{\partial T}\Big|_{T=T_0} T' = \Phi_0^a + G_{T_0} \cdot T', \tag{11.118}
$$

where $\Phi^a = \Phi^a(T)$ is the thermal flux on the water table in new conditions, $\Phi_0^a = \Phi^a(T_0)$ is the thermal flux on the water table at the natural temperature, that is in natural initial and boundary conditions, $G_{T_0} \cdot T'$ is the thermal flux exchange which arises due to an artificial source, G_{T_0} is the gradient of the thermal flux at the water table with natural temperature, depending on the natural temperature and meteorological conditions. By inserting Eq. (11.118) into Eq. (11.116) we obtain

$$
\underbrace{\frac{\partial}{\partial t}(\rho CAT_0) + \frac{\partial}{\partial l}(\rho CQT_0) - \frac{\partial}{\partial l}(\rho CDA\frac{\partial T_0}{\partial l}) + \Phi_0^a B +}_{(I)}
$$

$$
\underbrace{+ \frac{\partial}{\partial t}(\rho CAT') + \frac{\partial}{\partial l}(\rho CQT') - \frac{\partial}{\partial l}\left(\rho CDA\frac{\partial T'}{\partial l}\right) + G_{T_0} \cdot T' \cdot B = 0}_{(II)} \tag{11.119}
$$

As the natural temperature T_0 is also the solution of the differential equation (11.31) for some natural boundary and initial conditions and the known function Φ_0 of boundary thermal exchange: $\Phi_0 = \Phi(T_0)$, this part (I) equals zero in the (11.119) equation. So, the remaining part (II) of the equation determines the size of the temperature field disturbance

$$
\frac{\partial}{\partial t}(\rho CAT') + \frac{\partial}{\partial l}(\rho c CQT') - \frac{\partial}{\partial l}\left(\rho CDA\frac{\partial T'}{\partial l}\right) + G(T_0) \cdot T' \cdot B = 0. \tag{11.120}
$$

Equation (11.120) formally differs from Eq. (11.31) except in the interpretation of elements of the thermal flux through the water table.

The thermal flux consists of more components, see Eq. (11.113), the value $G(T_0)$ is influenced only by the components which depend on the watercourse temperature:

$$G(T_0) = \left.\frac{\partial \Phi^a}{\partial T}\right|_{T=T_0} = \left.\left\{\frac{\partial \Phi_{sn}}{\partial T} + \frac{\partial \Phi_{an}}{\partial T} - \frac{\partial \Phi_{br}}{\partial T} - \frac{\partial \Phi_e}{\partial T} - \frac{\partial \Phi_c}{\partial T} - \frac{\partial \Phi_a}{\partial T}\right\}\right|_{T=T_0}. \quad (11.121)$$

According to the expression (11.121), the first two elements depend on solar radiation, air temperature, and so on, which further leads to the conclusion that the solution to the problem of determining the extent of temperature disorder is possible without having all the meteorological data. The value of parameter $G(T_0)$ is easily calculated, by deriving appropriate expressions for thermal fluxes.

It should be noted (in avoiding a possible wrong conclusion) that solar and atmospheric radiation have a significant role in the formation of the watercourse natural temperature T_0, as well as in determining absolute values of the new temperature field T, formula (11.115).

References

Brocard, D.N. and Harleman, D.R.F. (1976) One-dimensional temperature predictions in unsteady flows, *Journal of the Hydraulics Division, HY3, ASCE*, March: 227–240.

Jobson, H.E. (1973) The dissipation of excess heat from water systems. *Journal of the Power Division, PO1, ASCE*, May: 89–103.

Taylor, G.I. (1954) The dispersion of matter in turbulent flow through a pipe. *Proceedings of the Royal Society London*, 233: 446–448.

Further reading

Abbott, M.B., Waren, I.R., Jensen, K.H., et al. (1979) Coupling of unsaturated and saturated zone models. *XVIII Congress IAHR.*, Cagliari, Italija, 5, p. D.c. 1.

Boericke, R.R. and Hall, D.W. (1974) Hydraulics and thermal dispersion in an irregular estuary. *Journal of the Hydraulics Division, HY1, ASCE*, January: 85–102.

Bowies, D.S., Fread., D.L., and Grenney, W.J. (1977) Coupled dynamic streamflow – temperature models. *Journal of the Hydraulics Division, HY5, ASCE*, May: 515–530.

Daily, J.W. and Harleman, D.R.F. (1966) *Fluid Dynamics*. Addison-Wesley, Reading, Massachusetts.

Eremenko, E.V. and Kolpak, V.Z. (1979) Mathematical modelling of pollutants in a stream recieiving non-point discharges. *XVIII Congress IAHR*, Cagliari, Italija, 5: p.D.b. 2.

Ford, D.E. and Stefan, H.G. (1980) Thermal predictions using integral energy model. *Journal of the Hydraulics Division, HY1, ASCE*, January: 39–55.

Holley, E.R. (1969) Unified view of diffusion and dispersion. *Journal of the Hydraulics Division, HY2, ASCE*, March: 621–631.

International Atomic Energy Agency (1979) *Hydrological Dispersion of Radioactive Material in Relation to Nuclear Power Plant Siting – A Safety Guide*.

Jović, V., Radelja, T., and Petrov, R. The simulation of the diffusion process on the non-steady shallow open flow, EUROMECH.

Jović, V. (1981) Formulacija modela simulacije nekih procesa disperzije u hidrodinamoci podzemnih voda, Simpozij: Hidrodinamički problemi zaštite voda. Akademija nauka i umjetnosti BiH, posebna izdanja – LII, knjiga 10, (in Croatian) Sarajevo.

Jović, V. (1993) *Introduction to Numerical Engineering Modelling* (in Croatian), Aquarius Engineering, Split.

Lin, J.D., Alfano, J.J., and Bock, P. (1979) A large-scale ground hydrologic model which interacts with the atmosphere. *XVIII Congress IAHR*, Cagliari, Italija, 3: p.B.d. 6.

Miller, A.C. and Richardson, E.V. (1974) Diffusion and dispersion in open channel flow. *Journal of the Hydraulics Division, HY1, ASCE*, January: 159–171.

Morris, F.W., Walton, R., and Christensen, B.A. (1979) Point and Non-point pollutant flushing in tidal canal networks. *XVIII Congress IAHR*, Cagliari, Italija, 5: p.D.b. 3.

Muller, H., and Visscher, G. (1979) experimental evaluation of heat exchange between water surface an atmosphere. *XVIII Congress IAHR*, Cagliari, Italija, 3: p.B.b. 7.

O'Neill, K. (1981) Highly efficient, oscillation free solution of the transport equation over long times and large spaces. *Water Resources*, 17: 1665–1675.

Paily, P.P. and Macagno, E.O. (1976) Numerical prediction of thermal-regime of rivers. *Journal of the Hydraulics Division, HY3, ASCE*, March: 255–274.

Panin, G.N., Volkova, S.V., and Foken, T.H. (1979) On heat exchange of surface layer of water reservoir with atmosphere. *XVIII Congress IAHR*, Cagliari, Italija, 3: p.B.b. 8.

Peterson, D.E., Sonnichsen, J.C., Engstrom, S.L., and Schrotke, P.M. (1973) Thermal capacity of our nation's waterways. *Journal of the Power Division*, 99: 193–204.

Pickens, J.F. and Gricak, G.E. (1981) Modelling of scale-dependent dispersion in hydrogeologic systems, *Water Resources Research*, 17: 1683–1693.

Shigemitsu, S. and Tsurumaki, Y. (1973) Experimental studies on thermal water pollution – especially for surface diffusion. *XV Congress of IAHR*, Istanbul, 1: 1–16.

Soerjadi, R. (1981) An alternative development of the dispersion equation. *Water Resources Research*, 17: 1611–1618.

Tang, D.H., Schwartz, F.W., and Smith, L. (1982) Stochastic modelling of mass transport in a random velocity field. *Water Resources Research*, 18: 231–244.

Tsal, Y.H. and Holley, E.R. (1978) Temporal moments for longitudinal dispersion. *Journal of the Hydraulics Division, HY12, ASCE*, December: 1617–1633.

Uzzell, J.C. and Ozisik, M.N. (1978) Three-dimensional temperature model for shallow lakes. *Journal of the Hydraulics Division, HY12, ASCE*, December: 1635–1645.

12

Hydraulic Vibrations in Networks

12.1 Introduction

Past experience has proved that implemented pressurized systems do not always behave as expected. The most common cause for this is the occurrence of hydraulic vibrations. These vibrations can be caused by various factors, the most frequent being the non-uniform operation of hydraulic engines. Mechanical rotations are never perfect due to certain drawbacks in their manufacture. Therefore periodic alternations of the discharge or pressure can be expected, in the spectrum of the basic frequencies or higher tones of the engine itself, the hydraulic components, or the system as a whole. Apart from these sources of vibrations, other causes may be stated, such as those caused by inappropriate manufacture of various pieces of hydromechanical equipment, the rhythmic separation of the boundary layer, and so on.

Various examples of damage created in hydraulic systems due to ordinary sources of vibration can be found in the literature.

Vibration sources are not themselves harmful to hydraulic systems, except when such excitation causes a resonance in the hydraulic system. Consequently, the study of vibrations is concerned with determining the intensity of forced vibrations for the known source of vibrations, that is excitation.

Due to the development of suitable models it is now possible to compute forced and free vibrations in very complex real pressurized systems for any spectrum of excitation frequency. Unfortunately, this does not solve the problem of vibrations, since the actual frequency of the possible source of vibration is not known in advance. When certain excitation frequencies are expected in the system, due to the operation of turbines or pumps, it is convenient to perform such an analysis, in order to avoid possible resonance by changing the characteristics of certain system elements before project evaluation.

The description of the fundamentals of hydraulic vibration analysis may be found in classical textbooks like V. L. Streeter and E. B. Wylie (Streeter and Wylie, 1967). Although the theory, based on the theory of telephone transition lines, has been known for a long time, there are still only a few complex softwares for vibration analysis.

This chapter describes the theory of vibration modelling in a complex hydraulic system, where we may find a network of pipes, valves, vessels, surge tanks, and so on, under natural operating conditions. The modelling of vibrations is an extension to the modelling of the non-steady flow, described in the previous chapters of the book.

Analysis and Modelling of Non-Steady Flow in Pipe and Channel Networks, First Edition. Vinko Jović.
© 2013 John Wiley & Sons, Ltd. Published 2013 by John Wiley & Sons, Ltd.

The analysis of forced vibrations is based upon certain assumptions:

- there is a static or dynamic equilibrium state of a hydraulic system where forced vibrations occur;
- the vibration amplitude is so small that hydrodynamic equations can be linearized in that interval of the variables changes without exerting significant influence upon the accuracy of the computation;
- the source of hydraulic vibrations, or excitations, has a periodic character, and lasts sufficiently long that steady harmonic vibrations in the entire system can be established.

Thus the system state, which is defined by a piezometric head $h(l, t)$ and by discharge $Q(l, t)$, can be expressed in the following form

$$h(l, t) = \bar{h}(l, t) + h'(l, t), \tag{12.1}$$

$$Q(l, t) = \bar{Q}(l, t) + Q'(l, t), \tag{12.2}$$

where $\bar{h}(l, t)$ is the average value of the piezometric head, $\bar{Q}(l, t)$ is the average value of discharge, $h'(l, t)$ is the piezometric head oscillation, $Q'(l, t)$ is the discharge oscillation at any point l of the system in time t.

Since the harmonic vibrations are well–developed, the piezometric head and discharge oscillations are expressed in a complex domain

$$h'(l, t) = \mathrm{Re}\underline{H}(l)e^{i\omega t}, \tag{12.3}$$

$$Q'(l, t) = \mathrm{Re}\underline{Q}(l)e^{i\omega t}, \tag{12.4}$$

where $\underline{H}(l)$ is the complex amplitude of piezometric head oscillations, $\underline{Q}(l)$ is the complex amplitude of the discharge oscillations, ω is the circular frequency of harmonic vibrations, t is time, and i is the imaginary unit of complex numbers.

It should be noted that, in order to avoid misunderstanding with common measurement units, circular frequency is defined by a period T of harmonic vibrations

$$\omega = \frac{2\pi}{T} \tag{12.5}$$

and is expressed in units $[s^{-1}]$. However, vibration frequencies are often measured in Hertz, which is the number of vibrations in a second. If harmonic vibrations are expressed in Hertz, then the circular frequency is computed by multiplying that value with 2π.

12.2   Vibration equations of a pipe element

Essentially, basic equations of unsteady flow in pipe elements are valid for vibrations as well, simply because vibrations are also unsteady phenomena. Consequently, vibrations could be generally modeled by any appropriate unsteady model, so that a sufficiently long computation is performed for a given periodic harmonic boundary condition. This approach would, however, be inefficient since the computation can be significantly shortened by a suitable transformation of the basic equations in a *frequency domain* for a developed harmonic state.

The efficiency of harmonic analysis compared to a direct approach can be particularly emphasized if the large frequencies spectrum of the excitation has to be analyzed, which is a common need in practice.

The basic equations of unsteady flow of a one-phase liquid in a pipe element, with constant characteristics of the cross-section along the pipe, may be written as follows

$$\text{continuity equation:} \quad gA\frac{\partial h}{\partial t} + c^2\frac{\partial Q}{\partial l} = 0, \tag{12.6}$$

$$\text{dynamic equation:} \quad \frac{\partial Q}{\partial t} + gA\frac{\partial}{\partial l}\left(h + \frac{Q^2}{2gA^2}\right) + gAJ_e = 0, \tag{12.7}$$

where A is cross-sectional area, c is water-hammer speed in the pipe, J_e is the energy line gradient, g is the gravity acceleration.

Equations (12.6) and (12.7) are actually valid for other pressurized elements which appear in the pressurized systems if resistances are generalized, so that the energy line gradient is expressed formally in the following way

$$J_e = C|Q|Q, \tag{12.8}$$

where parameter C is computed differently for each particular case. Introducing Eqs (12.1) and (12.2) into Eq. (12.6), we get

$$gA\frac{\partial \bar{h}}{\partial t} + c^2\frac{\partial \bar{Q}}{\partial l} + gA\frac{\partial h'}{\partial t} + c^2\frac{\partial Q'}{\partial l} = 0.$$

Since the mean values of both the piezometric head and discharge are at the same time the solutions of Eq. (12.6), which can be seen from the first two members of the previous expression, this means that the continuity equation is valid for the oscillatory values h' and Q'

$$gA\frac{\partial h'}{\partial t} + c^2\frac{\partial Q'}{\partial l} = 0. \tag{12.9}$$

Similarly, introducing Eqs (12.1), (12.2), and (12.8) into dynamic equation (12.7)

$$\frac{\partial(\bar{Q}+Q')}{\partial t} + gA\frac{\partial}{\partial l}\left((\bar{h}+h') + \frac{(\bar{Q}+Q')^2}{2gA^2}\right) + gAC|\bar{Q}+Q'|(\bar{Q}+Q') = 0.$$

In the preceding expression there are two non-linear members: the gradient of velocity head resistances, which should be linearized. The oscillatory part in the gradient of the velocity head can be neglected, since the pipe has a uniform cross-sectional area. The resistance member can be linearized by expanding into Taylor's series and by taking a constant and linear member

$$gAC|\bar{Q}|\bar{Q} + 2gAC|\bar{Q}|Q'.$$

Since the mean values of the piezometric head and discharge satisfy the dynamic equation, a dynamic equation for oscillatory values can be obtained

$$\frac{\partial Q'}{\partial t} + gA\frac{\partial h'}{\partial l} + RQ' = 0, \tag{12.10}$$

where $R = 2gAC|\bar{Q}|$.

12.3   Harmonic solution for the pipe element

Transferring the analysis into a complex domain of numbers, the oscillatory values of the piezometric head and discharge will be expressed in the form

$$h'(l, t) = \underline{H}(l)e^{i\omega t}, \tag{12.11}$$

$$Q'(l, t) = \underline{Q}(l)e^{i\omega t}. \tag{12.12}$$

By introducing Eqs (12.11) and (12.12) into Eqs (12.9) and (12.10) we obtain a system with two ordinary differential equations

$$\frac{d\underline{Q}}{dl} + \frac{i\omega}{c^2} gA\underline{H} = 0, \tag{12.13}$$

$$\frac{d\underline{H}}{dl} + \left(\frac{i\omega}{gA} + R\right)\underline{Q} = 0, \tag{12.14}$$

from which, introducing $\underline{Q}$ into the first and $\underline{H}$ from the first into the second, two ordinary differential equations of the second order can be obtained

$$\frac{d^2\underline{H}}{dl^2} = \gamma^2 \underline{H}, \tag{12.15}$$

$$\frac{d^2\underline{Q}}{dl^2} = \gamma^2 \underline{Q}, \tag{12.16}$$

where

$$\gamma^2 = \frac{igA\omega}{c^2}\left(\frac{i\omega}{gA} + R\right), \tag{12.17}$$

γ^2 is a complex constant which includes the characteristics of the pipe and a circular frequency of harmonic oscillations. It is called a transmission (propagation) constant, which is always in the first quadrant, that is the real and imaginary part are real positive numbers. Consequently, the square is in the second quadrant.

The solutions of Eqs (12.15) and (12.16) are independent and are solved for the given boundary conditions.

The boundary conditions can be the known values of the complex amplitude of discharge $\underline{Q}$ or piezometric heads $\underline{H}$, that is their combinations. Given a complex amplitude $\underline{Q}_1$ on the left, upstream end of the pipe, and the given $\underline{Q}_2$ on the right, downstream end, then the solution of Eq. (12.16) is

$$\underline{Q} = \frac{\left(\underline{Q}_2 - \underline{Q}_1 e^{-\gamma L}\right)e^{\gamma l} + \left(\underline{Q}_1 e^{\gamma L} - \underline{Q}_2\right)e^{-\gamma l}}{e^{\gamma L} - e^{-\gamma L}},$$

that is

$$\underline{Q} = \underline{Q}_1 \frac{\sinh \gamma(L - 1)}{\sinh \gamma L} + \underline{Q}_2 \frac{\sinh \gamma l}{\sinh \gamma L}. \tag{12.18}$$

In the same way we can obtain a solution for $\underline{H}$ for the known boundary values of complex amplitudes of the piezometric head

$$\underline{H} = \underline{H}_1 \frac{\sinh \gamma (L-1)}{\sinh \gamma L} + \underline{H}_2 \frac{\sinh \gamma l}{\sinh \gamma L}. \tag{12.19}$$

Furthermore, Eqs (12.13) and (12.14) can be written in the following form

$$\underline{H} = -\frac{Z_c}{\gamma} \frac{d\underline{Q}}{dl}, \tag{12.20}$$

$$\underline{Q} = -\frac{1}{\gamma Z_c} \frac{d\underline{H}}{dl}, \tag{12.21}$$

where the characteristic pipe impendance is

$$Z_c = \frac{\gamma a^2}{i \omega g A}. \tag{12.22}$$

Introducing Eqs (12.18) and (12.19) into Eqs (12.20) and (12.21) we obtain complex amplitudes of the discharge harmonic oscillations in the pipe for the given boundary values of a complex amplitude of harmonic oscillations of the piezometric head

$$\underline{Q} = \frac{1}{Z_c} \left(\underline{H}_1 \frac{\cosh \gamma (l-1)}{\sinh \gamma L} + \underline{H}_2 \frac{\cosh \gamma l}{\sinh \gamma L} \right), \tag{12.23}$$

that is complex amplitudes of the harmonic oscillation of the piezometric head for the given boundary values of a complex amplitude of the discharge harmonic oscillations

$$\underline{H} = Z_c \left(\underline{Q}_1 \frac{\cosh \gamma (l-1)}{\sinh \gamma L} + \underline{Q}_2 \frac{\cosh \gamma l}{\sinh \gamma L} \right). \tag{12.24}$$

Similarly, for the case of mixed upstream boundary conditions, we obtain

$$\underline{Q} = -\frac{\underline{H}_1}{Z_c} \sinh \gamma L + \underline{Q}_1 \cosh \gamma l, \tag{12.25}$$

$$\underline{H} = \underline{H}_1 \cosh \gamma l - \underline{Q}_1 Z_c \sinh \gamma L. \tag{12.26}$$

Actual values of the oscillations of piezometric head and discharge are obtained by introducing Eqs (12.23), (12.24), or (12.25), (12.26) into Eqs (12.3) and (12.4).

12.4 Harmonic solutions in the network

As the whole system consists of components of various types, the system model will be developed by connecting various corresponding finite elements into a model configuration.

Let M finite elements, connected into a network, define N nodes. Then the hydraulic state of vibrations can be uniquely described by N values of a complex amplitude of piezometric heads in the network nodes and by $2M$ boundary values of the complex amplitude of discharge on the elements. Naturally, this is possible since member $e^{i\omega t}$ from expressions (12.3) and (12.4) disappears from all equations, which is in accordance with the assumption of a developed steady harmonic state.

For M finite elements we write $2M$ equations of type (12.23). Taking Eq. (12.23) first from the upstream element end (index 1) and for the downstream element end (index 2), the result can be written in the matrix form

$$\begin{bmatrix} \underline{Q}_1 \\ \underline{Q}_2 \end{bmatrix} = \frac{1}{Z_c \sinh \gamma L} \begin{bmatrix} \cosh \gamma L & -1 \\ 1 & -\cosh \gamma L \end{bmatrix} \cdot \begin{bmatrix} \underline{H}_1 \\ \underline{H}_2 \end{bmatrix} \qquad (12.27)$$

that is, oscillations on the element are expressed by the nodal values of the oscillations of the piezometric head.

The remaining N equations are obtained from the N nodal continuity condition for m network nodal branches giving

$$\sum_{i=1}^{m} \underline{Q}_i e^{i\omega t} = 0 \quad \Rightarrow \quad \sum_{i=1}^{m} \underline{Q}_i = 0 \qquad (12.28)$$

which makes the system mathematically closed. The matrix of the finite element in the form ready assembling into the global system is obtained by changing the sign of the first row matrix (12.27)

$$\begin{bmatrix} A_1^e \\ A_2^e \end{bmatrix} = \frac{1}{Z_c \sinh \gamma L} \begin{bmatrix} -\cosh \gamma L & 1 \\ 1 & -\cosh \gamma L \end{bmatrix} \cdot \begin{bmatrix} \underline{H}_1 \\ \underline{H}_2 \end{bmatrix}. \qquad (12.29)$$

Subsequently, frequently occurring nodal boundary conditions are presented.

(a) **Node with a prescribed outflow**
    The continuity condition for m network nodal branches paths is given, so that

$$\sum_{i=1}^{m} \underline{Q}_i e^{i\omega t} = \underline{Q}^0 e^{i\omega t}, \qquad (12.30)$$

where $\underline{Q}^0$ is a complex amplitude of oscillations for a given outflow or external inflow. In practical modelling this node type is used as the source of vibration excitation.

(b) **Node with a prescribed piezometric head**
    This node cannot be the vibration source, due to the fact that the oscillation amplitude is equal to zero

$$\underline{H} e^{i\omega t} = 0. \qquad (12.31)$$

(c) **Surge tank at node**
    The continuity equation of the oscillatory part for node k can also be written in the form

$$\sum_{i=1}^{m} \underline{Q}'_i = A_k \frac{dh'_k}{dt}$$

from which it follows:

$$\sum_{i=1}^{m} \underline{Q}_i e^{i\omega t} = A_k \underline{H}_k i\omega e^{i\omega t}, \qquad (12.32)$$

where A_k is the area of the surge tank for the working water level and h'_k the water level oscillation in the surge tank.

Special attention should be paid to the difference between oscillations modelling of the hydro-power plant surge tank in the ordinary unsteady flow and vibration modelling. Namely, the surge tank is often constructed as a vertical cylindrical tank with upper and lower branched galleries. Standard unsteady computation implies the slow filling of the surge tank, so that the whole surge tank can be modeled using an area graph A_k dependent upon the water level. Consequently, the water hammer from the pressure pipeline is largely reflected upon the surge tank, whereas only a small part is transmitted to the headrace tunnel. This assumption is valid as long as the lengths of the tunnels in the tank are significantly shorter than the length of the penstock and headrace tunnel.

Unlike the water hammer, the vibrations occur much more quickly. Therefore, it is necessary, when the surge tank has a complex form, to model the tank tunnels as finite elements (usually as pipe elements), up to the position of the actual working water level in the tank.

The influence of the free water surface must be taken into account, which can be modeled according to Eq. (12.32), so here we place a node of the tank type, whose area A_k is equal to the area of the surface for the actual working water level.

According to practical experience it can be concluded that vibrations are transmitted both to the water filled tank tunnels and power tunnel. Relatively long galleries, which are "blind" ended branches, can be very harmful resonators, because vibrations are not well dampened because the water is steady in them for normal steady plant running conditions.

(d) **Air and exhaust vessels**

Like the standard unsteady modelling, the nodal equations can be reduced to surge tank equations.

(e) **Exhaust valves**

The continuity equation for nodes containing exhaust, for example exhaust valve, gate valve, Pelton turbine, and so on, can be expressed in the form

$$\sum_{i=1}^{m} \left(\bar{Q}_i + Q'_i \right) = a \left(\bar{h} - z + h' \right)^b,$$

where the left-hand side expresses the sum of fluxes in the node branches and the right side expresses the exhaust discharge. The exhaust discharge curve is determined by the net piezometric gradient $(h - z)$, where z is the exhaust level, and by parameters a and b which can occur, for the unsteady computation, as the time functions. These are considered constants for the vibration modelling of the considered working state.

The right hand side can be developed into a binomial series, for which only the first two members are written

$$a \left[\left(\bar{h} - z \right)^b + b \left(\bar{h} - z \right)^{b-1} \underline{H} e^{i\omega t} \right]$$

and the nodal conditions for vibrations follow as

$$\sum_{i=1}^{m} \underline{Q}_i e^{i\omega t} = ab \left(\bar{h} - z \right)^{b-1} \underline{H} e^{i\omega t}. \tag{12.33}$$

Finally, the obtained system of $(N + 2M)$ algebraic equations should be solved for the complex unknowns, that is N nodal values of the complex amplitude of the piezometric head and $2M$ values of the complex amplitude of element discharges.

The procedure of equation assembling and solution is similar to the procedure for standard frontal equation assembling and solution, described in Chapter 2. The only difference is the fact that here we deal with complex numbers.

12.5 Vibration source modelling

As stated in the introduction, the sources of vibration, that is excitations, can be caused by various mechanical factors. Without regard to the kind of the source vibrations, the transfer of vibrations into the hydraulic system is performed either by a change in pressure or in the local discharge, which does not affect further analyses.

In the described model, the natural method of vibrations modelling is to apply expression (12.30), that is by oscillations of the external discharge

$$\underline{Q}^0 e^{i\omega t} = \left|\underline{Q}^0\right| e^{i\varphi} e^{i\omega t} = \left|\underline{Q}^0\right| e^{i(\omega t + \varphi)}, \tag{12.34}$$

where $\left|\underline{Q}^0\right|$ is value of the real excitation amplitude and φ is excitation phase angle. The real excitation amplitude and phase angle are input data for the program, which is done interactively.

12.6 Hints to implementation in `SimpipCore`

Unlike other programming languages, Fortran has allowed work with complex numbers from the beginning. However, due to the Fortran implementation of a calculation with complex numbers, some hints for the realization of the task should be given.

In the main program `SimpipCore.f90` (see website www.wiley.com/go/jovic) the `Vibrations` subroutine call is required for the optional variable `runVibrations = .true.` It is the connecting subroutine between modelling of the hydraulic flow and module, which implements the entire calculation of hydraulic vibrations. Subroutine `Vibrations` is a simple interactive sequence that requires the name of the point where the source of vibration excitation is. After the logical function `InitVibra` performs the initialization of the `VibraModule` module for parameters:

- `SrcPnt` – index of vibration source point.
- `FromFreq` – initial frequency value.
- `StepFreq` – step in calculation.
- `PhaseFreq` – phase angle of source.
- `NoFreqSteps` – number of steps in calculation.

for every k frequency `logical function VibraSolver(k)` is called for solving vibrations.

```
subroutine Vibrations
use VibraModule
implicit none
integer k,cntSteps/1/
real FromFreq/0.001/,StepFreq/0.001/,PhaseFreq/0./
integer kItem,SrcPnt,NoFreqSteps/1000/
integer FindPoint
character*(OBJ_NAME_LEN) Key

  write(*,'(a\)') "Enter point name: "
  read(*,'(a)') Key
```

```
SrcPnt=FindPoint(trim(Key))
if(SrcPnt.eq.0) stop 'No such point'

if(InitVibra(SrcPnt,FromFreq,StepFreq,PhaseFreq,NoFreqSteps)) then
  do k=1,NoFreqSteps
    if(.not.VibraSolver(k)) then
      call OutputMsg("Not allowed elements in model!")
      call CloseVibra
      exit
    endif
  enddo
  call CloseVibra
endif
end subroutine Vibrations
```

It should be noted that the frequency is predetermined within the subroutine `Vibrations`, and that the calculation results are printed in a formatted form in the file `*.vib` for all points in the model configuration, which is appropriate for a smaller number of points only. This program solution is made as a textbook example of vibration for the requirements of tests in the next section. An interested reader can easily change the attached program sources at their own discretion, and link the precompiled `SimpipCore` project.

The entire calculation of hydraulic vibration is implemented in Fortran module `VibraModule`, logical function `VibraSolver(kFreq)` for which guidelines are shown here:

```
module VibraModule
use GlobalVars
implicit none
    private ch,sh,GetBDC,IncVar
    private GetBDC,IncVar
    ... other specifications and definitions
contains
    complex*16 function ch(x)
    ! complex function sinh(x)
    complex*16 x
          ch=(exp(x)+exp(-x))/2
    end function ch

    complex*16 function sh(x)
    ! complex function sinh(x)
    complex*16 x
          sh=(exp(x)-exp(-x))/2
    end function sh

    subroutine GetBDC
       ....
    end subroutine GetBDC

    subroutine InvVar
       ...
    end subroutine InvVar
```

```fortran
      logical function VibraSolver(kFreq)
      use CplxSimpipFront
      implicit none
          integer kFreq
          real Whs,Area,Resist,dL
          complex*16 elem_mtx(2,2),elem_rhs(2)
          real*8 freq,Omega
          integer ixelem,ielem
          integer iconnect(2)
          freq = StartFreq+(kFreq-1)*FreqStep
          Omega=2*Pi_*freq
          call CplxIniFrnt
          Cmplx_dU=cmplx(0,0)
          AmplitudaH=0
      call GetBDC(nPoints,Omega,Cmplx_Bdc)
      do ixelem=1,size(elems) !SizeOfElems()
          ielem = front_order(ixelem)
          if(ielem.eq.0) cycle
          if(Elems(ielem).infront==0) cycle
          select case(Elems(ielem).tip)
              case (PIPE_OBJ)
                  call GetPipeVibraInfo(ielem,Whs,Area,Resist,dL)
                  call PipeVibraMtx(Omega,Whs,Area,Resist,dL,elem_mtx)
              case (JOINT_OBJ)
                  ...
              case (VALVE_OBJ)
                  ...
              case default
                  VibraSolver=.false.
                  return
          endselect
          !  pass the element to the frontal procedure:
          iconnect(1)=Elems(ielem).n1
          iconnect(2)=Elems(ielem).n2
          call CplxFrontU(iconnect, ... ,Cmplx_Bdc,Cmplx_dU,LunFrn)
      enddo
!      Backward frontal procedure
call CplxFrontB(LunFrn,nPoints,Cmplx_dU)
kVibraStage=kFreq
call IncVar(freq,nPoints,Cmplx_dU,AmplitudaH)
VibraSolver=.true.
...
end function VibraSolver

... other procedures
...
end module VibraModule
```

The function `VibraSolver` uses module `CplxSimpipFront` to solve vibrations, where a generated equation system is solved by the frontal method of assembly and elimination in complex number form:

```
module CplxSimpipFront
implicit none
... specifications and definitions
    private flact,fldea
contains
    logical function OpenFrontalLun(FrontLen,lunfrn)
    ...
    end function OpenFrontalLun

    subroutine flact(noelnd,lelnod,nactiv,mxactv,kamo,lactiv)
    ...
    end subroutine flact

    subroutine fldea(nactiv,lactiv)
    ...
    end subroutine fldea

    subroutine CplxIniFrnt()
    ...
    end subroutine CplxIniFrnt

    subroutine CplxFrontU(iconnect, ... ,lun)
    ...
    end subroutine CplxFrontU

    subroutine CplxFrontB(lun, ... ,dU)
    ...
    end subroutine CplxFrontB
end module CplxSimpipFront
```

Well-known concepts and steps from the hydraulic solution are used and explained in the program solution `SimpipCore`. The names of previous functions and subroutines can also be used, for example `subroutine GetBDC` or subroutine `IncVar`, because they can be declared `private` by attribute in modules where they are located. All variables and procedures that are not declared with the attribute `private` are visible because they have the attribute `public` as implied.

12.7   Illustrative examples

Simple pipe vibration analysis. Let us consider a simple pipe of length L = 500 m and inner diameter D = 500 mm filled by water, see Figure 12.1. The source of vibration is located at the end of the pipe. The amplitude of the source is 1% of the prescribed Q value, that is 2 l/s, and the phase angle for analyzed frequency response is 0.

After we select `Option +Vibrations` analysis, the `SimpipCore` program will ask us to "Enter point name:" of the vibration source location and computation starts. The computation results will be stored in a file `Vibra test a).vib`.

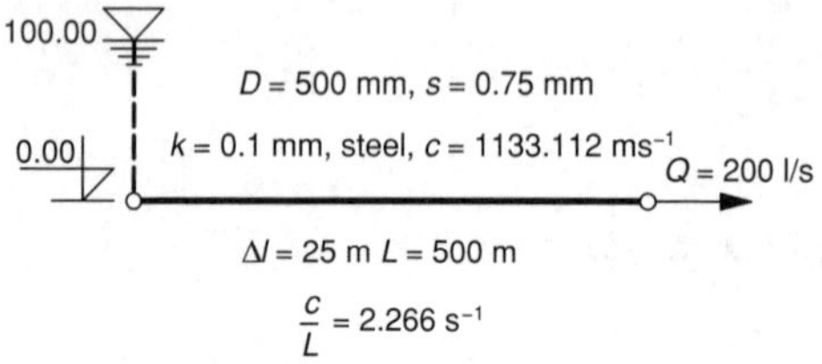

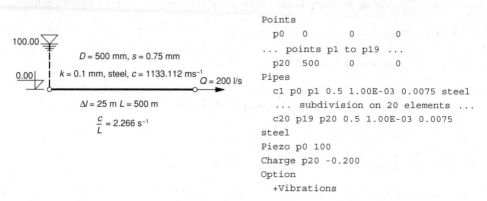

```
test file: Vibra test a).simpip

Points
  p0   0       0        0
... points p1 to p19 ...
  p20  500     0        0
Pipes
  c1 p0 p1 0.5 1.00E-03 0.0075 steel
   ... subdivision on 20 elements ...
  c20 p19 p20 0.5 1.00E-03 0.0075
steel
Piezo p0 100
Charge p20 -0.200
Option
  +Vibrations
```

Figure 12.1 Simple analysed pipe.

Figure 12.2 shows the piezometric head amplitudes for the location of the vibration source for a stage at time t = 0. It is obvious that the pipe has resonant frequencies.

For a simple pipeline the fundamental circular frequency may be computed as

$$\omega_0 = \frac{\pi c}{2L} = 3.56\left[\mathrm{s}^{-1}\right]$$

where $L = 500\,\mathrm{m}$ and $c = 1133.112\,\mathrm{m/s}$. For this simple system higher tones may be computed by multiplying the fundamental one by odd prime numbers.

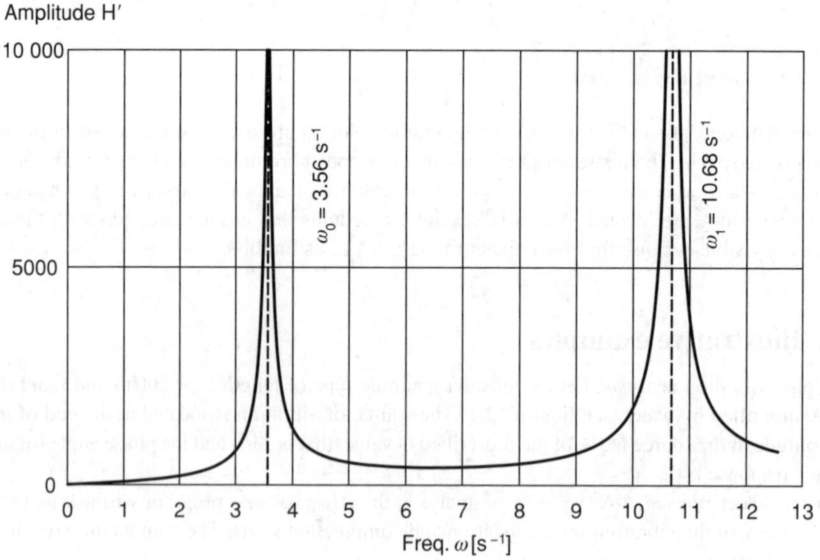

Figure 12.2 Nodal response vs. excitation frequency.

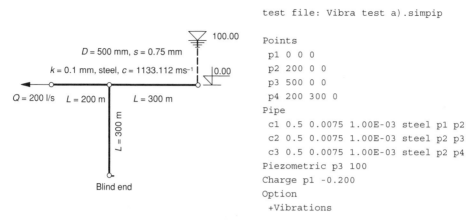

test file: Vibra test a).simpip

```
Points
  p1 0 0 0
  p2 200 0 0
  p3 500 0 0
  p4 200 300 0
Pipe
  c1 0.5 0.0075 1.00E-03 steel p1 p2
  c2 0.5 0.0075 1.00E-03 steel p2 p3
  c3 0.5 0.0075 1.00E-03 steel p2 p4
Piezometric p3 100
Charge p1 -0.200
Option
  +Vibrations
```

Figure 12.3   Simple network.

The analysis has been performed in steps of frequencies $0.001\ s^{-1}$. Using finer steps we may expect very high peaks of head responses, telling us that resonance will really occur. It may be noticed that the vibrations are dumped (flow resistance effect), but the resonance may occur. In order to make this clear, a computation with finer steps must be performed for a range of observed dangerous frequencies.

Simple network. Let us consider forced vibrations in a simple branched system, see Figure 12.3. The system consists of a pipeline and a blind ended branch. This example differs from the previous one because of the presence of the branch pipe of length 300 m.

Even for this simple branched system the resonant frequencies can be very hard to compute in advance. Figure 12.4 shows the nodal head's response vs. excitation frequencies for the steady flow conditions in

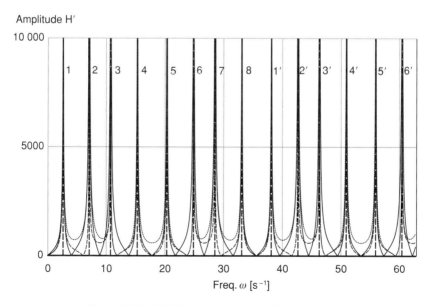

Figure 12.4   Nodal response vs. excitation frequency.

Table 12.1

ω_1	ω_2	ω_3	ω_4	ω_5	ω_6	ω_7	ω_8
2.53	7.06	10.74	15.27	20.32	24.86	28.54	33.07
ω_1'	ω_2'	ω_3'	ω_4'	ω_5'	ω_6'	ω_7'	ω_8'
38.12	42.65	46.34	50.87	55.92	60.45	...	...

the main pipe ($Q = 200\,l/s$ and no branch flow). The analysis has been performed in steps of circular frequencies $0.001 \times 2\pi\ s^{-1}$ from 0 to $10 \times 2\pi\ s^{-1}$.

It can be seen that this system has eight resonant circular frequencies that are repeated after the eighth step, see Table 12.1. It should be noted that the size of the pipe element is not limited for the determination of resonant frequencies. Unlike the first example where the pipe was subdivided into 20 elements, in the second example every pipe is considered a single pipe element.

This can be explained by the fact that time disappears as a variable from the developed harmonic state, that is it is a harmonic stationary state.

Reference

Streeter, V.L. and Wylie, E.B. (1967) *Hydraulic Transients*, McGraw–Hill Book Co., New York, London, Sydney.

Further reading

Fox, J.A. (1977) *Hydraulic Analysis of Unsteady Flow in Pipe Networks*. MacMillan Press, London.

Jović, V. (1987) Modelling of non–steady flow in pipe networks. *Proc. 2nd Int. Conf. NUMETA '87.* Martinus Nijhoff Pub. Swansea.

Jović, V. (1992) Modelling of hydraulic vibrations in network systems. *International Journal for Engineering Modelling.* 5: 11–17.

Watters, G.Z. (1984) *Analysis and Control of Unsteady Flow in Pipe Networks*. Butterworths, Boston.

Index